Waste Biomass Management – A Holistic Approach

Lakhveer Singh • Vipin Chandra Kalia
Editors

Waste Biomass Management – A Holistic Approach

Springer

Editors
Lakhveer Singh
Faculty of Engineering Technology
Universiti Malaysia Pahang (UMP)
Gambang, Kuantan, Malaysia

Vipin Chandra Kalia
CSIR - Institute of Genomics and Integrative Biology
New Delhi, India

ISBN 978-3-319-49594-1 ISBN 978-3-319-49595-8 (eBook)
DOI 10.1007/978-3-319-49595-8

Library of Congress Control Number: 2017935490

Printed on acid-free paper

This Springer imprint is published by Springer Nature
The registered company is Springer International Publishing AG
The registered company address is: Gewerbestrasse 11, 6330 Cham, Switzerland

Dedicated to my Parents

Preface

All living beings have quite efficient metabolisms. Plants are the basic producers of almost all things, which are available for consumption by human beings. Animals and human beings are able to digest most of the plant parts, however, a few components go undigested and are excreted out. A few other components of plant origin, which are not fit for human consumption, find their way to the landfills. Here, microbes play an active role in further degradation of wastes of biological origin. Microbes work either in isolation or in association with others. Anaerobic digestion is a wonderful process which involves different sets of microbes with highly specialized metabolic activities. These networks of different biochemical pathways result in almost complete degradation of organic matter present in the biowastes. The different intermediates can be either used as feed for the next set of microbes or may end up as bioproducts with unique applications. Although such processes are already operative in Nature, however, we need to work hard to elucidate the mechanisms and then exploit them for the benefit of human beings. Specialized pretreatments either physico-chemical or enzymatic help to make the process of waste degradation a bit easier. For enzymatic treatment, microbes with high hydrolytic process are required. Thermochemical conversion—liquefaction of biomass generates high value byproducts. These processes for waste biomass management result in producing oil, ethanol, biodiesel, sugars, biogas, activated carbon, etc. The last century has seen a transition from chemical to biological processes, primarily because of their eco-friendly nature. The Green Technologies are expected to save Earth. Scientific efforts are constantly needed to realize the significance of these approaches. Here, the rising stars of tomorrow—the young researchers and scientists need to make innovative contributions. This book intends to feed the young ignited minds and provoke them to contribute scientifically for the economic benefits of the society. In this book, scientists with strong academic backgrounds and practical experience have shared their research works carried out over the last few decades. This book is a reflection of the dedication of the scientific community towards human welfare. I am truly humbled by the contributions of each of the respected authors. I am extremely indebted to them for their efforts. Words are not sufficient enough to truly and adequately acknowledge the worthiness of their efforts. My true inspirations to write this book were bestowed in me and the constant faith and support of—my mother (Late Mrs. Kanta Kalia), who passed away, during the preparation of this book, and my father (Mr. R.B. Kalia), Amita (wife), Sunita, and Sangeeta (sisters), Ravi, Vinod, and Satyendra (brothers), Daksh and Bhrigu (sons), my teachers especially Dr. A.P. Joshi, my friends—Rup, Hemant, Yogendra, Rakesh, Atya, Jyoti, Malabika, Neeru, and Ritusree. I must also acknowledge the support of my young friends—Asha, Sadhana, Sanjay, Mamtesh, Subhasree, Shikha, Jyotsana, Ravi, Priyanka and Rahul.

Delhi, India Vipin Chandra Kalia

Contents

About the Editors

Vipin Chandra Kalia is presently working as Emeritus Scientist. He has been the Chief Scientist, and the Deputy Director, at Microbial Biotechnology and Genomics, CSIR-Institute of Genomics and Integrative Biology, Delhi. He is a Professor, AcSIR, who obtained his M.Sc. and Ph.D. in Genetics, from Indian Agricultural Research Institute, New Delhi. He has been elected as: (1) Fellow of the National Academy of Sciences (FNASc), (2) Fellow of the National Academy of Agricultural Sciences (FNAAS), and (3) Fellow of the Association of Microbiologists of India (FAMI), and. His main areas of research are Microbial biodiversity, Bioenergy, Biopolymers, Genomics, Microbial evolution, Quorum sensing, Quorum quenching, Drug discovery and Antimicrobials. He has published more than 100 papers in Scientific journals such as (1) Nature Biotechnology, (2) Biotechnology Advances, (3) Trends in Biotechnology, (4) Annual Review of Microbiology, (5) Critical Reviews in Microbiology, (6) Bioresource Technology, (7) PLoS One, (8) BMC Genomics, (9) International Journal of Hydrogen Energy, and (10) Gene. He has authored 14 book chapters. His works have been cited 3712 times with an h index of 32 and an i10 index of 71 (http://scholar.google.co.in/citations?hl=en&user=XaUw-VIAAAAJ). He has Edited three books: (i) Quorum Sensing versus Quorum Quenching: A Battle with No End in Sight (2015), http://link.springer.com/book/10.1007/978-81-322-1982-8, (ii) Microbial Factories Vol. 1: Biofuels, Waste treatment (2015) http://link.springer.com/book/10.1007%2F978-81-322-2598-0, and (iii) Microbial Factories Vol. 2: Biodiversity, Biopolymers, Bioactive Molecules (2015) http://link.springer.com/book/10.1007%2F978-81-322-2595-9. He is presently the Editor-in-Chief of the Indian Journal of Microbiology and editor of: (1) Journal of Microbiology and Biotechnology (Korea), (2) International Scholarly Res. Network Renewable Energy, (3) Dataset Papers in Microbiology, and (4) PLoS One. He is a life member of the following Scientific societies: (1) Society of Biological Chemists of India; (2) Society for Plant Biochemistry and Biotechnology, India; (3) Association of Microbiologists of India; (4) Indian Science Congress Association; (5) BioEnergy Society of India, and (6) the Biotech

Research Society of India (BRSI). He is also a member of the American Society for Microbiology. He can be contacted at: vckalia@igib.res.in; vc_kalia@yahoo.co.in.

Lakhveer Singh is presently working as senior lecturer at Faculty of Engineering Technology, Universiti Malaysia Pahang, Malaysia. He received his B.Sc. and M.Sc. in Chemistry from Himachal Pradesh Universiti Shimla and Punjab Universiti Chandigarh, India, in 2007 and 2009, respectively and Ph.D. in Industrial Chemistry from Universiti Malaysia Pahang (UMP) 2013. He has teaching, research and Industries experience with Universities and Industry in India. His main areas of interest are Bioenergy production, Bioreactors development, Wastewater treatment, MFC and Green Technology. In addition to research activities, he teaches undergraduate courses and guides doctoral students at Universiti Malaysia Pahang. He has a more than 30 number of international publications in high impact journals. His works have been cited 309 times with an h index of 10 and an i10 index of 10 (https://scholar.google.com/citations?user=TK3C9WEAAAAJ&hl=en). He has authored three book chapters and one book. He also holds a four patent filing application for his research. He is a reviewer of Elsevier, Wiley and Springer International journals. Presently he is the editor of Advance in Chemical and Biological Engineering journal, managing editor of International journal of engineering technology (IJETS) and guest editor of International Journal of Energy Engineering. He is also a member of International Water Association (IWA). He can be contacted at: lucki.chem09@gmail.com lakhveer@ump.edu.my.

Anaerobic Digestion

Sevcan Aydin

Abstract Identifying an alternative source of energy to those commonly in use has attracted increasing interest from researchers in recent years, and a particular focus has been placed on the identification of sustainable, renewable energy forms. Anaerobic digestion is a series of biochemical reactions by which different types of microorganisms break down biodegradable material to produce biogas, a clean form of energy. Almost all types of waste that have a high concentration of organic material can be degraded by a process of anaerobic biodegradation. Biogas, as a by-product of the anaerobic process, represents an ideal renewable energy source that can relieve some of the environmental problems associated with conventional energy sources. This chapter provides a brief overview of some of the basic concepts associated with the anaerobic digestion process and its complexities. It describes the steps involved in the anaerobic biodegradation process and provides insights into microbial relationships, substrata microorganism interactions, and some types of anaerobic digesters.

Keywords Anaerobic digestion • Microbial communities • Anaerobic bacterial diversity • Inhibition • Renewable energy

1 Introduction

The production of biogas from human and farm waste first emerged as a subject of interest in the latter part of the seventeenth century. John Dalton and Humphry Davy later independently documented the production of a flammable gas in the form of methane as a product of biodegradation. The microorganisms responsible for this phenomenon were isolated through different studies in the late nineteenth century. In 1881, French journal *Cosmos* published details of Louis M. Mouras' use

S. Aydin (✉)
Environmental Engineering Department, Istanbul Technical University, Maslak, Istanbul, Turkey

Catalan Institute for Water Research (ICRA), Science and Technology Park of the University of Girona, H2O Building, Girona, Spain
e-mail: sevcan_aydn@hotmail.com

L. Singh, V.C. Kalia (eds.), *Waste Biomass Management – A Holistic Approach*,
DOI 10.1007/978-3-319-49595-8_1

of this technology for wastewater treatment and referred to his innovation an "automatic scavenger." Cameron later developed the septic tank for use in treatment plants, the construction of which was based on a very similar approach to that suggested by Mouras (Pullen 2015). The first septic tank was used for the treatment of the city of Exeter's (UK) wastewater. The biogas that was produced by this plant was also used to provide heat and light during the treatment process. As such, Cameron's septic tank is largely considered to be the first anaerobic digestion plant to use biodegradation technology for the treatment of wastewater. The importance of the biogas that was produced during wastewater treatment as a potential renewable energy source gradually started to attract attention from both developed and undeveloped nations. In response to the increases in the world price of fossil fuels that resulted from World War II, India and China were two leading countries that applied biodegradation technology to generate biogas from food and human waste, and there was generally a larger interest in renewable energy during this period than there was in the second half of the twentieth century. Recently, the environmental problems associated with the utilization of fossil fuel, the elevated world oil prices, and the need for more sustainable disposal of domestic and agricultural waste have reignited interest in the anaerobic digestion process as a potential means of generating renewable energy (Massé et al. 2011).

One of the primary objectives of sustainable development is to reduce greenhouse gas emissions. One means by which this can be achieved is through the use of anaerobic digestion to dispose of livestock and agricultural waste (Massé et al. 2011). The digested manure obtained from the anaerobic process has less viable pathogens, low chemical oxygen demand, and diverse communities of psychrophilic, mesophilic, and thermophilic microorganisms (Massé et al. 1996; Sakar et al. 2009; Weiland 2010; Chen et al. 2012; Aydin et al. 2015a, b, c). Various studies have demonstrated that the amount and stability of biogas production in the treatment of agricultural and domestic waste are higher for co-digestion of manure than it is for mono-digestion. Furthermore, anaerobic digestion can be used as an effective means of disposing animal and plant waste. Plant materials, in particular, have a high methane production capacity; as such, anaerobic digestion represents a suitable option for the disposal of this form of waste. As plant materials have cellulosic and lignocelluloses in their structure, the efficient anaerobic digestion and hydrolysis of plant waste requires the use of a specific bacterial consortium (Sawatdeenarunat et al. 2015). Thermophilic conditions can enhance the degradation process; however, such conditions require higher levels of energy. Lignin cannot be broken down by anaerobic digestion; as such, the effective treatment of this type of waste relies on the use of an appropriate pretreatment phase that enhances the biodegradation process. Applying heat, steaming, and delignifying microorganisms are just some methods by which plant materials can be pretreated (Hendriks and Zeeman 2009; Frigon and Guiot 2010). Pretreatment methods require additional energy input; however, the energy potential from these residuals is somewhat similar to lignocellulosic waste matter (Chandra et al. 2012).

Over the course of the past 30 years, a number of studies have examined methods by which biogas production can be coupled with anaerobic digestion to increase the efficiency of the anaerobic process. Residuals from corn, cassava, cereal grains, and so on have been studied for this purpose (Mustafa et al. 2000; Zhang et al. 2010). In this coupled system, the bioethanol unit provides a continuous source of high COD waste, and the digestion process supplies energy to the bioethanol plant. As a result, overall energy production increases and the cost of bioethanol production decreases. The sustainability of agricultural production and the livestock industry can be improved through the use of anaerobic digestion in these sectors. To represent a viable component of the production system, it is important that anaerobic digestion is reliable and cost-effective (Massé et al. 2011; Aydin et al. 2015a, b, c).

2 Fundamentals in Anaerobic Digestion

The anaerobic process by which methane is produced involves four steps: hydrolysis, acidogenesis, acetogenesis, and methanogenesis. Various bacterial and archaeal populations participate in this process, and a fine balance needs to be achieved between these two groups as they perform biochemical activities. Anaerobic degradation starts with hydrolysis, through which the carbohydrates, proteins, and lipids as complex organic compounds are converted to simple monomers. The anaerobic process continues when acidogenic bacteria convert these monomers into volatile fatty acids (VFA). Acetate, propionate, and butyrate are the most abundant fatty acids involved in this step. The kinds of bacteria that participate in the hydrolysis step are determined by the composition of the substrate materials. *Clostridium* spp. are the most common strains, and these Gram-positive bacteria are capable of digesting cellulosic materials. *Bacteroidetes* and *Proteobacteria* can also perform this process; however, they are not as abundant as *Clostridium* spp. in cellulose-rich materials (Carballa et al. 2015). *Clostridium* spp. are also abundant in lipid-rich materials (Zakrzewski et al. 2012). Protein-rich materials are mostly degraded by *Clostridium* ssp., *Bacillus* spp., and *Proteobacteria* (Solli et al. 2014; Kovács et al. 2015).

The fatty acids that are produced during the anaerobic process are then converted to acetate by acetogenic bacteria. H_2 and CO_2 are produced during this process. The $\Delta G^{o\prime}$ values for oxidation of volatile fatty acids to acetate are +48.1 kJ/mol. The $\Delta G^{o\prime}$ amounts for butyrate and propionate are +48.1 kJ/mol and +76.1 kJ/mol. The $\Delta G^{o\prime}$ value for the conversion of acetate to H_2 and CO_2 is +104.6 kJ/mol (Hattori 2008; Aydin et al. 2015a, b). Acetogenic reactions are not energetically favorable, and special environmental conditions are required for this process to be successful. A very high concentration of volatile fatty acids is required. Furthermore, the low partial pressure of hydrogen is also necessary for the fulfillment of this process (Hattori 2008; Aydin et al. 2015a). Therefore, syntrophic relationships need to be formed between acetogenic bacteria and methanogenic Archaea to successfully produce an acetogenic reaction. The low partial pressure of H_2 can be realized by hydrogenotrophic Archaea, which converts the produced H_2 to methane. This event enables the growth of both syntrophic partners (Fig. 1). Any deficiency in this

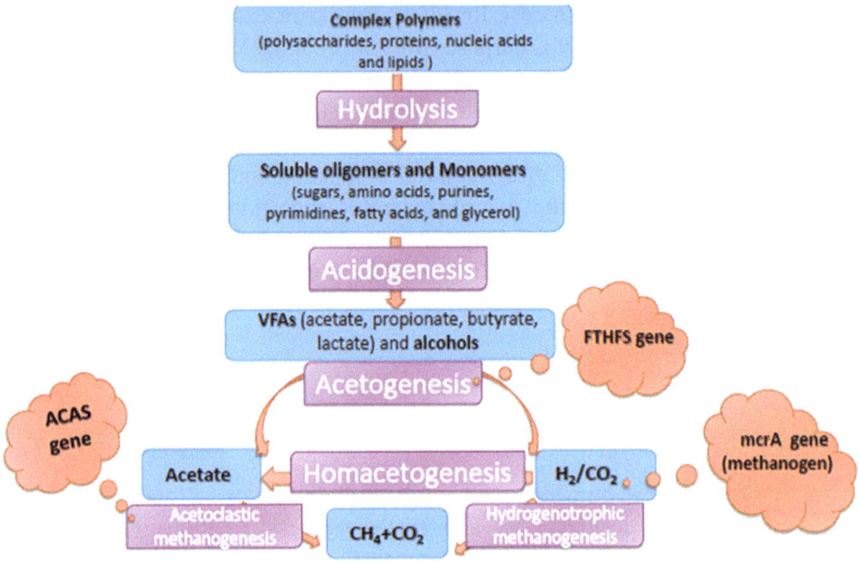

Fig. 1 Anaerobic process and targeting metabolic genes (Aydin et al. 2015c)

metabolic reaction will inhibit the catabolization of volatile fatty acids and reduce methane production. *Synergistetes* are active in the catabolizing acetate. *Syntrophomonas* spp. play the main role in the oxidation of propionate and butyrate. Syntrophic bacteria require specific culture to their demanding culture requirements and have slow growth rates. Therefore, the identification and characterization of these syntrophic bacteria are difficult on the culture-dependent methods (Carballa et al. 2015).

Methanogens belong to the phylum *Euryarchaeota* of Archaea. However, metabolic reactions within methanogens are very diverse. Even though methanogens require anaerobic conditions and low redox potential for growth, the aero-tolerance of *Methanosarcina* has been recently documented (Maeder et al. 2006). H_2 and CO_2, which are produced by the oxidation of fatty acids, are converted to methane by *Methanoculleus*, *Methanobacterium*, and *Methanospirillum*, which are hydrogenotrophic methanogens that are common in mesophilic and thermophilic digesters (Nelson et al. 2011). The low concentrations of sodium found in these environments can inhibit methanogenic activity. Furthermore, methanogenesis often takes place by sulfate-reducing bacteria, which demonstrates faster growth when the concentration of sulfate and nitrate is higher than 30 mM (Thauer et al. 2008). Acetate can be converted directly to methane without the requirement for additional bacteria. Due to the environmental condition of the biogas-producing reactor, *Methanosaeta* spp. and *Methanosarcina* spp. are the most prevalent in biogas reactors (Nelson et al. 2011; Aydin et al. 2015b). Acetate concentrations above 100 mM can readily inhibit the *Methanosaeta*; however, *Methanosarcina* are resistant to high concentrations of volatile fatty acids, low pH, and ammonia concentration (Hattori 2008). Because of their high diversity, *Methanosarcina* can perform methanogenesis through hydrogenotrophic, acetoclastic, and methylotrophic pathways (Maeder et al. 2006).

Preparing a specific culture for methanogenesis, and, therefore, their identification, is very difficult, and, in the case of syntrophic organisms, preparing a pure culture is almost impossible. Recent improvements in DNA sequencing technologies have made it possible for scientists to produce an in-depth analysis of the microbial communities that are responsible for anaerobic digestion. Microbial compositions, dynamics, and changes can now be readily investigated using this technique. Temperature and input material play the main roles in determining the richness and diversity of the microbial community (Levén et al. 2007). Increasing the abundance and composition of the species and operating at thermophilic temperatures can increase the reactor performance (Schauer-Gimenez et al. 2010; Werner et al. 2011). Amending the reactor with manure and compost has a greater effect on its performance and stability (Werner et al. 2011). The stability of the performance of the reactor can be enhanced by increasing the metabolic capacity and niche overlap in the reactor (Carballa et al. 2015).

3 The Characterization of Microbial Communities in Anaerobic Digestion

Environmental microbiology as a field of study has undergone significant developments in recent years as a result of the introduction of new molecular genomic tools. It is now much easier to observe the structural and functional diversity of environmental microbial communities using post-genomic molecular approaches. Traditionally, anaerobic digestion was assumed to operate as a "black box," where input and output alone played the main role in operational controls. The maintenance and enhancement of the diversity of the microbial population of this box were seldom assessed (Weiland 2010). However, more recent studies have considered microbial characterization and diversity when considering microbial characterization and diversity (Aydin et al. 2015c, d).

Fluorescence in situ hybridization (FISH) as a diagnostic method plays an important role in microbial characterization (Aydin et al. 2015c). The sample preparation associated with this technique requires minimal equipment, and specific fluorescent DNA probes are used to detect specific microbial targets. The results are viewed using an equipped microscope. This technique represents a fast and cost-effective method of monitoring microbial presence and development; however, it's not suitable for real-time analysis. Modern DNA-based technologies can present a complete image of the microbial and biochemical situation involved in the anaerobic digestion process. Recently integrated next-generation sequencing and quantitative PCR can be used as a basis for determining causal relationships between specific microorganisms and the performance of the digester (Aydin et al. 2015c).

At present, the various molecular approaches and techniques that are available for the purposes of community structure and function testing have their own inherent advantages and disadvantages, and none of them are currently able to offer a comprehensive understanding of the genetic and functional diversity of complex microbial communities. To improve the effectiveness and depth of study results, scientists should apply a combination of several techniques in their quest to understand the diversity, function, and ecology of microbial assemblages (Aydin et al. 2015b, c). The applications of new techniques, such as the metagenomic, metatranscriptomic, and proteogenomic methods, have improved understanding; however, a number of significant technical challenges remain, for example, extracting DNA, RNA, and proteins from environmental samples, mRNA instability, and scarcity of certain gene transcripts in total RNA. There is a specific need for further development of the bioinformatics methods that are available so that the significant amount of data that is generated through the whole-genome analysis and the application of metagenomic and metatranscriptomic tools can be better understood and applied. Producing an adequate quantitative understanding of microbial communities represents one of the most challenging hurdles because there are significant biases related to the nucleic acid isolation, PCR, and the more sophisticated DNA/RNA extraction methods. However, the approaches that are available are very much in a process of development, and further technical advancements are expected in the near future (Scherer and Neumann 2013; Aydin et al. 2015c).

4 The Environmental and Operational Factors that Affect Anaerobic Digestion

Suitable design of reactor and its operation is important for maximum performance and efficiency. Different reactor configurations have been operated successfully. Batch or continuous feeding manner, single or multiple reactors, and biomass or nutrient recycling are some of reactors types (Weiland 2010; Aydin et al. 2014, 2015b). Different parameters can be considered for having a maximum methane production and degradation of organic material with minimum maintenance and processing requirements. Floating cover, fixed dome, and balloon type were the earliest anaerobic digester types. These reactors which were adapted to continuous flow or batch operation had no mixing device and required long hydraulic retention time. With improvement of the microbial analysis methods which enable to study the microbial community evolved in the anaerobic digestion, more complicated reactors were constructed for maximum degradation of organic material and biogas production (Aydin et al. 2015b, c, d).

Batch reactor is the most common type of anaerobic reactor. In this reactor inputs are loaded inside the reactor, and the reactor then is sealed. Biogas is continually exit for the reactor. In this type of reactor, metabolic steps of anaerobic digestion occur in a sequence manner, starting with hydrolysis step and followed by acidogenic and acetogenic process and ends to methane production. In this process each group of microorganisms consume their desired substrate and produce the related products in maximum amount, and then if there is substrate depletion of substrate, their population decreases (Aydin et al. 2015b). The retention time is often longer compared to other reactor types. As the substance composition and microbial communities inside the batch reactor change significantly during the operational period, the modeling of this sort of reactors and prediction of their performance are so difficult (Donoso-Bravo et al. 2011; Aydin et al. 2015c). On the other hand, start-up phase is a very long period in these reactors and reaped every time that the reactor is loaded which reduces the system efficiency. However, the feeding manner has made this type of reactor so flexible for determining the composition of an amount of the input materials, and its changeover caused a minimum disruption in the microbial community in steady state (Aydin et al. 2015b). Continuous flow system is one of the other common types of anaerobic reactors. In this type, continuous loading of organic material and continuous removal of produced biogas and digestate take place. All steps of anaerobic reactions happen concurrently in this reactor. During the start-up phase, a gradual increase of the flow rate of organic material allows microbial community to reach to steady state. In this phase the activities of the acetogenic and methanogenic communities are approximately equal, methane is produced, and there is no fatty acid accumulation. In steady state, the phase equilibrium between acetogenic and methanogenic processes is maintained, which has been adjusted by the organic loading rate. Microbial behavior in the reactor can be quantified and predicted by Monod-type modeling of microorganism growth rate, but changes in microbial community degradation efficiency at different stages may lead to misleading results (Donoso-Bravo et al. 2011). The continuous flow reactor is a simple reactor system to design construction and operation; however, when its performance starts to lag, the adjusting of the system is difficult. The operation condition in steady-state phase is not the phase of maximum activity of the microorganisms involved; instead, it is a phase where microbial populations do not completely inhibit each other. Immobilization of microorganisms has been examined for increasing the stability of microbial populations and reducing microbial washout. Biofilm formation is an immobilization technique. Upflow anaerobic sludge blanket (UASB) reactor is reactor configuration which is a suitable base for microbial consortia which can naturally form biofilm. Microbial consortia develop in large granules and form biofilm layers in the reactor. Synthetic materials are also used for supporting microbial consortia and biofilm formation. Immobilization technique reduces the loss of microorganisms during high flow rate operation. However, the microbial diversity of these reactors is lower and may be less efficient when faced with various input materials and operational conditions such as pH and temperature (Nelson et al. 2011). Two-stage and plug flow reactors have been designed to optimize the different metabolic steps in anaerobic digestion (Nasr et al. 2012).

Two-phase reactor typically consists from two continuous flow reactors. The first reactor which is an acetogenic reactor is optimized for hydrolytic and fermentative communities, and the other one which is a methanogenic reactor is optimized for methanogenic community. Organic materials load into acetogenic reactor. The output of this reactor which contains high level of volatile fatty acids is then fed to the methanogenic reactor. As in this system, the loading of methanogenic reactor is under control, and the volatile fatty acid concentration and pH are in suitable level for methanogenic activity which allows the maximum degradation of organic material. Five to eighteenpercent increase in total energy production has been reported in the studies which have been done in two-stage systems (Luo et al. 2011; Nasr et al. 2012). The plug flow reactor is the other type of reactor which has been designed for the optimization of two different biochemical steps in anaerobic digestion. In this reactor fermentation and methanogenic stages have been separated temporally. In this reactor an incorporated continuous flow has been made it more flexible. Hydraulic retention time in this reactor is higher than two-stage digestion but lower than single-tank batch reactor (Weiland 2010). Microbial community changing pattern in this kind of reactor is so that hydrolytic, acetogenic, and methanogenic community profiles are obvious in different parts of the reactor. Physical partition of metabolic functions in this reactor allows for achieving a maximum microbial productivity. Temperature is among the most important operational factors in anaerobic biodegradation. All reactor configurations are successfully operated under psychrophilic (<25 °C), mesophilic (25–55 °C), and thermophilic (>55 °C) conditions; however, the stability of microbial community is different in each situation (Aydin et al. 2015b, c, d).

Psychrophilic digestion needs no energy input. This kind of digestion produces biogas less than mesophilic and thermophilic digestions when processing manure. Methanogenic archaea have slower growth in temperature lower than 25 °C. In this condition for preventing the increase of methane production, acclimatization of inoculums is required (Massé et al. 1996). Mesophilic digestion is a more prevalent anaerobic degradation form. This digestion is the most stable configuration with regard to the methane production. Maximum methane production is realized in thermophilic digestion; however, in terms of maintaining an equilibrium between acidogenic and methanogenic activities, mesophilic conditions are suitable than thermophilic ones. Thermophilic digestion on the other hand requires significant energy inputs which decreases the energy efficiency of the reactor. Different factors affect the maximum methane production of anaerobic digestion which is to be considered in the reactor design. Microbial community dynamics during anaerobic degradation process can be considered when reactor design for more waste breakdown and methane production (Aydin et al. 2014).

5 Inhibition of Anaerobic Digesters

Anaerobic biodegradation consists of different biochemical reactions that should be balanced among different microorganisms for effective degradation (Aydin et al. 2015a). Digester dysbiosis, such as volatile fatty acids and ammonia accumulation, pH decrease, and decrease in methane production, is well known, but the microbial fundamentals of dysbiosis are relatively undiscovered. A reduction in the diversity and abundance of methanogens is the most apparent sign of dysbiosis (Aydin et al. 2015c, d; Carballa et al. 2015). Reactors that contain healthy acetoclastic and hydrogenotrophic methanogen communities consistently produce methane. Acetoclastic methanogens are most sensitive to environmental variations, and any reduction in the number of this group of microorganisms is indicative of turbulence in the microbial community (Hattori 2008; Aydin et al. 2015a). In the methane production process, various bacterial strains are associated with Archaea; as such, differentiating between beneficial and harmful bacteria strains is very difficult. A correlation between specific bacteria and the hydrolytic, acidogenic, and acetogenic pathways has recently been found, and studies have identified taxonomic groups that perform specific metabolic reactions. Specific genes that can act as biomarkers for metabolic functions have also been detected. The difference in growth rate between the methanogenic Archaea and fermentative bacteria is caused by reactor dysbiosis. Growth performance in hydrolytic and fermentative bacteria is higher than that of Archaea (Hattori 2008; Aydin et al. 2015a). Bacterial populations have more ecological niches than Archaea. The high level of the metabolic community and the richness in ecological niches have made the bacterial community more productive than Archaea. Any changes in the metabolic environment beyond the Archaeal ecological niches will alter the bacterial/archaeal balance. Describing the healthy and unhealthy digester can be realized using information about bacterial and Archaeal niches and tendencies in microorganism compositions. An increase or decrease in the abundance of a certain microbial group can be a symptom of an imbalance between Archaeal and bacterial activities. In acetoclastic methanogens, *Methanosaeta* are more sensitive to a reduction in pH and ammonia accumulation. A shift from *Methanosaeta* to *Methanosarcina* can be a sign of condition deterioration (Aydin et al. 2015d; Carballa et al. 2015). In the suboptimal conditions that lead to volatile fatty acid and/or ammonia accumulation in the digester, *Methanomicrobiales* and *Methanobacteriales* are dominant (Carballa et al. 2015). A shift from *Firmicutes* to *Proteobacteria* has been reported in bacterial communities when an increase in the concentration of VFA and *Bacteroidetes* is detected in high concentrations of ammonia (De Vrieze et al. 2015). These shifts indicate that the condition of the reactor is deteriorating (Aydin et al. 2015a).

6 Bioaugmentation of Anaerobic Digesters

Bioaugmentation involves enhancing the biodegradation capabilities of digesters by adding microorganisms that have special metabolic abilities. It has also been tested successfully in terms of its propensity to increase methane production in an anaerobic digestion process (Werner et al. 2011). This method has been studied through different works, and the addition of microorganisms is performed before substrate materials are loaded. Pretreatment can increase hydrolytic and acidogenic activities. Because plant waste contains a high level of lignin and cellulose, it cannot be readily degraded by bacteria. Biological pretreatment prior to anaerobic digestion can increase the biodegradability of the plant materials and, therefore, overcome this problem. Fungi are able to degrade substrate containing lignin and cellulose into monosaccharides and can be used as a biologic pretreatment option. Manure amendments in digesters degrading plant and food wastes are the most common bioaugmentation applications. The diversity and even distribution of the microbial community are of critical importance to the performance and stability of the reactor (Schauer-Gimenez et al. 2010). A wide range of microorganisms with diverse metabolic properties increase the capabilities of the organisms present in the digesters when met with changes in operation conditions alongside substrate composition and loading rate. Adding manure and microorganism-containing materials to the digesters represents a suitable method of maintaining the reactor community. The addition of manure in digesters with substrate with very high carbohydrate or volatile solids increases performance stability and prohibits methanogenic collapse (Labatut and Scott 2008; Westerholm et al. 2012). Recycling the digestate can also increase and maintain the population of methanogens in the digester. This kind of recycling improves the performance of the digester and decreases the lag time between critical microorganisms and the commencement of methane production (Young et al. 2013). However, a constant recycling of digestate will increase the prevalence of selective adapted microorganisms in the reactor, and this reduces microbial richness that could develop in response to changes in input composition and operational conditions (Pervin et al. 2013; Aydin et al. 2015a).

Directly adding microorganisms to the reactor that have specific metabolic characteristics is the most common form of bioaugmentation, and it facilitates the identification of the microorganisms that are specifically responsible for a biochemical reaction. Very useful information can be obtained from molecular analysis of the effective biodegrader, and this can then be used to isolate and cultivate the studied microorganism. Isolation and cultivation will provide more information about the growth requirements of the studied microbial strains. Such studies will help to identify species that have an industrial value. *Methanosarcina* spp. play a unique role in methane production. They are among the few methanogens that can produce methane using hydrogenotrophic, acetoclastic, and methylotrophic pathways (Aydin et al. 2015a). This represents an example of how the capacity of the molecular analysis can be employed as a monitoring tool in the reactor to determine the most important species in the community.

If a suitable inoculum is provided, the system will have a shorter lag phase and will perform well. For this purpose, and to increase the performance of the reactor, specific species that are capable of producing specific metabolic reactions can be added to the system. Utilizing manure, which is a rich source of microorganisms, is an example of how microbial enriched input can be employed to increase the activity of the reactor. Successfully adding syntrophic bacteria and hydrogenotrophic Archaea as an exogenous culture to a psychrophilic digester system and increasing methane production by up to 50 % are another example (Akila and Chandra 2010). Exogenous culture amendment does not always successfully increase the performance and effectiveness of this method, which is mostly dependent on the establishment capability of microorganisms in the biodegradation system (Fotidis et al. 2013). As an enhanced treatment technology, bioaugmentation is very much in the early stages, and further research is required to better understand its potential in terms of bioremediation.

7 Perspectives

Ever-growing global concerns about environmental problems, such as global climate change and increased waste generation and its accompanying forms of pollution, have directly fueled the enthusiasm for research into anaerobic digestion. Anaerobic production is directly linked to biogas production. Unlike the processes employed to generate other biofuels, biogas production can be carried out using various substrates. However, certain substrates can be used to supply the necessary nutrients to the microorganisms. Bioenergy is as an economic renewable energy source that represents a promising alternative to fossil fuels. Finally, because of its low energy consumption, low sludge generation is preferred. Further advantages of anaerobic biodegradation are the production of manure as a soil biofertilizer and low waste generation.

Acknowledgments Authors thank TUBITAK (The Scientific and Technological Research Council of Turkey) for support of this project.

References

Akila G, Chandra TS (2010) Stimulation of biomethanation by *Clostridium* sp. PXYL1 in coculture with a *Methanosarcina* strain PMET1 at psychrophilic temperatures. J Appl Microbiol 108: 204–213. doi:10.1111/j.1365-2672.2009.04412.x

Aydin S, Ince B, Cetecioglu Z, Ozbayram EG, Shahi A, Okay O, Arikan O, Ince O (2014) Performance of anaerobic sequencing batch reactor in the treatment of pharmaceutical wastewater containing erythromycin and sulfamethoxazole mixture. Water Sci Technol 70: 1625. doi:10.2166/wst.2014.418

Aydin S, Cetecioglu Z, Arikan O, Ince B, Ozbayram EG, Ince O (2015a) Inhibitory effects of antibiotic combinations on syntrophic bacteria, homoacetogens and methanogens. Chemosphere 120:515–520. doi:10.1016/j.chemosphere.2014.09.045

Aydin S, Ince B, Cetecioglu Z, Arikan O, Ozbayram EG, Shahi A, Ince O (2015b) Combined effect of erythromycin, tetracycline and sulfamethoxazole on performance of anaerobic sequencing batch reactors. Bioresour Technol 186:207–214. doi:10.1016/j.biortech.2015.03.043

Aydin S, Ince B, Ince O (2015c) Application of real-time PCR to determination of combined effect of antibiotics on Bacteria, Methanogenic Archaea, Archaea in anaerobic sequencing batch reactors. Water Res 76:88–98. doi:10.1016/j.watres.2015.02.043

Aydin S, Shahi A, Ozbayram EG, Ince B, Ince O (2015d) Use of PCR-DGGE based molecular methods to assessment of microbial diversity during anaerobic treatment of antibiotic combinations. Bioresour Technol 192:735–740. doi:10.1016/j.biortech.2015.05.086

Carballa M, Regueiro L, Lema JM (2015) Microbial management of anaerobic digestion: exploiting the microbiome-functionality nexus. Curr Opin Biotechnol 33:103–111. doi:10.1016/j.copbio.2015.01.008

Chandra R, Takeuchi H, Hasegawa T (2012) Methane production from lignocellulosic agricultural crop wastes: a review in context to second generation of biofuel production. Renew Sustain Energy Rev 16:1462–1476. doi:10.1016/j.rser.2011.11.035

Chen Y, Fu B, Wang Y, Jiang Q, Liu H (2012) Reactor performance and bacterial pathogen removal in response to sludge retention time in a mesophilic anaerobic digester treating sewage sludge. Bioresour Technol 106:20–26. doi:10.1016/j.biortech.2011.11.093

De Vrieze J, Gildemyn S, Vilchez-Vargas R, Jáuregui R, Pieper D, Verstraete W, Boon N (2015) Inoculum selection is crucial to ensure operational stability in anaerobic digestion. Appl Microbiol Biotechnol 99:189–199. doi:10.1007/s00253-014-6046-3

Donoso-Bravo A, Mailier J, Martin C, Rodríguez J, Aceves-Lara CA, Wouwer AV (2011) Model selection, identification and validation in anaerobic digestion: a review. Wat Res 45: 5347–5364. doi:10.1016/j.watres.2011.08.059

Fotidis IA, Karakashev D, Angelidaki I (2013) Bioaugmentation with an acetate oxidising consortium as a tool to tackle ammonia inhibition of anaerobic digestion. Bioresour Technol 146: 57–62. doi:10.1016/j.biortech.2013.07.041

Frigon JC, Guiot SR (2010) Biomethane production from starch and lignocellulosic crops: a comparative review. Biofuels Bioprod Biorefin 4:447–458. doi:10.1002/bbb.229

Hattori S (2008) Syntrophic acetate-oxidizing microbes in methanogenic environments. Microbes Environ 23:118–127. doi:10.1264/jsme2.23.118

Hendriks ATWM, Zeeman G (2009) Pretreatments to enhance the digestibility of lignocellulosic biomass. Bioresour Technol 100:10–18. doi:10.1016/j.biortech.2008.05.027

Kovács E, Wirth R, Maróti G, Bagi Z, Nagy K, Minárovits J, Rákhely G, Kovács KL (2015) Augmented biogas production from protein-rich substrates and associated metagenomic changes. Bioresour Technol 178:254–261. doi:10.1016/j.biortech.2014.08.111

Labatut RA, Scott NR (2008) Experimental and predicted methane yields from the anaerobic co-digestion of animal manure with complex organic substrates. In: Providence, RI. pp. 6940–6957. doi:10.13031/2013.25178

Levén L, Eriksson ARB, Schnürer A (2007) Effect of process temperature on bacterial and archaeal communities in two methanogenic bioreactors treating organic household waste. FEMS Microbiol Ecol 59:683–693. doi:10.1111/j.1574-6941.2006.00263.x

Luo G, Xie L, Zhou Q, Angelidaki I (2011) Enhancement of bioenergy production from organic wastes by two-stage anaerobic hydrogen and methane production process. Bioresour Technol 102:8700–8706. doi:10.1016/j.biortech.2011.02.012

Maeder DL, Anderson I, Brettin TS, Bruce DC, Gilna P, Han CS, Lapidus A, Metcalf WW, Saunders E, Tapia R, Sowers KR (2006) The *Methanosarcina barkeri* genome: comparative analysis with *Methanosarcina acetivorans* and Methanosarcina mazei reveals extensive rearrangement within methanosarcinal genomes. J Bacteriol 188:7922–7931. doi:10.1128/JB.00810-06

Massé DI, Patni NK, Droste RL, Kennedy KJ (1996) Operation strategies for psychrophilic anaerobic digestion of swine manure slurry in sequencing batch reactors. Can J Civ Eng 23: 1285–1294. doi:10.1139/l96-937

Massé DI, Talbot G, Gilbert Y (2011) On farm biogas production: a method to reduce GHG emissions and develop more sustainable livestock operations. Anim Feed Sci Technol 166-167: 436–445. doi:10.1016/j.anifeedsci.2011.04.075

Mustafa AF, McKinnon JJ, Ingledew MW, Christensen DA (2000) The nutritive value for ruminants of thin stillage and distillers' grains derived from wheat, rye, triticale and barley. J Sci Food Agric 80:607–613. doi:10.1002/(SICI)1097-0010(200004)80:5<607::AID-JSFA582>3.3.CO;2-6

Nasr N, Elbeshbishy E, Hafez H, Nakhla G, Hesham El Naggar MH (2012) Comparative assessment of single-stage and two-stage anaerobic digestion for the treatment of thin stillage. Bioresour Technol 111:122–126. doi:10.1016/j.biortech.2012.02.019

Nelson MC, Morrison M, Yu Z (2011) A meta-analysis of the microbial diversity observed in anaerobic digesters. Bioresour Technol 102:3730–3739. doi:10.1016/j.biortech.2010.11.119

Pervin HM, Dennis PG, Lim HJ, Tyson GW, Batstone DJ, Bond PL (2013) Drivers of microbial community composition in mesophilic and thermophilic temperature-phased anaerobic digestion pre-treatment reactors. Wat Res 47:7098–7108. doi:10.1016/j.watres.2013.07.053

Pullen T (2015) Anaerobic digestion-making biogas-making energy. Routledge, New York, NY. doi:10.4324/9781315770772

Sakar S, Yetilmezsoy K, Kocak E (2009) Anaerobic digestion technology in poultry and livestock waste treatment: a literature review. Waste Manage Res 27:3–18. doi:10.1177/0734242x07079060

Sawatdeenarunat C, Surendra KC, Takara D, Oechsner H, Khanal SK (2015) Anaerobic digestion of lignocellulosic biomass: challenges and opportunities. Bioresour Technol 178:178–186. doi:10.1016/j.biortech.2014.09.103

Schauer-Gimenez AE, Zitomer DH, Maki JS, Struble CA (2010) Bioaugmentation for improved recovery of anaerobic digesters after toxicant exposure. Water Res 44:3555–3564. doi:10.1016/j.watres.2010.03.037

Solli L, Havelsrud O, Horn S, Rike A (2014) A metagenomic study of the microbial communities in four parallel biogas reactors. Biotechnol Biofuels 7:1–15. doi:10.1186/s13068-014-0146-2

Thauer RK, Kaster AK, Seedorf H, Buckel W, Hedderich R (2008) Methanogenic archaea: ecologically relevant differences in energy conservation. Nat Rev Microbiol 6:579–591. doi:10.1038/nrmicro1931

Weiland P (2010) Biogas production: current state and perspectives. Appl Microbiol Biotechnol 85:849–860. doi:10.1007/s00253-009-2246-7

Werner JJ, Knights D, Garcia ML, Scalfone NB, Smith S, Yarasheski K, Cummings TA, Beers AR, Knight R, Angenent LT (2011) Bacterial community structures are unique and resilient in full-scale bioenergy systems. Proc Natl Acad Sci USA 108:4158–4163. doi:10.1073/pnas.1015676108

Westerholm M, Hansson M, Schnürer A (2012) Improved biogas production from whole stillage by co-digestion with cattle manure. Bioresour Technol 114:314–319. doi:10.1016/j.biortech.2012.03.005

Young MN, Krajmalnik-Brown R, Liu W, Doyle ML, Rittmann BE (2013) The role of anaerobic sludge recycle in improving anaerobic digester performance. Bioresour Technol 128:731–737. doi:10.1016/j.biortech.2012.11.079

Zakrzewski M, Goesmann A, Jaenicke S, Jünemann S, Eikmeyer F, Szczepanowski R, Al-Soud WA, Sørensen S, Pühler A, Schlüter A (2012) Profiling of the metabolically active community from a production-scale biogas plant by means of high-throughput metatranscriptome sequencing. J Biotechnol 158:248–258. doi:10.1016/j.jbiotec.2012.01.020

Zhang C, Mao Z, Wang X, Zhang J, Sun F, Tang L, Zhang H (2010) Effective ethanol production by reutilizing waste distillage anaerobic digestion effluent in an integrated fermentation process coupled with both ethanol and methane fermentations. Bioprocess Biosyst Eng 33: 1067–1075. doi:10.1007/s00449-010-0432-8

Using Pretreatment and Enzymatic Saccharification Technologies to Produce Fermentable Sugars from Agricultural Wastes

Caoxing Huang, Ben Jeuck, and Qiang Yong

Abstract Every year, an abundance of agricultural wastes are produced after harvest, which are generally burned or discarded in the farmland. Agricultural wastes are an attractive lignocellulosic material for fermentable sugar production (glucose and xylose) since they are polysaccharide-rich resources. Unfortunately, the intimate associations between the main components of the cell wall create barriers for the enzymatic saccharification of cellulose and hemicellulose. Pretreatment plays a critical role in increasing enzymatic saccharification to obtain the glucose and xylose from pretreated agricultural wastes by different enzymes. The present review is a comprehensive evaluation to describe the advancements in pretreatment and enzymatic saccharification processes to produce the fermentable sugars from agricultural wastes. Using these agricultural wastes for the sugar production is ideal for developing bio-based chemicals while converting unwanted agricultural waste streams into valuable resources.

Keywords Agricultural waste • Pretreatment • Enzymatic saccharification • Sugars • Biorefinery

1 Introduction

For human societies, corn, wheat, and rice are the dominant grains for food consumption, due to their high content of starch and short growing seasons (Glover et al. 2007). After harvest, the grain seeds of crops are collected, while the residues,

C. Huang
College of Chemical Engineering, Nanjing Forestry University, Nanjing, Jiangsu, China

Department of Forest Biomaterials, North Carolina State University, Raleigh, NC, USA

B. Jeuck
Department of Forest Biomaterials, North Carolina State University, Raleigh, NC, USA

Q. Yong (✉)
College of Chemical Engineering, Nanjing Forestry University, Nanjing, Jiangsu, China
e-mail: swhx@njfu.com.cn; szxapy@163.com

L. Singh, V.C. Kalia (eds.), *Waste Biomass Management – A Holistic Approach*,
DOI 10.1007/978-3-319-49595-8_2

such as stover and straw, are discarded in the farmland. Maiorella (1985) stated that about 1 kg of harvested grain is accompanied by 1–1.5 kg of the straw. The amount of the corn stover, rice straw, and wheat straw are estimated about 830, 810, and 750 million tons, respectively. The produced stover and straw may be discarded in the farming fields for natural biodegradation, burned for heat, or used as cattle feed. In recent years, burning the straw has been challenged because the smoke from burning fields may be harmful to people and animals. In addition, some agronomists point out that leaving all the stover and straw in the field may actually negatively impact soil fertility and structures (López-Bellido et al. 2014). Hence, an immediate, convenient, and environmentally friendly way to dispose the agricultural crop wastes is needed.

As a natural composite material, the agricultural wastes are a chemical complex of cellulose, hemicellulose, lignin, extractives, and inorganic materials. In the cell wall, cellulose acts as the structural framework, and hemicellulose retains the water for the lignocellulosic biomass. Lignin plays a role as an adhesive between the various cell wall components. Additionally, lignin is important for mechanical support, water transport, and defense against natural invasion (Ragauskas et al. 2014). Although the composition of agricultural crop wastes is different, the typical percentage of the dominant chemical composition in corn stover, rice straw, and wheat straw is 35–50 % cellulose, 20–35 % hemicellulose, and 5–25 % lignin (Binod et al. 2010; Lomborg et al. 2010; Weiss et al. 2010). Possessing the property of high polysaccharide content, the agricultural wastes can be used as a resource for the production of C_6 sugar (glucose) and C_5 sugar (xylose), which are referred to fermentable sugars. These fermentable sugars are the building blocks of many chemical products, such as bio-based polymers, chemicals, and liquid fuels (Jäger and Büchs 2012). The US Department of Energy (DoE) reported that the following chemicals can be derived from the C_6 and C_5 fermentable sugars: 2,5-furan dicarboxylic, 3-hydroxy propionic, aspartic, levulinic, glucaric, itaconic, and glutamic acids, 1,4-diacids, 3-hydroxybutyrolactone, sorbitol, glycerol, and xylitol (Werpy et al. 2004). Technically, the biocatalytic conversion of lignocellulosic materials to these chemicals includes four major processing steps: pretreatment, enzymatic saccharification, fermentation, and product purification, as shown in Fig. 1 (Huber et al. 2006; Jäger and Büchs 2012; Zhang et al. 2014). Nowadays, production of bioethanol, lactic acid, and citric acid from fermentable sugars are well established and mature technologies (Cherubini and Strømman 2011; Hu et al. 2015).

Technically, the modern biotechnical process to convert agricultural wastes into fermentable sugars includes pretreatment and enzymatic hydrolysis (Fig. 1). It has been recognized that the architecture and component of lignocellulosic biomass such as lignin content, cellulose crystallinity, particle size, and pore volume have significant influence on enzymatic hydrolysis, making the polysaccharides resistant to be hydrolyzed, or recalcitrant (Hall et al. 2010; Yu et al. 2011; Leu and Zhu 2013). Hence, many novel and effective pretreatments have been proposed to make the agricultural wastes more susceptible to saccharify, by decreasing the cellulose crystallinity, increasing the biomass surface areas and pores, and degrading the lignin (Alvira et al. 2010; Gu et al. 2012; Yu et al. 2014; Min et al. 2015). Currently,

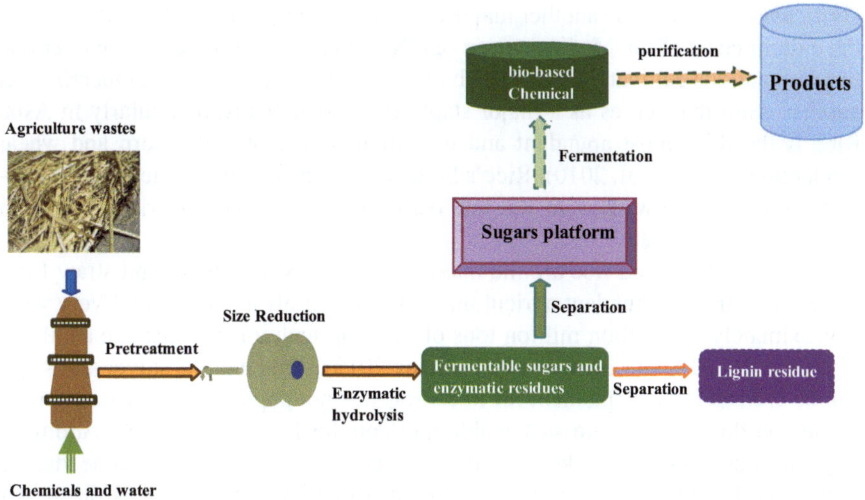

Fig. 1 Steps of sugar platform (fermentable sugars) and bio-based chemical production from agricultural wastes

the mechanisms of many pretreatments have been investigated in a fundamental level, and the researchers tend to seek more rational and effective standpoints to design new bioprocesses for the fermentable sugar production.

Enzymatic hydrolysis is the second process to obtain the fermentable sugars from the agricultural wastes. In this process, cellulase and hemicellulase enzymes are applied to cleave the polymers of cellulose and hemicellulose, respectively. Cellulase enzymes are critical in the saccharification process by catalyzing the hydrolysis of cellulose to glucose. Hemicellulases degrade hemicellulose and expose the cellulose fibers, which helps facilitate cellulose hydrolysis through increased accessibility. Hence, hemicellulase is supplemented with cellulase to raise the yield of enzymatic hydrolysis for a high sugar production (Menon and Rao 2012).

This article reviews the published investigations in the field of agricultural wastes (corn stover, wheat straw, and rice straw) for the fermentable sugar production of glucose and xylose. All recent advanced pretreatments and enzymatic hydrolysis technologies are presented, with special focus on the most recent and novel pretreatments.

2 Agricultural Waste

In prehistoric times, corn (*Zea mays L.*) was cultivated by the indigenous peoples of Mesoamerica and is now widely cultivated throughout the world. America is the top corn-producing country, followed by China, Brazil, and India (Khoo 2015). Wheat

(*Triticum aestivum L.*) is another major crop, originating from the Levant region or the eastern coast of the Mediterranean sea. Nowadays, it is estimated that over one hundred nations cultivate wheat (Talebnia et al. 2010). Rice (*Oryza glaberrima*) is another grain that serves as a major staple food for humans, particularly in Asia. Rice is the third most abundant and important food crop, after corn and wheat production (Binod et al. 2010). Rice's broad environmental parameters for cultivation allow it to grow. Due to the low demands on environment, rice have been cultivated worldwide.

As corn, wheat, and rice are the major global crops, the stover and straw from these crops are the abundant agricultural waste materials in the world. Every year, approximately 3–4 billion million tons of these agricultural residues are available when the annual grains are harvested (Zhang 2013). Although the stover and straw can be used as food supplement for livestock, it is not popular in many parts of the world and does not contain sustainable nutrients for livestock like hay. Additionally, the nutrients do not keep well in storage and decrease over time due to biological degradation. Overall, it is estimated that only about 5 % of the stover and straw is used as the livestock feed, while the rest tends to be burned in the fields after harvest (Glassner et al. 1999). The sheer volume of material being burned annually has caused serious problems with air pollution and consequently has shown negative impacts on human health (Mussatto and Roberto 2004).

Since lignocellulosic biomass can convert carbon dioxide and water into structural carbohydrate used as building blocks ($[CH_2O]_n$) and oxygen using natural light as its energy source, holocellulose is considered the most abundant source of biorenewable polysaccharides, with cellulose as the most abundant carbohydrate (Huber et al. 2006; Jäger and Büchs 2012). Cellulose and hemicellulose are the major potential candidates for the production of glucose and xylose, which function as fermentable sugars to be converted into fuels or other bio-based chemicals. According to Parfitt et al. (2010) and Lynd and Woods (2011), utilization of these agricultural wastes to produce bio-based chemicals can bring many benefits, which may be (1) an environment-friendly disposal of agricultural wastes, (2) a reduction of gas emissions responsible for the greenhouse effect and air pollution, (3) a provision of a sustainable approach to produce renewable and potentially carbon-neutral fuels and chemicals, (4) the creation of more high-paying manufacturing jobs, and (5) the reduction of the incentive for the human disease.

The three major polymers in agricultural wasters, cellulose, hemicellulose, and lignin, are organized into a three-dimensional structure, with the cellulose and hemicellulose form dense, anisotropic layers with lignin. In the cell wall of agricultural biomass, the subunit of cellulose is cellobiose, a disaccharide which is stabilized by hydrogen bonds (Hallac and Ragauskas 2011). The ordered arrangement of glucan makes cellulose, with its characteristic of high degree of polymerization and crystallinity. Hemicellulose is a random and amorphous complex, composed of xylan, glucan, arabinan, uronic acid, and acetyl group. All these carbohydrates may be covalently linked to the lignin, in a system referred to as lignin-carbohydrate complexes (LCCs) (Balakshin et al. 2011; Du et al. 2014; Huang et al. 2015a). Lignin is an amorphous polymer comprising with three types of units: guaiacyl (G), syringyl (S), and *p*-hydroxyphenyl (H) (Abdel-Hamid et al. 2013). In the cell wall, lignin functions as the cellular "glue" that yields rigidity to plant tissue and resist insects and pathogens (Ragauskas et al. 2014).

3 Pretreatment for the Fermentable Sugars from Agricultural Wasters

Technically, cellulose crystallinity, lignin hydrophobicity, and lignin-carbohydrate matrices carry the most relevant factors in lignocellulosic resistance to biodegradation or recalcitrance (Agbor et al. 2011; Barakat et al. 2013). To effectively convert the carbohydrate polymers into fermentable monosaccharides, pretreatment must first be applied on the agricultural wasters to break apart the firm structures. During the last few decades, various approaches have been used as the pretreatments for the fermentable sugars from agricultural wasters, including physical, chemical, physicochemical, and biological pretreatments (Alvira et al. 2010; Gu et al. 2012; Yu et al. 2014; Isikgor and Becer 2015; Min et al. 2015). Each pretreatment has shown to have unique advantages and disadvantages.

3.1 Physical Pretreatment

Physical pretreatments, such as milling, grinding, and irradiation, are used for increasing the surface accessible for enzyme action and reducing the degree of the polymerization of biomass (Binod et al. 2010; Talebnia et al. 2010; Hallac and Ragauskas 2011). Size reduction by milling or grinding is the first step prior to chemical pretreatments, which can promote efficient downstream processing. It is reported the enzymatic hydrolysis efficiency of wheat straw was increased significantly by ball milling, which went from 17.7 % (untreated sample) to 61.1 % (ball-milled sample) (Koullas et al. 1992). For rice straw, it is found that wet disk milling shows a better effect on glucose released and reduces energy demands, than that from ball milling (Hideno et al. 2009). Pedersen and Meyer (2009) investigated the effects of agricultural straw particle sizes on enzymatic hydrolysis. They found that the yield of glucan was increased by 39 % when the particle of the straw decreased from 2–4 cm to 53–149 μm. Different irradiation processes also improve the digestibility of agricultural wastes. Jin et al. (2009) used electron beam irradiation to pretreat the rice straw. The results showed that the 132 h enzymatic hydrolysis yield of the electron beam irradiated sample was 52.1 % higher than that of untreated rice straw (22.6 %). A microscopic study of the pretreated corn stover showed that the enzyme penetration increased from <1 % to 100 % of the thickness of secondary cell walls as the dilute acid pretreatment severity was increased from 20 min at 100 °C to 20 min at 150 °C (Donohoe et al. 2009). This was direct evidence supporting that the loosening of the plant cell wall structure through pretreatment improved the accessibility of cellulolytic enzymes. Recently, Peng et al. (2014) used the continuous microwave irradiation to pretreat corn stover and indicated that the glucose production was higher than that from the untreated material.

3.2 Chemical Pretreatment

Chemical pretreatments for agricultural wastes have been proven effectively in converting lignocelluloses into fermentable sugars. Among these methods, dilute acid and alkaline conditions are most widely used. The alkaline pretreatment shows a good performance in delignification, while the dilute acid pretreatment shows enhanced effectiveness in removing hemicellulose. For the agricultural wastes, the most promising pretreatment may be the alkaline and the alkaline-based pretreatment.

3.2.1 Acid Pretreatment

During the dilute acid pretreatment process, most of the hemicellulose can be degraded, creating more pores in the pretreated biomass and improving reaction areas for the enzymatic saccharification (Saha et al. 2005; Huang et al. 2015b). Dilute acid is the most widely reported pretreatments for corn stover, rice straw, and wheat straw. Technically, 0.5–1 % H_2SO_4 and high temperature are exploited. The monomeric sugars can be obtained from the liquors both in pretreatment and enzymatic hydrolysis processes. For rice straw, dilute acid is the excellent applied pretreatment in the laboratory scale. Chen et al. (2011) reported that a combination of dilute acid and steam explosion had a higher sugar yield (the total sugar in supernate of pretreatment and enzymatic hydrolysis), while less inhibitors were found in the hydrolyzate. After a consecutive pretreatment (at 165 °C for with 2 % H_2SO_4 for acid pretreatment and 180 °C for 20 min for steam explosion) and enzymatic saccharification process, the obtained sugar yield was 1.5-fold than that from the control pretreatment step. Saha et al. (2005) applied the dilute acid pretreatment with different temperatures and enzymatic saccharification to release the monomeric sugars from wheat straw. At the optimum pretreatment condition with 0.75 % acid (v/v), 0.57 mg sugars were released from 1 g wheat straw after enzymatic hydrolysis. Organic acids, such as fumaric or maleic acids, are also applied to improve cellulose hydrolysis for sugar production. Kootstra et al. (2009) reported saccharification yields and sugar degradation compounds from pretreatments with both acids, compared with sulfuric acid. It showed that high efficiency could be achieved by the fumaric acid. Meanwhile, less amounts of furfural were formed in the organic acid pretreatments than from sulfuric acid.

Xiang et al. (2003) demonstrated that dilute acid pretreatment can perform a quick reaction rates in a short time and improve the enzymatic saccharification efficiency, significantly. However, pretreatment conditions must be tailored to consider the unique chemical and structural composition of the biomass to be pretreated. Furthermore, corrosion of equipment is a significant drawback when applying the acid to pretreat biomass (Alvira et al. 2010). Additionally, many of the inhibitors (furfural, fumaric acid, and acetic acid) are generated during the pretreatment, which may inhibit the downstream enzymatic saccharification and fermentation process (Galbe and Zacchi 2012).

3.2.2 Alkaline Pretreatment

For the alkaline pretreatment of agricultural wastes, the agents of sodium, potassium, calcium, and ammonium hydroxides and green liquor (Na_2CO_3 and Na_2S) have been investigated (Kumar and Wyman 2009a; Gu et al. 2013). The alkaline pretreatment can cause lignin degradation, lignin structural alterations, partial cellulose decrystallization, and partial hemicellulose removal, resulting in a rapid increase in the yield of enzymatic hydrolysis (McIntosh and Vancov 2010; Ibrahim et al. 2011). In this paper, we review the most common sodium pretreatment and the novel green liquor pretreatment.

The primary effect of NaOH pretreatment is removing the lignin from lignocellulosic biomass, which occurs via ester bonds between lignin, hemicellulose, and cellulose being broken, consequently increasing more reaction areas for enzymatic saccharification (Huang et al. 2015c). Zhang and Cai (2008) reported that 20 % solid loading of rice straw pretreated by 2 % sodium hydroxide at 85 °C for 1 h decreased lignin composition by 36 %, resulting in a 68.21 % sugar yield by the *T. reesei* ZM4-F3. Sun et al. (1995) found that the 1.5 % sodium hydroxide was the optimal chemical loading for delignification and hemicellulose degradation in wheat straw at 20 °C, resulting in a 60 % degree of delignification and an 80 % release of hemicellulose. Chen et al. (2009) used NaOH to pretreat the corn stover, revealing that the pretreatment with 2 % sodium hydroxide significantly increased the degree of delignification and improved the enzymatic saccharification of cellulose. A yield with 81.2 % could be achieved from the NaOH-pretreated corn stover (8 % solid) with 20 FPU/(g substrate) cellulase. Although sodium pretreatment has potential in improving sugar production from cellulose in different agriculture wastes, the application is hindered by the high cost of sodium.

Recently, researchers at the North Carolina State University proposed an alkaline-based pretreatment utilizing green liquor (GL), the mixture of NaOH and Na_2S recovered from combustion of the black liquor in the recovery boiler from a kraft pulp mill (Wu et al. 2010; Jin et al. 2010). Gu et al. (2012) used this liquor to pretreat corn stover for improving sugar production. It showed that a maximum recovery yield with 70 % and a delignification degree with 45 % were obtained after this pretreatment with conditions as 8 % TTA, 40 % sulfidity, and 140 °C. After enzymatic hydrolysis, 70 % of the original carbohydrates were converted into monomeric sugars. According to Chu et al. (2014a), using the same pretreatment and enzymatic hydrolysis conditions, about 329.7 g glucose and 126.3 g xylose were obtained from 1000 g dry corn stover pretreated by the same condition. In 2013, Gu et al. (2013) also applied this pretreatment on the rice straw. Optimized GL pretreatment conditions were found to be 4 % TTA, 20 % sulfidity, and 140 °C. These conditions yielded 92.5 % of recovery glucan and 82.4 % of recovery xylan, with a 39.4 % degree of delignification. The sugar yields of glucan, xylan, and total sugar for this pretreated rice straw were 83.9 %, 69.6 %, and 78.0 %, respectively. All of the results indicated that GL pretreatment minimally reduces polysaccharides in the pretreated substrate before enzymatic hydrolysis, allowing almost all fermentable sugars to be released. In addition, all the equipment and processes in the GL pretreatment system have been industrially practiced for many decades in kraft pulp mills around the world, and the chemicals of green liquor, sodium carbonate, and sodium sulfide can be completely recovered from the chemical recovery system (Gu et al. 2013). Considering the capital costs, investment risk, technical feasibility, and efficiency for fermentable sugar production, the GL process shows a good performance in sugar production from agricultural wastes.

3.2.3 Ionic Liquid Pretreatment

Recently, using the ionic liquids as the solvents to pretreat the agricultural wastes has received much attention. At the room temperature, ionic liquids are a kind of fluid salt liquids, composed of organic cations and inorganic anions. The common inorganic anions in ionic liquids are Cl^-, Br^-, $AlCl_4{}^-$, $C_4F_9SO_3{}^-$, $(CF_3SO_2)N^-$, $CB_{11}H_{12}{}^-$, NO^{3-}, etc., whereas most of the organic cations are the derivative of imidazolium cation (Zakzeski et al. 2010). During pretreatment with ionic liquids, they act as solvent to dissolve (hemi)cellulose, lignin, or both disrupting the shield formed by these biopolymers. Ionic liquids with properties of the excellent dissolubility and good stability can quickly dissolve the network structure of lignocellulosic biomass, improving the accessibility and saccharification efficiency of cellulosic enzymes.

Li et al. (2009) found that after enzymatic hydrolysis (12 h), a 54.8 % yield of the sugars was achieved from the wheat straw pretreated by 1-ethyl-3-methyl imidazolium diethyl phosphate ($[Emim]^+$) liquid at 130 °C for 30 min. Using ammonia and [Emim]Ac to pretreat the rice straw, the pretreatment performed a synergistic effect on cellulose recovery and glucose released, with the yield of 82 % and 97 %, respectively. For corn stover, it reported that a high sugar yield with 70–80 % can be achieved when it was pretreated by 1-ethyl-3-methyl imidazolium chloride ([Emim]Cl) (Binder and Raines 2010). Sun et al. (2015) reported that 80 % of glucose could be released from the corn stover pretreated by 1-ethyl-3-methylimidazolium acetate ($[C_2C_1Im]$ [OAc]) after enzymatic saccharification.

Although application of ionic liquids can improve the released fermentable sugars from the agricultural wastes, many challenges prevail in applying these potential applications. The main challenges are high cost of ionic liquids, insufficient toxicological data, lack of basic physicochemical characteristic knowledge, mode of action on hemicellulose and/or lignin, and inhibitor generation (Menon and Rao 2012). Meanwhile, commercial ionic liquid recovery methods have not been fully developed.

3.3 *Physicochemical Pretreatment*

The physicochemical pretreatment refers to the combination of physical and chemical processes. Steam explosion, ammonia fiber explosion, and liquid hot water pretreatments are the major physicochemical pretreatments for the agricultural wastes to release the sugars.

3.3.1 Steam Explosion

Steam explosion is a cost-effective pretreatment widely used for agricultural wastes. During the steam explosion process, lignocellulosic material is treated by high-pressure saturated steam, with a sudden pressure drop causing rapid gas expansion within the material. The process can cause a large reduction of hemicellulose and a small amount of lignin degradation, due to the high temperature and acetic acid formation (Chandra et al. 2007). The removal of hemicellulose and lignin creates more pores in the pretreated biomass and improving reaction areas for the enzymatic saccharification.

Many factors can affect the efficiency of steam explosion, such as temperature, residence time, and particle size. The typically conditions for agricultural wastes are applied at a temperature of 160–260 °C and various residence times (Menon and Rao 2012). Beltrame et al. (1992) found that the optimal conditions for wheat straw was 230 °C with 1–2 residence time. After enzymatic hydrolysis, a maximum glucose yield of 93.5 % was achieved. Comparing that value with untreated wheat straw (11.8 %), it indicated that a great improvement in enzymatic saccharification efficiency was obtained after steam explosion. For the pretreatment technology, adding H_2SO_4 or SO_2 in steam explosion process can accelerate the removal of hemicellulose and lead to improved yields of enzymatic hydrolysis at lower pretreatment temperature. In 2006, Ballesteros et al. (2006) studied the effects of steam explosion on the acid-impregnated wheat straw at different temperatures and residence times. The highest yields of glucose and degree of hemicellulose dissolution were obtained from the steam explosion of acid-impregnated wheat straw at 180 °C for 10 min. For corn stover, Bondesson et al. (2013) investigated steam explosion pretreatment on the corn stover (with and without sulfuric acid catalyst), examining the effects of different residence times and temperatures on sugar released. After enzymatic hydrolysis, a maximum glucose yield with 86 % was obtained from the corn stover pretreated at 210 °C for 10 min, while the sugar yields maximized at 78 % from the sample pretreated by same conditions with sulfuric acid.

Steam explosion pretreatment has some attractive features, compared to other pretreatments. According to Avellar and Glasser (1998) report, these attractive features include the potential of significantly lower environmental impact, lower capital investment, more potential for energy efficiency, less hazardous process chemicals and higher carbohydrates recovery. However, there are still drawbacks of the steam explosion. The main disadvantages of the process are the partial hemicellulosic degradation and the generation of toxic compounds. These compounds can negatively affect the following enzymatic saccharification and fermentation steps (Galbe and Zacchi 2012).

3.3.2 Ammonia Fiber Explosion

Ammonia fiber explosion (AFEX) is an alkaline thermal pretreatment with liquid anhydrous ammonia at 60–100 °C for various amounts of time followed by a rapid pressure drop. The released pressure process causes an expansion of ammonia gas, which results in swelling, physical disruption of biomass fibers, and partial cellulose decrystallization (Alvira et al. 2010). Kumar and Wyman (2009b) reported that deacetylation of hemicellulose was also observed from the agricultural wastes, when using the AFEX pretreatment. After AFEX pretreatment, the digestibility of biomass is significantly increased and therefore improves the yield of released sugars. Mes-Hartree et al. (1988) conducted comparative analyses between steam explosion and AFEX, with the substrate of wheat straw. The results indicated that a highest released glucose with 0.38 g could be obtained from 1 g wheat straw pretreated by AFEX. Teymouri et al. (2004) observed that a maximum glucose yield with 98 % was reached for the pretreated corn stover, which was treated by the AFEX at 90 °C for 5 min with the ammonia/dry biomass ratio of 1:1. Meanwhile, AFEX has shown to be a promising pretreatment for rice straw, with only 3 % of sugars being lost during pretreatment process (Zhong et al. 2009). All the reported research indicates that enzymatic digestibility of AFEX-treated agricultural wastes can produce glucose and xylose with a yield more than 90 % and 80 %, respectively. Therefore, the AFEX can be an attractive pretreatment for fermentable sugar production from agricultural wastes. Meanwhile, a small amount of toxic inhibitors is generated during the AFEX process, while the pretreated sample contains higher lignin content and a part of hemicellulose is degraded during the process (Sun and Cheng 2002).

3.3.3 Liquid Hot Water

Liquid hot water (LHW) pretreatment is a hydrothermal process without any added chemicals. High pressure maintains the water in the liquid state above boiling temperatures, with the temperature from 170 °C to 230 °C (Sánchez and Cardona

2008; Yu et al. 2010; Min et al. 2015). By this pretreatment, most of the hemicellulose is removed from the raw biomass, making the cellulose more accessible to the enzymes. Pérez et al. (2007) reported that pretreatment temperature and time played the most important role on hemicellulose dissolution and enzymatic digestibility of pretreated wheat straw. The maximum yields of xylose recovery and glucose released were 80 % and 91 % under the optimum LHW pretreatment, respectively. Meanwhile, it has been shown that 80 % of the hemicellulose can be removed from the corn stover and improve the enzymatic digestibility of pretreated material. To optimize the recovery of hemicellulosic sugars and enhance the digestibility of pretreated biomass, two-step LHW pretreatment has been investigated.

It has been reported the LHW pretreatment has attractive and cost-saving potentials, requires no catalysts, uses a low-cost reactor construction, shows high sugar extraction and recovery, and forms little to no inhibitor (Kim et al. 2009; Alvira et al. 2010). Additionally, the particles are broken apart during pretreatment, eliminating the need for biomass size reduction. Therefore, LHW pretreatment shows a promising process for the fermentable sugar production in a large scale.

3.4 Biological Pretreatment

Previously, biological pretreatment has been studied for enhancing lignocellulosic materials for applications like feed and paper. This eco-friendly pretreatment has been applied to remove lignin from agricultural wastes for enhancing enzymatic saccharification for fermentable sugar production in ethanol processes (Alvira et al. 2010). Generally, the microorganisms of brown-, white-, and soft-rot fungi show the most promise in sugars recovery and delignification.

Hatakka (1983) investigated the effect of white-rot fungi on wheat straw and found 35 % of the carbohydrate was converted to fermentable sugars when the straw was pretreated by *Pleurotus ostreatus* for 35 d, with only 12 % sugar conversion for the un-pretreated straw. In 2007, Patel et al. (2007) obtained five different fungi through a screening process and applied these fungi to pretreat the wheat straw. The results showed that pretreatment with *Aspergillus niger* and *Aspergillus awamori* showed the best performance in obtaining the total fermentable sugars. For rice straw, Taniguchi et al. (2005) used four white-rot fungi to evaluate their effects on the structure changes and enzymatic digestibility. The results indicated the white-rot fungi of *Pleurotus ostreatus* showed the best performance. When rice straw was pretreated by *Pleurotus ostreatus* for 60 d, the degree of delignification was 41 %, and the degradation yields of cellulose and hemicellulose were only 17 % and 48 %, respectively. After 48 h enzymatic hydrolysis process, 44 % of the cellulose was converted into glucose in the pretreated rice straw.

The advantages from biological pretreatment are no additional chemical and energy, low capital cost, and mild environmental conditions. However, most ligninolytic microorganisms can not only degrade the lignin but also solubilize a part of cellulose and hemicellulose. Moreover, the degradation system of ligninolytic microorganisms is difficult to control so far (Alvira et al. 2010; Binod et al. 2010). To obtain a cost-competitive biological pretreatment for agricultural wastes, more screening studies need to be conducted to discover novel *Basidiomycota* fungi able to delignify agricultural wastes quickly and efficiently. Additionally, investigations into genetic modifications to known fungal species to increase their laccase and peroxidase productions are underway.

Different pretreatments for agricultural wastes have been widely investigated to improve the fermentable sugar production for downstream chemical production. An ideal pretreatment should have the following characteristics: decreasing the crystallinity of cellulose, increasing the accessible surface area by delignification, making the substrate digestible for enzymes, obtaining the maximum amount of available sugars from feedstocks, producing low toxic inhibitor, and being environmentally friendly (Alvira et al. 2010; Binod et al. 2010). Thus far, there has not been one single pretreatment that has been found to meet all requirements; every pretreatment has shown to have unique advantages and disadvantages. Table 1 shows these characteristics for the pretreatments mentioned previously.

Table 1 Advantages and disadvantages of pretreatment methods used for agricultural wastes

Pretreatment	Advantages	Disadvantages
Milling	1. Reduces cellulose crystallinity	1. High power and energy consumption
Alkaline	1. Removes hemicelluloses and lignin 2. Increases accessible surface area 3. High glucose and xylose yield 4. Low formation of inhibitors	1. Long residence times required 2. Irrecoverable salts formed and incorporated into biomass 3. Water pollution
Acid	1. High glucose yield 2. Obtaining the xylose without enzymes 3. Reduces hemicellulose content	1. Solvents need to be drained and recycled 2. Reactor corrosion problems 3. Formation of inhibitors 4. Generation of degradation products
Steam explosion Liquid hot water	1. Causes lignin transformation and hemicellulose solubilization 2. Higher yield of glucose and hemicellulose in the two-step method 3. Cost-effective	1. Generation of toxic compounds 2. Partial hemicellulose degradation
AFEX	1. Increases accessible surface area 2. Low formation of inhibitors	1. Not efficient for raw materials with high lignin content 2. High cost of large amount of ammonia
Biological	1. Degrades lignin and hemicellulose 2. Low energy consumption	1. Low rate of hydrolysis

4 Enzymatic Hydrolysis

Enzymatic hydrolysis is the second process to release the fermentable sugars from the lignocellulosic biomass. In this process, (hemi)cellulose-degrading enzymes are applied to cleave the polymers of (hemi)cellulose. Cellulases play a crucial role in the enzymatic process by initiating the catabolic hydrolysis of cellulose to glucose. Hemicellulases can degrade the hemicellulose covered on the cellulose fibers and expose more accessible surface area for cellulose hydrolysis (Menon and Rao 2012). Hence, hemicellulase is supplemented with cellulase to raise the yield of enzymatic hydrolysis for a high sugar production. The remaining lignin in the pretreated biomass can continue to limit the accessibility by a surface barrier and a nonproductive adsorption of enzymes, though this is greatly reduced through pretreatment. Yang et al. (2011) suggested that cellulase could be irreversibly adsorbed or entrapped within the three-dimensional matrix of lignin, which acts as a barrier that significantly reduces enzyme-cellulose interactions. The hydrophobic interaction, or ionic-type interaction, is thought to cause the adsorption of cellulases and hemicellulases onto lignin (Eriksson et al. 2002; Berlin et al. 2006). In response, many of the surfactants have been tested through addition in the enzymatic hydrolysis aqueous phase to reduce the electrostatic interactions that are believed to cause the enzymes to unproductively bind to lignin residues (Talebnia et al. 2010).

4.1 Cellulase and Hemicellulases

Due to the low growth rates of the bacteria, most of the commercial cellulase and hemicellulase are produced from fungi. As shown in Tables 2 and 3, cellulases can be produced by *Trichoderma reesei* and *Penicillium*, whereas hemicellulases can be

Table 2 The production of cellulases using microorganisms

Microorganism	Filter paper activity (FPU/ mL)	β-Glucosidase activity (IU/ mL)
Trichoderma reesei	2.49	2.17
Trichoderma reesei RUT C 30	6.2	0.39
Trichoderma species A-001	18	49
Trichoderma reesei ZU 02	0.25	Not detected
Trichoderma viride	0.88	0.33
Penicillium funiculosum	1.4	9.29
Penicillium pinophilum	2.0	10
Penicillium janthinellum	0.55	2.31
Penicillium decumbans	20.4	Not detected
Penicillium occitanis	23	Not detected

Table 3 The production of hemicellulase using microorganisms

Microorganism	Enzyme	Specific activity (IU/ mL)
Trichoderma longibrachiatum	Endo-1,4-β-xylanase	6630
Aspergillus nidulans	β-1,4-xylosidase	107
Aspergillus niger	Exo-β-1,4-mannosidase	188
	Feruloyl esterase	156
	Endo-*a*-1,5-arabinanase	90
	a-L-arabinofuranosidase	396
Schizophyllum commune	Acetyl xylan esterase	227

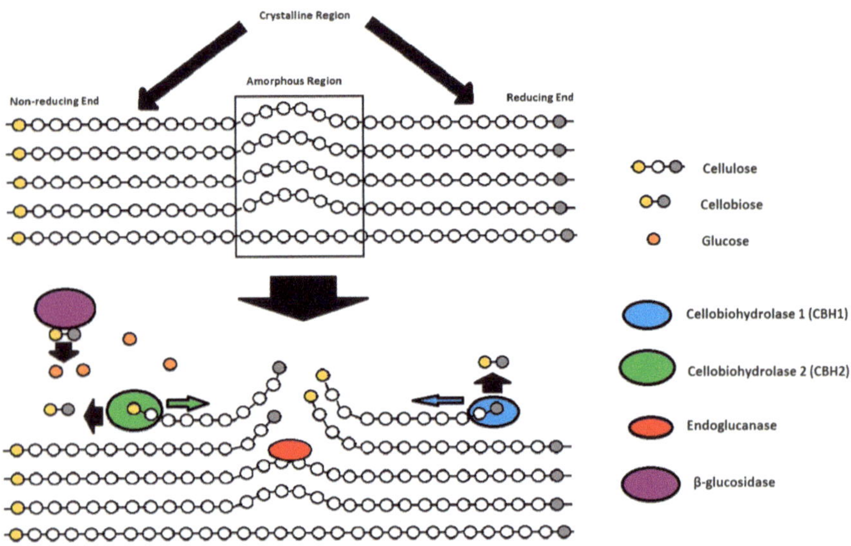

Fig. 2 The action of the different cellulases in the enzymatic hydrolysis process

produced from various organisms, such as *Aspergillus*, *Trichoderma longibrachiatum*, and *Schizophyllum commune*.

For synergistic action of carbohydrates to sugar during enzymatic hydrolysis process, cellulolytic enzyme cocktails should contain three majors group: endo-glucanase, exo-glucanase, and β-glucosidase. The endo-glucanase (EG) cleaves bonds from a random site along the cellulose chains, typically believed to attack amorphous regions of the cellulose bundles, whereas the exo-glucanase (CBH) cleaves cellobiose dimers from the cellulose chain ends, having two subtypes (I and II) with preferences toward reducing and nonreducing ends of the polysaccharide. Finally, β-glucosidase (β-G) catalyzes the production of the final monomeric glucose sugars from the solubilized cellobiose dimers (Henrissat et al. 1998; Lynd et al. 2002). The combined action of the cellulases is described in Fig. 2. It is reported that β-G plays a detoxifying role in the enzymatic hydrolysis process, and cellobiose has shown to be an end product inhibitor to both exo-glucanase and endo-glucanase (Rabinovich et al. 2002). Cellulase enzymes typically possess two protein domains, the catalytic domain and the cellulose-binding module (CBM). The structures of the catalytic domains in the main cellulase types have evolved toward their specific tasks. Endo-glucanase has a catalytic cleft optimal for the adsorption/desorption mechanism, while exo-glucanase has a tunnel-like catalytic domain, suitable for its progressive hydrolytic action along the polysaccharide chain (Teeri 1997).

The main glycosyl hydrolase to depolymerize the hemicellulose backbone is endo-1,4-β-D-xylanase. Endo-1,4-D-xylanase reduces the degree of polymerization of the substrate via initiating glycosidic bond cleavage in the xylan backbone (Biely 1993). The most abundant hemicellulose polymer in agricultural wastes is xylan consisting of *D*-xylose backbones, adorned with various side groups: *L*-arabinose, *D*-galactose, acetyl, feruloyl, *p*-coumaroyl, and glucuronic acid residues (Mazeau et al. 2005). Hence, hemicellulases, cooperating *a*-*L*-arabinofuranosidase, *a*-glucuronidase, acetyl xylan esterase, and hydroxycinnamic acid esterases with β-xylanase and β-xylosidase, are required for complete side chain residue cleavage in order to completely hydrolyze xylan to monosaccharides (Gilbert and Hazlewood, 1993).

4.2 *Hydrolysis of Cellulose and Hemicellulose*

Obtaining the fermentable sugars from carbohydrates in the agricultural wastes by hydrolysis has been explored since the early 1970s (Menon and Rao 2012). In the last century, obtaining the sugars was originally carried out through the application of mild reaction conditions (low temperature, acid charge, and long retention time) leading to the hydrolysis of cellulose and hemicellulose. For example, Yin et al. (1982) used the condition of 2 % H_2SO_4 and 110–120 °C to hydrolyze hemicellulose fractions in rice straw, giving a xylose yield over 70 %. Valkanas et al. (1998) pretreated rice straw by various acids at different acid concentrations. The results showed that with 3 h reaction time, the hemicellulose of rice straw was monomerized into the fermentable monosaccharides. Although acid hydrolysis can produce the fermentable sugars from the lignocellulosic biomass, many toxic inhibitors are generated during the process from the carbohydrates.

The hydrolysis technology for the fermentable sugar production from agricultural wastes has made significant progress. Now, most of the fermentable sugars are obtained from the agricultural wastes by the processes of pretreatment and enzymatic saccharification. In the laboratory studies, most of the cellulases are applied at temperatures of 45–55 °C and pH of 4–5, with a dosage of 10–30 (FPU/g glucan) and a reaction time of 48–72 h (Galbe and Zacchi 2002). Compared with acid-catalyzed hydrolysis, the application of enzymatic saccharification with different enzymes performs a better process for fermentable sugar production, since no degradation products are formed from glucose. However, it is a slower process than acid-catalyzed hydrolysis. Now, acid hydrolysis process is used as the pretreatment for the fermentable sugar production, combining with enzymatic hydrolysis process. For example, Abedinifar et al. (2009) applied the dilute H_2SO_4 and enzymatic hydrolysis on rice straw. The results showed that 720 mg/(1000 mg mass) sugar was obtained from the pretreated straw by dilute sulfuric acid after 48 h enzymatic hydrolysis, higher than that from untreated straw (460 mg/(1000 mg mass)). According to the previous reports, 0.56–0.67 g/(g mass) sugars can obtain from the wheat straw, after consecutive pretreatment and enzymatic saccharification process, with a glucose yield from 40 % to 90 % under the optimal reaction temperature and pH (Saha et al. 2005, Saha and Cotta 2007; Aderemi et al. 2008). For corn straw, it is reported that using different cellulase dosages, 15–90 % of the carbohydrates that remained in the pretreated solid residue could be converted to the sugars in subsequent enzymatic hydrolysis process (Lloyd and Wyman 2005). In 2009, NREL studied the whole slurry enzymatic saccharification of the acid-pretreated corn stover for fermentable sugar production. It reported that when 20 mg cellulase (per 1 g glucan) was applied in the enzymatic hydrolysis process, the cellulose-to-glucose conversion was approached 90 % at 20 wt % total acid-pretreated corn stover, also with the xylan-to-xylose conversion yield reached to 75 % (Humbird et al. 2011).

In the lignocellulosic material cell wall, hemicellulose coats the cellulose microfibrils and forms a barrier for efficient cellulose degradation. Meanwhile, the hemicelluloses can be covalently linked to the lignin, which is called lignin-carbohydrate complexes (LCCs) (Balakshin et al. 2011; Du et al. 2014; Huang et al. 2016). In the cell wall of agricultural wastes, the LCCs usually link via ester-ether bridges of ferulic acid to the C-5 hydroxyl position of arabinofuranoside in arabinoxylan (Chesson et al. 1983). Investigation by Min et al. (2014) showed that steric hindrance of multiple cross-linkages of LCCs could limit the action of cellulase to attack the cellulose. Hence, hemicellulases, like xylanase, α-L-arabinofuranosidase (AF), and feruloyl esterase, have been applied to degrade xylan and break the covalent bonds in the LCCs, facilitating the enzymatic saccharification of pretreated agricultural wastes (Kumar et al. 2009; Huang et al. 2015a).

Chu et al. (2014b) reported that when a specialized enzyme cocktail (cellulase, β-glucosidase, and xylanase with a ratio of 1:1.88:6.61) was applied to the corn stover pretreated by green liquor, a near-theoretical hydrolysis yield was achieved. The results showed that the glucan and xylan hydrolysis yields were 93.53 % and 86.00 %, respectively. 36.08 g of glucose and 14.06 g of xylose were obtained from 100 g of dry corn stover. Tabka et al. (2006) applied an enzyme cocktail, including xylanases, and feruloyl esterase to saccharify the pretreated wheat straw. The authors reported that applying feruloyl esterase with xylanases could cleave the diferulic bridges in xylan chains, which loosens and opens up the intrastructural matrix of the material. A highest glucose yield with 81 % was achieved when the cellulase, xylanase, and feruloyl esterases were applied simultaneously during the enzymatic hydrolysis process. Recently, Alvira et al. (2011) evaluated the effects of two recombinant hemicellulases (endoxylanase and AF) on the enzymatic saccharification of the wheat straw pretreated by steam explosion. The results showed the enzymatic hydrolysis yield was increased by 10 % when these two hemicellulases are applied with cellulase, indicating that accessory enzymes can improve the enzymatic digestibility of the pretreated biomass.

Recently, adding nonionic surfactants has shown to increase the enzymatic digestibility of lignocellulosic materials. PEG 6000 has been found to improve the enzymatic saccharification of the SO_2-catalyzed stream-exploded corn stover. An increased glucose yield with 17.41 % was achieved from the pretreated corn stover after enzymatic saccharification with 0.10 g/(g glucan) PEG 6000, comparing with no addition of surfactant (Ouyang et al. 2011). Saha et al. (2005) reported that Tween 20 has a stimulative effect on the enzymatic digestibility of H_2SO_4-pretreated wheat straw, increasing the sugar production from 0.49 g to 0.52 g/(g biomass). Kristensen et al. (2007) investigated the effects of four different types of nonionic surfactants (Berol, PEG, Tween 80, and bovine serum albumin) on the enzymatic saccharification of wheat straw pretreated by sulfuric acid. The results showed that the highest increase in enzymatic saccharification (70 %) was obtained from the pretreated straw with the addition of surfactant Berol 08 during the enzymatic saccharification process.

5 Conclusion

Biotechnologies designed to produce fermentable sugars from lignocellulosic biomass have been promoted worldwide. The utilization of agricultural waste is an excellent choice for bio-based chemical production, while converting unwanted agricultural waste streams into valuable resources. In this review, the biocatalytic conversion of agricultural wastes to the fermentable sugars, including the processes of pretreatments and enzymatic hydrolysis, was described comprehensively. Among the different pretreatments, chemical and thermochemical methods have shown to be the most effective and promising pretreatment for industrial applications. For the enzymatic hydrolysis, different enzymatic hydrolysis technologies have been applied to solve the issues in xylose and glucose productions from pretreated agricultural wastes. To further increase the amount of fermentable sugar production and decrease the cost of the whole process, the pretreatment should remove as little polysaccharides in the substrates as possible while minimizing sugar derivative formation in order to obtain near-theoretical sugar yields, while utilizing low enzyme loading during the enzymatic hydrolysis process.

6 Opinion

Although fermentable sugar production has seen many improvements through the invention of new technologies, few references exist that truly compare the economics of different pretreatments and enzymatic hydrolysis processes. Different lignocellulosic materials require unique pretreatment strategies, which likely require different pretreatment reactors and conditions. In addition, the costs of the biocatalytic enzymes are still regarded as major hurdles for economic viability of fermentable sugars. So, some economic analysis of the enzymatic hydrolysis process for pretreating agricultural wastes would be of great use in the field of pretreatment and fermentable sugar production research. To produce the fermentable sugars from agricultural waste in a commercial scale, further aspects of the economic analysis of the process need to be addresses.

Acknowledgments The research was supported by the National Natural Science Foundation of China (31570561) and the Priority Academic Program Development of Jiangsu Higher Education Institution (PAPD).

References

Abdel-Hamid AM, Solbiati JO, IK C (2013) Insights into lignin degradation and its potential industrial applications. Adv Appl Microbiol 82:1–28. doi:10.1016/B987-0-12-407679-2.00001-6

Abedinifar S, Karimi K, Khanahmadi M, Taherzadeh MJ (2009) Ethanol production by Mucor indicus and Rhizopus oryzae from rice straw by separate hydrolysis and fermentation. Biomass Bioenergy 33:828–833. doi:10.1016/j.biombioe.2009.01.003

Aderemi BO, Abu E, Highina BK (2008) The kinetics of glucose production from rice straw by *Aspergillus niger*. Afr J Biotechnol 7:1745–1752. doi:10.5897/AJB08.109

Agbor VB, Cicek N, Sparling R, Berlin A, Levin DB (2011) Biomass pretreatment: fundamentals toward application. Biotechnol Adv 29:675–685. doi:10.1016/j.biotechadv.2011.05.005

Alvira P, Tomás-Pejó E, Ballesteros M, Negro MJ (2010) Pretreatment technologies for an efficient bioethanol production process based on enzymatic hydrolysis: a review. Bioresour Technol 101:4851–4861. doi:10.1016/j.biortech.2009.11.093

Alvira P, Negro MJ, Ballesteros M (2011) Effect of endoxylanase and a-L-arabinofuranosidase supplementation on the enzymatic hydrolysis of steam exploded wheat straw. Bioresour Technol 102:4552–4558. doi:10.1016/j.biortech.2010.12.112

Avellar BK, Glasser WG (1998) Steam-assisted biomass fractionation. I. Process considerations and economic evaluation. Biomass Bioenergy 14:205–218. doi:10.1016/S0961-9534(97)10043-5

Balakshin M, Capanema E, Gracz H, Chang HM, Jameel H (2011) Quantification of lignin-carbohydrate linkages with high-resolution NMR spectroscopy. Planta 233:1097–1110. doi:10.1007/s00425-011-1359-2

Ballesteros I, Negro MJ, Oliva JM, Cabanas A, Manzanares P, Ballesteros M (2006) Ethanol production from steam-explosion pretreated wheat straw. Appl Biochem Biotechnol 130:496–508. doi:10.1007/978-1-59745-268-7_41

Barakat A, de Vries H, Rouau X (2013) Dry fractionation process as an important step in current and future lignocellulose biorefineries: a review. Bioresour Technol 134:362–373. doi:10.1016/j.biortech.2013.01.169

Beltrame PL, Carniti P, Visciglio A, Focher B, Marzetti A (1992) Fractionation and bioconversion of steam-exploded wheat straw. Bioresour Technol 39:165–171. doi:10.1016/0960-8524(92)90136-L

Berlin A, Balakshin M, Gilkes N, Kadla J, Maximenko V, Kubo S, Saddler J (2006) Inhibition of cellulase, xylanase and β-glucosidase activities by softwood lignin preparations. J Biotechnol 125:198–209. doi:10.1016/j.jbiotec.2006.02.021

Biely P (1993) Biochemical aspects of the production of microbial hemicellulases. In: Coughlan MP, Hazlewood GP (eds) Hemicellulose and hemicellulases. Portland Press, Cambridge

Binder JB, Raines RT (2010) Fermentable sugars by chemical hydrolysis of biomass. Proc Natl Acad Sci USA 107:4516–4521. doi:10.1073/pnas.0912073107

Binod P, Sindhu R, Singhania R, Vikram S, Devi L, Nagalakshmi S, Kurien N, Sukumaran RK, Pandey A (2010) Bioethanol production from rice straw: an overview. Bioresour Technol 101:4767–4774. doi:10.1016/j.biortech.2009.10.079

Bondesson PM, Galbe M, Zacchi G (2013) Ethanol and biogas production after steam pretreatment of corn stover with or without the addition of sulphuric acid. Biotechnol Biofuels 6. doi:10.1186/1754-6834-6-11

Chandra RP, Bura R, Mabee WE, Berlinl A, Pan X, Saddler JN (2007) Substrate pretreatment: the key to effective enzymatic hydrolysis of lignocellulosics? Biofuels 108:67–93. doi:10.1007/10_2007_064

Chen M, Zhao J, Xia L (2009) Comparison of four different chemical pretreatments of corn stover for enhancing enzymatic digestibility. Biomass Bioenergy 33:1381–1385. doi:10.1016/j.biombioe.2009.05.025

Chen WH, Pen BL, Yu CT, Hwang WS (2011) Pretreatment efficiency and structural characterization of rice straw by an integrated process of dilute-acid and steam explosion for bioethanol production. Bioresour Technol 102:2916–2924. doi:10.1016/j.biortech.2010.11.052

Cherubini F, Strømman AH (2011) Chemicals from lignocellulosic biomass: opportunities, perspectives, and potential of biorefinery systems. Biofuels Bioprod Biorefin 5:548–561. doi:10.1002/bbb.297

Chesson A, Gordon AH, Lomax JA (1983) Substituent groups linked by alkali-labile bonds to arabinose and xylose residues of legume, grass and cereal straw cell walls and their fate during digestion by rumen microorganisms. J Sci Food Agric 34:1330–1340. doi:10.1002/jsfa.2740341204

Chu QL, Li X, Yang DL, Xu Y, Ouyang J, Yu SY, Yong Q (2014a) Corn Stover bioconversion by green liquor pretreatment and a selected liquid fermentation strategy. BioResources 9:7681–7695. doi:10.15376/biores.9.4.7681-7695

Chu QL, Huang Y, Li X, Fan YM, Jin YC, Yu SY, Yong Q (2014b) Improved enzymatic hydrolysis of corn stover by green liquor pretreatment and a specialized enzyme cocktail. BioResources 9:4489–4502. doi:10.15376/biores.9.3.4489-4502

Donohoe BS, Selig MJ, Viamajala S, Vinzant TB, Adney WS, Himmel ME (2009) Detecting cellulase penetration into corn stover cell walls by immuno-electron microscopy. Biotechnol Bioeng 103:480–489. doi:10.1002/bit.22281

Du XY, Pérez-Boada M, Fernández C, Rencoret J, del Río J, Jiménez-Barbero J, Li JB, Gutiérrez A, Martínez AT (2014) Analysis of lignin-carbohydrate and lignin-lignin linkages after hydrolase treatment of xylan-lignin, glucomannan-lignin and glucan-lignin complexes from spruce wood. Planta 239:1079–1090. doi:10.1007/s00425-014-2037-y

Eriksson T, Börjesson J, Tjerneld F (2002) Mechanism of surfactant effect in enzymatic hydrolysis of lignocellulose. Enzyme Microb Technol 31:353–364. doi:10.1016/S0141-0229(02)00134-5

Galbe M, Zacchi G (2002) A review of the production of ethanol from softwood. Appl Microbiol Biotechnol 59:618–628. doi:10.1007/s00253-002-1058-9

Galbe M, Zacchi G (2012) Pretreatment: the key to efficient utilization of lignocellulosic materials. Biomass Bioenergy 46:70–78. doi:10.1016/j.biombioe.2012.03.026

Gilbert HJ, Hazlewood GP (1993) Bacterial cellulases and xylanases. J Gen Microbiol 139:187–194. doi:10.1099/00221287-139-2-187

Glassner D, Hettenhaus J, Schechinger T (1999) Corn stover potential: recasting the corn sweetener industry. In: Janick J (ed) Perspectives on new crops and new uses. ASHS, Alexandria, VA, pp 74–82

Glover JD, Cox CM, Reganold JP (2007) Future farming: a return to roots. Sci Am 297:82–89. doi:10.1038/scientificamerican0807-82

Gu F, Yang LF, Jin YC, Han Q, Chang HM, Jameel H, Phillips R (2012) Green liquor pretreatment for improving enzymatic hydrolysis of corn stover. Bioresour Technol 124:299–305. doi:10.1016/j.biortech.2012.08.054

Gu F, Wang WX, Jing L, Jin YC (2013) Effects of green liquor pretreatment on the chemical composition and enzymatic digestibility of rice straw. Bioresour Technol 149:375–382. doi:10.1016/j.biortech.2013.09.064

Hall M, Bansal P, Lee JH, Realff MJ, Bommarius AS (2010) Cellulose crystallinity-a key predictor of the enzymatic hydrolysis rate. FEBS J 277:1571–1582. doi:10.1111/j.1742-4658.2010.07585.x

Hallac BB, Ragauskas AJ (2011) Analyzing cellulose degree of polymerization and its relevancy to cellulosic ethanol. Biofuels Bioprod Bioref 5:215–225. doi:10.1002/bbb.269

Hatakka AI (1983) Pretreatment of wheat straw by white-rot fungi for enzymic saccharification of cellulose. Appl Microbiol Biotechnol 18:350–357. doi:10.1007/BF00504744

Henrissat B, Teeri TT, Warren RAJ (1998) A scheme for designating enzymes that hydrolyse the polysaccharides in the cell walls of plants. FEBS Lett 425:352–354. doi:10.1016/S0014-5793(98)00265-8

Hideno A, Inoue H, Tsukahara K, Minowa T, Inoue S, Endo T, Sawayama S (2009) Wet disk milling pretreatment without sulfuric acid for enzymatic hydrolysis of rice straw. Bioresour Technol 100:2706–2711. doi:10.1016/j.biortech.2008.12.057

Hu J, Zhang Z, Lin Y, Zhao S, Mei Y, Liang Y, Peng N (2015) High-titer lactic acid production from NaOH-pretreated corn stover by Bacillus coagulans LA204 using fed-batch simultaneous saccharification and fermentation under non-sterile condition. Bioresour Technol 182:251–257. doi:10.1016/j.biortech.2015.02.008

Huang CX, He J, Li X, Min DY, Yong Q (2015a) Facilitating the enzymatic saccharification of pulped bamboo residues by degrading the remained xylan and lignin-carbohydrates complexes. Bioresour Technol 192:471–477. doi:10.1016/j.biortech.2015.06.008

Huang CX, Chu QL, Xie YH, Li X, Jin YC, Min DY, Yong Q (2015b) Effect of kraft pulping pretreatment on the chemical composition, enzymatic digestibility, and sugar release of moso bamboo residues. BioResources 10:240–255. doi:10.15376/biores.10.1.240-255

Huang CX, He J, Wang Y, Min DY, Yong Q (2015c) Associating cooking additives with sodium hydroxide to pretreat bamboo residues for improving the enzymatic saccharification and monosaccharides production. Bioresour Technol 193:142–149. doi:10.1016/j.biortech.2015.06.073

Huang CX, He J, Du LT, Min DY, Yong Q (2016) Structural characterization of the lignins from the bamboo green and bamboo yellow of bamboo culm (*Phyllostachys pubescens*). J Wood Chem Technol 36:157–172. doi:10.1080/02773813.2015.1104544

Huber GW, Iborra S, Corma A (2006) Synthesis of transportation fuels from biomass: chemistry, catalysts, and engineering. Chem Rev 106:4044–4098. doi:10.1021/cr068360d

Humbird D, Davis R, Tao L, Kinchin C, Hsu D, Aden A, Schoen P, Lukas J, Olthof B, Worley M, Sexton D, Dudgeon D (2011) Process design and economics for biochemical conversion of lignocellulosic biomass to ethanol: dilute-acid pretreatment and enzymatic hydrolysis of corn stover. National Renewable Energy Laboratory Technical Report NREL/TP-5100-47764

Ibrahim MM, El-Zawawy WK, Abdel-Fattah YR, Soliman NA, Agblevor FA (2011) Comparison of alkaline pulping with steam explosion for glucose production from rice straw. Carbohyd Polym 83:720–726. doi:10.1016/j.carbpol.2010.08.046

Isikgor FH, Becer RC (2015) Lignocellulosic biomass: a sustainable platform for the production of bio-based chemicals and polymers. Polym Chem 6:4497–4559. doi:10.1039/c5py00263j

Jäger G, Büchs J (2012) Biocatalytic conversion of lignocellulose to platform chemicals. Biotechnol J 7:1122–1136. doi:10.1002/biot.201200033

Jin SB, Ja KK, Young HH, Byung CL, In-Geol C, Heon KK (2009) Improved enzymatic hydrolysis yield of rice straw using electron beam irradiation pretreatment. Bioresour Technol 100:1285–1290. doi:10.1016/j.biortech.2008.09.010

Jin YC, Jameel H, Chang HM, Phillips R (2010) Green liquor pretreatment of mixed hardwood for ethanol production in a repurposed kraft pulp mill. J Wood Chem Technol 30:86–104. doi:10.1080/02773810903578360

Khoo HH (2015) Review of bio-conversion pathways of lignocellulose-to-ethanol: Sustainability assessment based on land foot print projections. Renew Sust Energy Rev 46:100–119. doi:10.1016/j.rser.2015.02.027

Kim Y, Mosier NS, Ladisch MR (2009) Enzymatic digestion of liquid hot water pretreated hybrid poplar. Biotechnol Prog 25:340–348. doi:10.1002/btpr.137

Kootstra AMJ, Beeftink HH, Scott EL, Sanders JPM (2009) Comparison of dilute mineral and organic acid pretreatment for enzymatic hydrolysis of wheat straw. Biochem Eng J 46:126–131. doi:10.1016/j.bej.2009.04.020

Koullas DP, Christakopoulos P, Kekos D, Macris BJ, Koukios EG (1992) Correlating the effect of pretreatment on the enzymatic hydrolysis of straw. Biotechnol Bioeng 39:113–116. doi:10.1002/bit.260390116

Kristensen JB, Börjesson J, Bruun MH, Tjerneld F, Jorgensen H (2007) Use of surface active additives in enzymatic hydrolysis of wheat straw lignocellulose. Enzyme Microb Tech 40:888–895. doi:10.1016/j.enzmictec.2006.07.014

Kumar R, Wyman CE (2009a) Does change in accessibility with conversion depend on both the substrate and pretreatment technology? Bioresour Technol 100:4193–4202. doi:10.1016/j.biortech.2008.11.058

Kumar R, Wyman CE (2009b) Effect of xylanase supplementation of cellulase on digestion of corn stover solids prepared by leading pretreatment technologies. Bioresour Technol 100:4203–4213. doi:10.1016/j.biortech.2008.11.057

Kumar P, Barrett DM, Delwiche MJ, Stroeve P (2009) Methods for pretreatment of lignocellulosic biomass for efficient hydrolysis and biofuel production. Ind Eng Chem Res 48:3713–3729. doi:10.1021/ie801542g

Leu SY, Zhu JY (2013) Substrate-related factors affecting enzymatic saccharification of lignocelluloses: our recent understanding. Bioenergy Res 6:405–415. doi:10.1007/s12155-012-9276-1

Li Q, He YC, Xian M, Jun G, Xu X, Yang JM, Li LZ (2009) Improving enzymatic hydrolysis of wheat straw using ionic liquid 1-ethyl-3-methyl imidazolium diethyl phosphate pretreatment. Bioresour Technol 100:3570–3575. doi:10.1016/j.biortech.2009.02.040

Lloyd TA, Wyman CE (2005) Combined sugar yields for dilute sulfuric acid pretreatment of corn stover followed by enzymatic hydrolysis of the remaining solids. Bioresour Technol 96:1967–1977. doi:10.1016/j.biortech.2005.01.011

Lomborg CJ, Thomsen MH, Jensen ES, Esbensen KH (2010) Power plant intake quantification of wheat straw composition for 2nd generation bioethanol optimization-a near infrared spectroscopy (NIRS) feasibility study. Bioresour Technol 101:1199–1205. doi:10.1016/j.biortech.2009.09.027

López-Bellido L, Wery J, López-Bellido RJ (2014) Energy crops: prospects in the context of sustainable agriculture. Eur J Agron 60:1–12. doi:10.1016/j.eja.2014.07.001

Lynd LR, Woods J (2011) Perspective: a new hope for Africa. Nature 474:S20–S21. doi:10.1038/474S020a

Lynd LR, Weimer PJ, Zyl WH, Pretorius IS (2002) Microbial cellulose utilization: fundamentals and biotechnology. Microbiol Mol Biol R 66:506–577. doi:10.1128/MMBR.66.3.506-577.2002

Maiorella BL (1985) Ethanol. In: Moo-Young M (ed) Comprehensive biotechnology. Pergamon Press, Oxford, pp 861–914

Mazeau K, Moine C, Krausz P, Gloaguen V (2005) Conformational analysis of xylan chains. Carbohyd Res 340:2752–2760. doi:10.1016/j.carres.2005.09.023

McIntosh S, Vancov T (2010) Enhanced enzyme saccharification of *Sorghum bicolor* straw using dilute alkali pretreatment. Bioresour Technol 101:6718–6227. doi:10.1016/j.biortech.2010.03.116

Menon V, Rao M (2012) Trends in bioconversion of lignocellulose: biofuels, platform chemicals & biorefinery concept. Prog Energ Combust 38:522–550. doi:10.1016/j.pecs.2012.02.002

Mes-Hartree M, Dale BE, Craig WK (1988) Comparison of steam and ammonia pretreatment for enzymatic hydrolysis of cellulose. Appl Microbiol Biotechnol 29:462–468. doi:10.1007/BF00269069

Min DY, Li QZ, Chiang V, Jameel H, Chang HM, Lucia L (2014) The influence of lignin-carbohydrate complexes on the cellulase-mediated saccharification I: Transgenic black cottonwood (western balsam poplar, California poplar) *P. trichocarpa* including the xylan down-regulated and the lignin down-regulated lines. Fuel 119:207–213. doi:10.1016/j.fuel.2013.11.047

Min DY, Xu R, Hou Z, Lv JQ, Huang CX, Jin YC, Yong Q (2015) Minimizing inhibitors during pretreatment while maximizing sugar production in enzymatic hydrolysis through a two-stage hydrothermal pretreatment. Cellulose 22:1253–1261. doi:10.1007/s10570-015-0552-z

Mussatto SI, Roberto IC (2004) Optimal experimental condition for hemicellulosic hydrolyzate treatment with activated charcoal for xylitol production. Biotechnol Prog 20:134–139. doi:10.1021/bp034207i

Ouyang J, Ma R, Huang WT, Li X, Chen M, Yong Q (2011) Enhanced saccharification of SO_2 catalyzed steam-exploded corn stover by polyethylene glycol addition. Biomass Bioenergy 35:2053–2058. doi:10.1016/j.biombioe.2011.01.047

Parfitt J, Barthel M, Macnaughton S (2010) Food waste within food supply chains: quantification and potential for change to 2050. Philos Trans R Soc B Biol Sci 365:3065–3081. doi:10.1098/rstb.2010.0126

Patel SJ, Onkarappa R, Shobha KS (2007) Study of ethanol production from fungal pretreated wheat and rice straw. Internet J Microbiol 4. doi:10.5580/b7c

Pedersen M, Meyer AS (2009) Influence of substrate particle size and wet oxidation on physical surface structures and enzymatic hydrolysis of wheat straw. Biotechnol Prog 25:399–408. doi:10.1021/bp.141

Peng HD, Luo H, Jin SY, Li HQ, Xu J (2014) Improved bioethanol production from corn stover by alkali pretreatment with a novel pilot-scale continuous microwave irradiation reactor. Biotechnol Bioproc E 19:493–502. doi:10.1007/s12257-014-0014-8

Pérez JA, Gonzalez A, Oliva JM, Ballesteros I, Manzanares P (2007) Effect of process variables on liquid hot water pretreatment of wheat straw for bioconversion to fuel-ethanol in a batch reactor. J Chem Technol Biotechnol 82:929–938. doi:10.1002/jctb.1765

Rabinovich ML, Melnik MS, Boloboba AV (2002) Dedicated to the memory of I.V. Berezin and R.V. Feniksova microbial cellulases (review). Appl Biochem Microbiol 38:305–321. doi:10.1023/A:1016264219885

Ragauskas AJ, Beckham GT, Biddy MJ, Chandra R, Chen F, Davis MF, Dixon RA, Gilna P, Keller M, Langan P, Naskar AK, Saddler JN, Tschaplinski TJ, Tuskan GA, Wyman CE (2014) Lignin valorization: improving lignin processing in the biorefinery. Science 344:1246843. doi:10.1126/science.1246843

Saha BC, Cotta MA (2007) Enzymatic hydrolysis and fermentation of lime pretreated wheat straw to ethanol. J Chem Technol Biotechnol 82:913–919. doi:10.1002/jctb.1760

Saha BC, Iten LB, Cotta MA, Wu YV (2005) Dilute acid pretreatment, enzymatic saccharification and fermentation of wheat straw to ethanol. Process Biochem 40:3693–3700. doi:10.1016/j.procbio.2005.04.006

Sánchez ÓJ, Cardona CA (2008) Trends in biotechnological production of fuel ethanol from different feedstocks. Bioresour Technol 99:5270–5295. doi:10.1016/j.biortech.2007.11.013

Sun Y, Cheng J (2002) Hydrolysis of lignocellulosic materials for ethanol production: a review. Bioresour Technol 83:1–11. doi:10.1016/S0960-8524(01)00212-7

Sun R, Lawther JM, Banks WB (1995) Influence of alkaline pre-treatments on the cell wall components of wheat straw. Ind Crop Prod 4:127–145. doi:10.1016/0926-6690(95)00025-8

Sun N, Xu F, Sathitsuksanoh N, Thompson VS, Cafferty K, Li CL, Tanjore D, Narani A, Pray TR, Simmoms BA, Singh S (2015) Blending municipal solid waste with corn stover for sugar production using ionic liquid process. Bioresour Technol 186:200–206. doi:10.1016/j.biortech.2015.02.087

Tabka MG, Herpoël-Gimbert I, Monod F, Asther M, Sigoillot JC (2006) Enzymatic saccharification of wheat straw for bioethanol production by a combined cellulase xylanase and feruloyl esterase treatment. Enzyme Microbiol Tech 39:897–902. doi:10.1016/j.enzmictec.2006.01.021

Talebnia F, Karakashev D, Angelidaki I (2010) Production of bioethanol from wheat straw: an overview on pretreatment, hydrolysis and fermentation. Bioresour Technol 101:4744–4753. doi:10.1016/j.biortech.2009.11.080

Taniguchi M, Suzuki H, Watanabe D, Sakai K, Hoshino K, Tanaka T (2005) Evaluation of pretreatment with Pleurotus ostreatus for enzymatic hydrolysis of rice Straw. J Biosci Bioeng 100:637–643. doi:10.1263/jbb.100.637

Teeri TT (1997) Crystalline cellulose degradation: new insight into the function of cellobiohydrolases. Trends Biotechnol 15:160–167. doi:10.1016/S0167-7799(97)01032-9

Teymouri F, Laureano-Pérez L, Alizadeh H, Dale BE (2004) Ammonia fiber explosion treatment of corn stover. Appl Biochem Biotechnol 116:951–963. doi:10.1007/978-1-59259-837-3_77

Valkanas GN, Valkanas NP, Vlyssides AG, Theodoropoulos AG (1998) Method for production of ethyl alcohol. In: U.S.P. 5766895.

Weiss ND, Farmer JD, Schell DJ (2010) Impact of corn stover composition on hemicellulose conversion during dilute acid pretreatment and enzymatic cellulose digestibility of the pretreated solids. Bioresour Technol 101:674–678. doi:10.1016/j.biortech.2009.08.082

Werpy TA, Petersen G, Aden A, Bozell J, Holladay J, White J, Manheim A, Eliot D, Lasure L (2004) Top value added chemicals from biomass: I. Results of screening for potential candidates from sugars and synthesis gas. PNNL-14808. Pacific Northwest National Laboratory, Richland, WA

Wu SF, Chang HM, Jameel H, Phillips R (2010) Novel green liquor pretreatment of loblolly pine chips to facilitate enzymatic hydrolysis into fermentable sugars for ethanol production. J Wood Chem Technol 30:205–218. doi:10.1080/02773811003746717

Xiang Q, Kim JS, Lee YY (2003) A comprehensive kinetic model for dilute-acid hydrolysis of cellulose. Appl Biochem Biotechnol 3:337–352. doi:10.1007/978-1-4612-0057-4_27

Yang B, Dai Z, Ding SY, Wyman C (2011) Enzymatic hydrolysis of cellulosic biomass. Biofuels 2:421–449. doi:10.4155/bfs.11.116

Yin F, Hwang A, Yu N, Hao P (1982) Hydrolysis of agricultural wastes. In: Proceedings of resource recovery solid wastes conference, pp 447–456

Yu G, Yano S, Inoue H, Inoue S, Endo T, Sawayama S (2010) Pretreatment of rice straw by a hot-compressed water process for enzymatic hydrolysis. Appl Biochem Biotechnol 160:539–551. doi:10.1007/s12010-008-8420-z

Yu Z, Jameel H, Chang HM, Park S (2011) The effect of delignification of forest biomass on enzymatic hydrolysis. Bioresour Technol 102:9083–9089. doi:10.1016/j.biortech.2011.07.001

Yu H, Zhang M, Ouyang J, Shen Y (2014) Comparative study on four chemical pretreatment methods for an efficient saccharification of corn stover. Energy Fuels 28:4282–4287. doi:10.1021/ef5001612

Zakzeski J, Bruijnincx PCA, Jongerius AL, Weckhuysen BM (2010) The catalytic valorization of lignin for the production of renewable chemicals. Chem Rev 110:3552–3599. doi:10.1021/cr900354u

Zhang YHP (2013) Next generation biorefineries will solve the food, biofuels, and environmental trilemma in the energy-food-water nexus. Energy Sci Eng 1:27–41. doi:10.1002/ese3.2

Zhang QZ, Cai WM (2008) Enzymatic hydrolysis of alkali-pretreated rice straw by *Trichoderma reesei* ZM4-F3. Biomass Bioenergy 32:1130–1135. doi:10.1016/j.biombioe.2008.02.006

Zhang C, Lei X, Scott CT, Zhu JY (2014) Comparison of dilute acid and sulfite pretreatment for enzymatic saccharification of earlywood and latewood of Douglas fir. Bioenergy Res 7:362–370. doi:10.1007/s12155-013-9376-6

Zhong C, Lau MW, Balan V, Dale BE, Yuan YJ (2009) Optimization of enzymatic hydrolysis and ethanol fermentation from AFEX-treated rice straw. Appl Microbiol Biotechnol 84:667–676. doi:10.1007/s00253-009-2001-0

Various Sludge Pretreatments: Their Impact on Biogas Generation

J. Rajesh Banu and S. Kavitha

Abstract Biogas produced through anaerobic degradation of waste activated sludge (WAS) in wastewater treatment plants has gained much more attention as it is a renewable energy resource. An apparent advantage of anaerobic digestion is the generation of biogas that presents a renewable substitute for utilization of fossil fuels. It was observed that pretreatment has to be done prior to anaerobic degradation (AD) to enhance the biogas generation. This review deals with various pretreatments—chemical, physical, mechanical, combinative, phase-separated pretreatment, their mode of action, advantages and their impact on biogas generation. An extensive review of these pretreatment techniques could plausibly aid in increasing the biogas production and thus rises the sustainable energy generation.

Keywords Anaerobic digestion • Biogas • Physical pretreatment • Chemical pretreatment • Activated sludge flocs • Extracellular enzymes

1 Introduction

In recent years, wastewater treatment plants (WWTPs) have gradually increased, wide-reaching, as a result of swift urbanization. Superfluous sludge generated from different treatment plants is a massive environmental threat. The excess sludge disposal accounts up to 60 % of the overall operating cost of WWTPs. The common processes for treatment and disposal include biodegradation, heat treatment, dumping and so on. Among them, anaerobic digestion has been considered to be a sustainable option to current disposal strategies of sludge as the quantity of the organic waste is reduced and stabilized, and energy in the form of biogas is improved. Anaerobic degradation (AD) is applied at most municipal wastewater treatment plants (MWWTPs). It has numerous considerable merits over other existing methods. The anaerobic degradation of organics has dual benefits, i.e. treating unwanted materials and biogas which can be utilized as a substitute

J.R. Banu (✉) • S. Kavitha
Department of Civil Engineering, Regional Centre of Anna University, Tirunelveli 627007, Tamil Nadu, India
e-mail: rajeshces@gmail.com

L. Singh, V.C. Kalia (eds.), *Waste Biomass Management – A Holistic Approach*,
DOI 10.1007/978-3-319-49595-8_3

energy supply. During digestion 30–60 % of the degradable solids are transformed into biogas. The biogas generated through AD does not contain completely methane but also consist of CO_2 and lesser quantity of additional gases. Biogas upturn from sludge biomass by AD is a promising technology and has fascinated an escalating attention in current years due to the greater requirement of sustainable energy generation. The most extensively utilized conventional fuel is remnant fuel, which is transferred in the earth from plant or animal remnants, in a process (European Commission 2010). In up-to-date times, when fossil fuels are regularly diminishing in addition to growing costs and unsteadiness in the major producer countries, renewable energy has become one of the best substitutes for sustainable energy improvement. Renewable energy plays an imperative task in dropping the greenhouse gases; predominantly energy from biomass could contribute appreciably as it is a "carbon neutral" fuel. The need for alternative sources of energy for both decentralized and centralized power generation has led to the proliferation of research into alternative energy sources. Biogas is a combustible combination of gases formed by the inbuilt fermentation of wet biomass (anaerobic digestion/fermentation). The principal explosive element methane composed about 50–75 volume per cent (vol. %). Erstwhile molecules present in biogas comprise carbon dioxide, sulphide, oxygen and water vapour. Biogas generation happens obviously in wetlands when organic matter is decayed by anaerobic microbes to so-called swamp gas. A number of developing countries use biogas extensively. In India and China solitary, there are over 1 million diminutive, uncomplicated plants, all treating dissipate (sewage, animal manure, crop residues, etc.) from distinct domestic areas. The plants are excavated into the land and are unheated. The biogas is utilized in the housekeeping for cuisine, and the digested biomass is applied as a fertilizer. In AD process, the organic residues will be transformed by bacteria into the biogas through four sequential phases, i.e. hydrolysis, acidogenesis, acetogenesis and methanogenesis in anaerobic digestion process. The progression of anaerobic digestion has the prospective of transforming recyclable organics into biogas. On the other hand, in lack of a pretreatment, the biogas production potential of anaerobic digestion is inadequate due to the existence of refractory microbial cell walls and other organic materials present in the sludge. As a consequence to enhance the biogas augmentation, pretreatment is believed to be the plausible option. Sludge disintegration was extended to boost bioenergy generation by hastening the hydrolysis pace of AD.

To increase the biogas production, assorted pretreatment practices are applied. The recent development in AD process leads to improvement through pretreatment techniques which allows disintegration of waste. This in turn hasten the biodegrdability potential of AD. Pretreatment enhances liquefaction and its rate, which consequently enhance biogas production. All pretreatments bring about the breakdown of sludge biomass, thus solubilizing and discharging substance present inside the aqueous phase and converting obstinate organic substances into recyclable nature consequently creating substances easily accessible to microbes. These entire pretreatments are exposed to enhance the biogas generation in subsequent AD process. The various pretreatments include physiochemical (low-temperature thermochemical pretreatment), mechanical (ultrasonication, disperser pretreatment), chemical (acid, alkali, ozone pretreatment) and biological (enzymatic, microbial or biosurfactant pretreatment). This chapter focuses on review of various pretreatment processes for improving biogas generation.

2 Biogas

More than the preceding century, the incessant escalation of populace and industrialization has firmly impacted the environment. The two specifics in assistance can be directly coupled with two inconveniences in the authentic world: an amplified utilization of water and the inevitable requisite of elevated quantity of fuels. With the intention to lessen the upshot of increment in water utilization, wastewater treatment plants (WWTPs) were flattering, more sophisticated and more universal worldwide. An expansion in the number, extent and competence of the treatment plants has noteworthy affirmative collisions on the eminence and healthiness of marine surroundings. The preferred treatment process is connected with the type of waste generated. Sewage sludge is one among the derivatives in WWTPs, and its ultimate clearance is a crisis since it signifies up to 50 % of the functional expenditure during the set-up (Appels et al. 2008). In 1992, the quantity of municipal sludge generated in Europe was about 5.2×10^9 Kg/year, and in 1998 the volume became 7.2×10^9 Kg/year. The execution of the committee ordinance on urban wastewater treatment (91/171/EEC) resulted in numerous fresh treatment plants upcoming into function, and the overall quantity of sewage sludge was further amplified into twofold in 2006, when compared to 1998. A lately published statement primed for the European Commission estimated a production of 12×10^9 Kg/year of sludge in 2010 and expected an increment to roughly 13×10^9 Kg/year 2020 (European Commission 2010). So, there is an urgent need to dispose these huge quantities of sludge in an eco-friendly as well as cost-effective manner.

Biogas formed from the anaerobic digestion of these sludges at wastewater treatment plants (WWTPs) is considered as renewable energy. In topical years, the acquirement of fossil fuels has turn out to be pricier for economic and political purposes. The anxiety has been improved on the conservative furnish of energy owing to the continual obligation of energy that enhanced the significance of renewable and nonconventional source of energy. In contrast, due to the blazing of fossil fuels, the probability of global warming is moreover increased by which most of the countries pulled towards the significance of nonconventional resource of energy. Bioenergy production based on putrefaction of sludge substance positively is supportive in resolving the predicament of energy crisis in the household of staff/laboratory at site and to confiscate some stress from the conservative resources of energy.

In treatment plants, sewage is being treated, and the enduring residue (sludge) is called as slurry. The greater proportion of the slurry is having organic part and is estimated to be in the range of 70–80 %. The slurry is being condensed in thickener and propelled to digester tank. In the anaerobic digester tank, microbial action takes

Fig. 1 A photograph of semicontinuous anaerobic digester

place and generates 55–65 % of methane with the other gases (CO_2, H_2S). This is called as biomethanation process. On completion of digestion, the biogas is collected at the apex of the digester and transported to the gas burner (Metcalf and Eddy Inc 2006). A photograph of semicontinuous anaerobic digester is depicted in Fig. 1. Biogas is an eco-friendly, fresh and economical fuel. The present approach to prevail over these concerns is to keep on developing renewable energy from innate resources, such as geothermal, hydro, solar, tidal power and biomass.

It has been suggested that one of the accepted renewable carbon resources that is hefty enough to be utilized as an alternative for fossil fuel is biomass (Klass 2004). Biomass is the word employed to explain all biologically formed substance. Solid, gaseous or liquid fuels formed from biomass are biofuels. Biogas is one sort of biofuel that can be formed biologically by the untreated breakdown of organics. Biogas is formed by anaerobic digestion or fermentation of biodegradable biomass such as wastewater treatment sludge, municipal waste, agricultural and forestry waste, human and animal wastes and energy crops. This type of biogas is comprised principally of methane and carbon dioxide and is progressing to expand consideration right through the world. The methane produced from these organic transfer processes can be ignited or tarnished to discharge energy. Methane formed from biomass can be considered as a type of solar energy, as photosynthesis coalesces atmospheric carbon dioxide with water in the existence of sunlight to form the biomass. The process can be minimally obtainable by the equation

$$CO_2 + 2H_2O + 2H_2O \rightarrow ([CH_2O] + H_2O) + O_2$$

where $[CH_2O]$ symbolized the biomass (as carbohydrates) and oxygen is also produced as a side benefit of this reaction. The carbon dioxide bounce in the biomass, which is unconfined during the bioenergy upturn process, can be seen as the equilibrium of that exploited in photosynthesis and can be accessible to form fresh biomass (Peppley 2006). Thus, methane as a biogas formed from biomass during anaerobic digestion can be measured as an inexpensive and eco-friendly biofuel. Biogas is generally forced or compacted to the operation pressure needed for accurate appliance and then metred to the gas use apparatus. Previous to this, biogas may be treated to remove dampness, H_2S and CO_2, the main pollutants in biogas, in which case the biogas turn into methane. Elimination of pollutants, chiefly H_2S, will give up an intermediary product that has been described as partly upgraded biogas. Depending on the appliances, biogas may be stored either before or after processing, at "squat or lofty pressures". Collected biogas can be utilized by itself as fuel for burning, or it can be combusted in an apparatus to produce electricity or burst. If the biogas is improved to biomethane, extra uses may be probable. Biogas can be used for warmth production, production of current and heat or be upgraded to vehicle fuel. Production of heat and/or electricity in a gas reservoir, gas apparatus, gas turbine or fuel cell design can be consummated with the methane content generally reached in a digester 55–75 % (De Mes et al. 2003). For utilizing as vehicle fuel, the methane content has to be improved to at least 96–97 %. Biogas of vehicle fuel eminence has the similar methane concentration as ordinary gas and can be disseminated in an ordinary gas network.

Biogas, a fresh and renewable form, could be an excellent alternative for conservative energy reserve (remnant fuels, oil, etc.) that are the basis for ecological-environmental trouble and at the equivalent instant exhausting at a rapid speed. Regardless of its abundant benefits, the probability of biogas techniques might not be entirely exploited or spouted, as definite restraints are also allied with it. Most common amid these are the hydraulic retention time of 30–50 days, stumpy gas production in winter, etc. Consequently, attempts are obligatory to confiscate its diverse restrictions in order to initiate this technology in the pastoral region. Researchers have strived diverse practices to augment gas generation.

3 Anaerobic Digestion (AD)

Anaerobic digestion is the chiefly efficient practice for the management of sludge and generation of biogas (Banu et al. 2007a, b; Banu and Kaliappan 2007). AD can be an attractive option to convert the biomass into a biofuel. So a profound understanding about anaerobic digestion process to generate biogas is crucial. AD, as stated by Fraser (2010), is the upshot of a sequence of reactions between varieties of microorganisms. It takes place in four stages, namely, hydrolysis/liquefaction, acidogenesis, acetogenesis and methanogenesis. The first assemblage of microorganism oozes enzymes that lyse the polymeric materials to monomers such as glucose and amino acids. These are consequently transformed by the second assemblage (*Acetobacter*) of bacteria to volatile fatty acids such as acetic acid. Ultimately, the third assemblage of bacteria, *methanogens*, transfers H_2, CO_2 and acetate to CH_4. In the initial phase of hydrolysis, hydrolytic bacteria transfer the unhydrolyzable multifaceted organics such as proteins, carbohydrates and cellulose into soluble substances like amino acids, glucose and fatty acids (Appels et al. 2008). The hydrolytic activity is of noteworthy consequence in elevated organic waste and possibly will turn into a rate-restrictive step in anaerobic digestion (Banu et al. 2006; Rajakumar et al. 2010). Some industrialized functions beat this restriction by the utilization of diverse disintegration that augment hydrolysis. The purpose of pretreatments to boost up the initial step has been established to get a reduced digestion period, and this in turn affords a higher biogas yield.

3.1 Biochemical Methane Potential (BMP) Assay

An evaluation of pretreatment is necessary in order to scrutinize the potential of pretreated sludge for biogas production. Traditionally, biochemical methane potential (BMP) assay was used by many researchers to evaluate pretreatment efficiency. The BMP analysis is extensively employed to establish the anaerobic biodegradability of wastes. It is used to quantify the extent of methane production from given substrate (Angelidaki et al. 2009), and this aids the possibility of exploring the given substrate further for biofuel production. This assay can used to arrive at a biodegradability index of given substrate, which is subsequently used to scale up the process. It is a comparatively economical method for relative evaluation of anaerobic digestibility and probable biogas generation. It could be employed to establish relationship between organic content of a substance and its transformation into biogas. Thus, it was primarily recognized as effortless and economical process to scrutinize evaluation of AD of given substrates (Owen et al. 1979).

A photograph of BMP set-up and schematic diagram of BMP reactor are given for reference (Fig. 2). This simple experimentation can be carried out in batch reactors with inoculums substrate ratio as 3:1. After adding the substrates and the inoculums, the top portion over the liquor in the digesters are flushed by N_2 to remove O_2 present in the medium wholly to assure anaerobic environment. The digesters are closed by a lid to formulate it air stiff and swathed with silver sheet to afford a murky surrounding essential for development of microbes. The digesters are kept in orbital shaker (150–200 rpm). The gas is estimated by placing a needle in the lid. Gas build-up during the digestion period is collected with a help of a syringe. The syringe piston is pressed upwards owing to the increment in pressure within the digester when the gas is generated, and the exiled quantity is noted. The quantity of methane can be estimated through gas chromatograph.

Fig. 2 Experimental set-up of BMP assay

4 Pretreatment Technologies

Even though AD process has several benefits, several limitations also exist. AD process is frequently restricted by much extended retention times (20–30 days) and a stumpy decomposition of the organics (30–50 %). Restriction in spacing is also one of the restraining issues in anaerobic digestion at huge extent. By diminishing the stalk of the method, the flexibility and the usage of the process can be improved. The degradation of particulate organics into soluble substrates, known as hydrolysis, is the rate-limiting step during AD (Gerardi 2003; Aquino and Tuckey 2008; Deublein and Steinhauser 2008). The way to enhance AD performance is by pretreating the waste to convert insoluble organic polymers into soluble components (monomers). By doing this it is possible to enhance hydrolysis rate, which subsequently increase the biogas generation, as well as reduce the digestion time and the amount of final residuals (Mata Alvarez 2003).

4.1 Advantages of Pretreatment

The sludge pretreatment leads to the following advantages (Panter 2006):

- Increment in performance of digesters
- Hastened decomposition of sludge matter
- Increment in biogas generation
- Exclusion of pathogens
- Reduction of specific sludge production
- Lessening of bulking and foam formation
- Enhancement in dewaterability and settling properties
- Enhanced degradability of complicated substances and industrial remains
- High hasty digestion
- Enhanced solids reduction
- Elevated residuals stability
- More homogeneous end product

4.2 Mechanism of Pretreatment

To enhance digestion performance including elevated definite biogas generation, the main reasonable approach is to interrupt microbial cell walls in the waste activated sludge, and to allow degradable materials to release outside their cells is an important mechanism in the hydrolysis phase of digestion. As a result, modern technologies with a potentially advanced anaerobic digestion performance of sludge have emerged and were determined by sludge pretreatment before anaerobic digestion to hasten the rate-restricting step of sludge digestion. The purpose of pretreatment as disintegration of WAS is the discharge of cell content of microbes present in the sludge through distraction of their cell walls for the escalation of anaerobic digestion performance. The discharged cell substances of microbes are called as cell lysate which can hasten the method of anaerobic methane fermentation. It promotes quicker and enhanced degradation of organic substances and the comparable augmentation of biogas production. Through pretreatment, the bioavailability of sludge can be improved predominantly for anaerobic bacteria. As a result, enhanced methane production can be achieved.

4.3 Various Pretreatment Methods

There are lot of various pretreatment technologies which have high variabilities in their mode of action and outcome. In the following section, the outline of specific action of various pretreatment methods and their outcomes are summarized.

4.3.1 Physical Methods

Thermal Pretreatment

Thermal pretreatment is a physical technique that separates the liquid matter from solid organics and slackens the biomass arrangement of the residual organics (Mata-Alvarez et al. 2000; Ferrer et al. 2008). Thermal pretreatment is appropriate for the enhancement of stabilization, improvement of sludge dewatering and reduction of the pathogens (Uan et al. 2009, 2012; Uma et al. 2012a). As the stress and heat of thermal process increases, the organics subjected to treatment transformed to short chain fragments that are well appropriate for biodegradation by microbes (Wang et al. 2010).

Reaction Mechanism of Thermal Treatment

Thermal treatment is a fascinating option to decrease the surplus sludge production when linked with a usual bioprocess or anaerobic process (Paul et al. 2006). Thermal energy employed during heat pretreatment disrupts the chemical bonds of the cellular wall and the cellular covering, thus releasing the cellular materials into aqueous portion (Cacho rivero 2005; Uan et al. 2013a,b). In addition, thermal treatment causes the release of much water through disruption of sludge matrix (Bougrier et al. 2006; Esakki Raj et al. 2012, 2013). However this disintegration faces several major difficulties: odour problems, clogging of heat exchangers, corrosion and expensive. A photograph of thermal treatment set-up is given (Fig. 3).

High-Temperature Thermal Treatment

The heat treatment is considered to be low in the temperature range of less than 100 °C (Li and Noike 1992), medium in the temperature range of 100–175 °C and high in a temperature range of 175–225 °C. In these heat treatments, disintegration time with the general values in the range of 30–60 min is shown to have minimal effect than the temperature (Valo et al. 2004). Valo et al. (2004) reported solubilization increase of around 25 % and 60 % subsequent to thermal treatment of sludge at 130 and 170 °C, correspondingly. These authors also appraised the biogas generation of the heat-processed sludge in batch process and noted improvement of about 21 % and 45 % in biogas generation in contrast to unprocessed sludge. The most important shortcoming of this disintegration is elevated energy consumption.

Fig. 3 A photograph of thermal pretreatment set-up

Low-Temperature Thermal Treatment

This heat treatment has been considered as an effectual disintegration process for escalating biogas generation from both primary and secondary sludge (Uma et al. 2012a; Uan et al. 2013a). This disintegration entails lesser energy utilization. Some authors have reported that this disintegration employed at temperatures approximately 70 °C enhances microbial action of some thermophilic bacterial community in high thermophilic range (Ferrer et al. 2008).Thermal treatment at temperature less than 100 °C applied to conventional biological system led to noteworthy decrement of surplus sludge generation. In reality, a 50 % decrement in the surplus sludge production was noted when heat disintegration at 95 °C was employed to sludge from a principally settled urban wastewater treatment (Camocho et al. 2003). Only a few reports characterized the upshot of lower temperature on sludge biodegradation (Camocho 2001).

Microwave Pretreatment

Microwave pretreatment is a substitute option to usual heat treatment, and its prospective was highly renowned in sludge disintegration (Mudhoo and Sharma 2011; Ebenezer et al. 2015a, b). This radiation is electromagnetic having wavelength in the range of 1 mm–1 m equivalent to an oscillation frequency of about 300–0.3 GHz (Eskicioglu et al. 2007). Household microwave ovens and industrialized microwave producers are usually functioning at 2.45 GHz of oscillating frequency with a consequent wavelength of 12.24 cm and energy of 1.02×10^{-5} eV (Jones et al. 2002; Eskicioglu et al. 2007; Tang et al. 2010).

Reaction Mechanism of Microwave Treatment

The system of the microwave treatment involves thermal and an athermal (non-thermal) effect. The thermal effect is created by the impact of the fluctuating electrical field with bipolar compounds like H_2O, proteins, fats and other complex molecules. Potency (frequency), dosage, treatment time and dispersion intensity are the major parameters of microwave disintegration that affect the dielectric compounds (Park et al. 2010; Uma et al. 2013a). The substantial molecular rotary motion of the enduringly and provoked dipoles leads to abrasion that eventually heat the sludge (Dogan and Sanin 2009; Eskicioglu et al. 2007; Yu et al. 2010). The athermal effect is ascribed to the hastily shifting bipolar route in the polarized region of the cell membrane, which results in the rupture of hydrogen bonds, consequently paving the way for disintegration of the sludge matrix and alteration in the secondary and tertiary configuration of the proteins of the biomass (Eskicioglu et al. 2007; Tang et al. 2010; Park et al. 2010). Appels et al. (2013) reported about 50 % increment in biogas production of microwave pretreated sludge when compared to untreated sludge in their studies. Ebenezer et al. (2015a) reported about 0.385 (L/g VS) of biogas production potential in microwave pretreated sludge when compared to control (untreated sludge).

4.3.2 Chemical Methods

Chemical pretreatment is applied to attain the destruction of the organics through strong acids, alkalis or oxidants (Uan et al. 2013b). AD usually entails an alteration of the pH by rising alkalinity; thus alkali disintegration is the chosen chemical method (Li et al. 2012). Acidic pretreatments and oxidative methods such as ozonation are also employed to improve the biogas production and the hydrolysis pace. The upshots of chemical disintegration rely on the sort of method employed and the characteristics of the substrates. Chemical disintegration is inappropriate for simply recyclable substrates containing high quantity of carbohydrates, due to their hastened degradation and consequent build-up of VFA, which leads to letdown of the methanogenesis phase (Wang et al. 2012).

Alkaline Pretreatment

Alkaline pretreatment of sludge is a well-known method for rupturing the sludge biomass cell wall and membrane. It is known for field applicability as many of the full-scale treatment plants adopt alkaline pretreatment for sludge reduction. In addition, it was a cost-effective method compared to many other pretreatment techniques (Uma et al. 2012a, b). The various alkali agents used are NaOH, KOH, $Ca(OH)_2$, $Mg(OH)_2$, etc. Alkaline treatment of waste activated sludge using NaOH was thoroughly investigated by many researchers. Conversely, only some studies explored the efficiency of various alkali reagents for sludge pretreatment (Kim et al. 2003).

Reaction Mechanism of Alkaline Treatment

Alkaline treatment causes liquefaction or disintegration of flocs structure and cell membrane by hydroxyl radicals. During sludge pretreatment, the initial reaction involves salvation and saponification which provokes swelling of solids (Carlsson et al. 2012). As a result, the specific surface area is increased, and the substrates are easily accessible to anaerobic microbes (Hendriks and Zeeman 2009). An intense elevated pH induces the sludge protein to lose their usual forms, saponification of sludge lipid and hydrolysis of ribonucleic acid (RNA). Subsequent to the destruction of extracellular polymeric substances and gels, sludge biomass fails to maintain its turgor pressure. As a result, liquefaction of sludge happens which in turn discharges intracellular substances to the medium (Neyens et al. 2004; Li and Fang 2007). The alkali-mediated COD solubilization does not only correspond to high pH but also related to various other simultaneous actions such as saponification and neutralization. Neutralization effect is by acids, which are created during the degradation of the particulate organic matter (Kim et al. 2003). Lin et al. (2009) achieved an increment of 88 % of methane yield in NaOH pretreated sludge. Zhang et al. (2010) got accumulated methane yield of about 398 mL/g VSS in alkaline pretreated sludge.

Ozone Pretreatment

Ozone (O_3) is a potent oxidant and decontaminator. It has tough cell lytic action and acts with the polysaccharides, proteins, lipids and various multifaceted organic mixtures to change them into simpler ones. One more chemical pretreatment method is ozone treatment (Carrere et al. 2010; Banu et al. 2014; Sowmya et al. 2015), which does not create an increase of the salt quantity and devoid of chemical contamination when compared to other chemical pretreatment methods. In addition, it also eliminates the pathogens (Weemaes et al. 2000). Hence, ozone treatment has achieved vast attention for sludge pretreatment (Sri Bala Kameswari et al. 2011; Suresh Karthik et al. 2015). A photograph of ozone pretreatment is given (Fig. 4).

Reaction Mechanism of Ozone Treatment

Ozone is a strong oxidant, which decays itself into radicals and acts with organic substrates (Sri Bala kameswari et al. 2011) in two means: directly and indirectly. The direct reaction depends on the configuration of the reactant, whereas the indirect reaction depends on the hydroxyl radicals. Thus, the refractory substances turn into more recyclable and easily available to the anaerobic bacteria (Carballa et al. 2007). Among the techniques to collapse sludge, ozonation is one of the

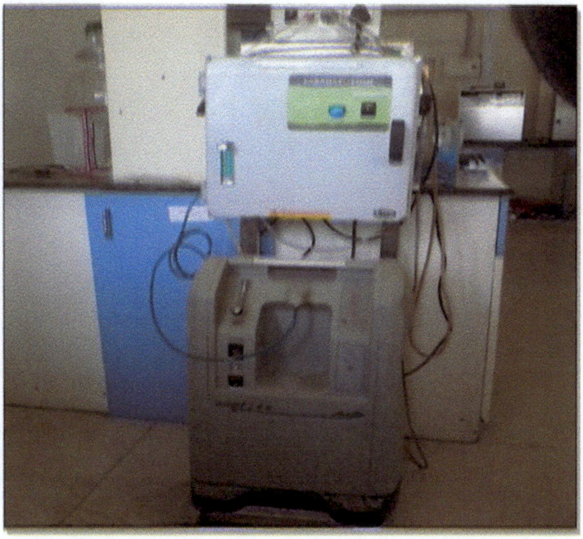

Fig. 4 A photograph of ozone pretreatment set-up

effectual ways and give up the utmost degree of dissolution. The flocs are wrecked down into very fine and minute, disseminated particles. Floc disruption (FD) and solubilization of sludge creates a large number of miniature particles diffused in the liquid part together with the soluble organic compounds. The sludge biodegradation is exaggerated if higher ozone dose is employed. Oxidation of organic molecules that are degradable for methanogenic bacteria will result in decrement of biogas production (Levlin 2010). The biogas production was augmented up to 80 % at ozone treatment with 0.1 g O_3/g COD, but the effect was not obvious at increased ozone concentration (Weemaes et al. 2000). By means of experimental test records, a comparable adverse effect for the utilization of O_3 was proposed (Vranitzky and Lahnsteiner 2002). If a process is carried out with an O_3 dosage of 0.06 g ozone/g dry solids, then AD extent enhanced from 45 % to 65 %, jointly by means of an increment in gas generation of 30–40%. Reactor action was unfavourably exaggerated as C and feed elimination competence diminished to some extent. The settle up period for executing O_3 processing at a 20,000 m^3reactor was calculated as 6 years.

4.3.3 Mechanical Methods

Ultrasonic Pretreatment

Ultrasonic pretreatment is a cell disruption technique which has been used in the biochemical field for more than four decades (Takatani et al. 1981).The crucial target of this treatment is to disintegrate the bacterial cell walls and to assist the accessibility of intracellular materials for subsequent degradation to digestion process. Ultrasonic irradiation has enhanced the process of degradability,

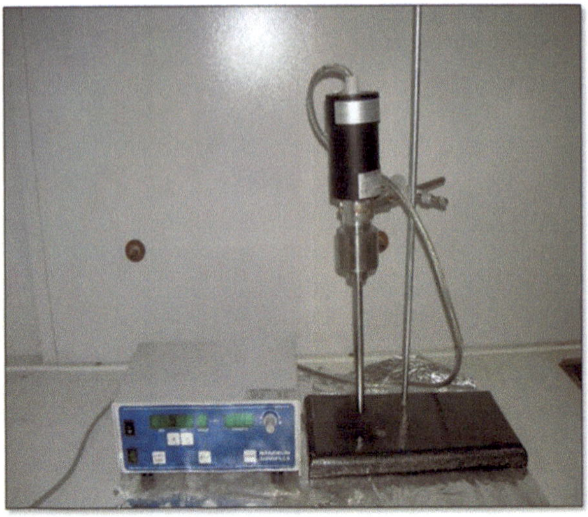

Fig. 5 A photograph of ultrasonic pretreatment set-up

liquefaction, disintegration of biomass, reduction of solids and decrement in floc size. This technique has the potential to enhance the digestion process and recovery of valuable products (Uma et al. 2013b).The efficiency of sonication lies on factors like time, reaction temperature, intensity of ultrasound, supply of energy, medium solid concentration, physiochemical characteristics of medium and ultrasonic device. A photograph ultrasonic pretreatment set-up is given (Fig. 5).

Reaction Mechanism of Ultrasonic Treatment

Sonication is the most effectual sludge pretreatment technique that has been studied comprehensively due to its/owing to its efficient and uncomplicated function (Tiehm et al. 2001; Chu et al. 2002; Bougrier et al. 2005; Uma et al. 2013b; Kavitha et al. 2015a). Ultrasound, when passing across the medium, creates stress emission that permeates through the particles and produces gas and vapour bubbles that develop and crumple forcefully at elevated speed; this effect is known as acoustic cavitation (Uma et al. 2013b). As this cavitation occurs, elevated shear stresses are generated and react on the materials in medium, which is primarily accountable for the sludge disintegration (Pilli et al. 2011). Cavitation takes place more rapidly at 20–40 kHz. Implosion and breakup of bubbles allow many smaller particles to be produced with high temperature of about 5000 K and pressures of 100 MPa, which could further induce efficiency of the process. The medium also experiences thermolytic process in the bubbles to liberate radical groups (H, OH and HOO) that collide with the materials in the medium (Zhang et al. 2008).The cavitation bubble was formed by ultrasound, and its radius was inversely proportional to the ultrasonic frequency (Tiehm et al. 2001). The ultrasonic cavitation bubble diameter can be calculated as

$$Rr \approx 3.28 fr^{-1} \quad (1)$$

where Rr (mm) is resonant bubble diagram and fr (kHz) is resonance intensity.

Cavitational crumple generates strong local temperature and elevated pressure on water-gas boundary, instability and elevated shearing event in the water. According to Bougrier et al. (2005), the overall effect of sonification is the sum of different mechanisms such as combustion, alkaline reactions by radicals, pyrolysis and shearing. On the other hand, it was observed that hydromechanical shear forces are chiefly accountable for ultrasonic dissolution of waste biomass. The localized elevated temperature and pressure provoke sludge pretreatment (Uma et al. 2013b). At elevated temperatures, lipids in the cell membrane become softer and cause puncture; this leads to the release of soluble organics in to exterior (Wang et al. 2005). Therefore, the quantity of dissolved solids is appreciably increased (Tiehm et al. 2001). Sonication mechanism is affected by three factors: specific energy, ultrasonic frequency and sludge characteristics (Khanal et al. 2007; Show et al. 2007). Energy input by ultrasound treatment has a crucial role during treatment in accordance with profitable analysis, and this will be a key parameter in choosing the well-suited ultrasound method for application. Ultrasonic-mediated sludge disintegration was effective only at high specific energy, and this increases pretreatment cost. For example, Bougrier et al. (2008) have used specific energy (SE) as high as 14,547 to achieve a biogas yield 1.84 times higher than the control. However, in recent years through phase-separated pretreatment, it is possible to achieve effective sludge disintegration in relatively low specific energy (Poornima et al. 2014). Working on phase-separated pretreatment, Gayathri et al. (2015) have achieved two times higher biogas production in sludge at relatively lower SE input of 7108. The following equation was used to calculate the specific energy (SE):

$$\mathrm{SE}\,(\mathrm{kJ/kg\,TS}) = P\,.\,t/V\,.\,\mathrm{TS}$$

where

P power in KW
t treatment time
V volume of sample
TS initial total solids concentration in kg

Among the various terminology used to represent ultrasonic dosage, SE is reliable and most widely used (Rai and Rao 2009). It can be applicable to measure energy requirement for disintegration for all types of sludge samples since input parameters for its measurement are simple and reliable.

Homogenization (Disperser) Pretreatment

Homogenization treatment was followed by dairy and foodstuff industries since the middle of the century. Even though the technique was applied to disperse food emulsions, the thorough understanding of the action mechanism was well established only after 1970 through basic research. More lately, in the early 1990s, a new invention of homogenizers, known as high-pressure homogenizers (HPH) also known as dispersers, has been developed.

Reaction Mechanism of Dispersion Treatment

Dispersion is also an eminent mechanical technique for biomass perturbance. Dispersion is the disbanding and dissemination of a solid, fluid or gaseous stage. Latterly, this has been suggested as a WAS disruption method, which transforms the speed as well as degree of sludge disintegration in the AD process (Rai and Rao 2009; Kavitha et al. 2016a). It has been chiefly used for making firm the stuffs from food and dairy suspensions and has paved way for progressively innovative relevance fields (Jacquel et al. 2008). It normally depends on cleaving microbial cell membrane and the discharge of inclusions into supernatant part by the utility of pressure gradient, cavitations and shear stresses (Zhang et al. 2008). This method has various merits in association to other disruption methods, i.e. no element alteration or denaturation during cleavage of cells, simple action and elevated lytic competence. The rich power requisite in dispersal pretreatment being diminished by joining with additional processing techniques lying on that diverse mode of reaction possibly will permit attaining elevated disruption. A photograph of disperser pretreatment set-up is depicted (Fig. 6). This new technique has a different treatment compartments that can achieve pressures 10–15 times greater than conventional dispersers. This process was usually based on the disintegration of sludge cell walls and the discharge of cytoplasm into the aqueous portion through the

Fig. 6 A photograph of disperser pretreatment set-up

effect of stress ascent, cavitation, instability, impingement, shear pressure and extended shear (Zhang et al. 2008; Kavitha et al. 2014a). Uma et al. (2012b) got 76 % of biogas production in dispersion alkaline pretreatment.

4.3.4 Biological Methods

Biological pretreatment can speed up hydrolysis rate via the catalytic activity of enzymes including lipases, proteinases, cellulases and hydrolases (Dohanyos et al. 2004; Gopi et al. 2015) and also through the addition of microbial cultures. In biological pretreatment, the pace of lytic action is mediated by hydrolases, and hence it is also a function of the action and amount of the microbial community in the digester. Literatures on the direct inoculation of bacteria or complex enzymes or a mixture of carbohydrases, peptidases and lipases have proved the improvement of fermentation (Veera Lakshmi et al. 2014). Furthermore, substantial consideration was given to the enzymatic disintegration of proteins for the enhancement of digestibility, since protein disintegration is considered as the rate-limiting step during WAS digestion (Haner et al. 1994).

Reaction Mechanism of Biological Treatment

Protease enzymes are universal, being found in all microbes and are indispensable for cell development and segregation. The exoproteases found outside the cells are of industrial value and discovered to have numerous uses in different industrial zone. Even though there are numerous microbial sources accessible for secreting proteolytic enzymes, barely some are renowned as industrial producers (Gupta et al. 2002a).Among them, strains of *Bacillus* sp. lead the industrialized zone (Gupta et al. 2002b). Priest in 1977 stated that the gram-positive, spore-forming bacterium *Bacillus* sp. secretes proteases, esterases and other kinds of extracellular enzymes. Additionally, numerous authors have explored about the secretion of proteolytic enzyme from *Bacillus subtilis* (Uchida et al. 1972).

Protease enzymes secreted by microbes may be positioned inside the cell (intracellular), cell membrane linked (periplasmic) or secreted into the system (extracelluar) (Kohlmann et al. 1991). Exoenzymes are typically proficient in degrading unsolvable materials such as cellulose, protein and starch, and the degraded materials are transformed inside the cell in which these materials are utilized as nutrients for development (Gibb and Strohl 1987; Oh et al. 2000).

The application of microbial enzymes for improvement of sludge reduction is the foundation to one more process called the enzymatic lysis (Kavitha et al. 2013). Prime advantages explained by the expanders of this method are eradicating the disease-causing microbes; nevertheless, an additional profit is the augmentation of gas generation in AD. A lipid-rich dairy industry wastewater using commercially obtainable inexpensive lipase enzyme pretreatment was also analyzed (Mendes et al. 2006). The high expenditure of commercially prepared enzymes makes the pretreatment reasonably impracticable. Therefore, bioaugmentation (augmentation of the endoenzymes or enzyme-secreting and biosurfactant-secreting microorganisms from the municipal sludge extracellular hydrolases) or addition of the superfluous enzyme-secreting bacterial stains is considered to be the appropriate means (Kavitha et al. 2014c; Kavitha et al. 2015b, c, d; Kavitha et al. 2016b, c; Godvin Sharmila et al. 2015). A biological pretreatment of WAS with the bacteria *Bacillus licheniformis* resulted in maximum 40 % solubilization which could further believed to enhance the biogas generation appreciably (Gopi et al. 2012). A bacterial pretreatment experiment along with anaerobic digestion yielded 57 % and 53 % increment in methane content and biogas production compared with the control (Merrylin et al. 2013a). A novel enzyme-secreting bacterial pretreatment was proposed in which the inoculated two mixed strains (*Bacillus jerish* 03 and *Bacillus jerish* 04) were the excellent enzyme (amylase and protease) producers. During this pretreatment the two strains together perform the lytic activity and enhanced the COD solubilization up to 17.33 % which could be expected to enhance the biogas generation economically (Kavitha et al. 2013). A bacterial pretreatment by two inoculated strains results in better SS reduction and COD solubilization. Further scrutinizing the BMP assay for the evaluation of pretreatment, the biogas production potential was found to be 0.467 L/ (g VS) by using modified Gompertz equation (Kavitha et al. 2014b).

4.3.5 Combinative Pretreatment

It is promising to unite more pretreatment techniques to promote the improvement and augmentation of biogas generation. Many works have emerged to perform a more relevant combinative technique without any tough concern being given to the amalgamated processes that could be oppressed to get the overall outcome. Others encompass additional considered approach combining techniques which pose balancing actions. The efficacy of low temperature thermo-chemical disintegration on biogas production in a semi-continuous anaerobic digester has yielded 103 % development in biogas generation than the blank reactor without pretreated sludge (Uma et al. 2012a). A chemical combined mechanical pretreatment is an effectual method for improvement of biodegradability and it laid the foundation to generate elevated biogas quantities (76 %), to progress clean energy generation from diary WAS (Uma et al. 2012b). A combined sono-alkalization pretreatment results in 83 % of fuel gas production during anaerobic digestion of dairy WAS (Uma et al. 2013a, b).

All the pretreatment processes have the ability to augment methane generation in anaerobic digestion process. The methods include thermal, oxidative, thermochemical, mechanical (ultrasonic, grinding, high-pressure homogenization) as well as other methods such as enzymatic hydrolysis and so forth. Prominence is chiefly focussed on their impact on biogas production. All these processes can enhance biogas production to some degree, but the energy required and the operation costs were varied. There is no ending about which method is the eminent. Heat action is progressively employed to eradicate disease-causing microbes and to influence the augmentation of biogas generation by high solubilization of sludge. The efficiency of this method on solubilization as well as gas generation was evidenced by amalgamation. Sonication disintegration was employed as a functional pace at a complete processing facility and accomplished a 1 % increment in COD solubilization, whereas a 19 % increment of solubilization has been accomplished by heat disintegration employing 70 °C for an hour (Uma et al. 2013b). This recommends that the effective sonication disintegration has been disrupting only biological flocs however not adequately demanding to cleave cell membranes. None of any advantages of solubilization were recovered from combined thermal disintegration in the combined action of sonication or from heat reaction of the sludge after it was formerly pretreated by means of ultrasonication. Nevertheless, gas generation from sonically pretreated sample was appreciably superior to the unprocessed sludge at any rate of stumpy SCOD release. On the contrary, the biogas generation from the pretreated sludge subjected to heat treatment has not been improved to the similar extent as the organic materials were solubilized representing that improved COD solubilization is not mechanically a fine pointer of augmented biogas generation.

5 Activated Sludge Flocs and Extracellular Polymeric Substances

Waste activated sludge refers to collection of microbial biomass that bounce to form floc matrix. The structure of a typical floc represents the presence of bacterial cells growing in dense, grape-shaped microcolonies, as filaments or as single cells entrenched in the matrix of EPS or attached to filamentous organisms (Monique et al. 2008). Filamentous bacteria are mostly known as “backbones” of flocs, responsible for their mechanical strength, as well as settling properties. The EPS matrix, composed of several fractions, is dense and sticky, gum-like fabric, responsible for a large extent of flocs and microcolony integrity. In the EPS matrix, many holes, cavities and channels are present, which make up for the large surface area of flocs and facilitate water and nutrient transport to the cells residing in the floc structure (Chu et al. 2002).

EPS are formed by natural processes and takes part in the organization of microbial consortia. Its basis is allied to the process relating dynamic secretion and peeling of biomass. EPS are the complex compounds located in the intercellular gap of bacterial attachment. The method of the floc generation is the outcome of the association of sludge components that are amply collected at the exterior side of sludge matrix at the time of endogenous respiration. EPS offers a principal connection system among the elements of sludge matrix—biomass, organic and inert molecules. Microbial flocs generated in treatment plants with activated sludge offer an effective protection system for bacterial colonies. The bacterial colonies are entrenched in an organization of EPS, which serve to be a trap for recyclable organics. Moreover, they behave as a meshwork having extracellular enzymes that show hydrolytic action (Kavitha et al. 2013). The agglomeration of tiny materials and colloids into well-built structures is a process that can be encountered naturally. The resultant aggregates that are formed are known as "flocs" which are greatly identified as being extremely porous, irregularly structured and slackly linked aggregates composed of smaller primary particles. Microbes, EPS or exocellular biopolymers and cations are the main features of flocs matrices. Numerous research reports indicated that the matrix pattern contains 60–80 % of the crude portion (Yang et al. 2010; Uma et al. 2012b).

In reality, sludge biomass goes on lysis or death at some point in which the intracellular components (substrates and nutrients) are expelled outside and offer foods that are later utilized in microbial metabolism (endogenous respiration). During this time, some part of the organics is discharged as resultant elements of respiration, ensuing in a diminished formation of new cells as a whole (lysis cryptic growth). This lysis process is mediated by extracellular enzymes. The cell lysis leads to the release of contents to the soluble phase in which the sludge biomass grows on cell lysate (released materials). This is known as cryptic growth. Therefore, an increment in cell lytic efficiency can thus contribute to a greater reduction of sludge and performs/fulfils a significant task in bringing down the energy expenditure and operational costs. The main objectives of this review were (1) to investigate the location of EPS and to study the distribution of extracellular enzymes in floc matrix, (2) to assess the importance of floc disruption prior to pretreatment and (3) to investigate various EPS removal methods and their specific roles.

6 Extracellular Enzymes

Proteins, polysaccharides and lipids are the major organic compounds in activated sludge, which should be broken down into minor substances using exocellular enzymes (Sheng and Yu 2006; Bala Subramanian et al. 2010). The sludge matrix contains paltry quantity of exoenzymes, which are entrapped in matrix and are not available for enzyme substrate action (Frolund et al. 1996). However, if entrapped enzymes are dispersed by some mechanism, then there is a possibility of improving degradation. Working on hydrolysis, Kavitha et al. (2013) have improved rate of reaction by releasing these trapped exoenzyme from sludge matrix. It is imperative to comprehend the location of enzymes in sludge flocs because the action of enzymes reveals their microbial activity while degrading soluble organics (Nybroe et al. 1992).

Guang et al. (2007) has reported that the enzymes alpha amylase and glucosidase are mainly entrapped in sludge pattern and that they are considered to be exoenzymes or extracellular enzymes. Guang et al. (2007) proposed that the protease action is chiefly restricted to the solid phase, but the least fraction is located in the tightly bound EPS (TB-EPS) area and barely appears in the loosely bound EPS (LB-EPS) part, while alpha amylase and glucosidase actions were mainly dispersed in the LB-EPS part, and slight quantities of those resided in the TB-EPS or solid phase. Thus, it was inferred that protease binds to the exterior of cell, i.e. ectoenzyme; at the same time, these enzymes are entrapped in the sludge matrix in detached form, i.e. exoenzymes. These enzymes are found to be imperative for the degradation of carbohydrates in sludge flocs. Majority of microbes residing in the floc matrix can straightly ingest certain elements that have merely lesser molecular weight (<1000) because of the selective permeability of cell walls (Cadoret et al. 2002). Therefore, most of the substrates (lysates) subjected to metabolic process of organism in the activated sludge floc have to be hydrolyzed before ingestion. This can be induced by liberating the enzymes that are entrapped in sludge flocs.

Proteolytic, lipolytic and cellulolytic enzymes synthesized inside the sludge cells are secreted into the extracellular surroundings and hydrolyze the entrapped complex materials into simpler tiny particles that can be transported transversely to the cell membrane and then get utilized for metabolic processes. Various diverse forms of enzymes were identified in sludge, such as amino peptidases, galactosidases, lipases and phosphatases (Nybroe et al. 1992; Frolund et al. 1995). Among these, leucine aminopeptidase and glucosidase showed the eminent activities (Watson et al. 2004). The hydrolyzing effects of these extracellular enzymes are regarded as the general rate-determining pace for the solubilization of particulate organics in the biological degradation of sludge.

7 Linkage Between EPS and Hydrolase

The structural pattern of the sludge floc relies on the linkage among the biopolymers and the different cations in the sludge. Minor alterations in ionic potency and quantity can consequently modify the morphological attributes of the sludge matrix (Merrylin et al. 2013a). Extraction of these ions from the matrix will disrupt the entrapped biopolymers of the inhabitant native microbes. The biopolymer arrangement is stabilized by lectin-like protein-binding polysaccharides that are cross-linked to adjoining proteins. The divalent cations promote structural pattern by linking the negatively charged surface of sludge components (Merrylin et al. 2013a, b; 2014a). EPS from activated sludge flocs are frequently split into two significant divisions: soluble and bound. Soluble EPS, also known as soluble microbial products (SMP), can shift liberally between the sludge matrix and bulk solution. Bound EPS possess an active two-layered organization comprising of LB-EPS and TB-EPS (Ramesh et al. 2006; Li and Yang 2007a, b; Merrylin et al. 2014b). The bound EPS is further subdivided into the following categories, specifically, "sheath" and "slime"; the former is intimately related to the biomass, whereas the latter has no straight connection with the biomass. The name of the former is also occasionally used for loosely bound EPS, whereas the name of the latter is used for steadily bound EPS (Yu et al. 2009). Both the soluble and bound EPS found in WAS are principally consisted of proteins and polysaccharides, and the former one have greater amount of polysaccharides than the latter (Pan et al. 2010). In accession to this fractionation, an enormous quantity of enzymes was immobilized in sludge matrix. These enzymes are indispensible to the hydrolysis of macromolecular substrates (Frolund et al. 1995). The EPS has a special mechanism, i.e. the linkage of enzyme with the substrate is stumpy, whereas substances within the flocs are sheltered from degradative activity of enzymes. However, the enzymes are trapped by or immobilized on the matrix; this immobilization reduces the natural process of enzymes on the sludge.

Even though microbes were integrated into the matrix, higher extent of disintegration took place in the exterior area of the matrix which contains greater amount of bacterial counts. Elements inside the flocs will be sheltered from the action of enzymes. Disturbance of these meshwork certainly pave the way to improved matrix dissolution (Watson et al. 2004). When the flocs get disrupted, complex substances that were sheltered from the action of enzymes get released and were degraded by hydrolases. On raising the incidence of matrix breakage, it is probable to make easy the process of disintegration by escalating the linkage among the enzyme and substrate. This disruption permits the lytic biomass and their allied enzymes to diffuse through the core, where the complex substances undergo disintegration (Watson et al. 2004).

Kavitha et al. (2013) proposed that sludge could be treated with complexing agents to convert the sludge as an improved substrate for enzyme action. Other reports investigated the methods involving improvement of the diffusion effectiveness of EPS by altering the EPS feature to enhance the disintegration of flocs matrix (Yuxuan et al. 2012).The extracellular enzymes played more significant roles in wastewater treatment process than intracellular enzymes (Confer and Logan 1998).

8 Phase-Separated Pretreatment and Its Benefits

Grounded along the above-cited reports, extracellular enzymes are found inevitable for hydrolysis processes. Since EPS reduces the connection between the substrate and extracellular enzymes, it is required to remove EPS and to disintegrate flocs so that the enzymes, proteins and carbohydrates immobilized on sludge matrix would get released and in turn enhance the sludge disintegration rate. A schematic representation of floc disruption and cell disintegration and the impact of phase-separated pretreatment on biogas production is shown in Fig. 7. It was divided into two phases (Cho et al. 2012). The initial phase is floc disruption (FD) (removal of EPS), and the final phase is cell disintegration (CD). FD meant for the removal of EPS from sludge matrix. CD denotes the release of cellular organics into liquid phase. EPS not only masks enzyme for substrate utilization but also retards

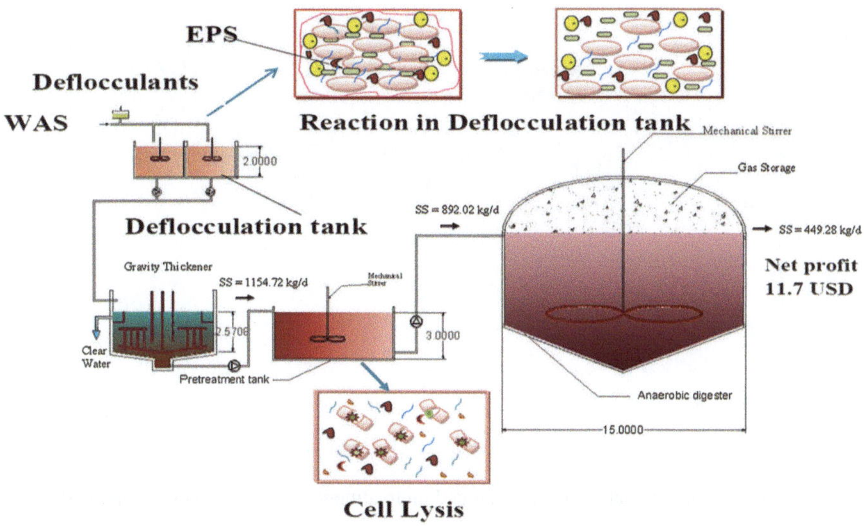

Fig. 7 A schematic representation of the impact of phase-separated pretreatment on biogas production

penetration competence of the substrate. The disruption of sludge matrix may lead to the accessibility of larger surface area for bacterial activity and the consequent discharge of trapped organics (substrates) to the bacterial action. Additionally, floc disruption (FD) improves the discharge of immobilized enzymes into the system consequently escalating biomass disintegration and liquefaction. Hence it is indispensable to release the EPS prior to disintegration to enhance the subsequent biodegradability. The floc disruption methods should be used prior to sludge pretreatments to enhance the biodegradability performance. The impact of phase-separated pretreatment on biodegradability and hydrolysis constant with confidence regions is depicted in Fig. 8. Figure 8 clearly depicts that compared to control (untreated sludge) and the pretreatment (bacterially disintegrated) alone, phase separated pretreatment with different chemicals (Ethylene diamine tetra acetic acid (EDTA), surfactants (SDS), Cation binding agent (Citric acid) $MgCl_2$ and NaCl—bacterially disintegrated sludges showed higher biodegradability. There are various EPS extraction methods employed to remove EPS and to enhance the activity of extracellular enzymes. This in turn improves the sludge disintegration. The extraction of EPS without cell cleavage is indispensable requirement of an extraction procedure. Many researchers have used DNA as a marker for indicating cell lysis (Kavitha et al. 2013; 2014b).

Energy requirement for sludge disintegration is crucial for profitable digestion of pretreated sludge. Economic evaluation of the pretreatment is required to demonstrate the practicability of employing pretreatment process. Floc disintegration prior to sludge pretreatment reduces the energy consumption for the pretreatment process. Poornima et al. (2014) proposed that the cost estimated for pretreated sludge

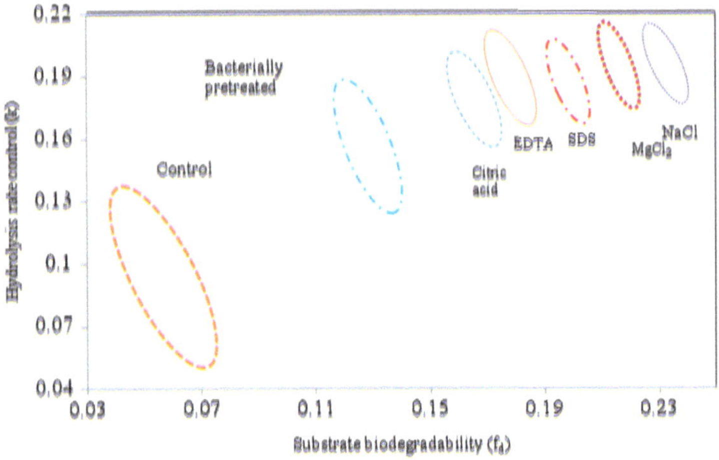

Fig. 8 Plot showing impact of phase-separated pretreatment on biodegradability and hydrolysis

(disperser alone) was found to be 46 USD/m^3, whereas the estimated cost for floc disrupted (floc disruption with sodium dodecyl sulphate) followed by disintegration with disperser which was found to be 26.4 USD/m^3. From the above, it was evident that a cost benefit of 43 % was possible with deflocculation-induced disperser pretreatment. Similarly, working on biogas generation from waste biomass, Gayathri et al. (2015) have reported a decrease in pretreatment cost from 370 Euro to 22 Euro, for phase-separated ultrasonic pretreatment, using citric acid as a deflocculant. Thus, the phase-separated disintegration was regarded to be inexpensive compared to the sonic treatment performed alone with net savings of 94.05 %.

9 Comparison of Different Pretreatments

All the disintegration processes have powerful effect on the improvement of hydrolysis and therefore can enhance the AD. The solubilization and biogas production efficiency of various pretreatment is depicted in Fig. 9. On evaluating all the pretreatments, the mechanical, chemical, thermal, thermochemical, ultrasonic, ozone, microwave and biological (enzymatic) methods are the most valuable, and universal pretreatment methods have been recommended for the escalation of biogas production in anaerobic digestion. Each method has revealed various benefits and merits for the diminution of organic waste mass, homogenization of samples, collapse of the cell structure and increment in the biodegradation competence. On the other hand, these diverse methods have hardly ever been joined together and evaluated openly with erstwhile pretreatment techniques. But it should

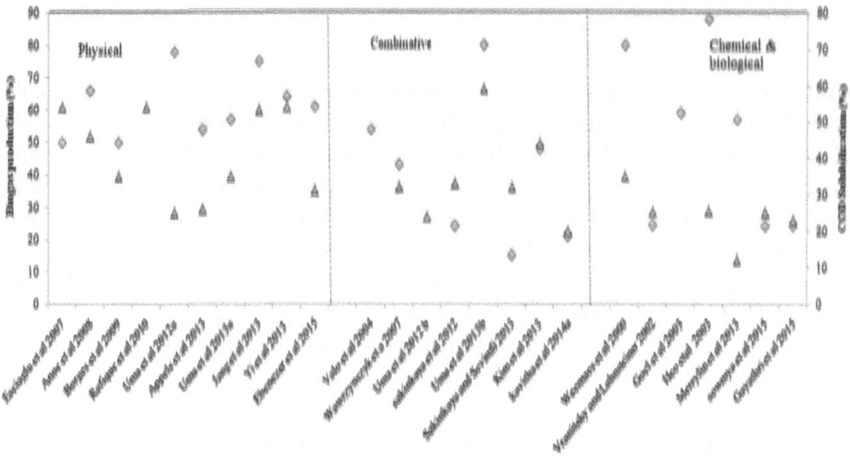

Fig. 9 Plot showing solubilization and biogas production efficiency of various pretreatments

be noticed that because of the variation in energy input, concentration of chemical, duration of reaction, the upshot of fermentation reaction, volatile solid degradation and biogas generation are not analogous in terms of operational cost versus reaction efficiency. Generally, mechanical, chemical and physical methods require shorter reaction time, from minutes to hours. Compared with the above-mentioned disintegration methods, biological treatment requires extended reaction duration, from hours to days.

In general, it is considered that the energy consumption for mechanical pretreatment is lesser than that of physical pretreatment; however it requires investment of mechanical equipments and maintenance cost. Thermal pretreatment consumes more heat energy, which is inexpensive and generally can be acquired from the blazing of biogas on the reaction spot. Since heat pretreatment at elevated temperatures usually produce inhibitory materials in the AD, temperatures lesser than 100 °C is generally employed to disintegrate the sludge. Chemo pretreatment by adding alkali and acids could be costly, and therefore it is obligatory to add foreign ions to the degradation medium. On comparing all the chemicals, the ozone is considered as the efficient one, since it does not lead to introduction of ions and in turn consumes only electric power for producing ozone. Although many pretreatments are available, such as mechanical, thermal and chemical, biological treatment is preferred based on the reduction of organic matter, removal of pathogen and control of odour problem (Goel et al. 1998). The energy consumption is also less when compared to the other methods. A noteworthy benefit of AD is the fairly elevated production of volatile fatty acids when compared to other disintegration methods. The factors which affect the bacterial pretreatment and decrease the activity of enzyme are extracellular polymeric substance, pH and temperature.

10 Perspectives

Compared with the widespread investigation about the disintegration methods, the fate of chemicals used for disintegration was still not analytically explored. The chemicals that are employed for disintegration have many negative aspects. Some are non-biodegradable agents and inhibit biological treatment by coupling with metabolic reactions. In the view of lack of reliable means for investigating these negative aspects, a resolution for these drawbacks is essential. The challenge for future research on disintegration should mainly focus on the safety of using chemicals. In addition to more comprehensive and possible analysis, a method taking into account all possible negative aspects must be investigated in order to obtain a valuable quantification procedure.

Acknowledgement The authors are thankful to the Department of Science and Technology, India, for providing financial support for this project (SR/FTP/ETA-0021/2010) under their Young Scientist Scheme.

References

Angelidaki I, Alves M, Bolzonella D, Borzacconi L, Campos JL (2009) Defining the Biomethane Potential (BMP) of solid organic wastes and energy crops: a proposed protocol for batch assays. Water Sci Technol 59:927–934. doi:10.2166/wst.2009.040

Appels L, Baeyens J, Degreve J, Dewil R (2008) Principles and potential of the anaerobic digestion of waste-activated sludge. Prog Energ Combust Sci 234:755–781. doi:10.1016/j.pecs.2008.06.002

Appels L, Houtmeyers S, Degreve J, Impe JV, Dewil R (2013) Influence of microwave pre-treatment on sludge solubilization and pilot scale semicontinuous anaerobic digestion. Bioresour Technol 128:598–603. doi:10.1016/j.biortech.2012.11.007

Aquino SF, Tuckey DC (2008) Integrated model of the production of soluble microbial products (SMP) and extracellular polymeric substances (EPS) in anaerobic chemostats during transient conditions. Biochem Eng J 38:138–146. doi:10.1016/j.bej.2007.06.010

Bala Subramanian S, Yan S, Tyagi RD, Surampalli RY (2010) Extracellular Polymeric Substances (EPS) producing bacterial strains of municipal wastewater sludge: isolation, molecular identification, EPS characterization and performance for sludge settling and dewatering. Water Res 44:2253–2266. doi:10.1016/j.watres.2009.12.046

Banu RJ, Kaliappan S (2007) Treatment of tannery wastewater using hybrid upflow anaerobic sludge blanket reactor. J Environ Eng Sci 6:415–421. doi:10.1139/s06-063

Banu RJ, Kaliappan S, Beck D (2006) High rate anaerobic treatment of Sago wastewater using HUASB with PUF as carrier. Int J Environ Sci Technol 3:69–77. doi:10.1007/BF03325909

Banu RJ, Kaliappan S, Yeom IT (2007a) Treatment of domestic wastewater using upflow anaerobic sludge blanket reactor. Int J Environ Sci Technol 4:363–370. doi:10.1007/BF03326295

Banu RJ, Kaliappan S, Yeom IT (2007b) Two-stage anaerobic treatment of dairy wastewater using HUASB with PUF and PVC carrier. Biotechnol Bioprocess Eng 12:257–264. doi:10.1007/BF02931101

Banu RJ, Arulazhagan P, Adish Kumar S, Kaliappan S, Lakshmi AM (2014) Anaerobic co-digestion of chemical- and ozone-pretreated sludge in hybrid upflow anaerobic sludge blanket reactor. Desalin Water Treat 54:3269–3278. doi:10.1080/19443994.2014.912156

Bougrier C, Carrere H, Delgenes JP (2005) Solubilisation of waste activated sludge by ultrasonic treatment. Chem Eng J 106:163–169. doi:10.1016/j.cej.2004.11.013

Bougrier C, Albasi C, Delgenes JP, Carrere H (2006) Effect of ultrasonic, thermal and ozone pre-treatments on waste activated sludge solubilisation and anaerobic biodegradability. Chem Eng Process Process Intensif 45:711–718. doi:10.1016/j.cep.2006.02.005

Bougrier C, Delgenes JP, Carrere H (2008) Effects of thermal treatments on five different waste activated sludge samples solubilization, physical properties and anaerobic digestion. J Chem Eng 139:236–244. doi:10.1016/j.cej.2007.07.099

Cacho Rivero JA (2005) Anaerobic digestion of excess municipal sludge, optimization for increased solid destruction. Ph.D. thesis. Department of Civil and Environmental Engineering of the University of Cincinnati, USA

Cadoret A, Conrad A, Block JC (2002) Availability of low and high molecular weight substrates to extracellular enzymes in whole and dispersed activated sludges. Enzyme Microb Technol 31:179–186. doi:http://dx.doi.org/10.1016/S0141-0229(02)00097-2

Camocho P (2001) Excess sludge process reduction by physical, alkaline or biological coupling. Ph.D thesis. Institute National des Sciences Appliquees, Toulouse, France

Camocho P, Ginestet P, Audic JM (2003) Pilot plant demonstration of reduction technology during activated sludge treatment of waste water WEFTEC'03, October 11-15, Los Angeles, USA

Carballa M, Manterola G, Larrea L, Ternes T, Omil F, Lema JM (2007) Influence of ozone pre-treatment on sludge anerobic digestion: removal of pharmaceutical and personal care products. Chemosphere 67:1444–1452. doi:10.1016/j.chemosphere.2006.10.004

Carlsson M, Lagerkvist A, Morgan-Sagastume F (2012) The effects of substrate pretreatment on anaerobic digestion: a review. Waste Manage 32:1634–50. doi:10.1016/j.wasman.2012.04.016

Carrere H, Dumas C, Battimelli A, Batsone DJ, Delgenes JP, Steyer JP (2010) Pretreatment methods to improve sludge anaerobic degradability: a review. J Hazard Mater 183:1–15. doi:10.1016/j.jhazmat.2010.06.129

Cho SK, Shin HS, Kim DH (2012) Waste activated sludge hydrolysis during ultrasonication: two-step disintegration. Bioresour Technol 121:480–483. doi:10.1016/j.biortech.2012.07.024

Chu CP, Lee DJ, Chang BV, You CS, Tay JH (2002) Weak ultrasonic pre-treatment on anaerobic digestion of flocculated activated biosolids. Water Res 36:2681–2688. doi:10.1016/S0043-1354(01)00515-2

Confer DR, Logan BE (1998) Location of protein and polysaccharide hydrolytic activity in suspended and biofilm wastewater cultures. Water Res 32:31–38. doi:10.1016/S0043-1354 (97)00194-2

De Mes TZD, Stams AJM, Reith JH, Zeeman G (2003) Methane production by anaerobic digestion of wastewater and solid wastes. In: Reith JH, Wijffels RH, Barten H (eds) Bio-methane and bio-hydrogen. Status and perspectives of biological methane and hydrogen production. Netherland Agency for Energy and the Environment (Novem), pp 58–102

Deublein D, Steinhauser A (2008) Biogas from waste and renewable resources, 1st edn. Wiley, Weinheim. doi:10.1002/9783527632794.

Dogan I, Sanin FD (2009) Alkaline solubilization and microwave irradiation as a combined sludge disintegration and minimization method. Water Res 43:2139–2148. doi:10.1016/j.watres.2009.02.023

Dohanyos M, Zabranska J, Kutil J, Jenıcek P (2004) Improvement of anaerobic digestion of sludge. Water Sci Technol 49:89–96. doi:10.1007/s11356-015-4677-2

Ebenezer VA, Kaliappan S, Adish Kumar S, Yeom IT, Banu JR (2015a) Influence of deflocculation on microwave disintegration and anaerobic biodegradability of waste activated sludge. Bioresour Technol 185:194–201. doi:10.1016/j.biortech.2015.02.102

Ebenezer VA, Arulazhagan P, Adish Kumar S, Yeom I-T, Banu JR (2015b) Effect of deflocculation on the efficiency of low-energy microwave pretreatment and anaerobic

biodegradation of waste activated sludge. Appl Energy 145:104–110. doi:10.1016/j.apenergy.2015.01.133
Esakki Raj S, Kaliappan S, Adish kumar S, Banu JR (2012) Combinative treatment (thermal-anaerobic) of EBPR sludge for the enhanced release and recovery of phosphorous. Int J Environ Eng 4:93–105. doi:10.1504/IJEE.2012.048097
Esakki Raj S, Banu JR, Adish Kumar S, Kaliappan S (2013) Effect of side stream thermal P recovery on the performance of AAO system integrated with sludge pretreatment. Bioresour Technol 140:376–384. doi:10.1016/j.biortech.2013.04.061
Eskicioglu C, Kennedy KJ, Droste RL (2007) Enhancement of batch waste activated sludge digestion by microwave pretreatment. Water Environ Res 11:2304–2317. doi:10.2175/106143007X184069
European Commission, Environmental, economic and social impacts of the use of sewage sludge on land (accessed 2010). http://ec.europa.eu/environment/waste/sludge/pdf/part_iii_report.pdf
Ferrer I, Ponsa S, Vazquez F, Font X (2008) Increasing biogas production by thermal (70°C) sludge pre-treatment prior to thermophilic anaerobic digestion. Biochem Eng J 42:186–192. doi:10.1016/j.bej.2008.06.020
Fraser KD (2010) Increased anaerobic digestion efficiencies via the use of thermal hydrolysis MSc thesis. Virginia Polytechnic Institute and State University Blacksburg
Frolund B, Griebe T, Nielsen PH (1995) Enzymatic activity in the activated-sludge floc matrix. Appl Microbiol Biotechnol 43:755–761. doi:10.1007/BF00164784
Frolund B, Palmgren R, Keiding K, Nielsen PH (1996) Extraction of extracellular polymers from activated sludge using a cation exchange resin. Water Res 30:749–1758. doi:10.1016/0043-1354(95)00323-1
Gayathri T, Kavitha S, Adish Kumar S, Kaliappan S, Yeom IT, Banu JR (2015) Effect of citric acid induced deflocculation on the ultrasonic pretreatment efficiency of dairy waste activated sludge. Ultrason Sonochem 22:333–340. doi:10.1016/j.ultsonch.2014.07.017
Gerardi MH (2003) The microbiology of anaerobic digesters. Wiley-Interscience, Wiley, pp 127–140. doi:10.1002/0471468967
Gibb GD, Strohl WR (1987) Physiological regulation of protease activity in *Streptomyces peucetius*. Can J Microbiol 34:187–190. doi:10.1139/m88-034
Godvin Sharmila V, Kavitha S, Rajashankar K, Yeom IT, Banu JR (2015) Effects of titanium dioxide mediated dairy waste activated sludge deflocculation on the efficiency of bacterial disintegration and cost of sludge management. Bioresour Technol 197:64–71. doi:10.1016/j.biortech.2015.08.038
Goel R, Mino T, Satoh H, Matsuo T (1998) Comparison of hydrolytic enzyme systems in pure culture and activated sludge under different electron acceptor conditions. Water Sci Technol 37:335–343. doi:10.1016/S0273-1223(98)00126-7
Gopi KS, Merrylin J, Kaliappan S, Adish Kumar S, Yeom IT, Banu JR (2012) Effect of cation binding agents on sludge solubilisation potential of bacteria. Biotechnol Bioprocess Eng 17:346–352. doi:10.1007/s12257-011-0465-0
Gopi KS, Arulazhagan P, Kavitha S, Adish Kumar S, Banu JR (2015) Evaluation of operational parameters for semi-continuous anaerobic digester treating pretreated waste activated sludge. Desalin Water Treatment 4:1–8. doi:10.1080/19443994.2015.1029526
Guang HY, He PJ, Shao LM, Lee DJ (2007) Enzyme activities in activated sludge flocs. Appl Microbiol Biotechnol 77:605–612. doi:10.1007/s00253-007-1204-5
Gupta R, Beeg QK, Loranz P (2002a) Bacterial alkaline proteases: molecular approaches and industrial applications. Appl Microbiol Biotechnol 59:15–32. doi:10.1007/s00253-002-0975-y
Gupta R, Beeg QK, Khan S, Chauhan B (2002b) An overview on fermentation, downstream processing and properties of microbial alkaline proteases. Appl Microbiol Biotechnol 60:381–395. doi:10.1007/s00253-002-1142-1
Haner A, Mason CA, Hamer G (1994) Death and lysis during aerobic thermophilic sludge treatment: Characterization of recalcitrant products. Water Res 28:863–869. doi:10.1016/0043-1354(94)90092-2

Hendriks ATWM, Zeeman G (2009) Pretreatments to enhance the digestibility of lignocellulosic biomass: a review. Bioresour Technol 100:10–18. doi:10.1016/j.biortech.2008.05.027

Jacquel N, Lo CW, Wei YH, Wu HS, Wang SS (2008) Isolation and purification of bacterial poly (3-hydroxyalkanoates). Biochem Eng J 39:15–27. doi:10.1016/j.bej.2007.11.029

Jones DA, Lelyveld TP, Mavrofidis SD, Kingman SW, Miles NJ (2002) Microwave heating applications in environmental engineering—a review. Resour Conserv Recyl 34:75–90. doi:10.1016/S0921-3449(01)00088-X

Kavitha S, Adish Kumar S, Yoga Lakshmi KN, Kaliappan S, Banu JR (2013) Effect of enzyme secreting bacterial pretreatment on enhancement of aerobic digestion potential of waste activated sludge interceded through EDTA. Bioresour Technol 150:210–219. doi:10.1016/j.biortech.2013.10.021

Kavitha S, Jayashree C, Adish Kumar S, Kaliappan S, Banu JR (2014a) Enhancing the functional and economical efficiency of a novel combined thermo chemical disperser disintegration of waste activated sludge for biogas production. Bioresour Technol 173:32–41. doi:10.1016/j.biortech.2014.09.078

Kavitha S, Jayashree C, Adish Kumar S, Yeom IT, Banu JR (2014b) The enhancement of anaerobic biodegradability of waste activated sludge by surfactant mediated biological pretreatment. Bioresour Technol 168:159–166. doi:10.1016/j.biortech.2014.01.118

Kavitha S, Adish Kumar S, Kaliappan S, Yeom IT, Banu JR (2014c) Improving the amenability of municipal waste activated sludge for biological pretreatment by phase-separated sludge disintegration. Bioresour Technol 169:700–706. doi:10.1016/j.biortech.2014.07.065

Kavitha S, Yukesh Kannah R, Yeom IT, Uan DK, Banu JR (2015a) Combined thermo-chemo-sonic disintegration of waste activated sludge for biogas production. Bioresour Technol 197:383–392. doi:10.1016/j.biortech.2015.08.131

Kavitha S, Kaliappan S, Adish Kumar S, Yeom IT, Banu JR (2015b) Effect of NaCl induced floc disruption on biological disintegration of sludge for enhanced biogas production. Bioresour Technol 192:807–811. doi:10.1016/j.biortech.2015.05.071

Kavitha S, Saranya T, Kaliappan S, Adish Kumar S, Yeom IT, Banu JR (2015c) Accelerating the sludge disintegration potential of a novel bacterial strain Planococcus jake 01 by CaCl2 induced deflocculation. Bioresour Technol 175:396–405. doi:10.1016/j.biortech.2014.10.122

Kavitha S, Kumar A, Kaliappan S, Yeom IT, Banu JR (2015d) Achieving profitable biological sludge disintegration through phase separation and predicting its anaerobic biodegradability by nonlinear regression model. Chem Eng J 279:478–487. doi:10.1016/j.cej.2015.05.051

Kavitha S, Jessin Brindha GM, Sally Gloriana A, Rajashankar K, Yeom IT, Banu JR (2016a) Enhancement of aerobic biodegradability potential of municipal waste activated sludge by ultrasonic aided bacterial disintegration. Bioresour Technol 200:161–169. doi:10.1016/j.biortech.2015.10.026

Kavitha S, Jessin Brindha GM, Sally Gloriana A, Rajashankar K, Yeom IT, Banu JR (2016b) Enhancement of aerobic biodegradability potential of municipal waste activated sludge by ultrasonic aided bacterial disintegration. Bioresour Technol 200:161–169. doi:10.1016/j.biortech.2015.10.026

Kavitha S, Saji Pray S, Yogalakshmi KN, Adish Kumar S, Yeom IT, Banu JR (2016c) Effect of chemo-mechanical disintegration on sludge anaerobic digestion for enhanced biogas production. Environ Sci Pollut Res 23:2402–2414. doi:10.1007/s11356-015-5461-z

Khanal SK, Grewell D, Sung S, Leeuwen JV (2007) Ultrasound application in wastewater sludge pre-treatment: a review. Critical Rev Environ Sci Technol 37:277–313. doi:10.1080/10643380600860249

Kim J, Park C, Kim TH, Lee M, Kim S, Kim K, Lee J (2003) Effects of various pretreatments for enhanced anaerobic digestion with waste activated sludge. J Biosci Bioeng 95:271–275. doi:10.1016/S1389-1723(03)80028-2

Klass D (2004) Biomass for renewable energy and fuels. In: Encyclopedia of energy, vol. 1. Elsevier, pp 193–212

Kohlmann KL, Nielsen SS, Stevenson LR, Ladisch MR (1991) Production of proteases by psychrotrophic microorganisms. J Dairy Sci 74:3275–3283. doi:10.3168/jds.S0022-0302(91)78513-5

Levlin E (2010) Maximizing sludge and biogas production for counteracting global warming. International scientific seminar, Research and application of new technologies in wastewater treatment and municipal solid waste disposal in Ukraine, Sweden and Poland 23–25 September 2009 Stockholm, Polish-Swedish, TRITA-LWR REPORT 3026, pp 95–10

Li CL, Fang HHP (2007) Inhibition of heavy metals on fermentative hydrogen production by granular sludge. Chemosphere 67:668–673. doi:10.1016/j.chemosphere.2006.11.005

Li YY, Noike T (1992) Upgrading of anaerobic digestion of waste activated sludge by thermal pretreatment. J Water Sci Technol 26:857–866. doi:10.2166/wst.2009.510

Li XY, Yang SF (2007a) Influence of loosely bound extracellular polymeric substances (EPS) on the flocculation, sedimentation and dewaterability of activated sludge. Water Res 41:1022–1030. doi:10.1016/j.watres.2006.06.037

Li XY, Yang SF (2007b) Influence of loosely bound extracellular polymeric substances (EPS) on the flocculation, sedimentation and dewaterability of activated sludge. Water Res 41:1022–1030. doi:10.1016/j.watres.2006.06.037

Li H, Li C, Liu W, Zou S (2012) Optimized alkaline pretreatment of sludge before anaerobic digestion. Bioresour Technol 123:189–194. doi:10.1016/j.biortech.2012.08.017

Lin Y, Wang D, Wu S, Wang C (2009) Alkali pretreatment enhances biogas production in the anaerobic digestion of pulp and paper sludge. J Hazard Mater 170:366–373. doi:http://dx.doi.org/10.1016/j.jhazmat.2009.04.086

Mata Alvarez J (2003) Biomethanization of the organic fraction municipal solid wastes. IWA Cornwall, London

Mata-Alvarez J, Macé S, Llabrés P (2000) Anaerobic digestion of organic solid wastes. An overview of research achievements and perspectives. Bioresour Technol 74:3–16. doi:10.1016/S0960-8524(00)00023-7

Mendes AA, Pereira EB, De Castro HF (2006) Effect of the enzymatic hydrolysis pre-treatment of lipids-rich wastewater on the anaerobic bio digestion. Biochem Eng J 32:185–190. doi:10.1016/j.bej.2006.09.021

Merrylin J, Adish Kumar S, Kaliappan S, Yeom IT, Banu JR (2013a) Biological pretreatment of non-flocculated sludge augments the biogas production in the anaerobic digestion of the pretreated waste activated sludge. Environ Technol 34:13–14. doi:10.1080/09593330.2013.810294

Merrylin J, Adish Kumar S, Kaliappan S, Yeom I-T, Banu JR (2013b) Effect of extracellular polymeric substances (EPS) on sludge reduction potential of *Bacillus licheniformis*. Int J Environ Sci Tech 8:85–92. doi:10.1007/s13762-012-0141-8

Merrylin J, Adish Kumar S, Kaliappan S, Yeom I-T, Banu JR (2014a) Effect of EPS removal on the sludge reduction potential of B. licheniformis on its optimized pH conditions. Water Environ J 28:95–103. doi:10.1111/wej.12014

Merrylin J, Adish Kumar S, Kalippan S, Yeom IT, Banu JR (2014b) Enhancing aerobic digestion potential of municipal waste activated sludge through removal of EPS. Environ Sci Pollut Res 21:1112–1123. doi:10.1007/s11356-013-1976-3

Metcalf and Eddy Inc (2006) Wastewater engineering: treatment and reuse. McGraw Hill series in civil and environmental engineering. 5th edn, McGraw-Hill, New York, NY

Monique R, Elisabeth GN, Etienne P, Dominique L (2008) A high yield multimethod extraction protocol for protein quantification in activated sludge. Bioresour Technol 99:7464–7471. doi:10.1016/j.biortech.2008.02.025

Mudhoo A, Sharma SK (2011) Microwave irradiation technology in waste sludge and wastewater treatment research. Crit Rev Environ Sci Technol 41:999–1066. doi:10.1080/10643380903392767

Neyens E, Baeyens J, Dewill R, Heyder B (2004) Advanced sludge treatment affects extracellular polymeric substances to improve activated sludge dewatering. J Hazard Mater 106:83–92. doi:10.1016/j.jhazmat.2003.11.014

Nybroe O, Jorgensen PE, Henze M (1992) Enzyme activities in waste water and activated sludge. Water Res 26:579–584. doi:10.1016/0043-1354(92)90230-2

Oh YS, Shih IL, Tzeng YM, Wang SL (2000) Protease produced by *Pesudomonas aeroginosa* K-187 and its application in the deproteinization of shrimp and crab shell wastes. Enzyme Microb Technol 27:3–10. doi:10.1016/S0141-0229(99)00172-6

Owen WF, Stuckey DC, Healy JB, Young LY, McCarty PL (1979) Bioassay for monitoring biochemical methane potential and anaerobic toxicity. Water Res 13:485–492. doi:10.1016/0043-1354(79)90043-5

Pan X, Liu J, Zhang D, Chen X, Li L, Song W, Yang JA (2010) Comparison of five extraction methods for extracellular polymeric substances (EPS) from biofilm by using three dimensional excitation-emission matrix (3DEEM) fluorescence spectroscopy. Water SA. 36:111–116. doi:10.4314/wsa.v36i1.50914

Panter K (2006) Energy prices rises fuel advanced digestion. World water and environmental engineering magazine, March/April 143

Park WJ, Ahn JH, Hwang S, Lee CK (2010) Effect of output power, target temperature, and solid concentration on the solubilization of waste activated sludge using microwave irradiation. Bioresour Technol 101:S13–S16. doi:10.1016/j.biortech.2009.02.062

Paul W, Ohlmeyer M, Leithoff H, Boonstra MJ, Pizzi A (2006) Optimising the properties of OSB by a onestep heat pre-treatment process. Holz als Roh- und Werkstoff 64:227–234. doi:10.1007/s00107-005-0073-9

Peppley BA (2006) Biomass for fuel cells: a technical and economic assessment. Int J Green Energy 3:201–218. doi:10.1080/01971520500439401

Pilli S, Bhunia P, Yan S, LeBlanc RJ, Tyagi RD, Surampalli RY (2011) Ultrasonic pretreatment of sludge: a review. Ultrason Sonochem 18:1–18. doi:10.1016/j.ultsonch.2010.02.014

Poornima DT, Vimala Ebenezer A, Adish Kumar S, Kaliappan S, Banu JR (2014) Effect of deflocculation on the efficiency of disperser induced dairy waste activated sludge disintegration and treatment cost. Bioresour Technol 167:151–158. doi:10.1016/j.biortech.2014.06.004

Rai CL, Rao PG (2009) Influence of sludge disintegration by high pressure homogenizer on microbial growth in sewage sludge: an approach for excess sludge reduction. Clean Technol Environ Policy 4:437–446. doi:10.1007/s10098-009-0202-y

Rajakumar R, Meenambal T, Banu JR, Yeom IT (2010) Treatment of poultry slaughterhouse wastewater in upflow anaerobic filter under low upflow velocity. Int J Environ Sci Technol 8:149–158. doi:10.1007/BF03326204

Ramesh A, Duu JL, Hong SG (2006) Soluble microbial products (SMP) and soluble extracellular polymeric substances (EPS) from wastewater sludge. J Environ Biotechnol 73:219–225. doi:10.1007/s00253-006-0446-y

Sheng GP, Yu HQ (2006) Characterization of extracellular polymeric substances of aerobic and anaerobic sludge using three-dimensional excitation and emission matrix fluorescence spectroscopy. Water Res 40:1233–1239. doi:10.1016/j.watres.2006.01.023

Show KY, Mao T, Lee DJ (2007) Optimization of sludge disruption by sonication. Water Res 41:4741–4747. doi:10.1016/j.watres.2007.07.017

Sowmya GP, Kavitha S, Adish Kumar S, Kaliappan S, Yeom IT, Banu JR (2015) Effect of sonically induced deflocculation on the efficiency of ozone mediated partial sludge disintegration for improved production of biogas. Ultrason Sonochem 26:241–248. doi:10.1016/j.ultsonch.2015.01.015

Sri Bala Kameswari K, Kalyanaraman C, Thanasekaran K (2011) Effect of ozonation and ultrasonication pretreatment processes on co-digestion of tannery solid wastes. Clean Technol Environ Policy 13:517–525. doi:10.1007/s-10098-010-0334-0

Suresh Karthik M, Krishna Kumar T, Arulazhagan P, Adish Kumar S, Yeom IT, Banu JR (2015) Effect of alkaline and ozone pretreatment on sludge reduction potential of a membrane bioreactor treating high-strength domestic wastewater. Desalin Wastewater Treatment 55:1127–1134. doi:10.1080/19443994.2014.923335

Tang B, Yu LF, Huang SS, Luo JZ, Zhuo Y (2010) Energy efficiency of pretreating excess sewage sludge with microwave irradiation. Bioresour Technol 101:5092–5097. doi:10.1016/j.biortech.2010.01.132

Tiehm A, Nickel K, Zellhorn M, Neis U (2001) Ultrasonic waste activated sludge disintegration for improving anaerobic stabilization. Water Res 35:2003–2009. doi:10.1016/S0043-1354(00)00468-1

Uan DK, Banu JR, Chung IJ, Yeom IT (2009) Effect of thermochemical sludge pretreatment on sludge reduction and on performances of anoxic-aerobic membrane bioreactor treating low strength domestic wastewater. J Chem Technol Biotechnol 84:1350–1355. doi:10.1002/jctb.2189

Uan DK, Banu RJ, Son DH, Yeom IT (2012) Influence of ferrous sulfate on thermochemical sludge disintegration and on performances of wastewater treatment in a new process: anoxic–oxic membrane bioreactor coupled with sludge disintegration step. Biochem Eng J 66:20–26. doi:10.1016/j.bej.2012.04.013

Uan DK, Banu JR, Kaliappan S, Yeom IT (2013a) Influence of the thermochemical sludge pretreatment on the nitrification of A/O reactor with the removal of phosphorus by simultaneous precipitation. Biotechnol Bioprocess Eng 18:313–320. doi:10.1007/s12257-012-0492-5

Uan DK, Yeom IT, Arulazhagan P, Banu JR (2013b) Effects of sludge pretreatment on sludge reduction in a lab-scale anaerobic/anoxic/oxic system treating domestic wastewater. Int J Environ Sci Technol 10:495–502. doi:10.1007/s13762-012-0120-0

Uchida K, Mogi T, Fukushima O (1972) Microbial production of protease, amylase and lipase (Japan.) 7250.389. C.F. Chem. Abst 79

Uma RR, Adish Kumar S, Kaliappan S, Yeom IT, Banu JR (2012a) Low temperature thermochemical pretreatment of dairy waste activated sludge for anaerobic digestion process. Bioresour Technol 103:415–424. doi:10.1016/j.biortech.2011.09.124

Uma RR, Adish Kumar S, Kaliappan S, Banu JR (2012b) Combined treatment of alkaline and disperser for improving solubilization and anaerobic biodegradability of dairy waste activated sludge. Bioresour Technol 126:107–116. doi:10.1016/j.biortech.2012.09.027

Uma RR, Adish Kumar S, Kaliappan S, Yeom IT, Banu JR (2013a) Impacts of microwave pretreatments on the semi-continuous anaerobic digestion of dairy waste activated sludge. Waste Manage 33:1119–1127. doi:10.1016/j.wasman.2013.01.016

Uma RR, Adish Kumar S, Kaliappan S, Yeom IT, Banu JR (2013b) Enhancing the anaerobic digestion potential of dairy waste activated sludge by two step sono-alkalization pretreatment. Ultrason Sonochem 21:1065–1074. doi:10.1016/j.ultsonch.2013.11.007

Valo A, Carrère H, Delgenès JP, Chem J (2004) Thermal, chemical and thermo-chemical pre-treatment of waste activated sludge for anaerobic digestion. Chem Technol Biotechnol 79:1197–1203. doi:10.1002/jctb.1106

Veera Lakshmi M, Merrylin J, Kavitha S, Adish Kumar S, Yeom IT, Banu JR (2014) Solubilisation of municipal sewage waste activated sludge by novel lytic bacterial strains. Environ Sci Pollut Res 21:2733–2743. doi:10.1007/s11356-013-2228-2

Vranitzky R, Lahnsteiner J (2002) Sewage sludge disintegration using ozone—a method of enhancing the anaerobic stabilisation of sewage sludge. In: Proceedings of the 7th European biosolids and organic residuals conference, Wakefield, UK

Wang F, Wang Y, Ji M (2005) Mechanism and kinetic models for ultrasonic waste activated sludge pretreatment. J Hazard Mat 123:145–150. doi:10.1016/j.jhazmat.2005.03.033

Wang XH, Gai LH, Sun XF, Xie HJ, Gao MM, Wang SG (2010) Effects of long term addition of Cu (II) and Ni (II) on the bioalkaline properties of aerobic granules in sequencing batch reactors. Appl Microbiol Biotechnol 86:1967–1975

Wang L, Mattsson M, Rundstedt J, Karlsson N (2012) Different pretreatments to enhance biogas production. Mas. Sci. thesis, Halmstad University

Watson SD, Akhurst T, Whiteley CG, Rose PD, Pletschke BI (2004) Primary sludge floc degradation is accelerated under biosulphidogenic conditions: enzymological aspects. Enzyme Microbial Technol 34:595–602. doi:10.1016/j.enzmictec.2004.01.004

Weemaes M, Grootaerd H, Simons F, Verstraete W (2000) Anaerobic digestion of ozonized biosolids. Water Sci Technol 34:2330–2336. doi:10.1016/S0043-1354(99)00373-5

Yang Q, Luo K, Li X, Wang D, Zheng W, Zeng G, Liu J (2010) Enhanced efficiency of biological excess sludge hydrolysis under anaerobic digestion by additional enzymes. Bioresour Technol 101:2924–2930. doi:10.1016/j.biortech.2009.11.012

Yu GH, He PJ, Shao LM (2009) Characteristics of extracellular polymeric substances (EPS) fractions from excess sludge and their effects on bioflocculability. Bioresour Technol 100:3193–3198. doi:10.1016/j.biortech.2009.02.009

Yu Q, Lei HY, Li Z, Li HL, Chen K, Zhang XH, Liang RL (2010) Physical and chemical properties of waste-activated sludge after microwave treatment. Water Res 44:2841–2849. doi:10.1016/j.watres.2009.11.057

Yuxuan Z, Panyue Z, Guangming Z, Weifang M, Hao W, Boqiang M (2012) Sewage sludge disintegration by combined treatment of alkaline + high pressure homogenization. Bioresour Technol 123:514–519. doi:10.1016/j.biortech.2012.07.078

Zhang GM, Zhang PY, Yang J, Liu HZ (2008) Energy efficient sludge sonication: power and sludge characteristics. Bioresour Technol 99:9029–9031. doi:10.1016/j.biortech.2008.04.021

Zhang D, Chen Y, Zhao Y, Zhu X (2010) New sludge pre-treatment method to improve methane production in waste activated sludge digestion. Environ Sci Technol 44:4802–4808. doi:10.1021/es1000209

Hydrolysis of Lignocellulosic Biomass for Recovering Hemicellulose: State of the Art

I.S.M. Rafiqul, A.M.M. Sakinah, and A.W. Zularisam

Abstract Hemicellulose, a heteropolysaccharide, is a second major component of lignocellulosic biomass (LCB). It is a potential source of various rare sugars, mainly xylose, because the biomass is cheap, renewable, and available globally. Xylose can be an economic and attractive substrate to produce numerous specialty chemicals, especially xylitol. It is particularly significant to depolymerize the complex composition of biomass to recover hemicellulosic sugars and to prepare cellulosic part available for efficient digestion. LCB hydrolysis by various techniques is an inevitable method for depolymerizing hemicellulose into xylose and other hemicellulosic sugars. Among the general methods of hemicellulose hydrolysis (such as acid, autohydrolysis, enzyme, combined acid-enzyme, and autohydrolysis-enzyme), dilute acid hydrolysis is the most investigated and extensively applied method due to its simplicity, effectiveness, and economic feasibility. A number of operating variables such as temperature, catalyst load, reaction time, and liquid to solid ratio significantly affect the kinetics of hemicellulose hydrolysis. Dilute acid catalyzes hemicellulose fractionation at elevated temperature and pressure within short residence time. This chapter reviews the current literature on hemicellulose hydrolysis methods and identifies the most suitable way to recover maximum hemicellulosic sugars (viz., xylose and arabinose) from LCB.

Keywords Hemicellulose • Hydrolysis • Kinetics • Lignocellulose • Mathematical modeling • Xylose

I.S.M. Rafiqul (✉)
Department of Genetic Engineering and Biotechnology, University of Chittagong, Chittagong 4331, Bangladesh
e-mail: drsmrafiqcu@gmail.com

A.M.M. Sakinah
Faculty of Chemical and Natural Resources Engineering, Universiti Malaysia Pahang, 26300 Kuantan, Pahang, Malaysia

A.W. Zularisam
Faculty of Engineering Technology, Universiti Malaysia Pahang, 26300 Kuantan, Pahang, Malaysia

L. Singh, V.C. Kalia (eds.), *Waste Biomass Management – A Holistic Approach*,
DOI 10.1007/978-3-319-49595-8_4

1 Introduction

Lignocellulosic biomass (LCB) is widespread, renewable, abundant, ecofriendly, and an inexpensive biosource of polysaccharides that offer sustainable production of various specialty chemicals, biomolecules, biofuels, and food ingredients having diverse industrial significance (Samanta et al. 2012; Rafiqul and Sakinah 2013; Singh et al. 2015). Lignocellulose is the principal constituent of plant cell walls. Cellulose is the main constituent of LCB followed by hemicellulose and lignin with smaller contents of extractives, ash, pectins, and proteins (Kumar et al. 2009; Rafiqul and Sakinah 2013). The integral benefit of LCB utilization can be obtained by its fractionation following the "biomass refining" philosophy, based on the selective separation of the major components to produce a variety of high-value products (Moure et al. 2006). However, biotransformation of LCB to specialty products is notably hindered by its chemical and structural complexity, which makes this material a challenge to be used as feedstock for industrial utilization. Cellulose is a linear homopolysaccharide composed of D-glucose units joined together via β-(1,4)-glycosidic bonds to form a skeleton (Kumar et al. 2009; Rafiqul and Sakinah 2013). The size of a cellulose molecule is generally expressed by its degree of polymerization (DP) (Park et al. 2010; Arvela et al. 2011). The cellulose chains are connected together by hydrogen linkages, and they are in turn joined together with hemicelluloses and lignin through hydrogen-bonding interaction and covalent interaction, respectively (Mosier et al. 2005). In LCB, cellulose is found in both crystalline and amorphous structures. About 50–90 % of the cellulose chain is joined laterally by hydrogen bonds to form a crystalline structure, whereas the remaining fraction is less ordered and is called amorphous cellulose (Park et al. 2010; Chundawat et al. 2011). The amorphous cellulose is rapidly solubilized by acid, whereas the decomposition of crystalline cellulose is restricted by the limited access of reactant to glycosidic linkages (Dhepe and Fukuoka 2008).

Hemicellulose is a second major component of LCB and is a heteropolysaccharide whose chemical composition varies from tissue to tissue within a single plant and from species to species. It is generally composed of different building block molecules including pentoses (e.g., xylose and arabinose), hexoses (e.g., glucose, galactose, and mannose), and uronic acids (e.g., galacturonic and 4-*O*-methyl-glucuronic acids) with acetyl side chains. It consists of 100–200 units of pentoses and hexoses in its molecular backbone (Park et al. 2010; Thomas et al. 2011). Hemicellulose is amorphous in structure, and a variety of branching, linkages, and monomer units contribute to its complex architecture and thereby its variety of conformations and function (Scheller and Ulvskov 2010; Pauly et al. 2013). Hemicellulose differs from cellulose in three main aspects. Firstly, it contains various types of sugars, whereas cellulose contains only β-(1,4)-D-glucopyranoses. Secondly, it exhibits a significant degree of chain branching, whereas cellulose is strictly a linear biopolymer. Thirdly, the DP of cellulose is 10–100 times higher than that of hemicellulose (Thomas et al. 2011). In biomass, hemicellulose is linked to cellulose through hydrogen-bonding interactions and builds a structural matrix, which is further bound to lignin to form a lignocellulosic complex (Mosier et al. 2005; Pauly et al. 2013). Hemicellulose, especially xylan, has attracted global attention due to its multipurpose uses including food and nonfood applications. It is extensively utilized as a raw material to produce numerous specialty chemicals such as xylitol, furfural, ethanol, erythritol, and lactic acid (Samanta et al. 2012; Rafiqul et al. 2015a).

Lignin, a third significant fraction in LCB, is an aromatic biopolymer of sinapyl-, coniferyl-, and coumaryl alcohol units linked by aryl-aryl, alkyl-alkyl, and alkyl-aryl ether linkages. Softwoods usually contain more lignin than hardwoods and herbaceous plants (Kumar et al. 2009; Rafiqul and Sakinah 2013; Eudes et al. 2014). Lignin is amorphous and hydrophobic in character and is not solubilized by acid but is hydrolizable with warm alkali, simply oxidized and condensed by phenol (Thomas et al. 2011). Cellulose, hemicellulose, and lignin are closely associated with each other; thus, biomasses can be regarded as composite materials in which lignin acts as a defensive agent that obstructs plant cell degradation using microorganisms (viz., fungi and bacteria) to bioconvert into chemicals and fuels (Rafiqul and Sakinah 2013; Eudes et al. 2014). Generally, hemicellulose accounts for 25–35 % of the dry wood mass (Spiridon and Popa 2008; Arvela et al. 2011). The predominant hemicelluloses in softwoods are galactoglucomannans and arabinoglucuronoxylans. A hardwood hemicellulose principally consists of *O*-acetyl-4-*O*-methylglucuronoxylan and glucomannans. *O*-acetyl-4-*O*-methylglucuronoxylan, simply called xylan, is the main hemicellulose in hardwoods (Spiridon and Popa 2008). Xylan is found in substantial amounts ranging from 11 % to 35 % (dry mass basis) in LCB such as hardwoods and agricultural residues. Softwoods (e.g., spruce, coniferous trees, and pine) and hardwoods (e.g., meranti, oak, and aspen) differ in structure and chemical constituents of the hemicellulose. Hemicelluloses in softwoods have higher quantities of mannose and glucose residues than hardwood hemicelluloses that typically hold larger amounts of xylose. Moreover, hemicelluloses are more highly acetylated in hardwoods than in softwoods (Palmqvist and Hahn-Hägerdal 2000a; Rafiqul and Sakinah 2013). During hydrolysis, xylose is the major sugar released from hardwoods, but arabinose can form a considerable quantity of pentoses generated from different agro- and forest-based wastes (Balat et al. 2008).

To make the feedstock suitable for bioconversion in bioprocesses, cellulose and hemicelluloses must be depolymerized into their constituents for use by microbes or enzymes. Hemicellulose is easily hydrolyzable owing to its branched and amorphous configuration compared to cellulose that requires harsh experimental condition because of its crystallinity, degree of polymerization, and accessible surface area (Chundawat et al. 2011; Rafiqul et al. 2014). Hemicellulose hydrolysis aimed to maximize the recovery of rare sugars (e.g., xylose, arabinose mannose, etc.) and minimize the coformation of toxic compounds. Additionally, hydrolysis of the hemicellulose makes the cellulosic part for succeeding transformation by acid or enzyme and shows potential opportunities for studies that could drastically diminish biomass treatment costs. Hemicellulose is normally hydrolyzed by either acidic, autohydrolytic, enzymatic, combined chemical-enzymatic, or combined autohydrolytic-enzymatic methods for manufacturing xylose and xylooligosaccharides (Teng et al. 2010; Hanim et al. 2011; Rafiqul and Sakinah 2013; Visser et al. 2015).

Dilute acid-catalyzed digestion is one of the most investigated and broadly utilized hydrolysis approaches as it is simple, low-cost, rapid, and effective (Swati et al. 2013; Rafiqul et al. 2014). This approach can be employed either as a pretreatment step preceding enzymatic digestion or as the main process of hydrolysis. Dilute acid hydrolysis is generally carried out at elevated temperature (160–230 °C) and pressure (~10 atm) using mineral acid (such as H_2SO_4 or HCl) concentration in the range of 2–5 %. The dilute sulfuric acid can effectively depolymerize hemicellulose into sugar monomers and soluble oligomers (Rafiqul and Sakinah 2013; Amirkhani et al. 2015). Compared to other hydrolytic techniques, dilute acid hydrolysis is principally important to degrade xylan in hemicellulose into xylose, which can be further transformed to bioxylitol or bioethanol (Sun and Cheng 2005; Tada et al. 2012; Rafiqul et al. 2015a, b). Furthermore, it is suitable for different biomasses such as hardwoods, softwoods, agricultural residues, municipal solid wastes, waste paper, and herbaceous crops (Mosier et al. 2005; Singh et al. 2015).

Autohydrolysis is an alternative technology to solubilize hemicellulose, with several benefits over the dilute acid hydrolysis, i.e., a more limited delignification and reduced contents of sugar decomposition compounds (Garrote et al. 2002; Parajó et al. 2004; Rafiqul and Sakinah 2013). Furthermore, it shows some environmental and technical advantages as no chemical (acid or alkali) other than water is employed (El Hage et al. 2010; Rafiqul and Sakinah 2013). However, this method needs special instrument, which can be run at elevated temperature. In enzymatic hydrolysis of LCB, the utility cost is lower compared to chemical treatment as the enzymatic hydrolysis is normally performed under gentle condition (pH ~5 and temperature 45–50 °C) and does not have a corrosion problem (Rabelo et al. 2011; Rafiqul and Sakinah 2013). Compared to acid and autohydrolysis, enzyme hydrolysis is environmental-friendly, milder, and more specific, but it requires pretreatment for enhancing enzymatic digestibility (Sun and Cheng 2005; Wu et al. 2011). Nevertheless, the main bottleneck of enzymatic treatment is that the presence of solid residuals (mainly lignin) and dissolution of enzymes in the hydrolysate makes it difficult to separate and recycle the enzymes in order to reduce the cost (Lu et al. 2007; Taherzadeh and Karimi 2007).

Hydrolysis of hemicellulose using a combined chemical and enzymatic treatment has been studied as an alternative method to produce mainly xylooligosaccharides (XOs). The access of enzymes to xylosidic linkages is restricted by the surrounding biomass constituents as well as by the substituents on its backbone (Chapla et al. 2012). Thus, pretreatment is needed before enzymatic digestion of LCB. In contrast to autohydrolysis, this approach is more attractive because it does not generate inhibitors or large amounts of monomeric sugars, and does not need special equipment. However, the combined chemical and enzymatic method is not favorable for xylose production. The two-stage method coupling with an autohydrolysis and enzyme treatment might also be effective in the production of XOs.

During acid hydrolysis, xylose was released as the predominant sugar from hemicellulose of hardwoods and agricultural residues together with other byproducts in varying amounts such as glucose, acetic acid, arabinose, lignin degradation products (LDPs), furfural, and hydroxymethylfurfural (HMF) (Rafiqul and Sakinah 2012a, b; Rafiqul et al. 2014). Some of these byproducts are toxic to organisms or enzymes. To assess treatment conditions in order to maximize xylose recovery, it is important to understand the mechanism and kinetics of LCB hydrolysis. All the inhibitory components need to be removed or their concentrations reduced so that the hemicellulosic hydrolysate can be effectively used in bioconversion processes (Marton et al. 2006).

The main problem related with efficient bioconversion of xylose to specialty products is that the hydrolysate contains a variety of byproducts, which are toxic to xylose-assimilating microbes (Villarreal et al. 2006). Xylitol is one of these products with broad applications in medical, pharmaceutical, odontological, cosmeceutical, and food industries (Rafiqul and Sakinah 2013). Several methods of detoxification including physical, chemical, biological, and combined treatments have been developed to reduce the concentration of inhibitors or to convert them into inactive compounds (Mussatto and Roberto 2004; Canilha et al. 2008). These approaches include neutralization, overliming, solvent extraction, charcoal adsorption, use of ion exchange resins, evaporation, microbial cell adaptation, and treatment with enzyme (viz., laccase and peroxidase) (Carvalheiro et al. 2005; Villarreal et al. 2006). When detoxification enhances the production costs, it is necessary either to bypass the detoxification steps or to develop efficient and cheap processes (Carvalheiro et al. 2005). This chapter aimed to review the state of art in selective recovery of hemicellulose, including a description of major structural composition of lignocellulosic biomass, application, mechanism, and kinetics of hydrolytic reaction and inhibitory compounds, and to identify the most efficient approach for maximum hemicellulose depolymerization into its monomeric constituents.

2 Lignocellulosic Biomass

Lignocellulosic biomass (LCB) is widespread, ecofriendly, renewable, cost-effective, and an inexpensive source of polysaccharides that can be utilized as substrate for the manufacture of various specialty chemicals, biofuels, and food ingredients (Samanta et al. 2012; Rafiqul and Sakinah 2013; Singh et al. 2015). A wide variety of LCB is available on earth for further conversion and utilization such as agricultural wastes (e.g., sugarcane bagasse, wheat and rice straw, corncobs, and corn stover), forest residues (e.g., wood sawdust and sawmill wastes), portions of municipal solid waste (e.g., waste paper), and woody (e.g., poplar trees) and herbaceous (e.g., switchgrass) crops (Kumar et al. 2009; Arvela et al. 2011; Rafiqul et al. 2014; Vithanage et al. 2016). The choice of LCB material bears great importance from both techno- and socioeconomical viewpoints. From ethical perspectives, the biomass should not compete with the food chain. Lignocellulosic waste with a low or even negative value such as agricultural and forest residues are preferred. Furthermore, it is also advantageous to select biomass sources that are locally available throughout the year. Lignocellulose is the principal constituent of plant cell walls. All LCBs are composed of three biopolymers cellulose (33–50 %), hemicellulose (18–35 %), lignin (10–30 %), and smaller contents of extractives, ash, pectins, and proteins. The amount of these constituents can vary according to plant species, growth stage, age, and other conditions (Rafiqul and Sakinah 2013). For example, hardwoods have greater contents of cellulose, whereas leaves, wheat straw, and grasses have more hemicellulose (Table 1). A detailed knowledge of the LCB structure is required to calculate the yields in polysaccharides and to assess the reactivity of components occurred in the biomass. Hence, the chemical composition of biomass, particularly cellulose, hemicellulose, and lignin, is detailed below.

Table 1 Composition of different lignocellulosic biomasses

Lignocellulosic biomass	Major components (%, w/w)			References
	Cellulose	Hemicellulose	Lignin	
Meranti wood sawdust (MWS)	41.06	30.64	25.22	Rafiqul and Sakinah (2012a), Rafiqul et al. (2014)
Oil palm empty fruit bunch (OPEFB)	30–40	20–30	20–30	Rahman et al. (2006), Zhang et al. (2012)
Oil palm frond	44	30.4	15.4	Hanim et al. (2011)
Brewer's spent grain (BSG)	16.8	28.4	27.8	Mussatto and Roberto (2005), (2008)
Corncobs	32–45	30-35	14–20	Kumar et al. (2009), Tada et al. (2012)
Sugarcane bagasse	40.5	23.9	26.4	Rabelo et al. (2011)
Rich straw	40	26.8	18.9	Sarkar and Aikat (2013)
Sunflower stalk	42.3	19.1	26.5	Akpinar et al. (2009)
Tobacco stalk	40.4	20.0	20.7	
Softwood stems	45–50	25–35	25–35	Kumar et al. (2009)
Leaves	15–20	80–85	0	
Nut shells	25–30	25–30	30–40	
Wheat straw	30	50	15	
Hardwood stems	40–55	24–40	18–25	
Grasses	25–40	35–50	10–30	
Cotton seed hairs	80–95	5–20	0	
Switchgrass	45	31.4	12	
Bermuda grass	25	35.7	6.4	
Solid cattle manure	1.6–4.7	1.4–3.3	2.7–5.7	
Waste newspaper	40–55	25–40	18–30	

2.1 Cellulose

Cellulose is the major constituent of LCB and is the most abundant biomaterial in nature. It is present in an organized fibrous form and usually comprised of 33–50 % (w/w) of the dry biomass. The chemical structure of a cellulose chain is shown in Fig. 1. Cellulose is a linear homopolysaccharide composed of D-glucopyranose units (5000–10,000 units) linked to each other by β-(1,4)-glycosidic bonds to form a skeleton (Kumar et al. 2009; Rafiqul and Sakinah 2013). The repeating unit of the cellulose molecule consists of two glucose anhydride residues, called cellobiose, which constitutes cellulose chains (Balat et al. 2008). The size of a cellulose molecule is generally expressed by its degree of polymerization (DP). The DP is defined as the number of anhydroglucose units that exist in a single chain, and it varies according to the biomass type in a range of 100–20,000 (Park et al. 2010; Arvela et al. 2011). The long cellulose chains in LCB are linked together via hydrogen linkages. They are in turn joined together with hemicellulose and lignin through hydrogen-bonding interaction and covalent interaction, respectively (Mosier et al. 2005).

Cellulose is present in both crystalline and amorphous forms. In LCBs, about 50–90 % of the cellulose chain is joined laterally by hydrogen bonds to form a crystalline (ordered) structure, whereas the remaining fraction is less ordered and is called amorphous cellulose (Park et al. 2010; Chundawat et al. 2011). In other words, the crystallinity is developed by the presence of inter- and intramolecular hydrogen linkages among several anhydroglucan chains. The crystalline form in cellulose is very stable. Moreover, H_2O is eliminated from crystalline part of cellulose (Dhepe and Fukuoka 2008; Deng et al. 2009). During acid treatment, the amorphous cellulose is rapidly solubilized, whereas the breakdown of crystalline cellulose is restricted by the limited access of reactant to β-(1,4)-glycosidic linkages. Cellulose is susceptible to enzyme digestion in its amorphous form. Glucose is released from cellulose via the action of either acid or enzyme by cleaving glycosidic linkages (Kumar et al. 2009). The resulting glucose can be transformed to sorbitol or bioethanol as a fuel additive.

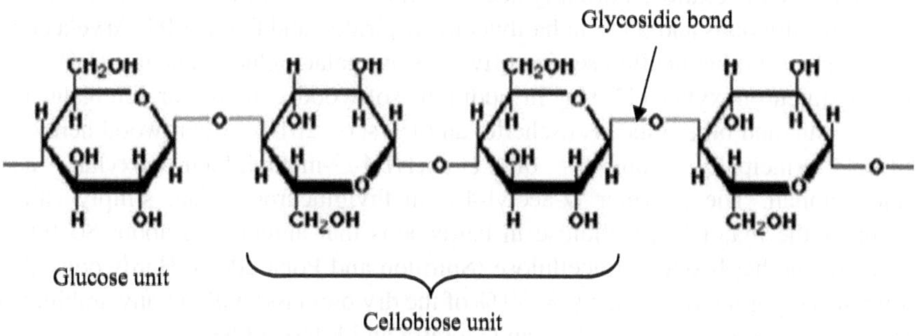

Fig. 1 Chemical structure of a cellulose chain

2.2 *Hemicellulose*

Hemicellulose, also known as polyose, is a second major constituent of LCB. It is a complex heteropolysaccharide whose chemical composition and structure vary from tissue to tissue within a single plant and from species to species. Hemicellulose is present in varying amounts in plants about 18–35 % of the total biomass, and it exists closely associated with cellulose and lignin. It is generally composed of different building block compounds including pentoses (e.g., xylose and arabinose), hexoses (e.g., glucose, galactose, and mannose), and uronic acids (e.g., 4-*O*-methylglucuronic and galacturonic acids) with acetyl side chains. It consists of 100–200 units of pentoses and hexoses in its molecular backbone (Park et al. 2010; Thomas et al. 2011). Hemicellulose is amorphous in structure, and a variety of branching, bonds, and monomers contribute to its complex architecture and thereby its variety of conformations and function (Scheller and Ulvskov 2010; Pauly et al. 2013). It is involved in the regulation of cell wall enlargement and modification and, unlike lignin, is thought to have a significant impact on the interaction among cellulose microfibrils. Hemicellulose differs from cellulose in three main aspects. Firstly, it contains various types of sugars, whereas cellulose contains only β-(1,4)-D-glucopyranoses. Secondly, it shows a significant degree of chain branching, whereas cellulose is firmly a linear biopolymer. Thirdly, the DP of cellulose is 10–100 times higher than that of hemicellulose (Thomas et al. 2011). In contrast to cellulose, the biopolymers found in hemicellulose are readily degradable. Hemicellulose in LCB is connected to cellulose through hydrogen-bonding interactions and builds a structural matrix (Mosier et al. 2005). This matrix is further bound to lignin for fabricating a lignocellulosic complex (Pauly et al. 2013). The composition of hemicellulose present in different lignocellulosics is described hereafter.

2.2.1 Hemicellulose in Wood

The prime hemicelluloses in woody materials are xylans, mannans, arabinans, and galactans. Hemicellulose normally accounts for 25–35 % of the mass of dry wood (27 % in softwoods and 35 % in hardwoods) (Spiridon and Popa 2008; Arvela et al. 2011). The vital hemicelluloses in softwoods are galactoglucomannan (21 %) and arabinoglucuronoxylan (11 %). In addition, softwood contains arabinogalactan, xyloglucan, and other glucans (Scheller and Ulvskov 2010). A hardwood hemicellulose principally consists of *O*-acetyl-4-*O*-methylglucuronoxylan and glucomannan. The polymer *O*-acetyl-4-*O*-methylglucuronoxylan, simply called xylan, is the major hemicellulose in hardwoods that amounts to about 80–90 % (w/w) of the hardwood hemicellulose (Spiridon and Popa 2008). Hardwoods also contain acetylglucuronoxylan (15–30 % of the dry biomass) without any arabinosyl branches opposite to softwood xylan (Moreira and Filho 2008).

Softwoods (such as spruce, coniferous trees, and pine) and hardwoods (such as meranti, oak, willow, and eucalyptus) differ in structure and chemical constituents of the hemicellulose. Hemicellulose in softwoods has a higher amount of mannose and glucose units than hardwoods hemicellulose, which typically holds a larger quantity of xylose residues. Moreover, it is more highly acetylated in hardwoods than in softwoods (Palmqvist and Hahn-Hägerdal 2000a; Rafiqul and Sakinah 2013). Arabinogalactan is more abundant hemicellulose in softwoods than in hardwoods with amounts ranging from 1–15 % to 1–5 %, respectively (Spiridon and Popa 2008). In arabinogalactan, galactose to arabinose ratio is typically 6:1 (Arvela et al. 2011). The long polymeric chain is greatly branched at β-(1,3) and β-(1,6) positions among galactose and small contents of glucuronate (Willför et al. 2005). During hydrolysis, xylose is the major sugar released from hardwood hemicelluloses, but arabinose can form a considerable quantity of pentoses liberated from different agricultural wastes and herbaceous crops (Balat et al. 2008). *Meranti* wood sawdust (MWS) is a valuable feedstock, since 71.7 % (w/w) of its dry mass is carbohydrate. This biomass is rich in xylan, about 29.22 % (Rafiqul and Sakinah 2012a; Rafiqul et al. 2014).

2.2.2 Hemicellulose in Agricultural Wastes

A number of agricultural wastes are utilized as feedstocks to produce rare sugars and specialty chemicals by acid hydrolysis, for example, wood sawdust, sugarcane bagasse, rice, wheat and sorghum straw, corncobs and stalk, sunflower stalk, cotton seeds, oil palm empty fruit bunch (OPEFB), brewer's spent grain (BSG), etc. Typically, agricultural residues contain 30–42 % (dry mass basis) cellulose, 11–35 % xylan, and 20–27 % lignin (Akpinar et al. 2009; Rafiqul and Sakinah 2013). Sugarcane bagasse is a potential feedstock because of its high abundance and xylan content (20.6 %). It contains hemicellulose with a xylan skeleton, which also holds arabinose units and glucuronic acids (Chong et al. 2004; Cheng et al. 2008). Additionally, the cellulose content in sugarcane bagasse is smaller than that in woods, being more prone to digestion. Nowadays, sugarcane bagasse is either utilized as boiler fuels or alternatively not used but considered as an environment pollution problem (Ewanick and Bura 2011). There is much interest in transforming it to sugar monomers (e.g., xylose) and further convert the sugars to xylitol and ethanol.

Rice straw, a side product from rice cultivation, contains 40.09 % cellulose, 26.8 % hemicellulose, and 18.9 % lignin (Sarkar and Aikat 2013). Its polymeric composition is as follows: glucan 42 %, xylan 21 %, lignin 16 %, ash 10 %, and arabinan 2.6 % (Roberto et al. 2003). This straw is currently incorporated into the soil or burnt. The later alternative is injurious to environment, generating CO_2, CO, and CH_4 emission. Moreover, it is suitable neither for animal feed nor for combustion due to its low protein content and energy (17.6 MJ/kg) value (Escribá and Porcar 2010). Rice straw is easily depolymerized with dilute acid because of its low lignin content. Wheat straw is available in huge amounts that contains arabinoxylan, in which xylose residues connected through β-(1,4)-xylosidic links and arabinose units are joined to the xylan backbone (Liavoga et al. 2007; Canilha et al. 2008). In general, wheat straw contains 24 % cellulose, 21 % hemicellulose, and 20 % lignin with a small amount of ash, waxes, and sugars. The remainder of constituents in wheat straw are protein and starch (Olsson 2006). The selective extraction of arabinoxylan from wheat straw via dilute acid treatment might be done at relatively low temperature due to its low content of cellulose (Liavoga et al. 2007; Arvela et al. 2011).

Corncobs and cornstalks are agricultural residues with low lignin content (Jørgensen et al. 2007). Typically, corncobs contain about 14 % lignin, but they have high contents of cellulose and hemicellulose with the values of 42 and 39 %, respectively (Kumer et al. 2009: Zhang et al. 2014). Corn stover normally consists of about 35 % cellulose, 20 % xylan, 17 % lignin, and 3 % arabinan (Aden and Foust 2009). Sunflower stalks are also residues generated during the production of sunflower seed, and they contain about 38 % cellulose, 30 % hemicellulose, and 11 % lignin. Normally, the ratio between stalk and seed weight is 2.3 (Arvela et al. 2011). Cotton seed residual cake is one of the promising sources of xylan. After delignification, it contains solely uronic acid and polymer xylan, and its sugar amount is about 75 %. Cotton seed was processed by steeping it in NaOCl solution for removing lignin followed by rinsing with H_2O (Sun et al. 2002).

Oil palm empty fruit bunch fiber (OPEFB) is a lignocellulosic waste from palm oil mills. Usually, the OPEFB is disposed in landfill areas or utilized as fuel in incineration plants that creates environmental pollution problem in nearby localities. It is composed of cellulose (30–40 %), hemicellulose (20–30 %), and lignin (20–30 %) with xylan being the principal constituent of hemicellulose. This biomass contains 18–26 % xylan, a xylose polymer (Rahman et al. 2006; Zhang et al. 2012). Brewer's spent grain (BSG) is the breweries wastes derived in enormous quantities, which corresponds to about 86 % of the overall production. Though BSG is generated in huge amounts, it has attracted little attention as a marketable commodity and is principally utilized as animal feeds. The BSG contains cellulose (17 %), hemicellulose (28.5 %), lignin (28 %), proteins (15.25 %), extractives (5.8 %), ash (4.6 %), and acetyl groups (1.35 %) (Mussatto and Roberto 2005; Mussatto et al. 2008).

2.2.3 Hemicellulose in Plant Gums

Arabinose is observed in large amounts in some plant gums like gum arabic (Kim et al. 2012). The main segment of gum arabic contains a highly branched polysaccharides β-(1,3)-connected galactose backbone with side chains of arabinose and rhamnose, and the chains end in glucuronate. Rhamnose, a deoxy sugar, has also been specifically extracted from gum arabic by means of hydrolysis (Dror et al. 2006; Kim et al. 2012). In addition to hemicellulose, rhamnose is also occurred in glycoside having a sugar residue together with a non-sugar residue. Rhamnose-containing glycosides (viz., hesperidin, naringin, and myricitin) are flavonoids, and hence their recovery is outside the scope of the current study.

Hemicellulose, particularly xylan, has gained much attention globally due to its multipurpose uses including food and nonfood applications. Hemicellulose is extensively utilized as a raw material for producing a wide variety of commodity chemicals such as xylose, xylitol, furfural, ethanol, erythritol, and lactic acid. Furthermore, it is industrially used as gelling material, viscosity modifier, tablet binder, and wet strength additive (Samanta et al. 2012). Several scientists have revealed the therapeutic application of xylan as antitumor, ulcer protective, and immunostimulatory agents (Barbat et al. 2008; Cipriani et al. 2008).

3 Lignin

Lignin, a third significant fraction in LCB, is an aromatic and rigid biopolymer. It is composed of monomers of phenyl propionic alcohol that is sinapyl, coniferyl, and coumaryl alcohol units linked by aryl-aryl, alkyl-alkyl, and alkyl-aryl ether linkages. Lignin is found in the plant cell walls, imparting structural support, impermeability, and resistance against microbial attack. It is covalently bonded to xylans and responsible for the rigidity and high level of compactness of the cell wall. The lignin fraction in biomasses varies considerably. In general, softwoods contain more lignin than hardwoods and herbaceous plants such as grasses, agricultural residues (Kumar et al. 2009; Rafiqul and Sakinah 2013; Eudes et al. 2014). Lignin contents on a dry basis in both hardwoods and softwoods usually ranged from 20 % to 40 % and from 10 % to 40 % in different herbaceous wastes, viz., straws, bagasses, corncobs, and rice husks (Balat et al. 2008; Rafiqul and Sakinah 2013). Lignin is amorphous and hydrophobic in character and is not solubilized by acid but is hydrolizable with warm alkali, simply oxidized and condensed by phenol (Thomas et al. 2011). Cellulose, hemicellulose, and lignin are closely associated with each other; thus, biomasses can be regarded as composite materials in which lignin acts as a defensive agent that obstructs plant cell degradation using microorganisms (viz., fungi and bacteria) to convert into chemicals and fuels (Rafiqul and Sakinah 2013; Eudes et al. 2014).

4 Hydrolysis Methods of Hemicellulose

In order to make the LCB suitable for bioconversion in bioprocesses, cellulose and hemicellulose must be depolymerized into their sugar constituents for use by microbes or enzymes. The most significant industrially used rare sugars including xylose, arabinose, mannose, galactose, and rhamnose that are manufactured by recovery of the hemicellulosic part of biomass. These compounds are widely utilized in the food, functional sweetener, and pharmaceutical industries. Hemicellulose is easily hydrolyzable owing to its amorphous, branched nature compared to cellulose (Rafiqul et al. 2014; Rennie and Scheller 2014), which requires harsh experimental condition because of its DP, crystallinity, and reachable surface region (Taherzadeh and Karimi, 2008; Chundawat et al. 2011). The purpose of hydrolysis of the hemicellulose portion of LCB, specifically hardwoods and agricultural residues, is to maximize the recovery of pentose sugars (e.g., xylose and arabinose) and minimize the coformation of toxic/inhibitory byproducts. Additionally, hydrolysis of the hemicellulose prepares the cellulosic part for subsequent transformation by acid or enzyme and shows particularly promising opportunities for studies that could drastically decrease biomass processing costs. Hemicellulose is generally hydrolyzed through the following methods to produce xylose and/or xylooligosaccharides: (1) acidic, (2) autohydrolytic, (3) enzymatic, (4) combined chemical-enzymatic, and (5) combined autohydrolytic-enzymatic methods (Yang et al. 2007; Teng et al. 2010; Hanim et al. 2011; Rafiqul and Sakinah 2013; Visser et al. 2015). The hydrolysis of cellulose and lignin is not focused in this work.

5 Acid Hydrolysis

Acid hydrolysis, a rapid and simple method, is one of the most popular approaches to get the highest yield of sugar from LCB. The hydrolysis conditions vary with raw material type, acid type and concentration, reaction temperature, and time (Rafiqul and Sakinah 2013; Singh et al. 2015). Xylose is the most abundant sugar released in the hydrolysate with a small amount of other sugars, while fully grown or aged hardwoods or agricultural wastes are utilized as feedstocks. Different acids such as sulfuric (Mussatto and Roberto 2005; Romero et al. 2010; Zhang et al. 2012; Rafiqul et al. 2014; Amirkhani et al. 2015), hydrochloric (Lavarack et al. 2002; Herrera et al. 2004; Sun et al. 2014), phosphoric (Lenihan et al. 2010), nitric (Chong et al. 2004; Salvi et al. 2010), acetic (Zhao et al. 2008; Chaturvedi and Verma 2013), and oxalic (Scordia et al. 2011) acids are commonly applied as catalysts in the hydrolysis process. Among these chemicals, the most extensively used and examined method is based on H_2SO_4 due to its high activity, low price, availability, low safety, and environmental concerns for industrial utilizations (Wang et al. 2011). Acid hydrolysis is mainly of two types according to the concentration of the acid applied: concentrated acid and dilute acid hydrolysis. The major decomposition pathway during acid-catalyzed hydrolysis of LCB is shown in Fig. 2.

5.1 *Concentrated Acid Hydrolysis*

Concentrated acid hydrolysis is a relatively old method. It is performed at low operating temperature (<50 °C) and atmospheric pressure with high acid concentration (50–70 %, v/v). Approximately 100 % cellulose is converted to glucose in this process (Kumar et al. 2009). In practice, concentrated acids such as H_2SO_4 and HCl have been utilized to treat LCB. Although the concentrated acids are strong

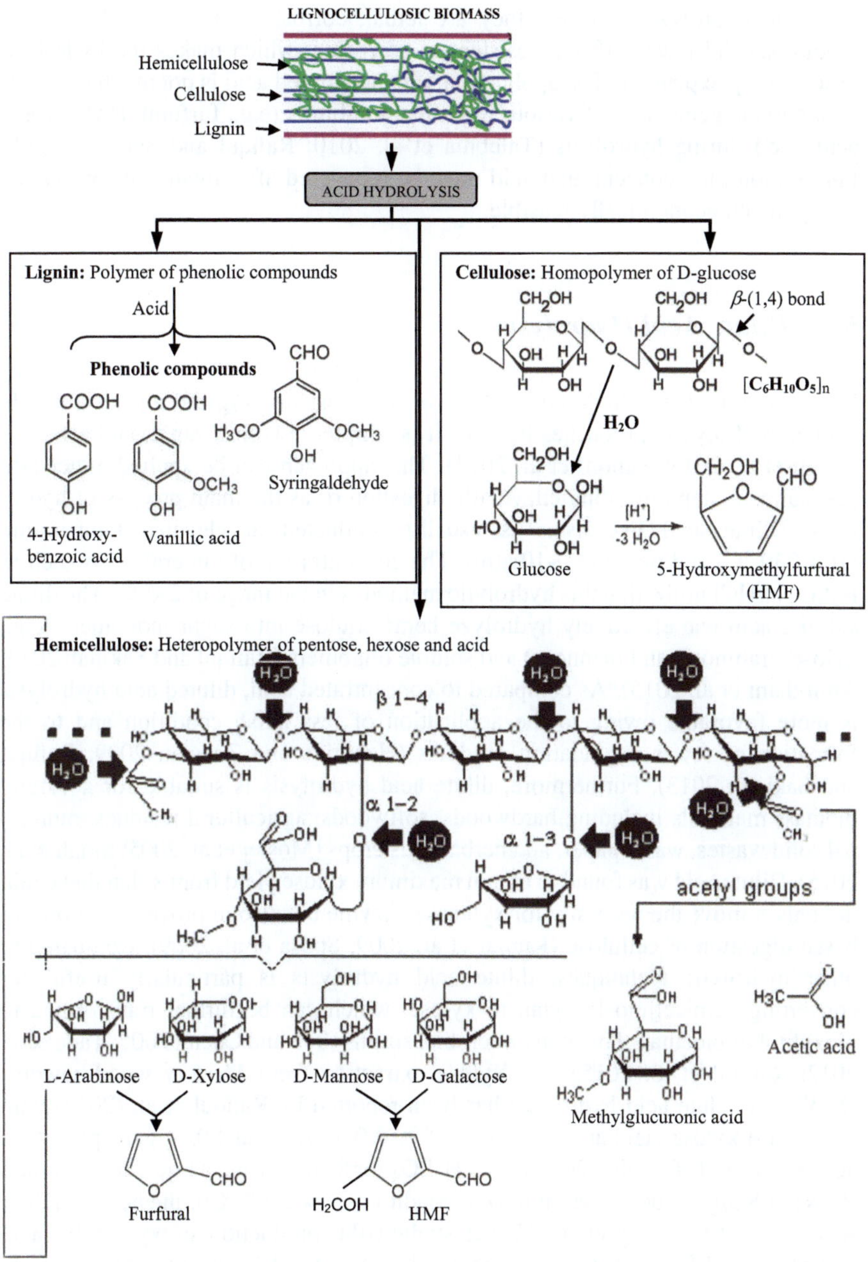

Fig. 2 Degradation of lignocellulosic biomass by acid hydrolysis

agents to hydrolyze cellulose, they are lethal, corrosive, and harmful and thus require special reactor that is resistant to corrosion, which makes the hydrolysis process very expensive. The application of concentrated acid is not much desirable owing to the generation of various microbial inhibitors (e.g., furfural, HMF, acetic acid, etc.) during hydrolysis (Talebnia et al. 2010; Rafiqul and Sakinah 2013). Furthermore, the concentrated acid must be recovered after treatment for making the approach economically feasible.

5.2 *Dilute Acid Hydrolysis*

Dilute acid-mediated hydrolysis is one of the most investigated and extensively applied hydrolysis approaches because it is simple, low-cost, rapid, and effective (Swati et al. 2013; Rafiqul et al. 2014). This approach can be applied either as a pretreatment step preceding enzymatic digestion or as the main process of hydrolysis. Dilute acid hydrolysis is usually conducted at elevated temperature (160–230 °C) and pressure (~10 atm). The concentration of mineral acids such as H_2SO_4 or HCl utilized in this hydrolytic method is in the range of 2–5 %. The dilute sulfuric acid can effectively hydrolyze hemicellulose into sugar monomers (e.g., xylose, arabinose, and mannose) and soluble oligomers (Rafiqul and Sakinah 2013; Amirkhani et al. 2015). As compared to concentrated acid, diluted acid hydrolysis is more favorable owing to the application of less harsh condition and to the formation of fewer degradation products (Hendriks and Zeeman 2009; Rafiqul and Sakinah 2013). Furthermore, dilute acid hydrolysis is suitable for a variety biomass materials including hardwoods, softwoods, agricultural residues, municipal solid wastes, waste paper, and herbaceous crops (Mosier et al. 2005; Singh et al. 2015). Dilute acid was found to obtain maximum xylose yield from xylan that could not only remove the necessity for xylanase enzyme but also improved the enzyme-based digestion of cellulose (Kumar et al. 2009; Sousa et al. 2009). Compared to other hydrolytic techniques, dilute acid hydrolysis is particularly useful for converting hemicellulosic xylan to xylose, which can be further transformed to bioxylitol or bioethanol by many microbial strains (Sun and Cheng 2005; Tada et al. 2012; Rafiqul et al. 2015a, b). Xylose extraction from *Meranti* wood sawdust (MWS) by dilute acid hydrolysis has been reported by Rafiqul et al. (2014) with the highest xylose yield and selectivity of 90.6 % (w/w) and 4.05 g/g, respectively, achieved at 124 °C with 3.26 % (w/w) H_2SO_4 for 80 min using a liquid to solid ratio (LSR) of 8 g/g. Under these optimized conditions, about 92 % of the hemicellulose was recovered. Zhang et al. (2012) studied the production of xylose by acid hydrolysis and found that the maximal xylose yield of 91.3 % was obtained from OPEFB hydrolysis catalyzed by 0.5 % H_2SO_4 and 0.2 % (w/v) H_3PO_4 at 160 °C, a LSR of 20 mL/g for 10 min. They pointed out that the combined utilization of H_2SO_4 and H_3PO_4 has a synergistic effect in enhancing xylose production compared to using H_2SO_4 alone. Table 2 presents the reported optimum conditions of acid hydrolysis of various LCBs along with the products yield.

Table 2 Reported data on the acid hydrolysis of different lignocellulosic biomasses

Lignocellulosic biomass	Optimum hydrolysis conditions	Products (g/L)	References
Wheat straw (particle size <1 mm)	1.85 % (w/v) H_2SO_4, 90 °C, 18 h, 20:1 w/w LSR (in reactor)	12.80 Xylose 1.70 Glucose 2.60 Arabinose 0.15 Furfural 2.70 Acetic acid	Nigam (2001)
Sorghum straw (particle size <0.5 mm)	6 % (w/w) HCl, 122 °C, 70 min, 10:1 w/w LSR (in autoclave)	16.20 Xylose 3.80 Glucose 2.0 Furfural 1.90 Acetic acid	Herrera et al. (2003)
Brewer's spent grain (particle size <0.5 mm)	3 % (w/w) H_2SO_4, 130 °C, 15 min, 8:1 w/w LSR (in autoclave)	26.70 Xylose 12.80 Arabinose 4.0 Glucose 1.50 Acetic acid 0.91 LDPs 0.29 Furfural 0.02 HMF	Carvalheiro et al. (2004a)
Oil palm empty fruit bunch (particle size <1 mm)	6 % H_2SO_4, 120 °C, 15 min, 8:1 w/w LSR (batch process)	29.40 Xylose 2.34 Glucose 1.25 Acetic acid 0.87 Furfural	Rahman et al. (2006)
Eucalyptus grandis wood (particle size <0.5 mm)	0.65 % H_2SO_4, 157 °C, 20 min, 8.6:1 w/w LSR (1.4 L pilot-scale reactor)	13.65 Xylose 3.10 Acetic acid 1.65 Glucose 1.55 Arabinose 1.23 Furfural 0.20 HMF	Canettieri et al. (2007)
Sugarcane bagasse (particle size 0.45–0.9 mm)	1.25 % (w/w) H_2SO_4, 121 °C, 2 h, 10:1 w/w LSR (in autoclave)	17.1 Xylose 7.2 Glucose 4.0 Acetic acid 2.0 Arabinose 1.4 Furfural 0.9 Galactose 0.5 Cellobiose 0.3 Mannose	Cheng et al. (2008)
Rice straw(particle size 10 mm)	2 % (w/v) H_2SO_4, 126 °C, 60 min, 10:1 v/w LSR (in pressure cooker)	13.30 Xylose 2.50 Glucose 2.30 Arabinose	Liaw et al. (2008)
Meranti wood sawdust	3.26 % (w/w) H_2SO_4, 124 °C, 80 min, 8:1 w/w LSR (in autoclave)	18.8 Xylose 4.64 Glucose 4.14 Acetic acid 2.55 Arabinose 1.55 LDPs 0.55 Furfural 0.08 HMF	Rafiqul and Sakinah (2012a); Rafiqul et al. (2014)

LDPs Lignin degradation products, *HMF* Hydroxymethylfurfural

5.3 Autohydrolysis

Autohydrolysis is defined as the pretreatment of biomass with hot water or saturated steam. It is given this name because the reaction is automatically initiated by the hydronium ions (H_3O^+) generated from the dissociation of water molecules. Subsequently, the dissolved acidic groups from hemicellulose liberate more H_3O^+ and accelerate the hydrolytic reaction. Autohydrolysis is also called hydrothermolysis, steam pretreatment, aqueous pretreatment, aquasolv, or hydrothermal treatment. Autohydrolysis is an alternative technology to solubilize hemicellulose, with different benefits over the dilute acid treatment, i.e., a slight delignification and reduced contents of sugar decomposition compounds (Garrote et al. 2002; Parajó et al. 2004; Rafiqul and Sakinah 2013). Furthermore, it shows some environmental and technical advantages as no chemical (acid or alkali) other than water is employed (El Hage et al. 2010; Rafiqul and Sakinah 2013). However, this method needs particular reactor that can be run at elevated temperature. Autohydrolysis is typically conducted at temperatures of 160–260 °C, which produces a high-molar mass of xylooligosaccharides (XOs) without altering substantially the structure of cellulose and lignin, allowing improved recovery during further processing (including enzyme hydrolysis of cellulose or chemical delignification) (Parajó et al. 2004; Garrote et al. 2007). The variables that affect the autohydrolysis efficiency and the hydrolysate composition are residence time, temperature, liquid to solid ratio, and structural integrity of raw material used (Parajó et al. 2004; Rafiqul and Sakinah 2013).

Garrote et al. (2002) studied XOs manufacturing from corncob by autohydrolysis and found that most of the cellulose was remained in the solid residue, whereas partial delignification (up to 25 % lignin removal) was attained. The authors reported that up to 95 % of xylan was recovered, forming XOs and other byproducts (mainly xylose and furfural). Production of XOs from brewery's spent grain (BSG) through autohydrolysis has been reported by Carvalheiro et al. (2004b) with a maximum yield of 61 % xylan achieved at 190 °C after 5 min of reaction. It was estimated that 62–78 % of the starting xylan was selectively depolymerized during autohydrolysis. Nabarlatz et al. (2007) examined six agricultural residues as raw materials for XOs production by autohydrolysis and found that the XOs yield followed xylan content and its accessibility and further progressively related to the contents of acetyl group in the feedstocks. Analyses of the hydrolysates revealed that they contained partially acetylated oligomeric and polymeric xylan fragments and a small amount of monosaccharides (e.g., xylose and arabinose) and degradation products (mainly furfural, HMF, and acid soluble lignin). However, the autohydrolysate is only partly fermented by organisms, while the carbohydrates are principally in the oligomeric forms and therefore a post-hydrolysis step (with dilute acid) is required for manufacturing sugar monomers. A two-step method (autohydrolysis followed by post-hydrolysis) is used for the generation of fermentable hydrolysate (Carvalheiro et al. 2005). Moreover, autohydrolysis is nonspecific, and other reactions than hemicellulose solubilization occur, which lead to a liquor having complex composition (Garrote et al. 2004).

Therefore, autohydrolysis is not feasible for the production of xylose to a satisfactory level.

5.4 Enzymatic Hydrolysis

Hydrolysis of LCB by enzyme technology has been justified as an alternative hydrolytic approach. In enzymatic hydrolysis, the utility cost is lower compared to chemical treatment as enzymatic hydrolysis is normally performed under gentle condition (pH ~5 and temperatures 45–50 °C) and does not have a corrosion problem (Rabelo et al. 2011; Rafiqul and Sakinah 2013). Compared to acid and autohydrolysis, enzyme hydrolysis is environmental-friendly, milder, and more specific, but it needs pretreatment for enhancing enzymatic digestibility (Sun and Cheng 2005; Wu et al. 2011). The yield and rate of enzyme-based hydrolysis of LCB are dependent on various parameters such as catalytic properties of enzymes, their loading concentrations, incubation time, raw material type, process variables, pretreatment process, and compounds released during pretreatment (Zhu et al. 2008; Wu et al. 2011). The pretreatment method removes lignin, reduces the crystallinity of cellulose, and increases the porosity of biomass (Sun and Cheng, 2005). An increase in surface area of cellulose/hemicellulose and enzymes loading improves the hydrolysis rate and time. Among all components in biomass, lignin is considered as a principal barrier to enzyme attack on cellulose, implying the significance of decreasing the structural integrity caused by lignin prior to hydrolysis (Rafiqul and Sakinah 2013).

Xylanases and cellulase are the important enzymes used in most of the enzyme treatment studies. The digestibility of LCB by enzymes is controlled by the surface area of raw materials, and the increase in surface area through pretreatment or reducing particle size increases the hydrolysis of materials (Zhu et al. 2008; Avci et al. 2013). Application of xylanases alone might not be sufficient due to the complex structure of LCB. Xylanases act synergistically and vary with microbial origin. They cleave the β-(1,4)-glycosidic bonds in the xylan molecule releasing short xylooligomers. The choice of a xylanase blend, consisting of xylosidases, arabinofuranosidases, mannanases, hemicellulolytic esterases, and glucuronidases, is one of the significant parameters for effective release of xylose from hemicellulosic material. The choice is also related to the nature of xylan structure, which differs with biomass type (viz., hardwood, softwood, agricultural residue, and grass). Pretreatment either by mild chemical agents at high temperature or by other specific enzymes would present the better hydrolysis method for the efficient recovery of xylose.

Though enzyme-based hydrolysis results in higher yield in the biotransformation of sugar from pretreated LCB, the price of enzymes is a crucial point in product costing. The use of a hemicellulolytic enzyme mixture is another choice. However, for each material, reaction conditions have to be identified and optimized for the specific enzyme blend. A pretreatment by wet oxidation is proved to be efficient for

LCBs, as a reduction in crystallinity was found along with the degradation of lignin to carboxylic acids, CO_2, and H_2O (Rafiqul and Sakinah 2013). Even with a number of advantages, the main shortcoming of enzymatic digestion is that the presence of solid residuals (mainly lignin) and dissolution of enzymes in the hydrolysate makes it difficult to separate and recycle the enzymes in order to reduce the cost (Lu et al. 2007; Taherzadeh and Karimi 2007). In addition, enzymes cannot freely penetrate the lignocellulosic matrix without pretreatment because of the lignin, and thus, the rate of enzyme hydrolysis is slower than acid hydrolysis (Lenihan et al. 2010; Avci et al. 2013). Therefore, the feasibility and applicability of enzymatic hydrolysis of hemicellulose requires further study.

5.5 *Chemical-Enzymatic Hydrolysis*

Hemicellulose hydrolysis via a combination of chemical and enzymatic treatments has been used as an alternative way for yielding xylooligosaccharides (XOs) and xylose from xylan-containing LCBs. Physical access of enzymes to xylosidic linkages in hemicellulose is limited by the surrounding biomass components as well as by the substituents on its backbone. The LCBs are usually pretreated before enzyme digestion. A number of pretreatment approaches are employed to expose the polysaccharide constituents for enzyme digestion such as alkaline extraction, cooking, and acidic treatment (Yang et al. 2007; Chapla et al. 2012). However, acidic treatment is not suitable for producing XOs because it can release numerous xylose in the liquor (Yang et al. 2007). For XOs production via chemical-enzymatic route, xylan is normally recovered with an alkali (e.g., NaOH or KOH) from biomasses, and then the xylan is hydrolyzed into XOs by enzyme xylanases containing low exo-xylanase and β-xylosidase activities (Akpinar et al. 2009). In contrast to autohydrolysis, this approach is more attractive as it does not generate undesired compounds or large quantities of sugars, and does not need special equipment. Hence, numerous reports are available on the production of XOs employing a two-stage method based on an alkali treatment followed by enzyme digestion of xylan from wheat straw (Swennen et al. 2005), corncobs (Aachary and Prapulla 2009, 2011), natural grass (Samanta et al. 2012), and hardwood (Falck et al. 2013). It can be pointed out that the chemical and enzymatic method is suitable for XOs production but not for xylose.

5.6 *Autohydrolytic-Enzymatic Hydrolysis*

A combination of autohydrolysis and enzymatic treatment was employed as a hydrolysis process for removing and recovering hemicellulose from LCBs. Hemicellulose in LCB could be a vital barrier to hydrolyze cellulose by enzymes. The elimination of hemicellulose from biomass, thus, would be a perquisite for entire digestion of cellulose. Autohydrolysis is an effective approach to decompose hemicellulose into XOs. The acids generated in autohydrolysis are week acids (viz., formic, acetic, and glucuronic acids) and are not strong enough to depolymerize cellulose. Autohydrolysis, thus, can selectively recover hemicelluloses from lignocellulosics. However, this approach generates XOs having broad range of degree of polymerization (usually DP 2–20) (Parajó et al. 2004; Makishima et al. 2009). Then, the high DP biomolecules are enzymatically hydrolyzed to smaller sugars (Moure et al. 2006). To avoid xylose coproduction, an enzyme preparation having low exo-xylanase and β-xylosidase activities is needed (Knob et al. 2010). XOs with DP of 2–4 are used as food and food ingredients (Samanta et al. 2012) since the sweetness of xylobiose (DP 2) is about 30 % of that of sugar and the sweetness of xylotriose (DP 3) and xylotetraose (DP 4) is moderate and they have no off-taste (Aachary and Prapulla 2011). To produce XOs from MWS biomass, a combined process involving an autohydrolysis and an enzymatic treatment was conducted. The enzymatic treatment of sawdust autohydrolysate using xylanases (from *Trichoderma viride*) produced a solution containing 15.61 % xylobiose and 10.4 % xylose as the prime oligomer and monomer, respectively. The endoxylanase from *Aspergillus oryzae* digested alkaline pretreated corncobs xylan to produce a reaction mixture containing 37.5 % xylobiose and 19 % xylose after 14 h (Aachary and Prapulla 2009). Thus, the two-stage process based on an autohydrolysis followed by enzymatic hydrolysis might be effective in the formation of XOs from LCBs for industrial applications.

Hemicellulose (xylan) hydrolysis principally produced xylose, which can be further utilized as a starting substrate for the manufacture of various specialty chemicals or biofuels via chemical and biotechnological routes (Mussatto and Roberto 2008; Rafiqul et al. 2015a). Xylitol is one of these chemicals with broad applications in pharmaceutical, odontological, cosmeceutical, and food industries (Rafiqul and Sakinah 2013). Moreover, xylitol is classified as one of the 12 high-valued bioproducts, which can be produced from lignocellulosic biomass (LCB). It can also act as a starting substrate to synthesize a wide variety of industrially important compounds, viz., glycols, lactic acid, and xylaric acid (Li et al. 2013; Rafiqul et al. 2015a).

6 Mechanism of Hemicellulose Hydrolysis

Hemicellulose hydrolysis occurs through the addition of proton either to glycosidic bonds or cyclic pyranic O_2. The development of cyclic intermediates need conformational change of the tetrahydropyran ring, and thus this way requires high energy compared to the acyclic means (Arvela et al. 2011). As a result, the pyranic oxygen way is not economically viable for the recovery of hemicellulose from biomass. Dilute acid-catalyzed hydrolysis of hemicellulose is very complex, mainly due to the fact that the substrate is in a solid phase and the catalyst in a liquid phase. It is a multistep reaction that takes place sequentially as follows (Herrera et al. 2003):

(1) protons diffusion in the wet lignocellulosic matrix; (2) protonation of a glycosidic bond oxygen that connects monomeric sugars; (3) cleavage of the gylcosidic linkage, which is considered as the rate-limiting step; (4) formation of a carbocation as intermediate complex; (5) solvation of the carbocation with water; (6) regeneration of the proton with cogeneration of the mono-, oligo-, or poly-saccharide based on the position of glycosidic bonds; (7) diffusion of the reaction products into the liquid phase; and (8) return of the protons to the second step to begin the cycle again. The initial and the rate-limiting steps are influenced by acid concentration, temperature, reaction time, and liquid to solid ratio (Liaw et al. 2008). It is necessary to select mild operating conditions that will improve xylose yield while decreasing the formation of byproducts such as acetic acid, lignin degradation products (LDPs), furfural, and hydroxymethylfurfural (HMF).

It is also important to determine the rate of hemicellulose hydrolysis that varies depending on its structure. Acid treatment of xylan was random, whereas in the acid-catalyzed extraction of arabinose in vegetable fiber was degraded particularly using dilute acid as catalyst (Sun et al. 2002). Analogously to arabinose recovery, rhamnose can be selectively released initiating from gum arabic with dilute H_2SO_4 (Dror et al. 2006), and rhamnose is present in the branches of galactose backbone. Furanoside hydrolyzes faster than pyranoside probably due to the faster hydrolysis rate for glycosidic bonds having nonreducing ends (Arvela et al. 2011). It can be concluded that different reaction groups in the hemicellulose show their own reactivities.

The mechanism of autohydrolysis reaction is similar to that of dilute acid hydrolysis. This reaction is catalyzed by hydronium ions (H_3O^+). In autohydrolysis, water is the only reactive agent added to substrate, and the reaction includes two stages. In the first stage, H_3O^+ deriving from water autoionization lead to hydrolyze hemicellulose by selective cleavage of both glycosidic links and acetyl groups. In the second stage, H_3O^+ coming from acetic acid ($C_2H_4O_2$) also acts as a catalyst, increasing reaction rate (Rafiqul and Sakinah 2013). The contribution of H_3O^+ from $C_2H_4O_2$ is larger than that from H_2O. During autohydrolysis, acids released from hydrolysis of uronic and acetyl groups present in hemicelluloses catalyze the hydrolysis of bonds between hemicellulose and lignin as well as that of carbohydrates. Although uronic acids are resistant to hydrolysis, they may also contribute to the formation of H_3O^+ although their role in hydrolysis is still not fully realized (Reyes et al. 2013).

The xylanolytic enzymes include xylanases, β-xylosidase, and debranching enzyme acetyl esterase. Xylanases (endo-β-1,4-D-xylan xylohydrolases, EC 3.2.1.8) is the key enzyme to digest hemicellulose. Xylanases randomly hydrolyze β-(1,4)-xylosidic bonds within the xylan polymer releasing XOs, and, in most reactions, the final compounds are xylobiose and xylotriose (Yang et al. 2007; Hanim et al. 2011). The enzyme β- xylosidase liberates xylose units from nonreducing ends of XOs. However, a complete decomposition needs the synergistic function of acetyl esterase to eliminate acetyl substituents from xylan backbone (Yang et al. 2007).

7 Kinetics of Acid Hydrolysis of Hemicellulose

To determine the kinetics of lignocellulose hydrolysis, it is important to analyze oligomers, sugars, and their decomposition products. Additionally, the kinetics of acetic acid formation and lignin dissolution can be attractive to find an overall scenery of the acid hydrolysis methods. The simple kinetic theory of hemicellulose extraction represents that xylan is initially hydrolyzed to xylose, which is then decomposed to furfural. Since differences in structure and composition of biomass affect reaction rate, it is important to examine the reaction kinetics of each raw material. The aim of the kinetic study was to evaluate the factors of kinetic model in order to predict the yield of xylose, glucose, furfural, and acetic acid during acid hydrolysis of lignocelluloses. It is crucial to determine the kinetic properties of compounds generated in the hydrolysis for assessing treatment conditions to optimize xylose yield.

Oven-dried *Meranti* wood sawdust (MWS) was mixed with H_2SO_4 solution at a LSR of 8 g/g. Hydrolysis was carried out at 130 °C in medium having 2–6 % (w/w) H_2SO_4 (Rafiqul and Sakinah 2012a). Samples were taken at different time intervals that ranged from 0 min to 120 min, and the hemicellulosic hydrolysate was processed. The hydrolysate was stored at 4 °C and examined by HPLC to estimate xylose, glucose, furfural, and acetic acid that are involved in kinetic studies on MWS hydrolysis. Numerous kinetic models of LCB hydrolysis have been reported (Saeman 1945; Carrasco and Roy 1992; Garrote et al. 2001; Rafiqul and Sakinah 2012a). The simple model of hemicellulose hydrolysis comprises a series of pseudohomogeneous irreversible first-order reactions that is expressed by Eq. (1):

$$\text{Xylan (s)} \xrightarrow{k_1} \text{Xylose (aq)} \xrightarrow{k_2} \text{Decomposit ion products} \tag{1}$$

where k_1 and k_2 are the rate of xylose production and degradation (min^{-1}), respectively. According to the above model and using differential equation, xylose concentration $[X]$ at time (t) can be presented as Eq. (2) (Saeman 1945; Rafiqul and Sakinah 2012a):

$$[X] = \frac{k_1 [Xn_\text{p}]}{k_2 - k_1} \left(e^{-k_1 t} - e^{-k_2 t}\right) + [X_0]\, e^{-k_2 t} \tag{2}$$

At time 0, the initial xylose and xylan amounts are defined as $[X_0]$ and $[Xn_\text{p}]$, respectively. It is assumed that $[X_0]$ is close to 0; Eq. (2) can be shortened through the following scheme that is known as Saeman's model in the present study. The kinetic constants k_1 and k_2 are estimated via the regression analysis of the observed data.

$$[X] = \frac{k_1\,[Xn_{\rm p}]}{k_2 - k_1}\left(e^{-k_1 t} - e^{-k_2 t}\right) \tag{3}$$

The potential concentration (PC) of xylose (g/L) is the quantitative conversion of xylan to xylose and is computed by Eq. (4). In which, CXn_0 denotes the initial xylan content in the biomass (e.g., 29.22 g xylan/100 g of MWS), LSR is the liquid to solid ratio (8 g/g), 150/132 is the stoichiometric factor, and D the dilution factor (e.g., $D = 5$).

$$[Xn_p] = \frac{150}{132} \times \frac{CXn_0}{LSR} \times D = 20.75\ \text{g xylose/L} \tag{4}$$

The kinetic model of cellulose hydrolysis also includes a sequence of pseudohomogeneous reactions. In this work, glucose degradation was negligible as the experimental conditions were not suitable for its breakdown. Thus, the model can be generalized as Eq. (5). The concentration of glucose $[G]$ as a function of "t" can be modified as Eq. (6). In Eq. (6), $[G_{\rm p}]$ and k_3 indicate the PC of glucose (g/L) and the rate of glucose production (min^{-1}), respectively.

$$\text{Glucan (s)} \xrightarrow{k_3} \text{Glucose (aq)} \tag{5}$$

$$[G] = [G_{\rm p}]\left(1 - e^{-k_3 t}\right) \tag{6}$$

Furfural concentration $[F]$ as a function of time can be presented by Eq. (7) in which k_4 and $[F_{\rm p}]$ are the furfural production rate (min^{-1}) and the PC of furfural (g/L), respectively.

$$[F] = [F_{\rm p}]\left(1 - e^{-k_4 t}\right) \tag{7}$$

The kinetic model for acetic acid formation and its concentration $[A]$ in the hydrolysate can be written as Eqs. (8) and (9), respectively.

$$\text{Acetyl groups} \xrightarrow{k_5} \text{Acetic acid} \tag{8}$$

$$[A] = [A_{\rm p}]\left(1 - e^{-k_5 t}\right) \tag{9}$$

where $[A_{\rm p}]$ is the PC of acetic acid (g/L) and k_5 is the rate of acetic acid formation (min^{-1}). $[G_{\rm p}]$, $[F_{\rm p}]$, and $[A_{\rm p}]$ were measured by regression. To develop models for MWS acid hydrolysis, Eqs. (3), (6), (7), and (9) were applied. Kinetic constants and factors were determined with iteration non-linear regression analysis and were statistically evaluated. The findings of kinetic studies on acid hydrolysis of MWS for the recovery of xylose are addressed in the following sections.

7.1 Kinetics of Xylose Formation

The xylan content $[Xn_p]$ in the MWS biomass was 20.75 g/L. The observed data were fitted to the Eq. (3) for determining kinetic and statistical parameters using regression analysis. The rate of xylose formation k_1 and decomposition k_2 increased with increasing H_2SO_4 concentration. The value of k_1 is about 26.5 fold higher than the k_2 values achieved by 6 % acid. This outcome implied that the enhancing H_2SO_4 concentration needed a smaller reaction time to recover maximum xylose in the liquor. For all regression parameters, the coefficient of determination R^2 (0.9943–0.9982) showed a good relation among the observed and model recommended data.

7.2 Kinetics of Glucose Formation

During acid hydrolysis of LCB, other sugars are produced, mainly glucose. The PC of glucose ($[G_p]$) could not be measured experimentally because it derived from both hemicellulose and cellulose. Kinetics of glucose release were performed via Eq. (6), and the kinetic parameters k_3 and $[G_p]$ were assessed through regression. The $[G_p]$ values achieved (8.02–9.15 g/L) in this study are consistent with other researcher (Téllez-Luis et al. 2002) who reported those ranging from 6.76 g/L to 7.22 g/L for H_2SO_4 hydrolysis of sorghum straw. The small amount of glucose in the hydrolysate pointed out that the cellulosic part remained almost unaltered while dilutes acid treatment of biomass. Hence, the glucose possibly is released from hemicellulose. The values of k_3 were determined ranging from 0.01134 to 0.03006 min^{-1}. It is mentioned that the values of $[G_p]$ and k_3 improved with increasing H_2SO_4 level. The values of R^2 obtained (0.9890–0.9992) were in good agreement with observed and expected results.

7.3 Kinetics of Furfural Formation

Furfural is formed from xylose as a prime decomposition product in the solubilization of LCB. The values of $[F_p]$ were observed within the range of 1.81–3.03 g/L. The obtained $[F_p]$ values are smaller than the values determined during acid treatment of sorghum straw (4 g/L) and OPEFB (4.88–6.57 g/L) (Téllez-Luis et al. 2002; Rahman et al. 2006). Furfural generation rate k_4 from both xylose and arabinose was somewhat larger than the rate of xylose breakdown to furfural k_2, denoting that the conversion of arabinose to furfural was insignificant. It was pointed out that k_4 value improved with increase in the H_2SO_4 concentration. The R^2 values (0.9894–0.9989) were well fitted with the kinetic equation of furfural production.

7.4 Kinetics of Acetic Acid Formation

During LCB hydrolysis, acetic acid is released from acetyl group present in the hemicellulose. The values of $[A_p]$ and k_5 were in the range of 4.55–5.1 g/L and 0.04188–0.08012 min^{-1}, respectively. The $[A_p]$ values are larger than the values reported for sorghum straw hydrolysis (1.47–1.56 g/L) using H_2SO_4 (Téllez-Luis et al. 2002). This finding indicated that H_2SO_4 is a suitable catalyst for liberating acetyl group in the sawdust. It was noticed that the value of k_5 increased with increasing H_2SO_4 level. The R^2 (0.9973–0.9985) showed that kinetic parameters attained for acetic acid was statistically satisfactory.

7.5 Generalized Kinetic Models

To predict all products in the hydrolysate, a generalized model was constructed by relating kinetic factors with acid concentration via the following empirical formula:

$$k_i = k_0 C_a^m \tag{10}$$

In Eq. (10), i is an integer ranged from 1 to 5, k_0 and m indicate regression variables, and C_a is the sulfuric acid level (%, w/w). In the generalized model, the rate of xylose release k_1 and decomposition k_2 was correlated with H_2SO_4 concentration (C_a) in empirical Eqs. (11), and (12), respectively. The values of m for k_1 and k_2 were found to be 1.2453 and 0.44, respectively. For both kinetic constants, the high values of R^2 (0.9918 and 0.9897, respectively) were obtained. By adding Eqs. (11) and (12) with Eq. (3), it is possible to calculate xylose content at any reaction time (0–120 min) and H_2SO_4 level (2–6 %).

$$k_1 = 0.01778\, C_a^{1.2453} \tag{11}$$

$$k_2 = 0.00276\, C_a^{0.44} \tag{12}$$

A generalized model was obtained by correlating kinetic constant k_3 with C_a to predict glucose production at any residence time and H_2SO_4 level applied in the experiment. The model of glucose production is expressed by Eq. (13), and the determination coefficient R^2 (0.9970) for k_3 fitted well. The value of m for k_3 was lower than the value of m for k_1 (Eq. 11). This result suggested that heterogeneous reactions might be taken place while glucan to glucose conversion. The impact of acid level on k_3 was different from that on k_1 due to the amorphous and crystalline structure of hemicellulose and cellulose, respectively. In addition, the solubilization of cellulose principally depends on its DP and accessible surface area (Rafiqul and Sakinah 2012a). Glucose generation can be predicted at each operating condition following Eq. (13) and (Eq. 6). Employing Eq. (14), the kinetic parameter k_4 was correlated with C_a, and the value of R^2 was very high (0.9932). Furfural is toxic to microbes, and thus, its concentration in the hemicellulosic liquor must be diminished for increasing the yield of various specialty chemicals, mainly xylitol.

$$k_3 = 0.00662\ C_a^{0.8135} \tag{13}$$

$$k_4 = 0.00288\ C_a^{0.59} \tag{14}$$

To predict acetic acid content, a generalized model was also created by correlating k_5 with C_a. The model of acetic acid is presented by Eq. (15), and the R^2 value fitted well (0.9880). The value of m for k_5 was 0.5411 that was smaller than the value of m (1.2453) for k_1. The effect of H_2SO_4 on the elimination of acetyl groups from hemicelluloses is lesser compared to that on xylan production that possibly demonstrated as the reason to obtain distinct m values (Rafiqul and Sakinah 2012a). To minimize acetic acid concentration, it is important to conduct biomass hydrolysis using high H_2SO_4 concentration and smaller residence time.

$$k_5 = 0.03003\ C_a^{0.5411} \tag{15}$$

8 Toxic Compounds in Hydrolysate

Treatment of lignocellulosic biomass (LCB) is required to produce xylose-rich hemicellulosic liquor that is applied as reaction media for the subsequent bioproduction of several high-value products, especially xylitol. A wide variety of compounds that are toxic to organisms are normally released or formed during acid or autohydrolysis of LCB. The nature of inhibitory components and their concentrations in hemicellulosic hydrolysate depend on both the type of raw materials and the hydrolysis condition used (Canilha et al. 2012; Rafiqul et al. 2015c). These compounds are usually divided into four main groups based on their origin: furan derivatives, weak acids, phenolic compounds, and heavy metals (viz., chromium, copper, iron, and nickel) released from the hydrolysis equipment. Furan derivatives such as furfural and HMF are generated from the breakdown of pentose and hexose sugars, respectively. Acetic acid, the major weak acid found in the hydrolysate, results from the deacetylation of hemicellulose. Phenolic and other aromatic compounds are produced from the partial decomposition of lignin (Palmqvist and Hahn-Hägerdal 2000b; Marton et al. 2006; Villarreal et al. 2006). HMF can be further decomposed to levulinic and formic acid, the latter also produced when furfural is broken down.

Normally, HMF is formed in less amounts compared to furfural by the decomposition of hexose sugars mainly as the small contents of hexoses in hemicelluloses. In addition, the condition employed in the depolymerization of hemicellulosic materials does not break down hexoses in huge amounts (Palmqvist and Hahn-Hägerdal 2000a). Other toxic compounds such as resin, terpene, vanillic, syringic, tannic, caprylic, caproic, palmitic, and pelargonic acids are also reported to be formed during chemical hydrolysis of biomass (Gírio et al. 2010). The inhibition intensity is directly associated with the initial concentrations of toxic compounds. A strong inhibition indicates high toxicity of the molecules and more severe negative impact on the bioconversion process (Palmqvist and Hahn-Hägerdal 2000b). The inhibitory effects have become one of the main bottlenecks to the commercial manufacture of xylitol and ethanol from LCBs (Marton et al. 2006; Villarreal et al. 2006; Canilha et al. 2008). All the inhibitory components need to be removed or their concentrations reduced so that the hemicellulosic hydrolysate can be effectively used in bioconversion processes (Marton et al. 2006).

9 Hydrolysate Detoxification Methods

The major and common problem related to the efficient bioconversion of xylose to value-added products (viz., xylitol and bioethanol) is that the hemicellulosic liquor contains a wide range of byproducts that are inhibitory to xylose-fermenting microorganisms (Villarreal et al. 2006). Four distinct approaches have been reported to minimize the presence of inhibitory components in lignocellulosic hydrolysates (Taherzadeh et al. 2000): (1) to use bioconversion-friendly hydrolysis methods to avoid the generation of inhibitors; (2) to detoxify the hydrolysate before use for conversion; (3) to construct and/or use inhibitor resistant microorganisms; and (4) to transform toxic compound into nontoxic product that does not interfere with microbial metabolism. When compared with the microbial conversion of detoxified hydrolysate or purified sugar, the conversion of non-detoxified hydrolysate is characterized by slow kinetics and very low yield and productivity. Hence, the lignocellulosic hydrolysate needs to be treated and neutralized to achieve the fermentation pH, thereby turning it to be more suitable for microbial assimilation. The selection of the best hydrolysate detoxification approach is significant to improve the efficiency of bioconversion processes. A variety of detoxification approaches including physical, chemical, biological, and combined treatments have been developed to remove the toxic compounds or to convert them into inactive compounds (Mussatto and Roberto 2004; Canilha et al. 2008, 2012). These approaches include neutralization, overliming, solvent extraction, charcoal adsorption, use of ion exchange resins, evaporation, microbial cell adaptation, and treatment with enzyme (e.g., laccase and peroxidase) (Carvalheiro et al. 2005; Villarreal et al. 2006). Among these approaches, charcoal adsorption and ion exchange resin have been pointed out as the most efficient (Palmqvist and Hahn-Hägerdal 2000a; Villarreal et al. 2006; Canilha et al. 2008).

The efficiency of a detoxification process depends both on the composition of hemicellulosic hydrolysate and on the species of microbes used because each type of hydrolysate has a discrete level of toxicity and every species of strain has a different degree of tolerance to inhibitors (Canilha et al. 2008). Before selecting a detoxification method, it is important to consider the hydrolysate composition, which varies with the raw materials and the operating conditions applied for hydrolysis (Mussatto and Roberto 2004). While detoxification increases the production costs, it is necessary either to bypass the hydrolysate treatment steps or to explore efficient and cheap processes (Carvalheiro et al. 2005). The construction of a genetically engineered new microorganism that tolerates inhibitors would be a better option because it would remove detoxification steps. It is important to identify the major toxic compounds in the hydrolysate before choosing a technique or a series of techniques for detoxification. This knowledge not only helps to select an efficient and low-cost detoxification method, but also to set up the hydrolysis conditions, which diminish the generation of inhibitors. Since this chapter particularly deals with the processes for hemicellulose hydrolysis from lignocellulose materials, information on detoxification methods are not detailed here.

10 Conclusion

Lignocellulosics are tough raw materials due to the compact packing of their constituents such as cellulose, hemicellulose, and lignin. Biomass hydrolysis is needed for disrupting its structure to extract hemicellulose and to prepare cellulose accessible for complete digestion by acid or enzyme. Hemicellulose, the second largest available polysaccharide on earth, is well suited to biomanufacture various specialty products (viz., xylitol and bioethanol) owing to its low cost, abundance, renewability, and environmental benign process. Till now, dilute acid treatment is preferable to other hydrolytic approaches because it is simple, cost-effective, rapid process, and broadly employed to hydrolyze LCB (Rafiqul et al. 2014). Nevertheless, the main drawback of acid treatment is that it produces a liquor, which has not only the sugars required for biotranformation but also toxic compounds (viz., acetic acid, LDPs, furfural, and HMF) that could hinder xylose biotranformation (Zhang et al. 2012; Rafiqul et al. 2014). It is, therefore, necessary to set faster hydrolytic conditions that can produce xylose-rich hydrolysate, which does not need detoxification. The efficient use of the hemicellulose of LCB offers an opportunity to decrease noticeably the cost of specialty chemicals production.

11 Opinion

In biorefinery, a fundamental method is the depolymerization of cellulose and hemicellulose to sugar monomers that is conducted in different routes. The selective hydrolysis of hemicellulose is an applicable approach to produce xylose and other pentoses, while it can be carried out at less severe conditions without disrupting cellulosic part. Acid hydrolysis of hemicellulose to recover rare sugars (e.g., xylose, arabinose, mannose, galactose, and rhamnose) can be optimized by selecting optimal variables (e.g., temperature, acid concentration, residence time, and LSR) of the method. At extreme experimental condition, the selectivity of sugars decreases mainly because of the sugar degradation and partial dissolution of cellulose. It is mentioned that the method of rare sugar production from lignocellulosics by acid hydrolysis is not yet completely understood, while it involves a number of separation and purification steps (viz., neutralization, filtration, precipitation, re-concentration, discoloration, and chromatographic separation) to obtain desirable products. In addition, since the dilute sugar solution is used, its re-concentration and reutilization are economically challenging. The separation and purification of rare sugar from its solution require further study on the process of using hemicellulose as a source of pentose sugars.

Acknowledgments The authors are grateful to the University of Chittagong, Bangladesh, Universiti Malaysia Pahang, and to the Ministry of Higher Education (MTUN-COE Research Grant No. RDU 121205), Malaysia, for providing necessary facilities and funds in order to conduct this study.

References

Aachary AA, Prapulla SG (2009) Value addition to corncob: production and characterization of xylooligosaccharides from alkali pretreated lignin-saccharide complex using *Aspergillus oryzae* MTCC 5154. Bioresour Technol 100:991–995. doi:10.1016/j.biortech.2008.06.050

Aachary AA, Prapulla SG (2011) Xylooligosaccharides (XOS) as an emerging prebiotic: microbial synthesis, utilization, structural characterization, bioactive properties, and applications. Compr Rev Food Sci Food Saf 10:2–16. doi:10.1111/j.1541-4337.2010.00135.x

Aden A, Foust T (2009) Technoeconomic analysis of the dilute sulfuric acid and enzymatic hydrolysis process for the conversion of corn stover to ethanol. Cellulose 16:535–545. doi:10.1007/s10570-009-9327-8

Akpinar O, Erdogan K, Bostanci S (2009) Production of xylooligosaccharides by controlled acid hydrolysis of lignocellulosic materials. Carbohyd Res 344:660–666. doi:10.1016/j.carres.2009.01.015

Amirkhani H, Yunus R, Rashid U, Salleh SF, Radhiah ABD, Syam S (2015) Low-temperature dilute acid hydrolysis of oil palm frond. Chem Eng Commun 202:1235–1244. doi:10.1080/00986445.2014.918881

Arvela PM, Salmi T, Holmbom B, Willför S, Murzin DY (2011) Synthesis of sugars by hydrolysis of hemicelluloses—a review. Chem Rev 111:5638–5666. doi:10.1021/cr2000042

Avci A, Saha BC, Kennedy GJ, Cotta MA (2013) Dilute sulfuric acid pretreatment of corn stover for enzymatic hydrolysis and efficient ethanol production by recombinant *Escherichia coli* FBR5 without detoxification. Bioresour Technol 142:312–319. doi:10.1016/j.biortech.2013.05.002

Balat M, Balat H, Öz C (2008) Progress in bioethanol processing. Prog Energy Combust 34:551–573. doi:10.1016/j.pecs.2007.11.001

Barbat A, Gloaguen V, Moine C, Catherine OS, Kraemer M, Rogniaux H, Ropartz D, Krausz P (2008) Structural characterization and cytotoxic properties of a 4-*O*-methylglucuronoxylan from Castanea sativa. 2. Evidence of a structure-activity relationship. J Nat Prod 71:1404–1409. doi:10.1021/np800207g

Canettieri EV, Rocha GJM, Carvalho JA Jr, Silva JBA (2007) Optimization of acid hydrolysis from the hemicellulosic fraction of *Eucalyptus grandis* residue using response surface methodology. Bioresour Technol 98:422–428. doi:10.1016/j.biortech.2005.12.012

Canilha L, Carvalho W, Felipe MGA, Silva JBA (2008) Xylitol production from wheat straw hemicellulosic hydrolysate: hydrolysate detoxification and carbon source used for inoculum preparation. Braz J Microbiol 39:333–336. doi:10.1590/S1517-83822008000200025

Canilha L, Chandel AK, Milessi TSS, Antunes FAF, Freitas WLC, Felipe MGA, Silva SS (2012) Bioconversion of sugarcane biomass into ethanol: an overview about composition, pretreatment methods, detoxification of hydrolysates, enzymatic Saccharification, and ethanol fermentation. J Biomed Biotechnol 2012:1–15. doi.org/10.1155/2012/989572

Carrasco F, Roy C (1992) Kinetic study of dilute-acid prehydrolysis of xylan-containing biomass. Wood Sci Technol 26:189–208. doi:10.1007/BF00224292

Carvalheiro F, Duarte LC, Medeiros R, Gírio FM (2004a) Optimization of brewery's spent grain dilute-acid hydrolysis for the production of pentose-rich culture media. Appl Biochem Biotechnol 113(116):1059–1072. doi:10.1385/ABAB:115:1-3:1059

Carvalheiro F, Esteves MP, Parajó JC, Pereira H, Gírio FM (2004b) Production of oligosaccharides by autohydrolysis of brewery's spent grain. Bioresour Technol 91:93–100. doi:10.1016/S0960-8524(03)00148-2

Carvalheiro F, Duarte LC, Lopes S, Parajó JC, Pereira H, Gírio FM (2005) Evaluation of the detoxification of brewery's spent grain hydrolysate for xylitol production by *Debaryomyces hansenii* CCMI 941. Process Biochem 40:1215–1223. doi:10.1016/j.procbio.2004.04.015

Chapla D, Pandit P, Shah A (2012) Production of xylooligosaccharides from corncob xylan by fungal xylanase and their utilization by probiotics. Bioresour Technol 115:215–221. doi:10.1016/j.biortech.2011.10.083

Chaturvedi V, Verma P (2013) An overview of key pretreatment processes employed for bioconversion of lignocellulosic biomass into biofuels and value added products. 3 Biotech 3:415–431. doi:10.1007/s13205-013-0167-8

Cheng KK, Cai BY, Zhang JA, Ling HZ, Zhou YJ, Ge JP, Xu JM (2008) Sugarcane bagasse hemicellulose hydrolysate for ethanol production by acid recovery process. Biochem Eng J 38:105–109. doi:10.1016/j.bej.2007.07.012

Chong AR, Ramírez JA, Garrote G, Vázquez M (2004) Hydrolysis of sugar cane bagasse using nitric acid: a kinetic assessment. J Food Eng 61:143–152. doi:10.1016/S0260-8774(03)00080-3

Chundawat SPS, Bellesia G, Uppugundla N, Sousa LC, Gao D, Cheh AM, Agarwal UP, Bianchetti CM, Phillips GN Jr, Langan P, Balan V, Gnanakaran S, Dale BE (2011) Restructuring the crystalline cellulose hydrogen bond network enhances its depolymerization rate. J Am Chem Soc 133:11163–11174. doi:10.1021/ja2011115

Cipriani TR, Mellinger CG, Souza LM, Baggio CH, Freitas CS, Marquez MCA, Gorin PAJ, Sassaki GL, Iacomini M (2008) Acidic heteroxylans from medicinal plants and their anti ulcer activity. Carbohyd Polym 74:274–278. doi:10.1016/j.carbpol.2008.02.012

Deng W, Tan X, Fang W, Zhang Q, Wang Y (2009) Conversion of cellulose into sorbitol over carbon nanotube-supported ruthenium catalyst. Catal Lett 133:167–174. doi:10.1007/s10562-009-0136-3

Dhepe PL, Fukuoka A (2008) Cellulose conversion under heterogeneous catalysis. ChemSusChem 1:969–975. doi:10.1002/cssc.200800129

Dror Y, Cohen Y, Rozen RY (2006) Structure of gum Arabic in aqueous solution. J Polym Sci Part B Polym Phys 44:3265–3271. doi:10.1002/polb.20970

El Hage R, Chrusciel L, Desharnais L, Brosse N (2010) Effect of autohydrolysis of *Miscanthus x giganteus* on lignin structure and organosolv delignification. Bioresour Technol 101:9321–9329. doi:10.1016/j.biortech.2010.06.143

Escribá LD, Porcar M (2010) Rice straw management: the big waste. Biofuels Bioprod Biorefin 4:154–159. doi:10.1002/bbb.196

Eudes A, Liang Y, Mitra P, Loque D (2014) Lignin bioengineering. Curr Opin Biotechnol 26:189–198. doi:10.1016/j.copbio.2014.01.002

Ewanick S, Bura R (2011) The effect of biomass moisture content on bioethanol yields from steam pretreated switchgrass and sugarcane bagasse. Bioresour Technol 102:2651–2658. doi:10.1016/j.biortech.2010.10.117

Falck P, Atsawanan SP, Grey C, Immerzeel P, Stalbrand H, Adlercreutz P, Karlsson EN (2013) Xylooligosaccharides from hardwood and cereal xylans produced by a thermostable xylanase as carbon sources for *Lactobacillus brevis* and *Bifidobacterium adolescentis*. J Agric Food Chem 61:7333–7340. doi:10.1021/jf401249g
Garrote G, Domínguez H, Parajó JC (2001) Kinetic modeling of corncob autohydrolysis. Process Biochem 36:571–578. doi:10.1016/S0032-9592(00)00253-3
Garrote G, Domínguez H, Parajó JC (2002) Autohydrolysis of corncob: study of non-isothermal operation for xylooligosaccharide production. J Food Eng 52:211–218. doi:10.1016/S0260-8774(01)00108-X
Garrote G, Domínguez H, Parajó JC (2004) Production of substituted oligosaccharides by hydrolytic processing of barley husks. Ind Eng Chem Res 43:1608–1614. doi:10.1021/ie0342762
Garrote G, Falqué E, Domínguez H, Parajó JC (2007) Autohydrolysis of agricultural residues: study of reaction byproducts. Bioresour Technol 98:1951–1957. doi:10.1016/j.biortech.2006.07.049
Gírio FM, Fonseca C, Carvalheiro F, Duarte LC, Marques S, Łukasik RB (2010) Hemicelluloses for fuel ethanol: a review. Bioresour Technol 101(13):4775–4800. doi:10.1016/j.biortech.2010.01.088
Hanim SS, Noor MAM, Rosma A (2011) Effect of autohydrolysis and enzymatic treatment on oil palm (*Elaeis guineensis* Jacq.) frond fibres for xylose and xylooligosaccharides production. Bioresour Technol 102:1234–1239. doi:10.1016/j.biortech.2010.08.017
Hendriks ATWM, Zeeman G (2009) Pretreatment to enhance the digestibility of lignocellulosic biomass. Bioresour Technol 100:10–18. doi:10.1016/j.biortech.2008.05.027
Herrera A, Téllez-Luis SJ, Ramírez JA, Vázquez M (2003) Production of xylose from sorghum straw using hydrochloric acid. J Cereal Sci 37:267–274. doi:10.1006/jcrs.2002.0510
Herrera A, Téllez-Luis SJ, González-Cabriales JJ, Ramírez JA, Vázquez M (2004) Effect of the hydrochloric acid concentration on the hydrolysis of sorghum straw at atmospheric pressure. J Food Eng 63:103–109. doi:10.1016/S0260-8774(03)00288-7
Jørgensen H, Kristensen JB, Felby C (2007) Enzymatic conversion of lignocellulose into fermentable sugars: challenges and opportunities. Biofuels Bioprod Bioref 1:119–134. doi:10.1002/bbb.4
Kim YS, Lim YR, Oh DK (2012) L-Arabinose production from sugar beet arabinan by immobilized endo- and exo-arabinanases from *Caldicellulosiruptor saccharolyticus* in a packed-bed reactor. J Biosci Bioeng 113:239–241. doi:10.1016/j.jbiosc.2011.10.008
Knob A, Terrasan CRF, Carmona EC (2010) β-Xylosidases from filamentous fungi: an overview. World J Microbiol Biotechnol 26(3):389–407. doi:10.1007/s11274-009-0190-4
Kumar P, Barrett DM, Delwiche MJ, Stroeve P (2009) Methods for pretreatment of lignocellulosic biomass for efficient hydrolysis and biofuel production. Ind Eng Chem Res 48:3713–3729. doi:10.1021/ie801542g
Lavarack BP, Griffin GJ, Rodman D (2002) The acid hydrolysis of sugarcane bagasse hemicellulose to produce xylose, arabinose, glucose and other products. Biomass Bioenergy 23:367–380. doi:10.1016/S0961-9534(02)00066-1
Lenihan P, Orozco A, O'Neill E, Ahmad MNM, Rooney DW, Walker GM (2010) Dilute acid hydrolysis of lignocellulosic biomass. Chem Eng J 156:395–403. doi:10.1016/j.cej.2009.10.061
Li Z, Qu H, Li C, Zhou X (2013) Direct and efficient xylitol production from xylan by *Saccharomyces cerevisiae* through transcriptional level and fermentation processing optimizations. Bioresour Technol 149:413–419. doi:10.1016/j.biortech.2013.09.101
Liavoga AB, Bian Y, Seib PA (2007) Release of D-xylose from wheat straw by acid and xylanase hydrolysis and purification of xylitol. J Agric Food Chem 55:7758–7766. doi:10.1021/jf070862k
Liaw WC, Chen CS, Chang WS, Chen KP (2008) Xylitol production from rice straw hemicellulose hydrolyzate by polyacrylic hydrogel thin films with immobilized *Candida subtropicalis* WF79. J Biosci Bioeng 105:97–105. doi:10.1263/jbb.105.97

Lu XB, Zhang YM, Yang J, Liang Y (2007) Enzymatic hydrolysis of corn stover after pretreatment with dilute sulfuric acid. Chem Eng Technol 30:938–944. doi:10.1002/ceat.200700035

Makishima S, Mizuno M, Sato N, Shinji K, Suzuki M, Nozaki K, Takahashi F, Kanda T, Amano Y (2009) Development of continuous flow type hydrothermal reactor for hemicellulose fraction recovery from corncob. Bioresour Technol 100:2842–2848. doi:10.1016/j.biortech.2008.12.023

Marton JM, Felipe MGA, Silva JBA, Pessoa A Jr (2006) Evaluation of the activated charcoals and adsorption conditions used in the treatment of sugarcane bagasse hydrolysate for xylitol production. Braz J Chem Eng 23:9–21. doi:10.1590/S0104-66322006000100002

Moreira LRS, Filho EXF (2008) An overview of mannan structure and mannan-degrading enzyme systems. Appl Microbiol Biotechnol 79:165–178. doi:10.1007/s00253-008-1423-4

Mosier N, Wyman C, Dale B, Elander R, Lee YY, Holtzapple M, Ladisch M (2005) Features of promising technologies for pretreatment of lignocellulosic biomass. Bioresour Technol 96:673–686. doi:10.1016/j.biortech.2004.06.025

Moure A, Gullón P, Domínguez H, Parajó JC (2006) Advances in the manufacture, purification and applications of xylo-oligosaccharides as food additives and nutraceuticals. Process Biochem 41:1913–1923. doi:10.1016/j.procbio.2006.05.011

Mussatto SI, Roberto IC (2004) Alternatives for detoxification of diluted-acid lignocellulosic hydrolyzates for use in fermentative processes: a review. Bioresour Technol 93:1–10. doi:10.1016/j.biortech.2003.10.005

Mussatto SI, Roberto IC (2005) Acid hydrolysis and fermentation of brewer's spent grain to produce xylitol. J Sci Food Agric 85:2453–2460. doi:10.1002/jsfa.2276

Mussatto SI, Roberto IC (2008) Establishment of the optimum initial xylose concentration and nutritional supplementation of brewer's spent grain hydrolysate for xylitol production by *Candida guilliermondii*. Process Biochem 43:540–546. doi:10.1016/j.procbio.2008.01.013

Mussatto SI, Fernandes M, Milagres AMF, Roberto IC (2008) Effect of hemicellulose and lignin on enzymatic hydrolysis of cellulose from brewer's spent grain. Enzyme Microb Technol 43:124–129. doi:10.1016/j.enzmictec.2007.11.006

Nabarlatz D, Ebringerová A, Montané D (2007) Autohydrolysis of agricultural by-products for the production of xylo-oligosaccharides. Carbohyd Polym 69:20–28. doi:10.1016/j.carbpol.2006.08.020

Nigam JN (2001) Ethanol production from wheat straw hemicellulose hydrolysate by *Pichia stipitis*. J Biotechnol 87:17–27. doi:10.1016/S0168-1656(00)00385-0

Olsson M (2006) Wheat straw and peat for fuel pellets–organic compounds from combustion. Biomass Bioenergy 30:555–564. doi:10.1016/j.biombioe.2006.01.005

Palmqvist E, Hahn-Hägerdal B (2000a) Fermentation of lignocellulosic hydrolysates. II: inhibitors and mechanisms of inhibition. Bioresour Technol 74:25–33. doi:10.1016/S0960-8524(99)00161-3

Palmqvist E, Hahn-Hägerdal B (2000b) Fermentation of lignocellulosic hydrolysates. I: inhibition and detoxification. Bioresour Technol 74:17–24. doi:10.1016/S0960-8524(99)00160-1

Parajó JC, Garrote G, Cruz JM, Domínguez H (2004) Production of xylooligosaccharides by autohydrolysis of lignocellulosic materials. Trends Food Sci Technol 15:115–120. doi:10.1016/j.tifs.2003.09.009

Park S, Baker JO, Himmel ME, Parilla PA, Johnson DK (2010) Cellulose crystallinity index: measurement techniques and their impact on interpreting cellulase performance. Biotechnol Biofuels 3:1–10. doi:10.1186/1754-6834-3-10

Pauly M, Gille S, Liu L, Mansoori N, Souza A, Schultink A, Xiong G (2013) Hemicellulose biosynthesis. Planta 238:627–642. doi:10.1007/s00425-013-1921-1

Rabelo SC, Fonseca NAA, Andrade RR, Filho RM, Costa AC (2011) Ethanol production from enzymatic hydrolysis of sugarcane bagasse pretreated with lime and alkaline hydrogen peroxide. Biomass Bioenergy 35:2600–2607. doi:10.1016/j.biombioe.2011.02.042

Rafiqul ISM, Sakinah AMM (2012a) Kinetic studies on acid hydrolysis of *Meranti* wood sawdust for xylose production. Chem Eng Sci 71:431–437. doi:10.1016/j.ces.2011.11.007

Rafiqul ISM, Sakinah AMM (2012b) Design of process parameters for the production of xylose from wood sawdust. Chem Eng Res Des 90:1307–1312. doi:10.1016/j.cherd.2011.12.009
Rafiqul ISM, Sakinah AMM (2013) Processes for the production of xylitol—a review. Food Rev Int 29:127–156. doi:10.1080/87559129.2012.714434
Rafiqul ISM, Sakinah AMM, Karim R (2014) Production of xylose from *Meranti* wood sawdust by dilute acid hydrolysis. Appl Biochem Biotechnol 174:542–555. doi:10.1007/s12010-014-1059-z
Rafiqul ISM, Sakinah AMM, Zularisam AW (2015a) Enzymatic production of bioxylitol from sawdust hydrolysate: screening of process parameters. Appl Biochem Biotechnol 176:1071–1083. doi:10.1007/s12010-015-1630-2
Rafiqul ISM, Sakinah AMM, Zularisam AW (2015b) Evaluation of sawdust hemicellulosic hydrolysate for bioproduction of xylitol by enzyme xylose reductase. Food Bioprod Process 94:82–89. doi:10.1016/j.fbp.2015.01.005
Rafiqul ISM, Sakinah AMM, Zularisam AW (2015c) Inhibition by toxic compounds in the hemicellulosic hydrolysates on the activity of xylose reductase from *Candida tropicalis*. Biotechnol Lett 37:191–196. doi:10.1007/s10529-014-1672-5
Rahman SHA, Choudhury JP, Ahmad AL (2006) Production of xylose from oil palm empty fruit bunch fiber using sulfuric acid. Biochem Eng J 30:97–103. doi:10.1016/j.bej.2006.02.009
Rennie EA, Scheller HV (2014) Xylan biosynthesis. Curr Opin Biotechnol 26:100–107. doi:10.1016/j.copbio.2013.11.013
Reyes P, Mendonça RT, Aguayo MG, Rodríguez J, Vega B, Fardim P (2013) Extraction and characterization of hemicelluloses from *Pinus radiata* and its feasibility for bioethanol production. Rev Árvore 37:175–180. doi:10.1590/S0100-67622013000100018
Roberto IC, Mussatto SI, Rodrigues RCLB (2003) Dilute-acid hydrolysis for optimization of xylose recovery from rice straw in a semi-pilot reactor. Ind Crop Prod 17:171–176. doi:10.1016/S0926-6690(02)00095-X
Romero I, Ruiz E, Castro E, Moya M (2010) Acid hydrolysis of olive tree biomass. Chem Eng Res Des 88:633–640. doi:10.1016/j.cherd.2009.10.007
Saeman JF (1945) Kinetics of wood saccharification: hydrolysis of cellulose and decomposition of sugars in dilute acid at high temperature. Ind Eng Chem 37:43–52. doi:10.1021/ie50421a009
Salvi DA, Aita GM, Robert D, Bazan V (2010) Ethanol production from sorghum by a dilute ammonia pretreatment. J Ind Microbiol Biotechnol 37:27–34. doi:10.1007/s10295-009-0645-5
Samanta AK, Jayapal N, Kolte AP, Senani S, Sridhar M, Suresh KP, Sampath KT (2012) Enzymatic production of xylooligosaccharides from alkali solubilized xylan of natural grass (*Sehima nervosum*). Bioresour Technol 112:199–205. doi:10.1016/j.biortech.2012.02.036
Sarkar N, Aikat K (2013) Kinetic study of acid hydrolysis of rice straw. ISRN Biotechnol ID 170615 :1–5. doi:10.5402/2013/170615
Scheller HV, Ulvskov P (2010) Hemicelluloses. Annu Rev Plant Biol 61:263–289. doi:10.1146/annurev-arplant-042809-112315
Scordia D, Cosentino SL, Lee JW, Jeffries TW (2011) Dilute oxalic acid pretreatment for biorefining gaint reed (*Arundo donax* L.). Biomass Bioenergy 35:3018–3024. doi:10.1016/j.biombioe.2011.03.046
Singh J, Suhag M, Dhaka A (2015) Augmented digestion of lignocellulose by steam explosion, acid and alkaline pretreatment methods: a review. Carbohyd Polym 117:624–631. doi:10.1016/j.carbpol.2014.10.012
Sousa LC, Chundawat SPS, Balan V, Dale BE (2009) 'Cradle-to-grave' assessment of existing lignocellulose pretreatment technologies. Curr Opin Biotechnol 20:339–347. doi:10.1016/j.copbio.2009.05.003
Spiridon I, Popa VI (2008) Hemicelluloses: major sources, properties and applications. In: Belgacem MN, Gandini A (eds) Monomers, polymers and composites from renewable resources. Elsevier, Amsterdam, pp 289–304. doi:10.1016/B978-0-08-045316-3.00013-2. isbn:9780080560519

Sun Y, Cheng JJ (2005) Dilute acid pretreatment of rye straw and bermudagrass for ethanol production. Bioresour Technol 96:1599–1606. doi:10.1016/j.biortech.2004.12.022

Sun HJ, Yoshida S, Park NH, Kusakabe I (2002) Preparation of (1→4)-β-D-xylooligosaccharides from an acid hydrolysate of cotton-seed xylan: suitability of cotton-seed xylan as a starting material for the preparation of (1→4)-β-D-xylooligosaccharides. Carbohyd Res 337:657–661. doi:10.1016/S0008-6215(02)00031-9

Sun Y, Yang G, Jia Z, Wen C, Zhang L (2014) Acid hydrolysis of corn stover using hydrochloric acid: kinetic modeling and statistical optimization. Chem Ind Chem Eng Q 20:531–539. doi:10.2298/CICEQ130911035S

Swati G, Haldar S, Ganguly A, Chatterjee PK (2013) Investigations on the kinetics and thermodynamics of dilute acid hydrolysis of *Parthenium hysterophorus* L. substrate. Chem Eng J 229:111–117. doi:10.1016/j.cej.2013.05.111

Swennen K, Courtin CM, Bruggen B, Vandecasteele C, Delcour JA (2005) Ultrafiltration and ethanol precipitation for isolation of arabinoxylooligosaccharides with different structures. Carbohyd Polym 62:283–292. doi:10.1016/j.carbpol.2005.08.001

Tada K, Kanno T, Horiuchi J (2012) Enhanced production of bioxylitol from corn cobs by *Candida magnolia*. Ind Eng Chem Res 51:10008–10014. doi:10.1021/ie202800h

Taherzadeh MJ, Karimi K (2007) Enzyme-based hydrolysis processes for ethanol from lignocellulosic materials: a review. BioResources 2:707–738. doi:10.15376/biores.10.1.1213-1223

Taherzadeh MJ, Karimi K (2008) Pretreatment of lignocellulosic wastes to improve ethanol and biogas production: a review. Int J Mol Sci 9:1621–1651. doi:10.3390/ijms9091621

Taherzadeh MJ, Niklasson C, Lidén G (2000) On-line control of fed-batch fermentation of dilute-acid hydrolyzates. Biotechnol Bioeng 69:330–338. doi:10.1002/1097-0290(20000805)69

Talebnia F, Karakashev D, Angelidaki I (2010) Production of bioethanol from wheat straw: an overview on pretreatment, hydrolysis and fermentation. Bioresour Technol 101:4744–4753. doi:10.1016/j.biortech.2009.11.080

Téllez-Luis SJ, Ramírez JA, Vázquez M (2002) Mathematical modelling of hemicellulosic sugar production from sorghum straw. J Food Eng 52:285–291. doi:10.1016/S0260-8774(01)00117-0

Teng C, Yan Q, Jiang Z, Fan G, Shi B (2010) Production of xylooligosaccharides from the steam explosion liquor of corncobs coupled with enzymatic hydrolysis using a thermostable xylanase. Bioresour Technol 101:7679–7682. doi:10.1016/j.biortech.2010.05.004

Thomas S, Paul SA, Pothan LA, Deepa B (2011) Natural fibres: structure, properties and applications. In: Kalia S, Kaith BS, Kaur I (eds) Cellulose fibers: bio- and nano-polymer composites: green chemistry and technology. Springer, New York, NY, pp 3–42. doi:10.1007/978-3-642-17370-7-1

Villarreal MLM, Prata AMR, Felipe MGA, Silva JBA (2006) Detoxification procedures of eucalyptus hemicellulose hydrolysate for xylitol production by *Candida guilliermondii*. Enzyme Microb Technol 40:17–24. doi:10.1016/j.enzmictec.2005.10.032

Visser EM, Leal TF, Almeida MN, Guimarães VM (2015) Increased enzymatic hydrolysis of sugarcane bagasse from enzyme recycling. Biotechnol Biofuels 8:1–9. doi:10.1186/s13068-014-0185-8

Vithanage LNG, Barbosa AM, Kankanamge GRN, Rakshit SK, Dekker RFH (2016) Valorization of hemicelluloses: production of bioxylitol from poplar wood prehydrolyzates by Candida guilliermondii FTI 20037. Bioenerg Res 9(1):181–197. doi:10.1007/s12155-015-9673-3

Wang GS, Lee JW, Zhu JY, Jeffries TW (2011) Dilute acid pretreatment of corncob for efficient sugar production. Appl Biochem Biotechnol 163:658–668. doi:10.1007/s12010-010-9071-4

Willför S, Sundberg K, Pranovich A, Holmbom B (2005) Polysaccharides in some industrially important hardwood species. Wood Sci Technol 39:601–617. doi:10.1007/s00226-005-0039-4

Wu L, Arakane M, Ike M, Wada M, Takai T, Gau M, Tokuyasu K (2011) Low temperature alkali pretreatment for improving enzymatic digestibility of sweet sorghum bagasse for ethanol production. Bioresour Technol 102:4793–4799. doi:10.1016/j.biortech.2011.01.023

Yang CH, Yang SF, Liu WH (2007) Production of xylooligosaccharides from xylans by extracellular xylanases from *Thermobifida fusca*. J Agric Food Chem 55:3955–3959. doi:10.1021/jf0635964

Zhang D, Ong YL, Li Z, Wu JC (2012) Optimization of dilute acid-catalyzed hydrolysis of oil palm empty fruit bunch fiber for high yield production of xylose. Chem Eng J 181(182):636–642. doi:10.1016/j.cej.2011.12.030

Zhang HJ, Fan XG, Qiu XL, Zhang QX, Wang WY, Li SX, Deng LH, Koffas MAG, Wei DS, Yuan QP (2014) A novel cleaning process for industrial production of xylose in pilot scale from corncob by using screw-steam-explosive extruder. Bioprocess Biosyst Eng 37:2425–2436. doi:10.1007/s00449-014-1219-0

Zhao XB, Wang L, Liu DH (2008) Peracetic acid pretreatment of sugarcane bagasse for enzymatic hydrolysis: a continued work. J Chem Technol Biotechnol 83:950–956. doi:10.1002/jctb.1889

Zhu L, O'Dwyer JP, Chang VS, Granda CB, Holtzapple MT (2008) Structural features affecting biomass enzymatic digestibility. Bioresour Technol 99:3817–3828. doi:10.1016/j.biortech.2007.07.033

Latent Potential of Microalgal Biomass: Research Efforts and Challenges

Pau Loke Show, Arumugasamy Senthil Kumar, Sue Wen Siow, Siti Sabariah Din, Vidya Sundaram, and Kulandai Arockia Rajesh Packiam

Abstract Microalgae, discovered to have the ability of contributing 60 % of oxygen in the atmosphere by using carbon dioxide for photosynthesis, are abundantly available in the environment. They can survive and grow rapidly even in inconsiderate conditions. Culturing microalgae is environment friendly because they do not require pesticides, whereas their residual biomass can be used as fertilizers. They are found to have the potential to produce biofuel: when carbohydrates of microalgae are fermented and further processed to produce bioethanol and biogas. Moreover, their lipids can be extracted to produce biodiesel. This present work focuses on how to recycle and reuse microalgal biomass, as spent microalgal biomass (SMAB) is constituent of 70 % of the entire microalgal biomass. SMAB is microalgal biomass which can be obtained from its primary use such as lipid extraction. Large-scale production of biodiesel from MAB will accumulate huge amount of waste de-oiled algal biomass (DAB) which can be further processed for other purposes like fertilizers, biofuel, food supply for microbes, etc.

Keywords Microalgae • Biomass • Waste • Recycling • Technology

---

P.L. Show (✉)
Department of Chemical and Environmental Engineering, Faculty of Engineering, University of Nottingham Malaysia Campus, Jalan Broga, Semenyih 43500, Selangor Darul Ehsan, Malaysia

Manufacturing and Industrial Processes Division, Faculty of Engineering, Centre for Food and Bioproduct Processing, University of Nottingham Malaysia Campus, Jalan Broga, Semenyih 43500, Selangor Darul Ehsan, Malaysia
e-mail: PauLoke.Show@nottingham.edu.my; showpauloke@gmail.com

A.S. Kumar • S.W. Siow
Department of Chemical and Environmental Engineering, Faculty of Engineering, University of Nottingham Malaysia Campus, Jalan Broga, Semenyih 43500, Selangor Darul Ehsan, Malaysia

S.S. Din
Chemical Engineering Technology (Biotechnology), University Tun Hussein Onn Malaysia, Batu Pahat, Malaysia

V. Sundaram • K.A.R. Packiam
Department of Biotechnology, Bannari Amman Institute of Technology, Erode Dist, Tamil Nadu, India

L. Singh, V.C. Kalia (eds.), *Waste Biomass Management – A Holistic Approach*,
DOI 10.1007/978-3-319-49595-8_5

1 Introduction

Standing tall among the basic and industrial needs of people in the current day earth is electricity, the pivotal and inevitable discovery of science. The past decades have seen tremendous increase in population vis-a-vis huge utilization of electrical power. Concomitant to the exploding demand and utilization, the world now faces the downside of the same through pollution. Air pollution is the direct result of huge amounts of carbon dioxide which gets emitted from the combustion of fossil fuel in ever-expanding industries. Global warming, the much debated topic of the increase in the temperature of the surface of the earth, is also an immediate result of excessive carbon dioxide emission. Hence it has become mandatory to resolve these problems in order to conserve the environment. Steps to conserve the environment at the cost of decreasing the electricity generation is nevertheless a solution and is not feasible. Microalgal biomass may be a single promising answer to all the abovesaid problems (Fig. 1). Their applications range from biofuel generation while at the same time acting as carbon dioxide sequesters reducing greenhouse gas emissions. Apart from this, they act as key players in waste water bioremediation and are also exploited in the production of methane, biohydrogen and eicosapentaenoic acid.

2 Oilgae: Algae as Biofuel

As push has come to shove, the research community's major focus is on fuel generation utilizing renewable energy resources, thereby limiting harmful gas emissions. One such potent fuel generator is the microalgae, a group which has obtained immense scientific interest at recent times. Microalgal research is spearheaded towards the effective utilization of algal biomass and maximal biofuel generation. Ghayal and Pandya (2013) have propounded that microalgal biomass

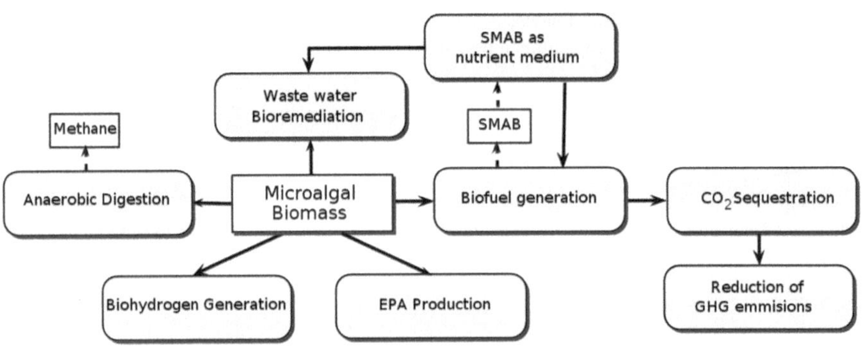

Fig. 1 Various applications of the microalgal biomass

(MAB) are excellent performers in the generation of biofuels. Their research has focused on maximal extraction of lipids from specific species of microalgae. The work has also aimed at replacing coal by mixing algal biomass with waste water and carbon dioxide from the industrial site as an efficient source of generating power.

Algal biomass has a high oil-producing capacity due to their high lipid content and calorific value. In a study by Liu et al. (2015), various species of green and blue-green algae were collected from different sites and tested for their energy-generating capacity. C1strain, a variant of *Chlorella vulgaris* among all the other species of algae examined, had the highest oil content of 11.45 %. Energy recovery from this particular strain was 29.35 % with highest calorific values. Because of the rich calorific value, C1 could serve as an acceptable blend with coal and be used as a solid fuel for generation of electricity. In both the photobioreactor and open ponding system, C1 exhibited comparably good growth rate and also used chimney fumes from the plant as a source of nutrient for growth (Matsuda et al. 2016).

Down the pipeline, photobioreactors and ponding systems could yield enough biomass which can undergo further process of drying and supplement coal. Industries in need of continuous and affordable energy supply can largely rely on algae as the ultimate tailor-made generator of power. Looking at the downstream of the algal cultivation, harvesting of the biomass is a challenge. Methods to decrease the energy consumption of processing algal biomass to obtain products are now being highly debated.

Liu et al. (2015) proposed that algal biomass could be flocculated and recovered by methodically increasing the pH of the medium. Increasing the pH, increases the flocculation efficiency of the microalgae due to the formation of magnesium hydroxide precipitates from magnesium ions present in the medium. Magnesium hydroxide precipitates has been found to induce the coagulation of microalgal cells. Also the study reveals that flocculation efficiency directly depends upon the concentration of microalgal biomass and the released polysaccharides (RPS). The work is effective in the way which suggests easier harvesting of algal cells through a feasible technique with an added advantage of medium reuse. There is an improvement of the current pilot-scale study design to arrive at a profitable industrial production of biodiesel from algae. The experimentation was carried out only using few algal strains. Further explorations into other algal groups will yield a novel maximal producer of biodiesel at larger scales. The supply of availability of nitrogen and phosphorus, nutrients of high demand and microalgae limits the overall process of biodiesel generation. In an analysis, Rösch et al. (2012) stated that the nutrient requirement of the entire process can be reduced by recycling the spent microbial algal biomass (SMAB) after oil extraction. Through hydrothermal gasification of SMAB, maximal recycling values of nitrogen and phosphorus were achieved.

3 Salvaged Algal Biomass: SMAB

Microalgal biomass, after the production of primary fuels, is still rich in nutrients and is collectively called as spent microalgal biomass (SMAB). Recycling and reuse of spent microalgal biomass can produce sustainable biofuels. The usage of spent microalgal biomass (SMAB) and the microalgal waste for the large-scale production of biofuel demonstrated by Rashid and co-workers in 2013 proved the sustainability of algal biofuel. The major nutrients required for efficient microalgal cultivation are carbon and nitrogen (Van Den Hende et al. 2011). Spent microalgal biomass (SMAB) recovered after its utilization as a source of biofuel or waste water remediation contains an initial energy level of up to 70 % and is also found to contain nutrients like complex lipids, proteins and carbohydrates (Table 1). These nutrients in SMAB could be pretreated to breakdown the complex lipids and proteins to simple assimilable carbon and nitrogen sources and further provided for the microalgae, thereby reducing the cost of culture medium for the algal growth (Rashid et al. 2013).

3.1 *Processing of SMAB*

SMAB is a complex source composed of macromolecules of carbohydrates and proteins which are to be broken down into simple fermentable monomers by the process of pretreatment. There are two types of pretreatment listed below.

3.1.1 Substrate Pretreatment

Substrates are substances including organic waste, waste water, sludge and ligno-cellulosic or cellulosic biomass that can enhance digestion of SMAB by providing enough of nutrients. Pretreatment is needed to increase the solubility of substrate by

Table 1 Composition of spent microbial algal biomass (SMAB)

Microalgal species	Protein (%)	Carbohydrates (%)	Lipids (%)	References
Scenedesmus spp.	32.4	6.5	24.7	Yang et al. (2011)
Chlorella vulgaris[a]	54.5	18.0	14.5	Becker (2004), Brown et al. (1997)
Chlamydomonas reinhardtii	18.0	21.0	17.0	
Spirulina platensis[a]	54.5	11.0	7.5	
Dunaliella bioculata	49.0	4.0	8.0	
Anabaena cylindrica[a]	49.5	27.5	5.5	

[a]Shows averaged values for simplicity

removing the unnecessary bacteria and to enrich the useful bacteria. Activated sludge is the most common substrate in co-digestion of SMAB (Passos et al. 2015).

3.1.2 SMAB Pretreatment

Depending on the species of algae and the desired secondary product, pretreatment of SMAB can be categorized into either of the three methods—mechanical, chemical or enzymatic. Mechanical pretreatments are carried out by using autoclaves, homogenizers, microwaves or sonicators. Chemical method falls broadly into either acid or alkaline treatment, although alkaline-based process is especially preferred so as to enhance the solubility of hemicellulose. Basically, chemical pretreatments are generally suitable for SMAB with high content of hemicellulose and to increase the availability of substances that can hinder anaerobic hydrolysis. A combinatorial approach of physiochemical treatment via acid or alkaline catalysis at elevated temperatures is often exploited for SMAB. Such a strategy increases the solubility of COD and the amount of hydrogen produced. Enzymatic pretreatment acts on cell walls, destroys it and increases the amount of compound extracted from SMAB. A two-step enzymatic treatment is now common instead of physiochemical because chemical usage is investigated to produce inhibitors that restrict the production of hydrogen. The first step of enzymatic treatment breaks down carbohydrates into simple sugars, while proteins will be hydrolysed in the second step. These hydrolysates can be used as nutrients for algae (Carrere et al. 2016). Most of the abovesaid pretreatments are applicable to dry biomass. Wet biomass should be dried before proceeding to pretreatment, and the process of drying consumes additional revenue. Energy consumption of pretreatment should be minimal in order to reduce the energy demand of the overall process and thereby balancing global energy need (Mendez et al. 2014).

The need for technoeconomic analysis to put forth a futuristic stratagem for the effective utilization of SMAB is higher. By completely understanding the potential of SMAB, generation of energy at minimal economy is possible. Pretreatment of SMAB, being the critical process which determines the amount of simple compounds available for the desired product, requires significant improvements to enhance SMAB digestibility. It is to be noted that both upstream and downstream processing for the utilization of SMAB are highly energy consuming. Hence comprehensive studies driving the reaction towards the maximum generation of energy by complete usage of substrate algal biomass but with limited feed power for these upstream and downstream processes are to be carried out to arrive at the eco-friendly industrial-scale methodology. Recycling of nutrients from waste should be the focus of optimization procedures to avoid nutrient losses. Thus it is of paramount importance to concentrate on some efficient treatment technologies for use in algal cultivation in order to ably recycle the residual nutrient spent waste (Rothlisberger-Lewis et al. 2016).

4 Biofilter: Algae for CO_2 sequestration and Greenhouse Gas Mitigation

The nutritional inorganic carbon requirement of high-rate algal cultures could be satiated by the 6–12 % carbon dioxide-rich power plant flue gas which is otherwise not possible with just the atmospheric sources because of the exponential growth of the algal biomass. Microalgae sequest and bioconvert carbon dioxide from both atmospheric and flue gas origin (Li et al. 2015) leading to reduction of greenhouse gas emission and preservation of ecosystem. The innovation of microalgal-based carbon dioxide biosequestration is required to bring significant advancement in carbon dioxide biomitigation aimed as a solution for global warming.

In March 2008, California Public Utilities Commission and California Energy Commission proposed a policy to be adopted by California Air Resources Board in which it was recommended that it is obligatory to reduce greenhouse gas (GHG) emissions. The main trigger for the aforesaid recommendations was the California Global Warming Solutions Act 2006 which established statewide targets of GHG emissions for the next five decades (Chandrashekeran et al. 2015). As per the executive order, GHG emissions has to be reduced to levels 2000, 1990 and to 80 % below 1990 by the years of 2010, 2020 and 2050, respectively. Possibilities are more that similar drivers would soon affect the United States, especially in all economic sectors and GHG sources. Clearly, the United States and the world as a whole is looking at GHG emissions as a threat and hunting for ways to decrease emissions for a better tomorrow. These restrictive regulations on GHG emissions in California motivated a team to undertake a microalgal research near the Salton Sea, in Southern California.

In an effort to put forth a realistic technology for reducing the emission of GHG, Brune et al. (2009) constructed a 50-MW natural gas-fired electrical generation plant. His team has analyzed the possibility of reducing GHG emissions using an algal biomass system twined with multiple purpose flue gas CO_2 trapping and waste sludge and animal manure utilization. In a power plant producing 216 MkW and operating at 240-day season, the algal production system was so modelled to sequest almost 70 % of flue gas carbon dioxide and in the process producing 42,000 ton of protein-rich algal biomass which again becomes a source of algal oil-producing biodiesel, animal feed supplement and the leftover residues being used as manure (Bai et al. 2015). The effort predicts the carbon dioxide sequestration of 60–80 % and a 22–30 % reduction in GHG emissions. The undertaking will pave way to an integrated approach of combined carbon dioxide sequestration, fossil fuel replacement and animal feed supplementation with the greater good of reduction in GHG emissions. It is currently viewed that the microalgae mass culture GHG remediation strategy is one of the feasible solutions to the reductions of power plant emissions.

5 Bioremediation: Algal Waste Water Treatment

Microalgae are being currently exploited for sustainable biofuel production and phycoremediation of domestic waste water (Rawat et al. 2011). The benefits of waste water bioremediation by microalgae are threefold. Waste water is biologically treated, biomass is generated for biofuel production and the harmful carbon dioxide is sequestered (Fig. 1). The overall process is eco-friendly with zilch pollution effects on the atmosphere as long as the biomass and nutrients are reused and recycled. Waste water is one of the major sources from which biofuels are being produced by exploiting the properties of algal biomass. Biodiesel and bioethanol from waste water with excellent results are produced by using microalgae. Thus, a combinatorial approach of bioremediation with biofuel production must be judiciously expanded (Hernández et al. 2016). Expansion should focus on several important parameters such as effluent volume and composition, species of microalgae under consideration, temperature and regional light conditions. The commercial initiatives for this purpose must be framed taking into consideration the regional interest for a particular biofuel (Lauersen et al. 2015). Therefore, in-depth market analysis based on the individual need and ethnic demand is very much needed, and, unfortunately, there is no single model to accomplish the same.

6 Anaerobic Digestion: Algae as Methane Producer

Methane considered to be an ideal clean fuel is produced from biomass by a mature and stable technology. It is an environment-friendly fuel produced by the process of anaerobic digestion. Anaerobic digestion is a sequential process encompassing the prominent steps of hydrolysis, acidogenesis, acetogenesis and methanogenesis (Fig. 2). Any organic material consisting of insoluble complex suspended compounds are initially hydrolysed or broken down into their simplest monomeric molecules. Acetates, alcohols, hydrogens and other volatile fatty acids obtained by initial hydrolysis are directly utilized by the methanogens. Remaining components are further catabolized by acidogenic organisms ending the process with the creation of volatile fatty acids. Acetogenesis further digests the simple molecules of acidogenesis and produces acetic acid along carbon dioxide and hydrogen. The process ends with methanogenesis which use the intermediate products of the preceding steps and converts them into methane, carbon dioxide and water.

SMAB can be anaerobically digested to yield methane (Tijani et al. 2015). There are several ideal characteristics of microalgae fitting for AD which are thin cell wall, resistive to natural contaminants and majorly made up of carbohydrates and large cell size (few millimetres to a few 100 micrometres) and high growth rate in non-sterile medium. The characteristic of thin cell wall is of utmost importance because it is this feature that causes the contact of AD bacteria with the microalgae

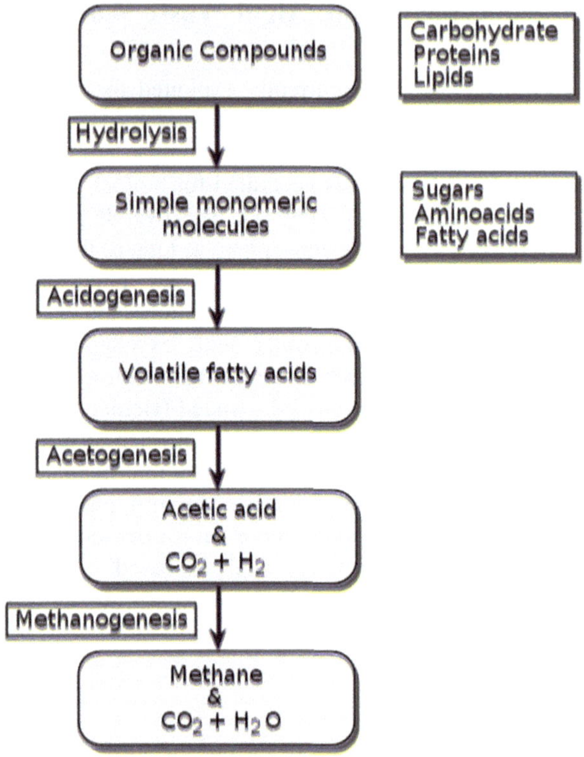

Fig. 2 Methane production via anaerobic digestion

easier. It is also to be noted that marine microalgae with strong cell walls need aggressive treatment, whereas it is sufficient for fresh water microalgae to be handled mildly. Also the composition of the cell wall has been found to influence and affect the efficiency of the process of AD. Cell wall composition directly has effect on the extent of its degradability, and cell wall degradability plays a crucial role in affecting AD. A protein-rich cell wall with a less cellulose content is highly degradable (Torres et al. 2015). The higher the degradability of the cell wall, the higher is the kinetics of AD and the yield of methane resulting in the very less indigestible SMAB.

Biopolymers present in the outer layer of the cell wall of the microalgae are reported to be resistant to the process of AD. These biopolymers form a protective trilaminar outer wall known as trilaminar sheath (TLS) which slows down the AD. Moreover, the presence of some inhibitors like ammonia and salt hinder efficient microalgal degradation. Ammonium exists in both protonated (NH^{4+}) and deprotonated (NH_3) forms. Deprotonated form is alone the inhibitor of anaerobic digestion because of its active permeability through the cell wall. The presence and distribution of the two forms of ammonium is affected by the pH and temperature. Production of the deprotonated form is elevated at higher pH, thereby lowering the methanogenic activity which generates a decrease in pH and amount of ammonia (NH_3).

Likewise temperature is another indispensable factor drastically affecting the performance of AD by changing the physical and physicochemical properties of the media which in turn alter the thermodynamics of the biological process. Various factors such as substrate loading rate, C/N ratio, biodegradability and quality of effluent needed also affect the thermodynamics of the material. Thermophilic (49–57 °C) rather than mesophilic (30–38 °C) conditions favour AD as it provides better waste stabilization, more sludge dewatering, higher production of methane, greater hydrolysis rate, lower formation of foam and better volatile organic reduction. Operating at thermophilic conditions has few disadvantages like high costs due to increased consumption of energy; longer period of sludge adaptation; poor stability; vulnerability to ammonia, potassium and sodium inhibition; and the effect on medium pH due to higher generation of volatile fatty acid. Hydrolysis is the rate-limiting step throughout the anaerobic digestion of SMAB because the complex compounds consume much time to be degraded (Gruber-Brunhumer et al. 2015). More notably, hydrolysis rate can be accelerated by thermophilic condition because when the overall biochemical reaction rate is increased to enhance the performance of organic matter degradation, it will directly result in the higher yield of biogas. Thus and so, as evidently recorded by the world studies, the entire rate of anaerobic digestion can be improved by changing the rate of hydrolysis (González-Fernández et al. 2016; Hernández et al. 2016). Salt is also one critical component required in low concentration for ethanoic bacteria. However excessive salt will cause the bacteria to be dehydrated by osmosis, and hence, the bacteria will be deactivated (Skorupskaitė et al. 2015).

AD is a technology which is acknowledged worldwide for its sustainability and stability. The advantages of AD are production of biogas, reduction of greenhouse gases, combination of organic wastes, best in removing pathogens, low requirement of nutrients and fewer solids to be dealt with. Nutrients such as phosphorus and ammonium in the form of phosphate and ammonia, respectively, are produced during the AD of SMAB (Mendez et al. 2014). These nutrients can be reused as substrate for cultivating microalgae. However, it has high specification on the methanogenetic bacteria used for abundance of chemical compounds.

7 Generation of Biohydrogen

Another significant application of microalgal biomass, apart from methane production, is its use as feedstock for biohydrogen as well as EPA production (Table 2). Biohydrogen can be produced through SMAB by converting carbohydrates and proteins into hydrogen via bacterial fermentation. It is proposed that SMAB rich in nitrogen and lower C/N contents gives a good biohydrogen yield. Other parameters playing vital role in the biohydrogen production are algal solubility, substrate-algae ratio, pH of the culture medium and availability of light. Two-stage processes to produce hydrogen and methane in successive steps have been suggested in order for the advantages of multiple productions (Rashid et al. 2013). This has been relatively

Table 2 Key microalgal products

Microalgal product	Application	Production method	References
Methane	Vehicle fuel	Anaerobic digestion	Wang et al. (2013), Tijani et al. (2015)
Biohydrogen	Bioenergy	Fermentation	Rashid et al. (2013)
Eicosapentaenoic acid (EPA)	Medicine, dietary component	Fermentation	Belarbi et al. (2000)

a new arena of research in the current days which involves optimization of operational condition using statistical experimental design.

8 Niche Product: Algal EPA

Eicosapentaenoic acid (EPA) , an omega-3 fatty acid, has found its application in medicine and is considered an important part of human diet. Although this polyunsaturated fatty acid is obtained by consuming fish, an industrial process to produce the EPA of high purity from a widely available source has been intensively explored. Microalgae apart from being the sources of biofuels and a sturdy reducer of greenhouse gases have also found their use as a producer of a value-added product such as EPA. Belarbi et al. (2000) have demonstrated a simple yet scalable technique for the recovery of EPA ester from microalgae. The ultimate focus of the investigation is to deduce a cost-efficient process for the recovery of esterified EPA at purities greater than 90 % from crude extracts of microalgae. Both moist and freeze-dried crude algal biomasses along with fish oil were tested for their ability to produce high-purity EPA. Algal biomass from the species of *P. tricornutum* and *M. subterraneus* were simultaneously extracted and transesterified followed by silica gel (argenated) chromatographical purification. Extracts of the algal biomass proved to be top-notch producers of 96 % pure EPA than when compared to conventional fish oil which yielded EPA of 70 % purity (Suganya et al. 2016). Among the two algal species researched, P. *tricornutum* had economic advantages over *M. subterraneus* because the latter could be transesterified only as a dried biomass, but the former work well in both dry and paste conditions.

9 Prospects and Challenge

The rise in the price of fossil fuels and depletion of non-renewable resources has caused a shift in the focus towards plant-based biofuels. Nowadays corn starch sugarcane or sugar beet is used to produce bioethanol, while palm and oil seed rape are harvested for biodiesel. The disadvantage to the utilization of plants as sources of biofuel are competition with food crops for land to grow, and the energy contained in the biofuel of the plants is less than the energy produced from fuel. Microalgae have emerged as efficacious biofuel sources overcoming the aforesaid disadvantages of plant-based biofuels in terms of ease of cultivation and energy efficiency and are also the sources of other value-added products. The vast diversity of microalgae including 300,000 species makes it possible to be cultivated on various lands and less stringent conditions for growth in abundance. High lipid content of algae is conducive for biodiesel production. The spent microalgae, since being rich in nutrients, can be further reused. The sustainability of reusing the SMAB depends upon the cost-effectiveness and eco-friendly nature of the technology involved. Pretreatment of SMAB is a crucial step which needs to be explored to reduce the usage of harsh chemicals and for the economy. Microalgal farming supercedes traditional farming in a way of utilizing the by-products within the process chains provided the obstacles behind the technical operations of waste recycling are surmounted. Algal biomass propels the generation of biofuels and electricity; the waste being used as animal feed and most importantly bioremediation of waste water by microalgae have become the centre of many research interests. Given that the choice of the species of microalgae is multiple and the amplitude with which research is taking place, the possibility of a condition for viable and maximal microalgae production is not too far. Decisively, microalgae with its million species have proven to be an industrial treasure trove of economically important products of immense need to the society.

References

Bai X, Schenk PM, Yuan Z, Lant PA, Pratt S (2015) Enhanced triacylglyceride extraction from microalgae using free nitrous acid pre-treatment. Appl Energy 154:183–189. doi:10.1016/j.apenergy.2015.04.045

Becker EW (2004) Microalgae in human and animal nutrition. In: Richmond A (ed) Handbook of microalgal culture. Blackwell, Oxford, pp 312–351. ISBN:978-04-709-9528-0. doi:10.1002/9780470995280.ch18

Belarbi EH, Molina E, Chisti Y (2000) A process for high yield and scaleable recovery of high purity eicosapentaenoic acid esters from microalgae and fish oil. Enzyme Microb Technol 26:516–529

Brown MR, Jeffrey SW, Volkman JK, Dunstan GA (1997) Nutritional properties of microalgae for mariculture. Aquaculture 151:315–331

Brune D, Lundquist T, Benemann J (2009) microalgal biomass for greenhouse gas reductions: potential for replacement of fossil fuels and animal feeds. J Environ Eng 135:1136–1144. doi:10.1061/(ASCE)EE.1943-7870.0000100

Carrere H, Antonopoulou G, Affes R, Passos F, Battimelli A, Lyberatos G, Ferrer I (2016) Review of feedstock pretreatment strategies for improved anaerobic digestion: from lab-scale research to full-scale application. Bioresour Technol 199:386–397. doi:10.1016/j.biortech.2015.09.007

Chandrashekeran S, Zuckerman J, Deason J (2015) Raising the stakes for energy efficiency: a qualitative case study of California's risk/reward incentive mechanism. Util Policy 36:79–90. doi:10.1016/j.jup.2015.09.006

Ghayal MS, Pandya MT (2013) Microalgae biomass: a renewable source of energy. Energy Procedia 32:242–250. doi:10.1016/j.egypro.2013.05.031

González-Fernández C, Mahdy A, Ballesteros I, Ballesteros M (2016) Impact of temperature and photoperiod on anaerobic biodegradability of microalgae grown in urban wastewater. Int Biodeterior Biodegrad 106:16–23. doi:10.1016/j.ibiod.2015.09.016

Gruber-Brunhumer MR, Jerney J, Zohar E, Nussbaumer M, Hieger C, Bochmann G, Schagerl M, Obbard JP, Fuchs W, Drosg B (2015) Acutodesmus obliquus as a benchmark strain for evaluating methane production from microalgae: influence of different storage and pretreatment methods on biogas yield. Algal Res 12:230–238. doi:10.1016/j.algal.2015.08.022

Hernández D, Riaño B, Coca M, Solana M, Bertucco A, García-González MC (2016) Microalgae cultivation in high rate algal ponds using slaughterhouse wastewater for biofuel applications. Chem Eng J 285:449–458. doi:10.1016/j.cej.2015.09.072

Lauersen KJ, Huber I, Wichmann J, Baier T, Leiter A, Gaukel V, Kartushin V, Rattenholl A, Steinweg C, von Riesen L, Posten C, Gudermann F, Lütkemeyer D, Mussgnug JH, Kruse O (2015) Investigating the dynamics of recombinant protein secretion from a microalgal host. J Biotechnol 215:62–71. doi:10.1016/j.jbiotec.2015.05.001

Li J, Liu Y, Cheng JJ, Mos M, Daroch M (2015) Biological potential of microalgae in China for biorefinery-based production of biofuels and high value compounds. New Biotechnol 32:588–596. doi:10.1016/j.nbt.2015.02.001

Liu W, Zhao C, Jiang J, Lu Q, Hao Y, Wang L, Liu C (2015) Bioflocculant production from untreated corn stover using *Cellulosimicrobium cellulans* L804 isolate and its application to harvesting microalgae. Biotechnol Biofuels 8. doi:10.1186/s13068-015-0354-4

Matsuda S, Durney AR, He L, Mukaibo H (2016) Sedimentation-induced detachment of magnetite nanoparticles from microalgal flocs. Bioresour Technol 200:914–920. doi:10.1016/j.biortech.2015.11.006

Mendez L, Mahdy A, Ballesteros M, González-Fernández C (2014) Methane production of thermally pretreated *Chlorella vulgaris* and *Scenedesmus* sp. biomass at increasing biomass loads. Appl Energy 129:238–242. doi:10.1016/j.apenergy.2014.04.110

Passos F, Carretero J, Ferrer I (2015) Comparing pretreatment methods for improving microalgae anaerobic digestion: thermal, hydrothermal, microwave and ultrasound. Chem Eng J 279:667–672. doi:10.1016/j.cej.2015.05.065

Rashid N, Rehman MSU, Han JI (2013) Recycling and reuse of spent microalgal biomass for sustainable biofuels. Biochem Eng J 75:101–107. doi:10.1016/j.bej.2013.04.001

Rawat I, Ranjith Kumar R, Mutanda T, Bux F (2011) Dual role of microalgae: Phycoremediation of domestic wastewater and biomass production for sustainable biofuels production. Appl Energy 88:3411–3424. doi:10.1016/j.apenergy.2010.11.025

Rösch C, Skarka J, Wegerer N (2012) Materials flow modeling of nutrient recycling in biodiesel production from microalgae. Bioresour Technol 107:191–199. doi:10.1016/j.biortech.2011.12.016

Rothlisberger-Lewis KL, Foster JL, Hons FM (2016) Soil carbon and nitrogen dynamics as affected by lipid-extracted algae application. Geoderma 262:140–146. doi:10.1016/j.geoderma.2015.08.018

Skorupskaitė V, Makarevičienė V, Šiaudinis G, Zajančauskaitė V (2015) Green energy from different feedstock processed under anaerobic conditions. Agron Res 13:420–429

Suganya T, Varman M, Masjuki HH, Renganathan S (2016) Macroalgae and microalgae as a potential source for commercial applications along with biofuels production: a biorefinery approach. Renew Sust Energy Rev 55:909–941. doi:10.1016/j.rser.2015.11.026

Tijani H, Abdullah N, Yuzir A (2015) Integration of microalgae biomass in biomethanation systems. Renew Sustain Energy Rev 52:1610–1622. doi:10.1016/j.rser.2015.07.179

Torres A, Fermoso FG, Neumann P, Azocar L, Jeison Nuñez D (2015) Anaerobic digestion as a tool for resource recovery from a biodiesel production process from microalgae. J Biobased Mater Bioenergy 9:342–349. doi:10.1166/jbmb.2015.1522

Van Den Hende S, Vervaeren H, Saveyn H, Maes G, Boon N (2011) Microalgal bacterial floc properties are improved by a balanced inorganic/organic carbon ratio. Biotechnol Bioeng 108:549–558. doi:10.1002/bit.22985

Wang X, Nordlander E, Thorin E, Yan J (2013) Microalgal biomethane production integrated with an existing biogas plant: a case study in Sweden. Appl Energy 112:478–484. doi:10.1016/j.apenergy.2013.04.087

Yang Z, Rongbo G, Xiaohui X, Xiaolei F, Shengjun L (2011) Hydrogen and methane production from lipid-extracted microalgal biomass residues. Int J Hydrogen Energy 36:3465–3470. doi:10.1016/j.ijhydene.2010.12.018

Treatment of Dye Wastewater for Water Reuse Using Membrane Bioreactor and Biofouling Control

Muhammad Faisal Siddiqui, Lakhveer Singh, and Zularisam Ab Wahid

Abstract Wastewater treatment for water reuse and membrane biofouling control is of significant value to sustainable performance of a membrane bioreactor system. Different treatment techniques have been employed to treat dye wastewater. In recent studies, membrane bioreactor was employed to treat dye wastewater; however, membrane bioreactors are facing biofouling problem. Biofouling (is a process of membrane surface colonization by microbial cells via adhesion and production of extracellular polymeric substances (EPSs)) is almost always a major hitch for membrane bioreactors (MBRs) and membrane systems. Biofouling of membrane reactors results in higher operational expenses and reduced stability and operational performance. In this chapter, biological treatment of membrane biofouling is demonstrated. Furthermore, major causes of biofouling and biological control strategies are discussed. Lastly, conclusions on wastewater treatment and membrane biofouling are presented.

Keywords Dye wastewater • Membrane biofouling • Extracellular polymeric substances • Biofouling control

List of Abbreviations

AHL	Acyl homoserine lactone
AIP	Autoinducer peptide
COD	Chemical oxygen demand
DNA	Deoxyribonucleic acid
eDNA	Extracellular deoxyribonucleic acid
EPS	Extracellular polymeric substance

M.F. Siddiqui (✉)
Department of Microbiology, Abdul Wali Khan University Mardan, Mardan, Pakistan

Faculty of Chemical and Natural Resources Engineering, Universiti Malaysia Pahang, Gambang 26300, Malaysia
e-mail: send2biotech@gmail.com

L. Singh • Z.A. Wahid
Faculty of Engineering Technology, Universiti Malaysia Pahang, Gambang 26300, Malaysia

L. Singh, V.C. Kalia (eds.), *Waste Biomass Management – A Holistic Approach*,
DOI 10.1007/978-3-319-49595-8_6

MBRs	Membrane bioreactors
NO	Nitric oxide
QS	Quorum sensing
SMBRs	Submerged membrane bioreactors
SMP	Soluble microbial products
SNP	S-Nitroso-N-acetylpenicillamine
UF	Ultrafiltration

1 Introduction

Textile wastewater is often treated with physiochemical methods, but these methods are generally very expensive (Flemming et al. 1997). Moreover, the complex molecular structure of dyes makes them more resistant to degradation via biological methods. Thus, there is an urgent need to develop a suitable technology for treatment of dyes in textile effluent.

In recent years, submerged membrane bioreactors (SMBRs) have received significant interest, because they eliminate the requisite of a secondary sedimentation tank. Because SMBRs have small reactor space and usually generate very less amount of sludge waste (Yun et al. 2006), the biofouling of membrane prevents its large-scale application, as it is a major hitch (Flemming et al. 1997). Membrane fouling is of different kinds (e.g. biofouling, organic and inorganic). Among various types of fouling, "biofouling" due to microorganism and from EPS is a challenging work (Yu et al. 2010). Membrane bioreactors affected with biofouling results in higher operational expenses and reduced stability and operational performance (Yu et al. 2010).

2 Textile Dye Wastewater and Treatment

2.1 *Textile Dye Wastewater and Its Effects*

The occurrence of colour and its intermediates has been undesirable for use in the domestic or industrial sector. The coloured effluents affect aesthetic merits; hence, it is clear that colour is a pollutant. Therefore, colour in wastewater has to be treated before discharge. Various agents, i.e. pigments, dyes, tannins, etc., are often used as a colouring. Among different colour wastewaters with complex nature, dye wastewater is the leading waste. Almost 8000 various types of dyes have been produced, and these dyes are mostly used by paper and pulp industries (Ali and Sreekrishnan 2001).

It is of great interest to treat textile effluents owing to aesthetic and hazardous effects on receiving waters. Many research works have been carried out to develop an effective treatment technology, but owing to the mixture of different dyes in dye wastewater, no solo technique is effective for the removal of dyes from dye wastewater. In this context, ecological and human health has provoked the government to discharge the coloured effluents in lower colour level. Although most of the textile industries are aware of the hazards of dyes, but still these textile industries have failed to treat the dye wastewater (Lewis 1999).

2.2 Nature of Textile Effluent

There is a major concern about the discharge of textile dye effluent after textile processing owing to consumption of the large volumes of water. For the discharge of colour effluent, discharge of colour is also included in the water quality standards in Malaysia. The limits for the discharge of colour effluents are 100 and 200 platinum-cobalt units according to standards A and B, respectively (Awang and Aziz 2012).

A large amount of mineral salts, reactive dyes and water is used for cotton processing in textile refining processes. Hence, they discharge a huge amount of wastewater to water streams, which includes different nature of contaminants. The effluent discharge after every textile process usually contains various dyes, solvents, salts and detergents (Barredo-Damas et al. 2006). Textile effluent contains various chemicals of different concentrations (Marcucci et al. 2003). The dye wastewater affects the visual, gas solubility and water transparency in water streams, and it can be hazardous to aquatic plants and animals. Moreover, almost all dyes, which are produced synthetically and their metabolites are mutagenic, are carcinogenic and toxic to the human health (Nilsson et al. 1993).

2.3 Batik Wastewater

Batik is a cloth which is conventionally produced manually by wax-resistant dyeing method. The term has been expanded due to modern trends in the textile sector to include fabrics, which integrate conventional batik designs, even if they are not made using the wax-resistant dyeing methods. There are almost 1500 textile factories in Malaysia, most of these factories work as local factories making the batik. These homemade textile factories are well known in Malaysia. These industries are traditionally inherited from generation to generation (Ahmad et al. 2007).

Batik-making process consists of several processes before the outcome looks attractive and eye catching. Generally, the batik-making process can be categorized in four processes, including cloth preparation, application of wax, dying of cloth and removal of wax in boiling water. A detailed process of making "batik" is shown in Fig. 1 (Ahmad et al. 2007).

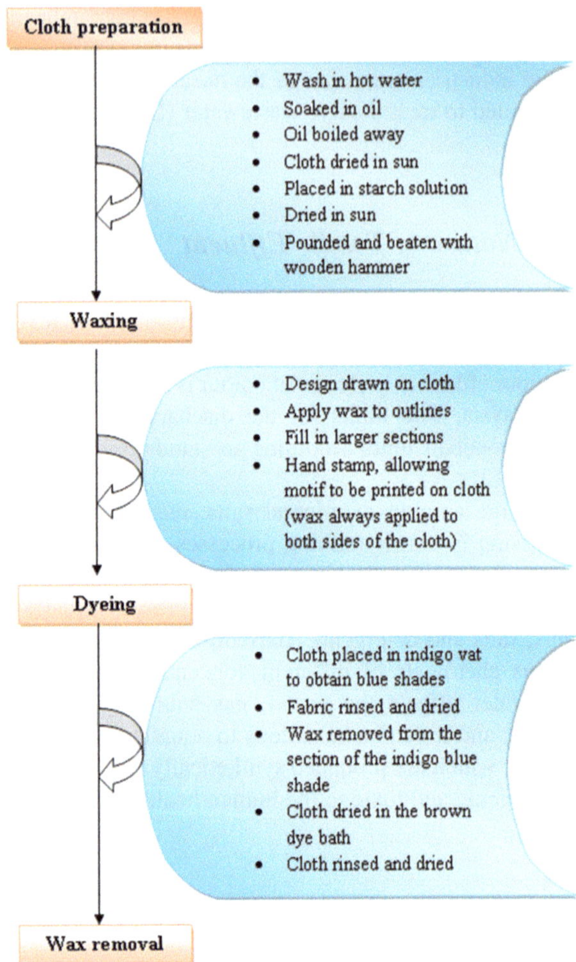

Fig. 1 Flow chart shows the process of making batik (Drawn after Ahmad et al. 2007)

2.4 *Dye Treatment Technologies*

Textile dye wastewater treatment is still a main concern owing to the synthetic nature of dyes, which is very difficult to treat with conventional treatment techniques (Zhang et al. 2004). Most of the biological, physical and chemical techniques used for dye removal work either via adsorption on solid supports, into sludge, or complete degradation of the dye molecule. Recently, the main techniques used to treat textile wastewaters are chemical or chemical-physical as membrane filtration, adsorption, ion exchange, electrolysis, advanced chemical oxidation and chemical reduction (Gogate and Pandit 2004).

Recently, there are two fundamental problems. (1) How to dispose or treat a large amount of wastes, which are continuously producing by different industries?

(2) How can we remove those chemicals which are dumped or accumulating in soil or river bodies? Biological treatment methods can be employed in this regard, which includes aerobic, anaerobic and combined aerobic/anaerobic treatment with fungi (Wang and Yu 1998; Ali et al. 2009; Siddiqui et al. 2009, 2010), yeasts, bacterial, enzymes, algae and plants (Shrivastava et al. 2005).

2.5 *Membrane Bioreactors in Treatment of Dyes*

Bioreactors are the core of any biotechnology-based production processes be it for vaccine, proteins, enzymatic or microbial biotransformation, bioremediation and biodegradation (Chisti and Moo-Young 1999). Nowadays, sound and efficient treatment technologies are needed to face the recent challenges. In this regard, treatment of wastewater using membrane bioreactor is one of the promising alternative methods for effective and efficient technology. Though, this membrane technology needs an initial high setup cost, and it is overcome via the reuse of permeate.

There are a number of studies that have been conducted for treating various types of wastewater by using membrane technology for treating, recovering and recycling of water. Koyuncu (2002) studied nanofiltration to treat Reactive Orange 16 solutions where the experiments were operated at crossflow filtration mode. They succeeded in achieving rejection coefficient higher than 99 % in optimal conditions of pressure, feed concentration, and crossflow velocity. Tang and Chen (2002) studied nanofiltration to treat CI Reactive Black 5 contained in textile wastewater. In this study, at low pressures, high fluxes were achieved, with NaCl rejections of less than 14 % and average dye removal of 98 %. Yun et al. (2006) studied the treatment of dye wastewater in aerobic submerged membrane bioreactor. They found that the average chemical oxygen demand (COD) and dye removal was 94.8 % and 86.6 %, respectively.

With the environment regulation tightened, ultrafiltration (UF) membranes are employed in different areas of a textile sector. In the textile industry, ultrafiltration can be used for treating size and latex contaminant effluents, treating the effluents from wool washing and treating wastewater from a dying process. Ultrafiltration can completely remove many dyes such as acid, disperse, direct, azo and vat. Although ultrafiltration achieves complete colour removal for usually all dyes, however, it's important to mitigate fouling cause by the dyes or their intermediates (Woerner 2003).

3 Membrane Biofouling and Biological Treatment of Biofouling for Wastewater

3.1 Fundamentals of Membrane

Membrane can simply refer to a thin film-like structure, which separates two fluids. Membrane acts as a selective barrier, which allows some molecules to pass through and not allow others. There are many benefits of using membrane technology, which include low energy consumption, easy scale up, continuous separation, no additive used and easy combination with other technologies. Based on the particle size, the membrane filtration is divided into different ranges as follows.

Microfiltration It is a membrane process which removes contaminants from the fluid using microporous membrane. This membrane is usually used to remove suspended solid particles. A typical microfiltration membrane is classified based on its pore size range from of 0.1 to 10 micrometre (μm) (Cheryan 1998).

Ultrafiltration It is a membrane filtration system in which the hydrostatic pressure forces a liquid against a membrane. Ultrafiltration is employed for separation of macromolecules (10^3–10^6Da), especially starches, proteins and microbes such as virus and bacteria. Ultrafiltration membrane prevents molecules in the range from 1000 to 500,000 daltons in molecular weight (Cheryan 1998).

Nanofiltration This membrane system prevents molecules ranging from nanofiltration membrane 100 to 1000 daltons in molecular weight. These membranes are also classified as ultrafiltration membrane based on molecular weight cut-off. This membrane can also prevent smaller (0.001 μm) contaminants (Taylor and Jacobs 1996).

Reverse Osmosis These membranes have very small pore size compared to other membranes, and it can reject even a very small (0.0001 μm) particle (Taylor and Jacobs 1996). These membranes are usually classified by rejection of NaCl in a solution, which is ranged from 95 % to 99.4 %. Its solute efficiency depends on many factors like pressure, water flux rate and solute concentration.

3.2 Membrane Bioreactor

The membrane bioreactor system was first introduced in 1970s for the treatment of wastewater and comprised of a membrane unit combined with suspended growth biological reactor into a single process. There are different configurations of membrane bioreactors. The membrane operation can be either submerged in which the membrane is installed as internal or side-stream operation in which the membrane is independent of bioreactor. MBR is fed with feed wastewater, which

Table 1 Advantages and disadvantages of side-stream and submerged MBR

Side-stream MBR	Submerged MBR
Advantages • Small footprint • Effluent disinfection • Complete solids removal • High loading rate compatibility • Sludge bulking not a problem • Rapid start-up • Combined COD, solids and nutrient removal in a single unit	Advantages • Small footprint • Less frequent cleaning required • Lower operating costs • Feed-forward control of O_2 demand • Low liquid pumping costs • Low energy consumption
Disadvantages • Aeration limitations • Membrane costs • Membrane fouling • High operating costs • High pumping costs • High cleaning required • Process complexity	Disadvantages • Susceptible to membrane fouling • High aeration costs

then comes into contact with the biomass present in the reactor. The mixture of biomass and feed wastewater is then pumped to the membrane unit, and the permeate is discharged into a separation tank, and the rententate is recycled. To maintain the constant age of sludge, excess sludge is wasted. Normally, chemical cleaning or back washing is used to clean the membrane. In submerged membrane bioreactor, the costs are reduced because there is no recirculation (Chua et al. 2002). In a submerged system, the pressure is applied either by pressurizing the reactor or by suction through the membrane. The costs for both reactors are discussed by Gander et al. (2000). Table 1 shows advantages and disadvantages of the submerged and side-stream MBRs.

3.3 *Membrane Biofouling*

Membrane biofouling control is a challenging and demanding problem nowadays (Wang et al. 2005). Biofouling is very difficult to mitigate. Even if few microbes are reduced, still, remaining microbes can grow on limited nutrients (Flemming et al. 1997). Microbes present in water will always invade and colonize the surface of membranes. Hence, if microbes are removed up to 99.9 %, still, some microbes can grow on very limited nutrients. In membrane bioreactor systems, biofouling is very challenging to control (Flemming et al. 1997). Biofouling of membrane reactors results in higher operational expenses and reduced stability and operational performance (Flemming et al. 1997).

Extracellular polymeric substances (EPSs) are found to be key substances to cause membrane biofouling (Rosenberger and Kraume 2002). These compounds are high molecular weights that are produced by bacterial cells. These compounds (EPSs) comprised of polysaccharides, lipoproteins, proteins and deoxyribonucleic acid (DNA). Recently, studies on EPS in either soluble microbial product (SMP) or bound form have attained growing interest (Yeon et al. 2005). EPS shows an essential component of biofilm development and structure, particularly mechanical strength, attachment and protection against environmental deleterious effects (Tansel et al. 2008). It is necessary to remove EPS (soluble) from the activated sludge, because they can pose internal fouling and hence decrease the membrane flux (Lee et al. 2003).

3.4 Treatment of Membrane Biofouling by Biological Methods

So far, various physical and chemical methods have been used to treat the biofouled membrane, and these methods are energy efficient and not effective (Le-Clech et al. 2006). Biofouling is hard to control, even by reducing the number of microorganisms, they can multiply, and they will again grow if nutrients are available (Flemming et al. 1997).Thus, it seems that biological mitigation of membrane biofouling is a novel approach that needs further research (Siddiqui et al. 2015a; Xiong and Liu 2010).

Recently, different biofouling control strategies (Siddiqui et al. 2012a, 2015a, 2015c) have been used for mitigation of membrane biofouling, and there are some reviews on biofouling mitigation based on quorum sensing (QS) (Dobretsov et al. 2009; Siddiqui et al. 2015b). In this review paper, biological methods (Fig. 2) to control biofouling of membrane bioreactors such as enzymatic, nitric oxide, bacteriophage and quorum sensing inhibitions are discussed. A summary of biological methods to control membrane biofouling have been given in Table 2.

Enzymatic Control. It is already reported that bacterial EPS facilitates the bioflocculation process in activated sludge and has a role in its surface charge and structural and settling properties (Houghton et al. 2001). Bioflocs are formed

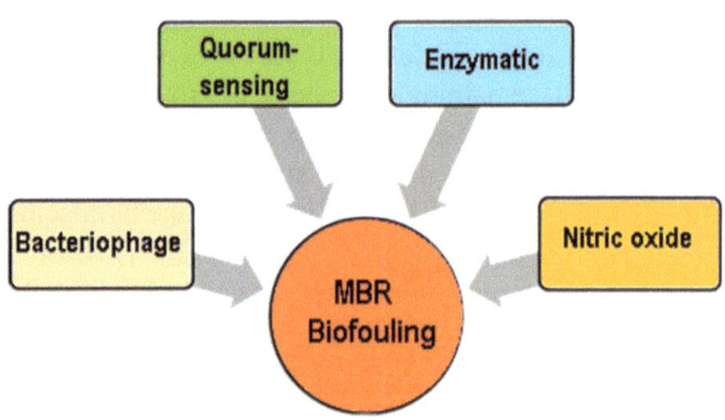

Fig. 2 Different biological methods to control membrane biofouling

Table 2 Capabilities, advantages and limitations of current biological methods to control membrane biofouling

Biological method	Capabilities	Advantages	Limitations	Reference
Enzymatic				
Proteinase K	Disperse the established biofilm	Reduce biofilm and environment friendly	Limit its large-scale application	Chaignon et al. (2007)
Trypsin	Hydrolyzes lysine peptides and can also remove the mature *S. aureus* 383 biofilm	Reduced bacterial resistance	Require optimum conditions	Chaignon et al. (2007)
Protease	Control biofouling on UF membrane for wastewater treatment	High removal of foulants and high permeate recovery	Sensitive to elevated salt concentration	Leroy et al. (2008)
Antimycotic lysozyme	Candida biofilm formation control and attachment	Environment friendly and acts more specific than biocides	Limits widespread application	Caro et al. (2009)
Nitric Oxide *SNP*	Pretreatment of biofilm grown on the RO membrane led to a twofold increase in the biofilm removal efficiency	Attenuates the signalling mechanism rather than by toxic effects	Low solubility in water and not stable as it can be easily oxidized	Barraud et al. (2009)
Bacteriophage	UF membrane biofouling reduction of 40 % on average	Continuous infection and multiplication as long as the host is present	The phages detected in permeative solution	Goldman et al. (2009)

through the interaction of bacterial cells with organic and inorganic particles, which are held together by EPS (Subramanian et al. 2008). It is not easy to remove EPS from fouled membrane via chemical and physical methods. However, enzymes could be used to hydrolyze the EPS, which suggests a new approach to mitigate membrane biofouling (Leroy et al. 2008).

Various enzymes have already been applied to mitigate biofouling. The already formed biofilm was dispersed using proteinase K, which hydrolyzes peptides and can also disperse the biofilm formed by *S. aureus* 383 (Chaignon et al. 2007). Leroy et al. (2008) studied various enzymes, such as papain, subtilisin and amino protease A in the attachment of *Pseudoalteromonas* sp. D41. The most efficient enzyme was subtilisin, which not only detached bacterial cells but also inhibited the microbial attachment (Poele and Van der Graaf 2005). Various enzymes were applied to mitigate membrane biofouling for wastewater treatment. It was found that the cleaning of biofouling using protease showed higher removal efficiency, with increased flux recovery. Yu et al. (2010) investigated to remove fouling of UF membrane using four different enzymes, including lipase, alpha amylase, protease and cellulase. They found that when these enzymes were used in combination with NaOH and citric acid, about 90 % cleaning was achieved. Moreover, they supposed that "enzymes may offer an additional advantage due to their biodegradability as to the adverse effect expected from chemical cleaning.

It is clear from the above studies that enzymatic treatment of membrane biofouling appears to be a novel alternative compared to traditional methods. Enzymatic methods disrupt the EPS to mitigate the microbial attachment.

It has also been found that extracellular DNA (eDNA) helps in the development of biofilm and also provides a structural support to biofilm (Wu and Xi 2009). An investigation on *S. epidermidis* revealed that eDNA was vital in its biofilm formation on the surface of polystyrene (Qin et al. 2007). It is already known that lysozyme can disrupt the bacterial cell wall. Therefore, lysozymes have been used to treat bacterial biofilms. Caro et al. (2009) found that with the coating of stainless steel with lysozyme, the microbial attachment was completely inhibited. The *Candida* biofilm was controlled using antimycotic protein lysozyme, which acts more specifically than biocides, and it is eco-friendly (Samaranayake et al. 2009). Thus, it can be seen from the different studies that lysozyme could be a nontoxic and effective for the mitigation of biofilm. Moreover, the instability of this enzyme is already known in the environment, and it depends on the ionic strength, pH and temperature.

These investigations indicated that it would be better to mitigate the initial microbial attachment than mature biofilm. Furthermore, the enzymatic control of EPS has limitations, e.g. the activity of an enzyme alters with the change of pH, temperature and salt concentration. However, the enzymatic method is eco-friendly and nontoxic compared to the chemical methods.

Control by Nitric Oxide Nitric oxide (NO) is as a naturally universal molecule, which can regulate the proliferation of useless cells (Liu et al. 2012). It attenuates the microbial cells via signalling mechanisms. Liu et al. (2012) reported that NO decreased the capacity of *Pseudomonas putida* and other multispecies bacteria to develop biofilm. It has been found that the genes responsible for encoding regulators are usually present in bacterial cells (Radhakrishnan and Viollier 2012). There are many factors linked to the NO response system and production, which show the conservation across microbial cells (Gusarov et al. 2008). It has been found that NO could decrease an attachment of *S. epidermidis*, *Escherichia coli* and *S. aureus* with a reduction rate of 48 %, 88 % and 96 %, respectively (Charville et al. 2008). It is suggested from these reports that various bacterial biofilms can be inhibited using NO. The biofilms formed by *S. epidermidis* and *S. aureus* were inhibited using NO (Schlag et al. 2007). Additionally, the attachment and removal of mix species biofilms can also be inhibited using various NO donors from water treatment systems (Barraud et al. 2009).

Antimicrobial agents in combination with a low dose of NO can enhance the biofilm dispersal. A twofold increase was found in the removal of multispecies biofilm grown on reverse osmosis (RO) membrane, when NO was used in combination with S-nitroso-N-acetylpenicillamine (SNP) (Barraud et al. 2009).

As discussed above, NO can disperse biofilm and membrane biofouling. However, it should be noted that NO can easily be oxidized, and it has low solubility in water. Thus, it would be less effective if NO is directly added to mitigate biofouling (Cai et al. 2005).

Control by Bacteriophage An overview of research on "phages treatment" exposes its role in combating a number of bacteria in different environments, such as foam formation reduction (Petrovski et al. 2011), slime and biofilm control and plant diseases, medicine and foodborne pathogen control and detection (Górski et al. 2009).

The application of bacteriophages has also been studied to mitigate membrane biofouling. Goldman et al. (2009) first time applied bacteriophages to mitigate UF membrane biofouling. They found that with the addition of phages, the microbial attachment was reduced by 40 %, and membrane filterability was significantly improved. According to Goldman et al. (2009), the use of phages has the benefit of continuous infection and multiplication in the presence of bacterial cells. Goldman et al. (2009) observed that there were insufficient (1–10 PFU/100 ml) phages in permeate; it could be the result of penetration of phage through membrane. Goldman et al. (2009) supposed that the presence of phage in permeate would not be an issue owing to the absence of bacteria in permeate. Goldman et al. (2009) suggested phage as a potential agent to control membrane biofouling with continuous multiplication. However, the specific features of phage would ultimately pose a challenge on its large-scale usage.

Control by Quorum Sensing Recently, mitigation of membrane biofouling based on quorum sensing (QS) and quorum quenching (QQ) (Fig. 3) was employed as a new biofouling mitigation paradigm (Yeon et al. 2008). QS based on cell-to-cell signalling mechanism which is used by cells to assess local densities and is a mechanism of cell-to-cell communication that is used by microbial cells to control gene expression (Bassler and Losick 2006; Thiel et al. 2009). QS mechanism is based on secretion, production and sense of signalling molecules (Fig. 4) (Bassler and Losick 2006; Thiel et al. 2009). Figure 4a represented AHL-mediated QS system in Gram-negative bacteria. When the population density is low, AHLs

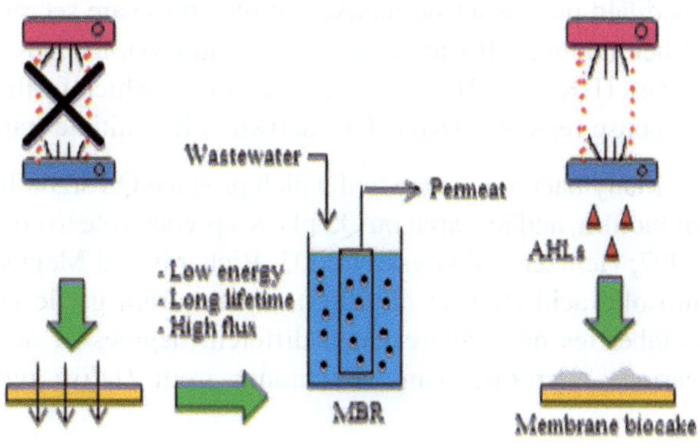

Fig. 3 Concept of membrane biofouling control based on QS

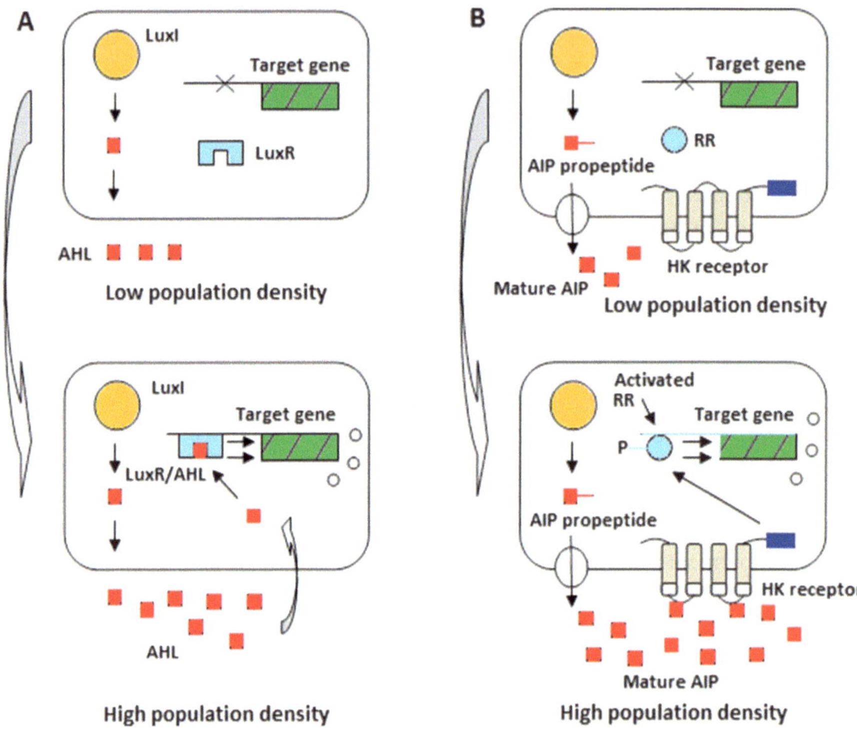

Fig. 4 Bacterial QS systems. (**a**) AHL-mediated QS system in Gram-negative bacteria. (**b**) Autoinducer peptide (AIP) QS in Gram-positive bacteria (Bassler and Losick 2006; Thiel et al. 2009)

present in low concentration and the LuxR receptor is quickly degraded. When the AHL concentration reaches a specific concentration, the AHL signalling molecules bind with LuxR and make an AHL/LuxR complex, hence activated. This activated complex starts transcription of the specific genes required. Figure 4b represented autoinducer peptide (AIP) quorum sensing in Gram-positive bacteria. In this system, small precursors of peptides are produced, and these molecules are generally modified postranscriptionally, and then these are secreted via specific transporters. When mature AIPs are in high concentration, it binds to transmembrane histidine kinase (HK), and HK receptor is activated, which further activates the downstream response regulator (RR). This activated RR initiate transcription of specific genes.

Many bacteria are studied which produce QS signalling molecules for formation of biofilm, and research on QS plays a pivotal role to control biofilm (McClean et al. 1997; Hentzer and Givskov 2003; Richards and Melander 2009). Compounds like urosolic acid (Ren et al. 2005), extract from garlic (Bjarnsholt et al. 2005) and cranberries have all revealed different degrees of antibiofilm properties against various microorganisms. Furanones from *Delisa pulchera* are widely studied classes of natural compounds, which inhibit biofilm formation. The potential application of QS mitigation appears to be a novel way to mitigate membrane biofouling. Therefore, different approaches to target QS systems to control membrane biofouling are used (Siddiqui et al. 2012b, 2015a).

4 Conclusions

This chapter discussed research on dye wastewater treatment, membrane biofouling and some biofouling control strategies. In view of the critical function of bound EPS in the sludge and biofouling, it was considered to be restrained to control membrane biofouling. Different studies exhibited the presence of EPS in MBRs, which affect the membrane fouling substantially, EPS concentration and EPS composition to establish the biofouling tendency. Several strategies are currently used to control and treat biofouling problem. Biological methods discussed in this chapter have a great potential in mitigating membrane biofouling. Compared with the other chemical and physical methods, biological methods to mitigate membrane biofouling would present benefits of less bacterial resistance, low toxicity and high efficiency.

Acknowledgements The authors wish to thank (Grant No. GRS100308) the Faculty of Chemical and Natural Resource Engineering, University of Malaysia Pahang, Malaysia, for the support.

References

Ahmad A, Puasa S, Abiding S (2007) Cross flow ultrafiltration for removing direct-15 dye from waste water of textile industry. ASEAN J Sci Technol Devel 23:207–216. doi:10.3125/asean.v23i3.420

Ali M, Sreekrishnan T (2001) Aquatic toxicity from pulp and paper mill effluents: a review. Adv Env Res 5:175–196. doi:10.1016/S1093-0191(00)00055-1

Ali N, Hameed A, Siddiqui M, Ghumro P, Ahmed S (2009) Application of *Aspergillus niger* SA1 for the enhanced bioremoval of azo dyes in simulated textile effluent. Afr J Biotechnol 8:3839–3845. www.ajol.info/index.php/ajb/article/download/62069/50111

Awang NA, Aziz HA (2012) *Hibiscus rosa-sinensis* leaf extract as coagulant aid in leachate treatment. Appl Water Sci 2:293–298. doi:10.1007/s13201-012-0049-y

Bala Subramanian S, Yan S, Tyagi R, Surampalli R, Lohani B (2008) Isolation and molecular identification of extracellular polymeric substances (EPS) producing bacterial strains for sludge settling and dewatering. J Environ Sci Health A Tox Hazard Subst Environ Eng 43:1495–1503. doi:10.1080/10934520802293602

Barraud N, Storey MV, Moore ZP, Webb JS, Rice SA, Kjelleberg S (2009) Nitric oxide-mediated dispersal in single-and multi-species biofilms of clinically and industrially relevant microorganisms. Microb Biotechnol 2:370–378. doi:10.1111/j.1751-7915.2009.00098.x

Barredo-Damas S, Alcaina-Miranda M, Iborra-Clar M, Bes-Pia A, Mendoza-Roca J, Iborra-Clar A (2006) Study of the UF process as pretreatment of NF membranes for textile wastewater reuse. Desalination 200:745–747. doi:10.1016/j.desal.2006.03.497

Bassler BL, Losick R (2006) Bacterially speaking. Cell 125:237–246. doi:10.1016/j.cell.2006.04.001

Bjarnsholt T, Jensen PØ, Rasmussen TB, Christophersen L, Calum H, Hentzer M, Eberl L (2005) Garlic blocks quorum sensing and promotes rapid clearing of pulmonary *Pseudomonas aeruginosa* infections. Microbiology 151:3873–3880. doi:10.1099/mic.0.27955-0

Cai TB, Wang PG, Holder AA (2005) NO and NO donors. In: Wang PG, Cai TB, Taniguchi N (eds) Nitric oxide donors for pharmaceutical and biological applications. Wiley VCH, Weinheim. doi:10.1002/3527603751

Caro A, Humblot V, Méthivier C, Minier M, Salmain M, Pradier CM (2009) Grafting of lysozyme and/or poly (ethylene glycol) to prevent biofilm growth on stainless steel surfaces. J Phys Chem B 113:2101–2109. doi:10.1021/jp805284s

Chaignon P, Sadovskaya I, Ragunah C, Ramasubbu N, Kaplan J, Jabbouri S (2007) Susceptibility of staphylococcal biofilms to enzymatic treatments depends on their chemical composition. Appl Microbiol Biotechnol 75:125–132. doi:10.1007/s00253-006-0790-y

Charville GW, Hetrick EM, Geer CB, Schoenfisch MH (2008) Reduced bacterial adhesion to fibrinogen-coated substrates via nitric oxide release. Biomaterials 29:4039–4044. doi:10.1016/j.biomaterials.2008.07.005

Cheryan M (1998) Ultrafiltration and microfiltration handbook. CRC Press. https://www.crcpress.com/Ultrafiltration-and-Microfiltration-Handbook/Cheryan/9781566765985

Chisti Y, Moo-Young M (1999) Fermentation technology, bioprocessing, scale-up and manufacture. Biotechnology-The Science and the Business, Harwood Academic Publishers, New York, NY, pp. 177–222

Chua H, Arnot T, Howell J (2002) Controlling fouling in membrane bioreactors operated with a variable throughput. Desalination 149:225–229. doi:10.1016/S0011-9164(02)00764-6

Dobretsov S, Teplitski M, Paul V (2009) Mini-review: quorum sensing in the marine environment and its relationship to biofouling. Biofouling 25:413–427. doi:10.1080/08927010902853516

Flemming H-C, Schaule G, Griebe T, Schmitt J, Tamachkiarowa A (1997) Biofouling—the Achilles heel of membrane processes. Desalination 113:215–225. doi:10.1016/S0011-9164(97)00132-X

Gander M, Jefferson B, Judd S (2000) Aerobic MBRs for domestic wastewater treatment: a review with cost considerations. Sep Purif Technol 18:119–130. doi:10.1016/S1383-5866(99)00056-8

Gogate PR, Pandit AB (2004) A review of imperative technologies for wastewater treatment II: hybrid methods. Adv Environ Res 8:553–597. doi:10.1016/S1093-0191(03)00031-5

Goldman G, Starosvetsky J, Armon R (2009) Inhibition of biofilm formation on UF membrane by use of specific bacteriophages. J Membr Sci 342:145–152. doi:10.1016/j.memsci.2009.06.036

Górski A, Miedzybrodzki R, Borysowski J, Weber-Dabrowska B, Lobocka M, Fortuna W, Filby G (2009) Bacteriophage therapy for the treatment of infections. Curr Opin Investig Drugs 10:766–774. PMID:19649921

Gusarov I, Starodubtseva M, Wang Z-Q, McQuade L, Lippard SJ, Stuehr DJ, Nudler E (2008) Bacterial nitric-oxide synthases operate without a dedicated redox partner. J Biol Chem 283:13140–13147. doi:10.1074/jbc.M710178200

Hentzer M, Givskov M (2003) Pharmacological inhibition of quorum sensing for the treatment of chronic bacterial infections. J Clin Invest 112:1300–1307. doi:10.1172/JCI20074

Houghton J, Quarmby J, Stephenson T (2001) Municipal wastewater sludge dewaterability and the presence of microbial extracellular polymer. Water Sci Technol 44:373–379 http://www.ncbi.nlm.nih.gov/pubmed/11548008

Koyuncu I (2002) Reactive dye removal in dye/salt mixtures by nanofiltration membranes containing vinylsulphone dyes: effects of feed concentration and cross flow velocity. Desalination 143:243–253. doi:10.1016/S0011-9164(02)00263-1

Le-Clech P, Chen V, Fane TA (2006) Fouling in membrane bioreactors used in wastewater treatment. J Membr Sci 284:17–53. doi:10.1016/j.memsci.2006.08.019

Lee W, Kang S, Shin H (2003) Sludge characteristics and their contribution to microfiltration in submerged membrane bioreactors. J Membr Sci 216:217–227. doi:10.1016/S0376-7388(03)00073-5

Leroy C, Delbarr C, Ghillebaert F, Compere C, Combes D (2008) Effects of commercial enzymes on the adhesion of a marine biofilm-forming bacterium. Biofouling 24:11–22. doi:10.1080/08927010701784912

Lewis DM (1999) Coloration in the next century. Rev Prog Color Relat Top 29:23–28. doi:10.1111/j.1478-4408.1999.tb00124.x

Liu P, Huang Q, Chen W (2012) Heterologous expression of bacterial nitric oxide synthase gene: a potential biological method to control biofilm development in the environment. Can J Microbiol 58:336–344. doi:10.1139/w11-141

Marcucci M, Ciabatti I, Matteucci A, Vernaglione G (2003) Membrane technologies applied to textile wastewater treatment. Ann N Y Acad Sci 984:53–64. doi:10.1111/j.1749-6632.2003.tb05992.x

McClean KH, Winson MK, Fish L, Taylor A, Chhabra SR, Camara M, Bycroft BW (1997) Quorum sensing and *Chromobacteriumviolaceum*: exploitation of violacein production and inhibition for the detection of N-acylhomoserine lactones. Microbiology 143:3703–3711. doi:10.1099/00221287-143-12-3703

Nilsson R, Nordlinder R, Wass U, Meding B, Belin L (1993) Asthma, rhinitis, and dermatitis in workers exposed to reactive dyes. Br J Ind Med 50:65–70. doi:10.1136/oem.50.1.65

Petrovski S, Seviour RJ, Tillett D (2011) Prevention of Gordonia and Nocardia stabilized foam formation by using bacteriophage GTE7. Appl Environ Microbiol 77:7864–7867. doi:10.1128/AEM.05692-11

Qin Z, Ou Y, Yang L, Zhu Y, Tolker-Nielsen T, Molin S, Qu D (2007) Role of autolysin-mediated DNA release in biofilm formation of *Staphylococcus epidermidis*. Microbiology 153:2083–2092. doi:10.1099/mic.0.2007/006031-0

Radhakrishnan SK, Viollier P (2012) Two-in-one: bifunctional regulators synchronizing developmental events in bacteria. Trends Cell Biol 22:14–21. doi:10.1016/j.tcb.2011.09.004

Ren D, Zuo R, Barrios AFG, Bedzyk LA, Eldridge GR, Pasmore ME, Wood TK (2005) Differential gene expression for investigation of *Escherichia coli* biofilm inhibition by plant extract ursolic acid. Appl Environ Microbiol 71:4022–4034. doi:10.1128/AEM.71.7.4022-4034.2005

Richards JJ, Melander C (2009) Controlling bacterial biofilms. Chem Bio Chem 10:2287–2294. doi:10.1002/cbic.200900317

Rosenberger S, Kraume M (2002) Filterability of activated sludge in membrane bioreactors. Desalination 146:373–379. doi:10.1016/S0011-9164(02)00515-5

Samaranayake Y, Cheung B, Parahitiyawa N, Seneviratne C, Yau J, Yeung K, Samaranayake L (2009) Synergistic activity of lysozyme and antifungal agents against *Candida albicans* biofilms on denture acrylic surfaces. Arc Oral Biol 54:115–126. doi:10.1016/j.archoralbio.2008.09.015

Schlag S, Nerz C, Birkenstock TA, Altenberend F, Götz F (2007) Inhibition of staphylococcal biofilm formation by nitrite. J Bacteriol 189:7911–7919. doi:10.1128/JB.00598-07

Shrivastava R, Christian V, Vyas B (2005) Enzymatic decolorization of sulfonphthalein dyes. Enzyme Microb Technol 36:333–337. doi:10.1016/j.enzmictec.2004.09.004

Siddiqui MF, Andleeb S, Ali N, Ghumro PB, Ahmed S (2009) Up-flow immobilized fungal column reactor for the treatment of anthraquinone dye drimarene blue K2RL. Afr J Biotechnol 8:5570–5577. www.ajol.info/index.php/ajb/article/download/66007/53725

Siddiqui MF, Andleeb S, Ali N, Ghumro PB, Ahmed S (2010) Biotreatment of anthraquinone dye Drimarene Blue K 2 RL. Afr J Environ Sci Technol 4:45–50 http://www.ajol.info/index.php/ajest/article/view/56319

Siddiqui MF, Oh H-S, Rzechowicz M, Winters H, Chong TH, Fane AG (2015a) Biofouling control potential of tannic acid, ellagic acid, and epigallocatechin against *Pseudomonas aeruginosa* and reverse osmosis membrane multispecies community. J Indust Eng Chem 30:204–211. doi:10.1016/j.jiec.2015.05.023

Siddiqui MF, Rzechowicz M, Harvey W, Zularisam A, Anthony GF (2015b) Quorum sensing based membrane biofouling control for water treatment: a review. J Water Proc Eng 7:112–122. doi:10.1016/j.jwpe.2015.06.003

Siddiqui MF, Rzechowicz M, Oh H-S, Saeidi N, Hui LJ, Winters H, Anthony GF, Chong TH (2015c) The efficacy of tannic acid in controlling biofouling by *Pseudomonas aeruginosa*is dependent on nutrient conditions and bacterial density. Int Biodeterior Biodegrad 104:74–82. doi:10.1016/j.ibiod.2015.05.004

Siddiqui MF, Sakinah M, Ismail AF, Matsuura T, Zularisam A (2012a) The anti-biofouling effect of *Piper betle* extract against *Pseudomonas aeruginosa* and bacterial consortium. Desalination 288:24–30. doi:10.1016/j.desal.2011.11.060

Siddiqui MF, Sakinah M, Singh L, Zularisam A (2012b) Targeting N-acyl-homoserine-lactones to mitigate membrane biofouling based on quorum sensing using a biofouling reducer. J Biotechnol 161:190–197. doi:10.1016/j.jbiotec.2012.06.029

Tang C, Chen V (2002) Nanofiltration of textile wastewater for water reuse. Desalination 143:11–20. doi:10.1016/S0011-9164(02)00216-3

Tansel B, Sager J, Garland J, Xu S, Levine L, Bisbee P (2008) Biofouling affinity of membrane surfaces under quiescent conditions. Desalination 227:264–273. doi:10.1016/j.desal.2007.06.030

Taylor J, Jacobs EP (1996) Water treatment membrane processes. Reverse osmosis and nanofiltration. McGraw-Hill, New York, NY

TePoele S, Van der Graaf J (2005) Enzymatic cleaning in ultrafiltration of wastewater treatment plant effluent. Desalination 179:73–81. doi:10.1016/j.desal.2004.11.056

Thiel V, Kunze B, Verma P, Wagner-Döbler I, Schulz S (2009) New structural variants of homoserine lactones in bacteria. Chem Bio Chem 10:1861–1868. doi:10.1002/cbic.200900126

Wang S, Guillen G, Hoek EM (2005) Direct observation of microbial adhesion to membranes. Environ Sci Technol 39:6461–6469. doi:10.1021/es050188s

Wang Y, Yu J (1998) Adsorption and degradation of Synthetic dyes on the mycelium of *Trametes versicolor*. Water Sci Technol 38:233–238. doi:10.1016/S0273-1223(98)00532-0

Woerner DL (2003) Membrane technology in textile operations. Koch Membrane Systems, Wilmington, MA. cdam.minam.gob.pe/multimedia/contaminacion/TextileMembrane.pdf

Wu J, Xi C (2009) Evaluation of different methods for extracting extracellular DNA from the biofilm matrix. Appl Environ Microbiol 75:5390–5395. doi:10.1128/AEM.00400-09

Xiong Y, Liu Y (2010) Biological control of microbial attachment: a promising alternative for mitigating membrane biofouling. Appl Microbiol Biotechnol 86:825–837. doi:10.1007/s00253-010-2463-0

Yeon K-M, Cheong W-S, Oh H-S, Lee W-N, Hwang B-K, Lee C-H, Lewandowski Z (2008) Quorum sensing: a new biofouling control paradigm in a membrane bioreactor for advanced wastewater treatment. Environ Sci Technol 43:380–385. doi:10.1021/es8019275

Yeon K-M, Park J-S, Lee C-H, Kim S-M (2005) Membrane coupled high-performance compact reactor: a new MBR system for advanced wastewater treatment. Water Res 39:1954–1961. doi:10.1016/j.watres.2005.03.006

Yu H, Fang L-C, Lateef SK, Wu C-H, Lin C-F (2010) Enzymatic treatment for controlling irreversible membrane fouling in cross-flow humic acid-fed ultrafiltration. J Hazard Mater 177:1153–1158. doi:10.1016/j.jhazmat.2010.01.022

Yun M-A, Yeon K-M, Park J-S, Lee C-H, Chun J, Lim DJ (2006) Characterization of biofilm structure and its effect on membrane permeability in MBR for dye wastewater treatment. Water Res 40:45–52. doi:10.1016/j.watres.2005.10.035

Zhang F, Yediler A, Liang X, Kettrup A (2004) Effects of dye additives on the ozonation process and oxidation by-products: a comparative study using hydrolyzed CI reactive red 120. Dyes and Pigm 60:1–7. doi:10.1016/S0143-7208(03)00111-6

Economic and Market Value of Biogas Technology

Abu Yousuf, Maksudur Rahman Khan, Domenico Pirozzi, Zularisam Ab Wahid, and Samson Mekbib Atnaw

Abstract The demand and prospect of biogas technology as a renewable energy source in terms of market value have not been adequately addressed. This paper focuses on the status of legal framework and future market situation related to biogas technology in order to facilitate its improvements. Biogas technology, known as biofuel production process through fermentation of biological wastes, is a well-established technique to improve lives, livelihoods, health, and ecosystem. This approach generates a large revenue opportunity that supports the socioeconomic development in rural areas. However, very little initiative has been introduced specially in the developing world to gear up the biogas technology. For more sustainable development of this technology, policy-makers should reform the existing institutional framework by reorganizing subsidies, motivating and attracting investor with flexible financial conditions, liberalizing the management of gas grids, and involving farmers in local projects. Therefore, it is a great challenge to find a proper mode of marketing policy, business models and multi-profit options, and a sustainable financing mechanism. This paper covers the state-of-the-art enlargements and future consequences of the hastily emerging biogas market, starting with a universal viewpoint and going through the market characteristics of Europe, the USA, Africa, and Asia Pacific.

Keywords Biogas market potentiality • Economic policy • Market barriers and solutions • Biogas sustainability

A. Yousuf (✉) • Z.A. Wahid • S.M. Atnaw
Faculty of Engineering Technology, University Malaysia Pahang, Ganmbang 26300, Kuantan, Malaysia
e-mail: ayousufcep@yahoo.com; ayousuf@ump.edu.my

M.R. Khan
Faculty of Chemical and Natural Resources Engineering, University Malaysia Pahang, Ganmbang, Kuantan, Malaysia

D. Pirozzi
Department of Chemical Engineering, Materials and Industrial Production, University Naples Federico II, Naples, Italy

L. Singh, V.C. Kalia (eds.), *Waste Biomass Management – A Holistic Approach*,
DOI 10.1007/978-3-319-49595-8_7

Nomenclature

BDT	Bangladeshi taka
bln m^3/a	Billion meter^3/annum
BMW	Bioorganic municipal waste
CAGR	Compound annual growth rate
CHP	Combined heat and power plants
EU	European Union
GHG	Greenhouse gas
GWh	Gigawatt-hour
GWth	Gigawatt thermal
kg/c/a	Kilogram/capita/annum
kWh	Kilowatt-hour
MWel	Megawatt electric
MWh	Megawatt-hour
MWth	Megawatt thermal
NGO	Nongovernmental organization
THB	Thai bath
tpa	Tonnes per annum

1 Introduction

Biogas technology offers multiple benefits simultaneously corresponding to health, food, agriculture, energy, environment, and sanitation. It also bears the potential to facilitate an independent economic and energy development especially of the third world. The overall practice of biogas technology proves its simplicity and ease adaptability in terms of construction capacity and management (Amigun and Blottnitz 2010). Biogas is produced through anaerobic fermentation process from biodegradable organic wastes, such as cow manure, food residues, agricultural residues, city garbage, etc. The basic design of a biogas digester is illustrated in Fig. 1. Biogas is a mix of methane, carbon dioxide, and other trace gases which are able to generate heat, electricity, or light. In the least developed countries, biogas technology is mainly used by small-scale farmers, when these farmers have enough livestock for running a small-scale biogas plant (Buysman and Mol 2013). Major constraints of the biogas technology are related to the lack of financial attractiveness, although highly subsidized, and to the conditions to be met by households: they require a minimum amount of cattle, access to water, and a financial contribution to construction cost. Unfortunately, the majority of households are excluded from these programs by default due to their inability to meet the conditions.

Methane content, which lies between 50 % and 75 % of biogas, defines its energy value, and it depends on the composition of biomass substrate and the fermentation process (Iqbal et al. 2014; Yousuf et al. 2014). One cubic meter of methane has an energy content of about 10 kilowatt-hours. The one cubic meter of biogas,

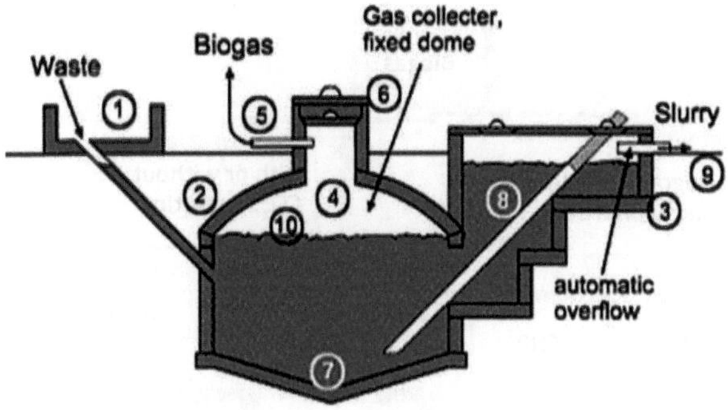

Fig. 1 Common structure of biogas digester: (1) feed inlet (mixing tank), (2) digester, (3) compensation and removal tank, (4) gas chamber, (5) biogas outlet pipe, (6) removal cover for annual discharging, (7) undigested residues, (8) undigested residues outlet pipe, (9) reference level slurry, and (10) supernatant scum

containing 60 % methane, will produce 6 kilowatt-hours or 1.5 to 3 kWh of electrical energy. Therefore, 1 m^3 of biogas could replace 0.56 m^3 of natural gas.

It is hoped that from the 2012–2016 period, the worldwide biogas plant installation capacity will increase to 2700MWel. By this time, additional 3800 new plants will be built. The growth of this number will shift away from Germany, where the revolution in biogas technology happened during the last 5 years. The dynamic biogas market will be expanded shortly to other European states, North America, and Asia. Most countries in those areas already have developed or going to develop a regulatory framework for biogas technology and marketing policy. It is predicted that the world biogas market will reach over 50 billion USD $ by 2030 for plant constructor, operator, service provider, and supplier. The stronger biogas market will ensure the development for the cities and rural areas offering new scope of income, job, investment, and profit. Therefore, in future biogas will play an important and efficient role to local energy (electricity, heat, cooling, and fuel) distribution policies.

The market analysis of biogas technology from global aspect is absent in the research arena. Very few studies have been done on the market-oriented biogas business and profit analysis in regional or local basis (Buysman and Mol 2013). This paper, therefore, focuses on the future market potential of biogas technology and its profit options which will insist the policy-maker, investor, and entrepreneurs to confirm social, economic, and environmental returns from biogas technology. It also gives an overview of opportunities, as well as weak points and potential barriers for biogas market development.

2 Global Biogas Market

To meet up the growing demand of heating, power, and fuel markets, biogas already showed its potentiality as a versatile energy carrier. As well as it clamps a strategic position in terms of resource availability, efficiency, and from an environmental

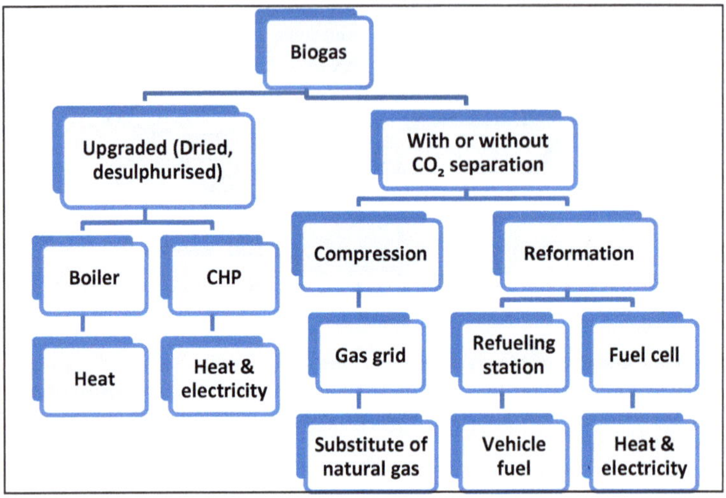

Fig. 2 Versatile uses of biogas

and socioeconomic viewpoint. A number of technologies already developed to purify and upgrade biogas to that of natural gas, so that it can be input into natural gas grids. Therefore, biogas (desulfurized and cleaned) can be used as substitute of natural gas in many different purposes. Figure 2 shows the available options for the use of this very versatile source of energy. Recently, biogas is widely used in combined heat and power plants (CHPs) where it replaced oil and gas.

Recent analysis by Global Information, Inc. (GII) reported that by the next 12 years, global biogas installation capacity will increase moderately, and it will reach to 22.040 MW by 2025 making a compound annual growth rate (CAGR) of 7.2 %. Mainly the world biogas market has grown considerably between 2001 and 2011 having capacity from 2.388 MW to 8.377 MW, respectively. This equals a CAGR of 13.4 %. As it is mentioned, the leading nations in the global biogas market are Germany, the USA, and the UK, accounting for over 50 % of the total installed capacity worldwide (Maghanaki et al. 2013).

The report "Global Biogas Market 2014–2018" projected that the global biogas market will grow at a CAGR of 8.64 % over the 2013–2018 period. The industry experts prepared the report on the basis of in-depth market analysis. The report covers the Americas and the Europe, the Middle East, and Africa (EMEA) and Asia Pacific (APAC) regions; it also focuses the comprehensive biogas market scenery and its growth projections in the coming years. According to the report, the advancement of the global biogas market is driven by numerous factors. A smart solution to sanitation, offered by biogas technology, is one of the major drivers since biogas is produced from garbage, rotten food, manure, and wastewater. Depletion of fossil fuel, increasing of energy demand, and changing of global climate are also influencing the expansion of global biogas market.

In 2012, a research group, Penn State Information, Knowledge, and wEb (PIKE), has published an attractive outcome of their research that may encourage the biogas investors (Pike 2012). PIKE Research is a consulting firm and that delivers in-depth

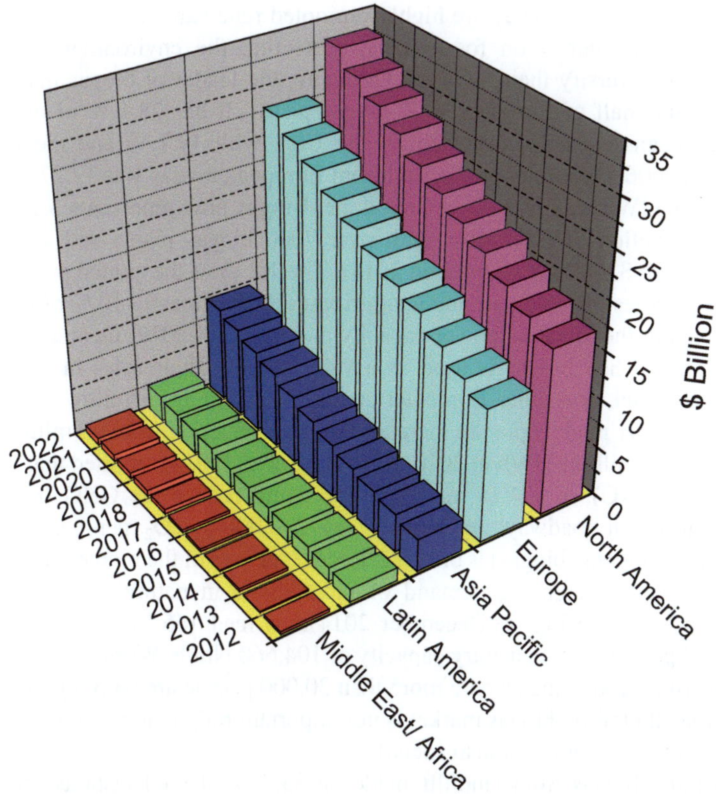

Fig. 3 Biogas world market value ($ in billion) by region (2012–2022)

analysis of global clean technology markets. It suggests that the global commercial biogas market will more than double in the next 10 years as their findings is presented in Fig. 3. The figure illustrated that global revenue of this fast-growing market was $17.3 billion in 2012, and it will be almost double in 2022, reaching $33.1 billion. PIKE report analyzes the prospective global market of biogas demand considering four key industrial segments: municipal solid waste, agriculture, industrial, and sewage treatment. Another report estimated the value of the world market for anaerobic digesters and landfill gas equipment, and the value was at nearly $4.5 billion for 2013 and will reach to $7 billion by 2018.

3 Biogas Market Characteristics of Developed World

3.1 EU Countries

Biogas is considered as the sustainable solution in EU member countries in both sectors to fulfill the renewable energy target by 2020 and meet organic waste management directive. They are highly promoted renewable energy production to reduce the dependency on fossil fuels, protecting the environment from CO_2 addition and diversify their incomes. Germany, the leader of biogas market, successfully runs half of Europe's 9000 biogas plants. It has 28.1 % of the world's installed biogas capacity, followed by the USA with 19.7 % and the UK with 15.5 %. In 2006, electricity production from European biogas was 17,272 GWh, of which 7338 GWh was from Germany alone. Biogas now represents 3.9 % of the annual production of electricity, with over 7000 biogas plants and an installed power close to 3300 MW (Findeisen 2014). The 1.2 % of the annual production of electricity comes from biogas, and it contributes approximately 10 % of renewable energy having the installed power capacity close to 1500 MW. Germany does not impose a tax when biogas is used as a vehicle fuel. Their roles in biogas uses become a model, and other European countries are going to follow the German model. Germany made a plan to increase biogas production from 90 million MWh in 2013 to 130 million MWh in 2020. Rosenstiel et al. (2014) reported that liquefied natural gas (LNG), at a 4 % market share, could reduce the GHG from goods transportation on roadway in Germany by 240.000 t CO_2 per year, if 20 % biomethane (upgrade biogas) is blended. Recent surveys of the biomethane industry barometer show increasing demand of biomethane in natural gas grid line in Germany. As of the Fig. 4, December 2014 thus feed 165 biomethane plants in the natural gas grid with an entry capacity of 104,660 Nm/h. World biogas sector is led by German companies where more than 30,000 people are working in this field of business; therefore, biogas market is not important only in terms of environmental or monetary benefits but also socially.

In Britain, biogas from landfills made up 33.7 % of total installed bio-power capacity in 2011. Sewage sludge gas also contributed 6 % to the UK's total installed bio-power capacity in 2011. Global biogas capacity is expected to reach

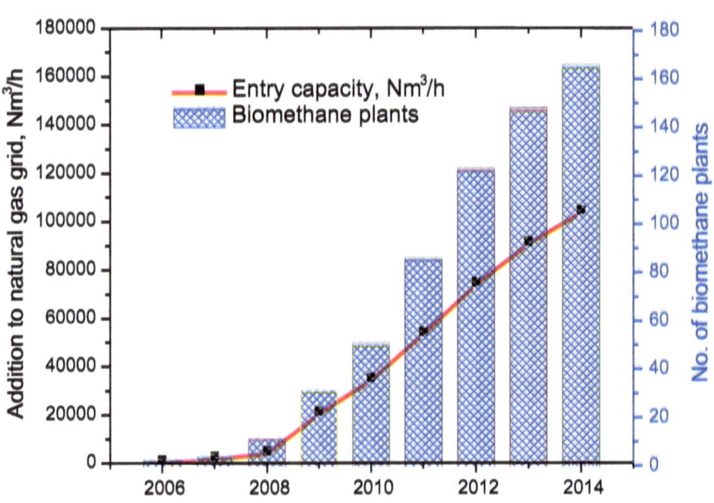

Fig. 4 Market development of biomethane in Germany

22,000 MW by 2025. The UK is generating approximately 30 % electricity from coal and nuclear power. As these sources are old age and causing serious emissions, they need to be close down in the near future. Corporations are looking for new coal-fired stations, but the government has denied unless the facilities have scrubbers to clean emissions. Therefore, biogas would be a good alternative and sustainable solution in the energy sector over there.

The biogas markets are already established in Italy due to the favorable subsidization schemes. The UK, the Czech Republic, and the Netherlands will continue to be potential markets in the next 5 years. France is one of the new strong and growing European biogas markets. Sweden is already running city buses on natural gas, and Italy is a promising growing market in favor of biogas utilization in transportation. The UK, France, Spain, and Italy are adopting parts of the German model. Poland has lots of plans and potential, but in practice not so much action is visible. Denmark and Norway are growing, so Sweden is at a high level. However, EU has the bright future to expand their biogas market because of their new trend of energy uses (Fig. 5).

Liquefied methane gas, from either biogas or natural gas, is a new vehicle fuel in Sweden. Recently, Sweden has opened the world's first plants for the production of liquefied biogas (LBG) in the town of Linköping, southern Sweden. The plant designed by Swedish Biogas International will have a production capacity of 7.5 MWth with an annual target of 60 GWth. There were 47 factories in 2010, which were upgrading biogas to biomethane, and seven of them were linked to public grid line. The most common route for the use of biomethane is the transport sector. In 2010, 608 GWh was produced as biomethane, where 572 GWh (94 %) was sold as vehicle fuel. Sweden is one of the pioneer countries, which follows a

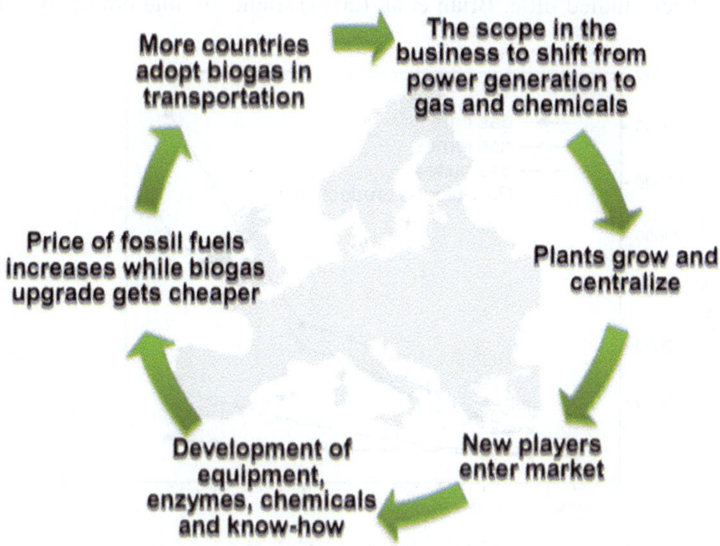

Fig. 5 Reasons behind the biogas market development in European high-tech countries

national standard for biogas as a transportation fuel. The Swedish government has a plan to have a vehicle fleet that would be independent of fossil fuels by 2030. In fact, in 2050, Sweden wishes to supply clean energy without carbon emissions.

Since biogas is more beneficial to the environment compared to diesel fuel due to reduction of air pollution, the use of biogas as vehicle fuel is increasing (Olsson and Fallde 2015). Figure 6 shows the increasing trend of gas-fueled vehicles in Sweden that implies the prospect of the future biogas market. It is reported that biogas buses have increased significantly in 2011, and 11 % of buses were gas fueled in Sweden. A number of filling stations are also growing up, while there were very few filling stations in 1996, and the number increased to 122 in 2010. Additionally, in Linköping, 1.8 to 2.8 million Nm^3 raw biogas is produced annually through digestion of sewage sludge. In 2013, the total amount of raw biogas produced in Linköping exceeded 17.5 million Nm^3. All biogas is upgraded to vehicle gas (Fallde and Eklund 2015).

3.2 *The USA*

In the USA, still, it is a great challenge to control the greenhouse gas (GHG) emissions from energy- and fuel-oriented fields. The multipurpose use of biogas as a substitute of traditional fuel can mitigate the addition of GHG to the atmosphere. In a carbon-constrained future, biogas is a very promising source of renewable energy, which does not offer only energy and environmental benefits, but rather it plays roles to socioeconomic welfares. Although pipeline-quality biogas can increase the supply of natural gas market, its potential and sustainability has long been studied little. Brian et al. (2014) studied to find out the way how to

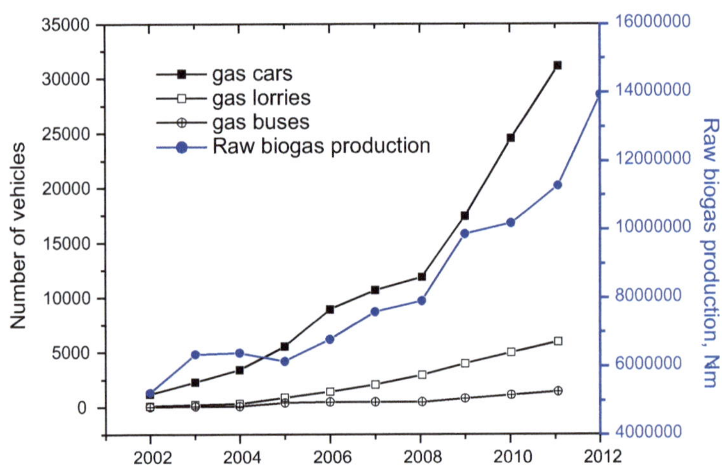

Fig. 6 Number of gas-fueled vehicles in Sweden and raw biogas production only in Linköping, Sweden

disperse domestic biogas market throughout the USA by 2040. Their report suggested a number of crucial points that will support to boost up biogas technology in the future such as the following:

(1) Agriculture and forest residual biomass should be considered as major sources of biogas.
(2) Around 3–5% of the total natural gas market could be replaced by biogas.
(3) Electricity generation from biogas may be more profitable than supplying through pipeline because it needs high pressure and needs to be oriented with system loss.
(4) Energy market is yet not ready to welcome biogas technology; therefore, new policy and financial encouragements are essential to spread up its use.

Natural gas (including biogas) consumption in the USA increased by four times from 1950 to 2012, while between 1950 and 1970 (Fig. 7), the consumption rates were very sharp as the economy became strong and more natural gas was discovered in the postwar era. This growth was followed by a decrease since new gas findings declined, prices rose, and replacement occurred during the period of 1970 and 1990. New gas withdrawal technologies such as hydraulic fracturing were invented in the late 1990s, and plentiful resources of shale gas was opened which was economically accessible. Between 2011 and 2040, US natural gas intake is expected to upsurge by 0.7 % per year under the projected baseline as reported in the US Energy Information Administration's Annual Energy Outlook 2013 with projections to 2040.

The USA is full of biodegradable organic wastes that could be feed for biogas systems, feedstock include city wastes, food wastes, biosolids, livestock manure, landfill gas, and residues of agro-food industries. Energy production from food

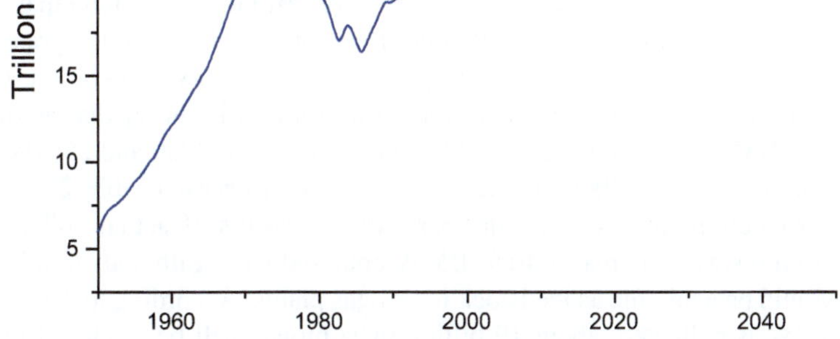

Fig. 7 Natural gas consumption in the USA from 1950 to 2012, with projections to 2040

Table 1 Biogas production capacity and its energy potential in the USA

Energy potential of biogas	Biogas sources			Total
	Livestock manure	Landfill gas	Water resource recovery	
Currently operational biogas systems	239	636	1241	2116
Total number of biogas systems	8241	1086	3681	13,008
Biogas production (billion cubic feet/year)	257	284	113	654
Annual electricity (billion kWh/year)	13.1	22.5	5.6	41.2
Equivalent residential electricity use (1000 homes/year)	1089	1864	539	3492

industry wastes in a biogas system is recognized as ecologically healthy waste managing system. As reported by the US Federal government and industry sources, the USA has more than 2000 functioning biogas systems and more than 13,000 prospective sites. The potentiality of these systems to generate energy is summarized in the following Table 1.

4 Biogas Market Characteristics of Developing World

4.1 China

China is the world's leader in energy demand, coal consumption, municipal solid waste production, CO_2 and CH_4 emissions, and N-fertilizer consumption. As well as, the country has to survive with one of the faster period of urbanization in history. To solve these environmental issues, China government has initiated a number of strategies such as huge investment in renewable energy sectors, increase primary energy consumption from renewable sources and international collaboration to boost up bioenergy production. In this circumstance, biogas technology can play a vital role that has a plenty of feedstock throughout the country.

The biogas cooperation platform has been built between Germany and China. A Sino-German project—Resource Recovery of Bioorganic Municipal Waste (RR-BMW)—is announced to elaborate forecasts on how to integrate better waste treatment and renewable energy production via biogas generation, through a joint approach between urban and agricultural areas. It shows that as much as 6 GW_{tot} of BMW-derived biogas could be generated until 2020 and 25 GW_{tot} until 2050. Raninger et al. (2006) made a conservative forecast (Table 2) of biogas energy production for China on the basis of the results of actual collections of BMW from residential area. If 40 % BMW potential were gathered, 28 million tpa BMW would be available as feedstock for biogas plants. According to their study, if 75 % BMW is collected, about 19 billion m^3/a biogas will be produced by 2050, which is equivalent to 11.8 GW_{tot} energy potential (Table 2).

Table 2 Estimates of biogas building potential from MSW/BMW in China until 2050

Description	2005	2010	2020	2050
Urbanization rate in China (%)	28	36	51	70
Citizens (million)	360	470	650	940
MSW collection rate (%)	53.5	60	65	75
MSW from cities (million tpa)	155	240	380	700
MSW organic matter (% FM)	66	62	55	45
Organic matter in MSW (tpa)	102	150	210	315
Citizen participation[a] (%)	70[a]		80	
BMW overall potential (million tpa)	71	105	147	252
BMW collection area (%)	40	25	55	75
BMW net potential (million tpa)[a]	>28	>47	>80	>189
Biogas potential (billion m^3/a)	>2.8	>4.7	>8.0	>19
Biogas energy potential (GW)	>1.8	>2.9	>5.0	>11.8

[a]According to RR-BMW field test results, participation rate 70 % and net-BMW 70 kg/c a (2005/2006)

According to the medium- and long-term development plan for renewable energy in China by 2020, 10,000 major projects of biogas production on farms' breeding and 6,000 projects of biogas production using organic industrial effluent will be built with a production total annual output of 14 billion m^3 biogas and the total installed capacity of 3 GW. One analysis reported that only MSW can produce about 25 billion m^3 of biogas, derived from municipal wastewater (sewage sludge), and agricultural residues and livestock waste can produce 70 billion m of biogas. In total, there are about 1.5–2.0 billion tpa of biomass available, and the existing landfill sites may produce at least 155 billion m biogas per year, or about 96 GW. It is estimated that one-third (43 billion m^3) might be produced until 2020. The Chinese biogas energy potential forecast information is presented in Fig. 8.

4.2 India

Among the developing world, India has proved its potentiality to be a fast-growing economy. Therefore, to continue the socioeconomic development, energy is considered as a critical input. The country's energy strategy has been renewed and aims to include efficient and environment-friendly energy generation technology, to ensure energy security and provide energy access to all and optimum mixing of primary and renewable energy generation. The 95 % need of transportation fuels meet by fossil-based oil in India and demand is still rising gradually.

More than 40 % population of India does not have the access to the energy of national grid line. Another important challenge for India is to fulfill the demand of rural cooking needs. The report published by the Ministry of New and Renewable Energy (MNRE 2011) stated that the Family Size Biogas Plants Programme covers

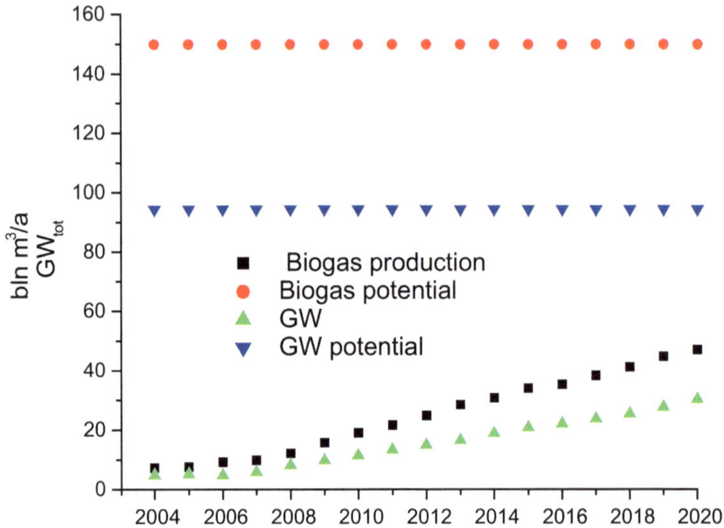

Fig. 8 The future trends of biogas production (billion m^3/a) and its energy potential (GW_{tot}) in China

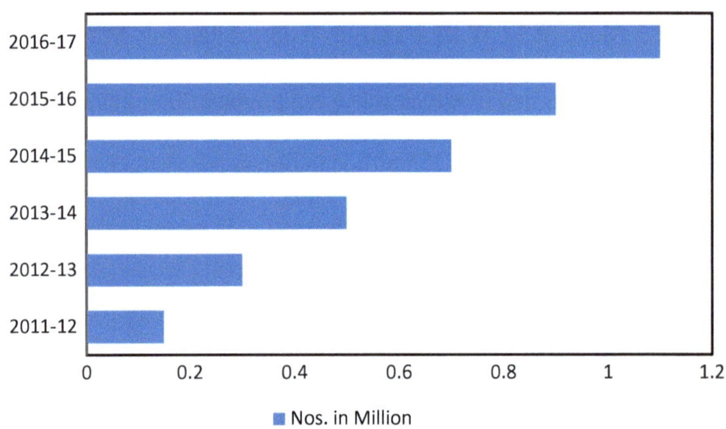

Fig. 9 Family size biogas plant set up targets (cumulative) for the period of 2011–2017, data adopted from

the issues of cook stoves for both the individual household and the community levels. About 0.1 million family size biogas digesters have been establishing by the ministry annually. Although the annual target is increasing by 0.15 million, the ministry would like to cover more than 2 million by 2022. Year-wise targets for family size biogas plant setup are summarized in Fig. 9. Indian renewable energy market, which is fastly growing up, offers lucrative business opportunities to local, foreign, and international entrepreneurs, and the government also provides a favorable business environment to boost up the renewable energy sector. India is the third country with the most business-friendly environment in order to finance the renewable energy sector. It earns annual revenues of nearly $ 185 billion and has the most attractive and developing renewable energy markets in South Asia.

4.3 Thailand

Instead of the lavishness of biodegradable organic waste in the urban markets of Thailand, dynamic practices of bioenergy generation are missing there. Green waste conversion to renewable energy will not only benefit the environment but also financially rewarding. Large amounts of organic wastes come from fruit and vegetable markets each year, and their disposal is very costly, both financially and environmentally. Therefore, it has a bright future to produce biogas from green waste generated from the fruit and vegetable markets, also known as green markets. With proper utilization of green market waste, the market itself can easily fulfill their energy needs. However, intensive research is required to face the challenges and achieve the sustainability goal. Ali et al. (2012) showed that production of biogas from the waste generated is environmentally friendly as well as cost-effective. The authors also studied benefits-cost analysis of biogas production and proposed a scenario (Table 3). Several factors were included in their cost analysis procedure such as equipment and machinery, construction cost, working capital, and contingency but excluded land purchasing cost. According to the current market price, around THB 4 million needs to be invested to set up a biogas plant.

4.4 Malaysia

Energy demand for the transportation and industrial sectors in Malaysia is increasing hastily, and the increasing rate is about 5–7 % per year during the next 20 years

Table 3 Benefit-cost analysis

Investment cost[a]		Functional cost		Benefits	
Construction	30,000	Cost per ton	75	Cost per ton	360
Working capital and contingency	30,000	Total waste quantity (tonnes)	29.4	Total biogas quantity (tonnes)	15.1
Process equipment and machinery	45,000	Total cost per day	2205	Total income per day	5436
Fees[b]	15,000	Total cost per year	26,460	Total benefits per year	65,232
Total	120,000				

[a]All figures are given in $USD
[b]Fees of local consultant, authority permission, etc.

Table 4 Economic viability for biogas generation and compost with and without CDM

	Biogas generation without CDM	Biogas generation with CDM	Compost without CDM	Compost with CDM
NPV (RM million)	−0.03	12.39	1.31	11.87
IRR (%)	5	25	8	31
PBP (year)	8	3.5	6.5	2.9

NPV net present value, *IRR* internal rate of return, *PBP* payback period

from 2004 (Lau et al. 2009). Recently the petroleum price has become volatile in world market, and the Malaysian government has been paying many subsidies to control the fossil fuel price. Therefore, Sustainable Energy Development Authority (SEDA) Malaysia has taken National Renewable Energy Policy and Action Plan (2010). According to SEDA, the Malaysian government made a plan to generate power from small hydro plant 197.79 MW, biomass 124.24 MW, biogas (landfill/agri-waste) 98.22 MW, and solar PV 87.98 MW for the period of 2012–2017. Thus, RE generation from biogas technology is the third largest source in Malaysia.

From the last decades, the palm oil industry has been considered as a main source of income for the rural people. The derivatives of palm oil industry such as monocarp, empty fruit bunch, and shell and mill effluent could be an emerging source to produce biogas (Hosseini and Wahid 2013). Yoshizaki et al. (2012) reported that a palm oil mill with a capacity of 54 t fresh fruit bunch per hour has the potential to produce 6.9 GWh of electricity from biogas. In their case study, they compared the economic viabilities of biogas generation and compost projects in a palm oil mill in Malaysia with and without clean development mechanism (CDM). Their core findings are presented in Table 4.

4.5 Bangladesh

Bangladesh is one of the most densely populated countries in the world. Most of the people, especially in rural areas, either living as a family or in a group, where every house joins another house, usually have relatives that live very close to each other. It is a quite favorable condition for biogas generation. The biogas sector in Bangladesh is today characterized by competition of subsidy levels, and not by market-oriented competition of companies that search to win customers by delivering high-quality products and by providing reliable services. Moreover, in Bangladesh only 3 % of people enjoy the convenience of natural gas that come to their homes through pipelines. Most of the rural population of Bangladesh depends on biomass, plant debris, fragments of plants, animal dung, and wood for fuel, and this fueling system creates deforestation, flooding, and soil erosion. Women and children feel the burden of collecting fuel and suffer the most. They are the worst victims of indoor air pollution like smoke in the kitchen.

Table 5 Financial savings from family size biogas plant in Bangladesh

Amount saved (in BDT)/month	No. of households	% households
Zero saving	9	14
Less than BDT 200	5	9
BDT 200–500	29	44
BDT 501–1000	18	27
More than BDT 1000	4	6
Total	65	100

Bangladesh has been suffering a lot of energy crisis, and, to fulfill the demand, it has signed an agreement with Russia to establish a nuclear power plant. Most recently, 7 June 2015, Bangladesh government has signed a G-to-G contract with Indian government to set up new power plant by Indian companies. Therefore, energy market has still a vacuum in Bangladesh, and renewable energy is a most viable option to fill up the vacancy. Ghimire (2005) reported the survey result on existing family size biogas plant in terms of financial benefits, and it is summarized in Table 5. It clearly shows that 84.5 % of the biogas households are experiencing financial benefits from their plants. The average financial savings of biogas plants was calculated to be BDT 4947.10 per year/household, which is a significant amount.

Another report (Lohan et al. 2015) states that a family size biogas plant (2 m^3) can replace kerosene 316 L, 5535 kg of firewood, and 4400 kg cow dung cakes, which will reduce emissions of NO_x, SO_x, CO, and volatile organic compounds into the air by 16.4, 11.3, 69.7, and 987.0 kg per year, respectively. Therefore, biogas would be a sustainable alternative of natural gas in Bangladesh.

4.6 *African Countries*

The most powerful future biogas market is in Africa. Biogas technology in Africa appears to be held by an oligopoly of technology-driven economic situation where there are only a few suppliers of a particular product. The actions of one supplier can have a major impact on prices and competition (Amigun and Blottnitz 2010). The low number of suppliers, the less competition and thus promote a higher charge. This concept is presented in Eq. (1). Higher capital costs experienced in the biogas industry in Africa is compounded by the fact that the current biogas market in Africa is slow.

$$Q = \left(\frac{S}{n + S \times b \times \bar{P}}\right) - S \times b \times P \tag{1}$$

where Q is the firm sales, S is the total sales of the industry, n is the number of firms in the industry, b is the constant (a firm's sales to its price), P is the price charged by the firm itself, and $\bar{P}$ is the average price charged by its competitors.

Biogas technology was introduced for the first time in 1950 in South Africa and Kenya; however, recently many sub-Saharan countries are practicing this technology such as Burundi, Botswana, Burkina Faso, Cote d'Ivoire, Ethiopia, Ghana, Guinea, Lesotho, Namibia, Nigeria, Rwanda, South Africa, Uganda, and Zimbabwe. Africa has gathered mixed experience from biogas practice. Previous experience has shown that a technology focusing on engineering and capacity building does not always function properly and will not essentially create a market. There are also success stories in Africa which learned experiences from Nepal and Vietnam that with dedicated market approach, well-defined roles between participants (government, business, suppliers, contractors, consumers and credit institutions) proved to be the successful way forward. However, a number of global organizations and foreign aid agencies (Parawira 2009) are inspiring the biogas technology in Africa proving technical and financial support. Their efforts are becoming a success, since it is accepted as a simple and easily adaptable technology for Africa to meet up their energy needs.

Netherlands Development Organisation (SNV) has started to install biogas plant in Africa in 2007, and installation rate is slower compared to Asia as reported by (Mwirigi et al. 2014). Biogas plant installation by SNV is presented in Fig. 10 with their average installation cost. It illustrates that without 2–3 exceptions, cost becomes lower with the higher number of plants. Which implies that if more number of suppliers or entrepreneurs is allowed to contribute in biogas plant

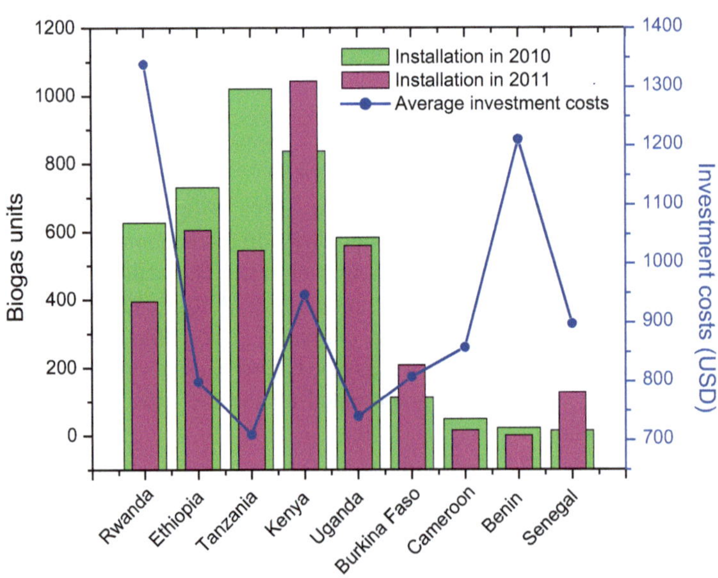

Fig. 10 Biogas units supported by SNV program in Africa

installation, after-sales service, maintenance, affordable biogas appliances (valves, stoves, lamps), it is possible to reduce the overall per unit operation cost.

There is a lot of scope to expand biogas market in Africa. An AGAMA Energy fact sheet (Brown 2006) reported that in South Africa, there are 400,000 homes with two or more cattle and having no electricity that could be provided by biogas digesters. The statistical report also describes that lack of electricity is suffered by 45 % of schools, 66 % have poor sanitation facilities, 27 % have no clean water, and 12 % have no sanitation at all in South Africa. Only one technology, that is, biogas, would be a potential solution to mitigate all of these problems. Therefore, this scenario is proposing a bright biogas market in Africa.

5 Biogas Market Barriers and Opportunities

Tsoutsos and Stamboulis (2005) reported eight types of obstacles that may hinder the deployment of renewable energy technologies: technological, regulatory, cultural, demand, production, infrastructure, environmental, and socioeconomic. Street and Miles (1996) outlined three broad categories: basic, technical, and nontechnical. Jacobsson and Johnson (2000) cited three types of factors such as market barriers, networks, and institutions. To overcome the primary barriers by realizing the full potential of biogas market, a better understanding between investors, policy-makers, and the public is essential.

5.1 Lack of Consciousness of Biogas Benefits

The main barrier that can be highlighted is the general low level of awareness of biogas utilization possibility and possible benefits to the local community. Investors, policy-makers, and the public should deeply understand what types of benefits they could gain from the investment in biogas sectors. They should realize that biogas technology is such a solution which offers multiple applications and can be adopted in both urban and rural areas. The sociocultural constraints can be overcome through an intensive educational campaign to raise awareness about the importance of this technology. This may include the issuance and distribution of manual tools and illustrated various mechanisms of propaganda, especially in rural areas through mass media, and organizing training courses and seminars.

5.2 Lack of Awareness of Waste Classification

Mixed waste (organic, plastic, glass, and metal) collection is a great barrier to introduce biogas technology in urban areas or in municipal territory. Most of the cities of developing countries are not practicing separate containers or dust bins for categorize wastes. As a result, biogas plants need to set up an extra unit to classify the wastes, which increases the operation cost.

5.3 Volatile Conditions of Biogas Market

Fluctuations of the renewable energy market are a major barrier toward greater investment in biogas systems. In fact, rural consumers have no knowledge or are not motivated to commercialize biogas and spin it. Market insecurity arises when people recognize the irregularity in the quality or quantity of biogas scheme.

5.4 Market Immaturity

A significant barrier toward biogas systems adoption is posed from infantile markets for greenhouse gas reduction benefits. Immature markets or unpopularity of non-energy products such as nutrient rich soil adjustments, pelletized and pumpable organic fertilizers, and feedstock for plastics reduce encouragements to invest. Any remarkable step is invisible to popularize and commercialize the non-energy products coming from biogas process, although organic fertilizer has a vast potential market around the world, particularly in agro-based counties. Commonly, uneducated farmer in developing countries are mostly depending on chemical fertilizer rather than organic fertilizer. Therefore, the biofertilizer market is not expanding promptly.

5.5 Lack of Full Assessment

Small-scale generators are insufficient to interconnect to the grid and to evaluate a fair market price and full environmental value that imitates the real contribution of biogas. In addition, high project costs, inaccurate assessment, and lack of proper financial appreciation and administrative bureaucracy related to energy and non-energy services create barriers to widespread investment in biogas systems.

5.6 *Lack of Coordination Across Central and Local Governments*

Many countries do not have linkup between local and central governments, public and private research organizations (NGOs), and research and business institutions to commercialize the advanced products and process. The disintegration of current resources, authority, and jurisdiction at the state and local levels prejudices to the implementation of the biogas system. Additional barriers arise in financing of biogas systems due to scattered efforts initiated by investors, governments, local agencies, and policy-makers.

5.7 *Lack of Technical Information, Analysis, and Research*

Most of the biogas producers' countries currently lack sufficient environmental, methodical, and monetary performance data related to biogas system. Actual information of biogas with the by-product, greenhouse gas emissions, and benefit of the water quality is also absent in the government's database. Data-consolidation can help market analysis, policy-making, and planning and underwriting the future. Advanced research is essential to fully utilize the biogas potential. Very few countries are upgrading biogas to supply in the national gas grid.

6 Solutions to Enhance Biogas Potential

The barriers identified in this study can probably be overcome if biogas provides a satisfactory financial return to the stakeholders when they deal with biogas feedstock, by-product, biogas-related stuffs (stove, balloon, generator, etc.), technology, networks, and infrastructure. All activities related to biogas production and the trading system should be closely monitored during the initial period. Data collection and evaluation will help to establish a sustainable business. Some factors significantly influence the economics of biogas such as initial installation cost of the digester, the price of natural gas and CO_2, the upgradation of biogas to natural gas, and the liquefaction of upgraded biogas. The technology of upgradation and liquefaction of biogas is still now developing and not cost-effective. To overcome this financial obstacle, governmental subsidies can make up a big proportion of the revenue for producers.

Furthermore, the government should take necessary step to uplift biogas utilization through existing programs and facilities and improve harmonization and communication between public and private sectors. The education and science-related ministry should allocate special research fund to encourage researcher for further development of biogas technology. Combined efforts of these activities will speed up the dissemination of biogas technology from rural to urban areas, reduce the dependency on fossil fuel, strengthen the economy, meet our renewable energy targets, reduce the methane emissions, and advance to make a green environment.

The proper realization of the potentiality of biogas market will attract massive investment from project developers, food manufacturers, municipal authorities, livestock producers, energy entrepreneurs, and private waste collectors. A collective line of action of the government, policy-makers, investors, and researchers is urgent to overcome the barriers of limiting growth of the biogas industry. However, to attract the investor in biogas sector, the government should offer some relaxations in this division, like:

1. Providing nonarable land for free of cost or with minimum cost.
2. Bank loan with flexible conditions.
3. Low cost insurances to guarantee plant performance.
4. Tax-free capital gains in an Enterprise Investment Scheme for biogas plants.
5. Dividends also should be tax-free.
6. To avoid waste and sewage fees, chain supermarket, schools, restaurant, hospitals, etc. need to support for installing anaerobic digesters.
7. Support to build a biogas market and create linkages to farmers through promotion and marketing and networking through women associations, local authorities, and NGOs/associations.
8. Media/advertising support to popularize the technology.

7 Conclusion

Biogas technology is molded by a set of sociotechnical features. Though most of the policies related to biogas technology are biased toward energy production; however, it is highly influenced by environmental and agricultural factors. Unfortunately, when any biogas project is designed, agricultural and environmental policies are integrated very poorly with that. Therefore, to meet the future energy demand and effectively utilize the biogas market opportunities, policy-makers should highlight integrated benefits of this technology. Authors have additional implications for policy-makers to overcome biogas market barriers:

– Integration of socio-territorial factors to rewarding the use of heat and other by-products of biogas digester as well as diversify the biogas uses—not only for electricity.

– Energy policy has to move away from family size digester to large-scale programs or strict regulatory frameworks in which only certain types of biogas providers (e.g., NGOs or foreign investors) are targeted.

– Biogas subsidy policy should be reformed such as biogas technology training for households, and commercial and financial supports for digester maintenance might be considered in future.

– Financing should be relaxed for a market-based biogas model.

– It is necessary to cut off the existing inefficient strategies that are creating obstacles against developing biogas marketing and to diversify types of plants, modes of supply, and uses and destinations of the biogas.

References

Ali G, Nitivattananon V, Abbas S, Sabir M (2012) Green waste to biogas: renewable energy possibilities for Thailand's green markets. Renew Sustain Energy Rev 16:5423–5429. doi:10.1016/j.rser.2012.05.021

Amigun B, Blottnitz HV (2010) Capacity-cost and location-cost analyses for biogas plants in Africa. Resour Conserv Recycl 55:63–73. doi:10.1016/j.resconrec.2010.07.004

Brian CM, Christopher SG, Tibor V (2014) Biogas in the United States: an assessment of market potential in a carbon-constrained future NI R 14-02 Durham. Duke University, Durham, NC

Brown VJ (2006) Biogas: a bright idea for Africa. Environ Health Perspect 114:A300

Buysman E, Mol APJ (2013) Market-based biogas sector development in least developed countries—the case of Cambodia. Energy Policy 63:44–51. doi:10.1016/j.enpol.2013.05.071

Fallde M, Eklund M (2015) Towards a sustainable socio-technical system of biogas for transport: the case of the city of Linköping in Sweden. J Clean Prod 98:17–28. doi:10.1016/j.jclepro.2014.05.089

Findeisen C (2014) Biogas as a key in future energy systems—trends in Germany and opportunities in developing and emerging countries. Paper presented at the 3rd Biogas Asia Pacific Forum, Kuala Lampur, Malaysia

Ghimire PC (2005) Technical study of biogas plants installed in Bangladesh. Development Partners, Dhaka

Hosseini SE, Wahid MA (2013) Feasibility study of biogas production and utilization as a source of renewable energy in Malaysia. Renew Sustain Energy Rev 19:454–462. doi:10.1016/j.rser.2012.11.008

Iqbal SA, Rahaman S, Rahman M, Yousuf A (2014) Anaerobic digestion of kitchen waste to produce biogas. Proc Eng 90:657–662. doi:1016/j.proeng.2014.11.787

Jacobsson S, Johnson A (2000) The diffusion of renewable energy technology: an analytical framework and key issues for research. Energy Policy 28:625–640. doi:10.1016/S0301-4215(00)00041-0

Lau LC, Tan KT, Lee KT, Mohamed AR (2009) A comparative study on the energy policies in Japan and Malaysia in fulfilling their nations' obligations towards the Kyoto Protocol. Energy Policy 37:4771–4778. doi:10.1016/j.enpol.2009.06.034

Lohan SK, Dixit J, Kumar R, Pandey Y, Khan J, Ishaq M, Modasir S, Kumar D (2015) Biogas: a boon for sustainable energy development in India's cold climate. Renew Sustain Energy Rev 43:95–101. doi:10.1016/j.rser.2014.11.028

Maghanaki MM, Ghobadian B, Najafi G, Galogah RJ (2013) Potential of biogas production in Iran. Renew Sustain Energy Rev 28:702–714. doi:10.1016/j.rser.2013.08.021

MNRE (2011) Ministry of New and Renewable Energy, Strategic plan for new and renewable energy sector for the period 2011–17

Mwirigi J, Balana BB, Mugisha J, Walekhwa P, Melamu R, Nakami S, Makenzi P (2014) Socio-economic hurdles to widespread adoption of small-scale biogas digesters in Sub-Saharan Africa: a review. Biomass Bioenergy 70:17–25. doi:10.1016/j.biombioe.2014.02.018

Olsson L, Fallde M (2015) Waste (d) potential: a socio-technical analysis of biogas production and use in Sweden. J Clean Prod 98:107–115. doi:10.1016/j.jclepro.2014.02.015

Parawira W (2009) Biogas technology in sub-Saharan Africa: status, prospects and constraints. Rev Environ Sci Biotechnol 8:187–200. doi:10.1007/s11157-009-9148-0

Pike R (2012) Methane recovery and utilization in landfills and anaerobic digesters: municipal solid waste, agricultural, industrial, and wastewater market analysis and forecasts. http://www.navigantresearch.com/research/renewable-biogas

Raninger B, Li R, Bidlingmaier W (2006) Bioorganic municipal waste resource recovery and utilisation project in Shenyang-RRU-BMW summary report. In: Shenyang: Proceedings of the 1st ICEEE Int Conference on bioorganic municipal waste and landfill management Shenyang, PR China, pp 1–120

Rosenstiel DP, Bünger U, Schmidt PR, Weindorf W, Wurster R, Zerhusen J (2014) LNG in Germany: Liquefied natural gas and renewable methane in heavy-duty road transport. Berlin, Germany

Street P, Miles I (1996) Transition to alternative energy supply technologies: the case of windpower. Energy Policy 24:413–425. doi:10.1016/0301-4215(96)00013-4

Tsoutsos TD, Stamboulis YA (2005) The sustainable diffusion of renewable energy technologies as an example of an innovation-focused policy. Technovation 25:753–761. doi:10.1016/j.technovation.2003.12.003

Yoshizaki T, Shirai Y, Hassan MA, Baharuddin AS, Abdullah NMR, Sulaiman A, Busu Z (2012) Economic analysis of biogas and compost projects in a palm oil mill with clean development mechanism in Malaysia. Env Dev Sustain 14:1065–1079. doi:10.1007/s10668-012-9371-7

Yousuf A, Iqbal SA, Sarker NC, Hasan MN, Sarker MSH (2014) Optimization and fabrication of a portable biogas reactor. J Chem Eng 27:36–40. doi:10.3329/jce.v27i2.17799

Biomass Gasification

Samson Mekbib Atnaw, Shaharin Anwar Sulaiman, and Suzana Yusup

Abstract This chapter discusses the state of the art in biomass gasification studies. It initially gives a brief account on the energy potential and characterization of different biomass fuels. A review of the mechanisms of biomass gasification process and influence of major operating conditions on syngas composition and heating value (HV) was presented. Consideration of syngas quality requirements for different downstream applications and the means of achieving the same through optimum operation were highlighted. The theoretical studies of gasification process mainly focused on prediction of syngas composition and investigating influence of various operating conditions on process output. The equilibrium modeling assumes conditions of the ideal, well-stirred reactor with sufficiently long residence time to allow the reactions to reach equilibrium. Kinetic models present detailed information on the residence time and kinetic limitations; however, equilibrium models were widely used as a valuable tool in predicting the thermodynamic limits of chemical processes.

Keywords Syngas • Biomass • Gasification

1 Introduction

Biomass is a solid form of fuel with a complex mixture of organic compounds and polymers. Common biomass sources range from woody and herbaceous plants, forest residues, wood wastes generated from furniture and carpentry plants,

S.M. Atnaw (✉)
Energy and Environmental Engineering Technology Program, Faculty of Engineering Technology, Universiti Malaysia Pahang, 26300 Gambang, Kuantan, Malaysia
e-mail: mekbibsams@ump.edu.my

S.A. Sulaiman
Department of Mechanical Engineering, Universiti Teknologi PETRONAS, 31750 Tronoh, Perak, Malaysia

S. Yusup
Department of Chemical Engineering, Universiti Teknologi PETRONAS, 31750 Tronoh, Perak, Malaysia

L. Singh, V.C. Kalia (eds.), *Waste Biomass Management – A Holistic Approach*,
DOI 10.1007/978-3-319-49595-8_8

agricultural residues and by-products, animal waste, food processing waste, and more recently organic components from municipal solid waste (MSW). In addition, industrial wastes like plastics and rubber are currently being used as biomass waste (Larsen et al. 2006; Mitta et al. 2006; Ponzio et al. 2006; Jarana et al. 2008; Mastellone et al. 2010; Donatelli et al. 2012). Biomass resources from industrial and energy crops (such as tires, oil palm wastes, sugarcane, etc.) are of higher advantage because of their abundant availability and sustainability. As biomass includes a vast scope of various solid organic wastes, their physical and chemical composition and characteristics vary widely (Hanaoka et al. 2005). Biomass materials are renewable with a sustainable generation in shorter span of time.

Biomass materials are highly reactive compared to conventional solid fuels like coal. More than 80 % by mass of dry biomass materials is composed of volatile matter (VM) compared to only 20 % for that of coal (Atnaw et al. 2013). This higher volatile matter and oxygen composition of biomass allows a large portion of the biomass to undergo rapid devolatilized at relatively lower temperature. More interestingly the biomass char left after devolatilization is mostly composed of pure carbon which is highly reactive allowing for further gasification processes to take place. Furthermore, the lower ash and sulfur content of biomass materials as compared to fossil fuels like coal makes them more environment-friendly sources of energy. Though the relative compositions as well as physical and chemical characteristics of different biomass materials vary widely, in general it is deem to note that biomass is the only renewable energy source, which contains hydrocarbon compounds that can substitute existing fossil fuels. Therefore, the use of such renewable energy sources which does not contribute for environmental deterioration is in accordance with the current global energy policy reforms.

Energy fuels derived from biomass conversion are gaining more attention due to the current global energy crisis and the need for energy security. Recent scientific studies unarguably ascertained that environmental problems like global warming are not only occurring but also increasing at an alarming rate. In addition, the scarcity and high prices of fossil fuels like coal and crude oil and their startling depletion rate are the major reasons for expanding the search for renewable and sustainable energy sources such as biomass. Further to being the only renewable form of energy-producing carbon-containing gaseous and liquid fuels, the utilization of biomass materials as energy source has the following advantages (McKendry 2002a):

1. Widespread availability locally and globally at low or no cost
2. Decreases reliance on imported and expensive energy sources like crude oil
3. No net contribution to greenhouse gases like CO_2, as the CO_2 produced during biomass conversion will be recycled as a carbon source for growth of plantation as long as sustainability of utilization and generation is kept in balance
4. High volatile matter composition and resulting higher reactivity
5. Reducing disposal problems of waste materials

Biomass is a fairly distributed resource which can effectively support medium-scale, decentralized energy systems with only moderate transportation distance from the feed to the conversion plants.

2 General Characterization of Biomass Materials

Biomass materials have initial moisture content usually reported to be as high as 70 % on mass basis. Therefore, it is customary to compare various properties of biomass materials at a given moisture content or more conveniently on dry basis. The major characteristics of biomass materials that play a significant role in their utilization as energy fuel include initial moisture content, bulk density, proportion of fixed carbon (FC) and volatiles, ash content, alkali metal (such as K, Cl) content, and heating value of the solid fuel. Such physical and chemical composition and properties which play a major role for utilization of biomass materials for energy use are known to widely vary with the specific biomass material. Knowledge of the structure, nature, and properties of the specific biomass materials is a crucial point in determining their potential and feasibility for energy utilization. Proximate and ultimate analysis of biomass materials helps to determine the chemical and physical properties of the solid fuel, respectively.

2.1 *Proximate Analysis*

Proximate analysis test basically determines the relative percentage of volatile matter (VM), fixed carbon (FC), and ash in the biomass material. Shown in Table 1 is a comparison of proximate analysis results of different biomass materials and that of coal (Kelly Yong et al. 2007; Saidur et al. 2011). Higher ash content of biomass obviously decreases the energy content of the fuel as the ash components do not have heating value. Typically woody biomass materials have a lower ash content of about 1 %, while the ash contents of different crop residues might range as high as 20 %. High ash content of solid fuels causes problems of ash slagging in gasifiers working at high temperature. As shown in Table 1, the ash contents of woody biomass types (beechwood and sawdust) were lower, while that of rice straw and coal were high. Studies showed that biomass materials with ash content lower than 6 % are categorized as non-slagging fuels with no serious ash slagging problems during thermochemical conversion. Severe slagging is much common for biomass fuels with ash content higher than 10 %, and ash content of less than 6 % is generally recommended for downdraft gasification. The ash contents of OPF and other oil palm biomass derivatives (except that of mesocarp fibers) were shown to be lower than 5 %, indicating their suitability as a gasification fuel.

The volatile matter (VM) content of a solid fuel can be defined as the amount of volatile content including moisture content that would be released during heating of the fuel to 950 °C in a period of 7 min (McKendry 2002a). Whereas, the fixed carbon (FC) content of a solid fuel accounts for the mass remaining after all the volatiles are completely released from the fuel excluding the ash content (McKendry 2002a). The proximate analysis results (Table 1) showed that more than 65 % by mass of biomass materials consisted of volatile matter (VM), while coal has a lower VM composition and the highest percentage of fixed carbon.

Table 1 Proximate analysis and heating value of biomass fuels (Saidur et al. 2011)

Biomass type	Proximate analysis (%)		
	Volatile matter	Ash	Fixed carbon
OPF	85.1	3.4	11.5
Oil palm tree trunk	86.7	3.4	9.9
EFB	87.1	4.6	8.4
Mesocarp fibers	84.9	6.1	9.0
Palm kernel shells	83.5	3.0	13.5
Hazelnut shell	69.3	1.4	28.3
Switch grass	76.7	9.0	14.3
Wheat straw	75.2	7.0	17.7
Rice straw	65.5	18.7	15.8
Beechwood	82.5	0.5	17.0
Sawdust	82.2	2.8	15.0
Coal	33.9	10.3	55.8

The volatile matter (VM) content of solid fuels is very important in determining the reactivity of solid fuels, as it represents the amount of organic hydrocarbon concentration of the biomass material and it indicates the ease with which the biomass could be ignited and undergo thermal conversion (McKendry 2002a). As shown in Table 1, the volatile matter contents oil palm-derived fuels is higher than all other biomass types including woody biomass fuels. In addition, the fixed carbon content of the fronds and trunks of oil palm fronds, EFBs, and fibers from oil palm biomass derivatives are shown to be lower as compared to other biomass materials like wheat straw, beechwood, and sawdust. Studies showed that the fixed carbon content of solid fuels mainly contributes to the unconverted carbon in thermochemical conversion (McKendry 2002a; Hanaoka et al. 2005). This showed that the lower fixed carbon of OPF and the other oil palm biomass fuels imply their high potential as a gasification fuel.

2.2 *Ultimate Analysis*

The ultimate analysis which involves advanced chemical analysis techniques is a useful method for determining element-based chemical composition (C, H, N, O, S, and Cl) of solid fuels. The ultimate analysis results for selected biomass fuels are shown in Table 2 (Kelly Yong et al. 2007; Saidur et al. 2011).

2.3 *Heating Value*

The heating value (HV) of a solid fuel which is measured using a bomb calorimeter indicates the amount of heating value released per unit mass of the fuel during

Table 2 Ultimate analysis of biomass fuels (Kelly Yong et al. 2007; Saidur et al. 2011)

Type of biomass	C	H	N	O	S
OPF	42.4	5.80	3.60	48.20	–
Oil palm tree trunk	41.9	6.00	3.80	48.20	–
EFB	48.8	7.33	0.00	36.30	0.7
Mesocarp fibers	50.3	7.07	0.42	36.28	–
Palm kernel shells	53.8	7.20	0.00	36.30	–
Hazelnut shell	50.8	5.60	1.00	41.10	0.1
Switch grass	46.7	5.90	0.80	37.40	0.2
Wheat straw	41.8	5.50	0.70	35.50	–
Rice straw	38.5	5.28	0.88	–	–
Coconut shell	51.1	5.60	0.10	43.10	0.1
Beechwood	49.5	6.20	0.40	41.20	–
Sawdust	46.9	5.20	0.10	37.80	0.1
Coal	75.5	5.00	1.20	4.90	3.1

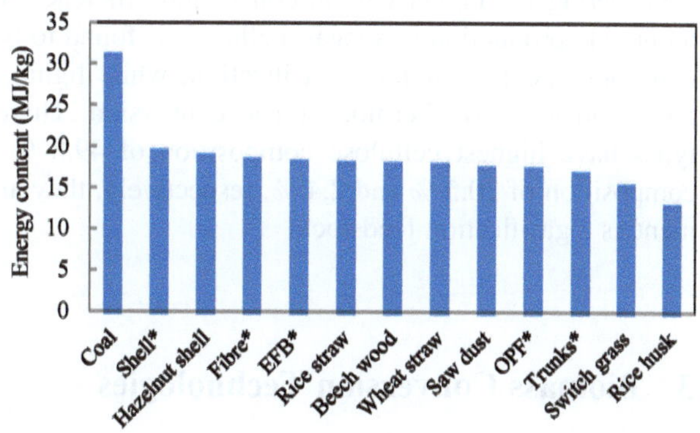

Fig. 1 Heating value of different biomass type (Saidur et al. 2011), * oil palm biomass waste derivatives

complete burning with the supply of air (McKendry 2002a). Shown in Fig. 1 is heating value of a range of biomass materials (Kelly Yong et al. 2007; Saidur et al. 2011).

2.4 *Chemical Composition*

In general, the major constituents of biomass include cellulose, hemicelluloses, lignin, and ash. Usually biomass material contains 40–65 % cellulose, 20–40 % hemicelluloses, and 5–30 % lignin on dry basis (McKendry 2002a). The chemical composition in weight percent of the different oil palm biomass types, reported in the work of Mohammed et al. (2011a) and Kelly Yong et al. (2007), is summarized in Table 3, in comparison with literature values for other biomass types. It can be seen that (Table 3) oil palm biomass derivatives have the highest cellulose composition and the lowest lignin percentage compared with other woody biomass

Table 3 Chemical compositions of biomass materials (Kelly Yong et al. 2007; Mohammed et al. 2011a; Saidur et al. 2011)

Component	OPF[a]	EFB[a]	Shell[a]	Fiber[a]	Trunk[a]	Hardwood	Wood Bark	Beach Wood
Cellulose	49.8	38.3	20.8	34.5	37.1	45.8	24.8	45.8
Hemicellulose	40.4	35.3	22.7	31.8	31.8	31.3	29.8	31.8
Lignin	20.5	22.1	50.7	25.7	22.3	21.7	43.8	21.9
Ash	2.4	1.6	1.0	3.5	4.3	2.7	1.6	0.4

[a]Oil palm biomass

materials. A study carried out by Hanaoka et al. (2005), on effect of woody biomass with air–water gasification, reported that the carbon conversions in cellulose, hemicellulose, and lignin are 97.9 %, 92.2 %, and 52.8 %, respectively. Furthermore, syngas composition in cellulose was reported to have higher carbon monoxide composition (35.5 % mole), compared to that of hemicellulose and lignin. Considering the higher carbon conversion efficiency and resulting higher carbon monoxide production in syngas, cellulose is found to be the most favorable chemical component in biomass gasification, while lignin components show inferior quality in terms of thermochemical conversion. Therefore, as oil palm biomass types have highest cellulose composition of 49.8 % and lower lignin and ash composition of 20.5 % and 2.4 %, respectively, they are found to be more convenient as a gasification feedstock.

3 Biomass Conversion Technologies

Conversion of solid biomass to intermediate and final fuel products that can be used as a direct fuel or used in further synthesizing for production of useful chemicals is commonly used. The major biomass conversion technologies for energy use include thermochemical conversion processes and biochemical conversion. Thermochemical conversion has the highest throughputs compared to other biomass conversion processes. Unlike biochemical conversion process which acts on only part of the biomass, specifically cellulose, thermochemical conversion operates on almost all components of the biomass material. Thermochemical conversion is further classified into pyrolysis, gasification, and liquefaction. There is a limited use in the application of biomass pyrolysis due to difficulty in downstream processing (Wang et al. 2008). Hence, gasification has a higher advantage over pyrolysis and liquefaction thermal conversion methods, as the pyrolysis liquid produced is needed to be further processed, while on the other hand syngas from gasification could readily be used with little or minor cleaning.

Comparison of the various thermochemical conversion processes in terms of equivalence ratio (ER), products, and end-use application is summarized in Table 4. Generation of syngas of higher energy value and better homogeneity that could

Table 4 Comparison of various thermochemical conversion processes

ER	Process	Reaction	Products	Applications
0	Pyrolysis	Endothermic	Liquid hydrocarbons	Chemical energy
$0 < ER < 1$	Gasification	Endothermic/ exothermic	Product gas (CO, H_2)	Chemical and sensible energy
$ER > 1$	Combustion	Exothermic	Heat	Sensible energy

Table 5 Products of thermochemical conversion and application

Process	Product	Application
Pyrolysis	Chemical	Slurry fuel reductant
	Bio-oil	Chemical
		Ammonia
	Syngas	Chemicals
		Methane
		Ammonia
		Electricity
Liquefaction	Bio-oil	Chemical
		Ammonia
Gasification	Syngas	Chemicals
		Methane
		Ammonia
		Electricity
Combustion	Heat	Electricity

more easily be stored and transported with multifaceted applications is possible through gasification of lower-quality solid fuels. In addition to being used as gaseous fuel, the syngas produced from gasification could also be used as a raw material for producing various useful chemical compounds via synthesizing. This versatility of syngas is a major advantage sought for wide range of industrial applications.

Schematic representation of the major products and application of fuel products from thermochemical conversion are shown in Table 5. Some of the major advantages obtained via conversion of the solid fuel to syngas through gasification instead of direct burning include:

1. Production of a homogeneous fuel gas from a highly nonhomogeneous solid fuel
2. Efficient and controlled heat production
3. Improved power production efficiency
4. The use of syngas as a raw material for synthesizing other chemicals

4 Overview of Gasification Technology

The main purpose of gasifier reactors is the thermochemical conversion of solid biomass fuel into a secondary fuel in a gaseous state. The synthetic fuel gas produced through gasification is often referred to as syngas, wood gas, or producer gas. A gasifier unit is basically a cylindrical vessel lined with a refractory material for heat insulation. A simpler description of the mechanism of the gasification process is production of further combustible syngas through partial combustion of part of the feed. Unlike the incombustible gaseous products from complete combustion like nitrogen, water vapor, carbon monoxide, sulfur, etc., the partial combustion of hydrocarbon content of fuel during gasification produces combustible gases like carbon monoxide (CO), hydrogen (H_2), and trace amount of methane (CH_4). Hence, the process of gasification converts the solid biomass waste of no value or negative value to useful carbon- and hydrogen-rich fuel gas which is more suitable to handle and utilize in various thermal and energy applications. Unlike the highly heterogeneous solid fuel, the syngas produced through gasification could be used for running IC engines and turbines or could be used as a source for synthesizing various useful chemical compounds (Demirbas 2009).

The various oxidizing agents used for partial combustion of biomass during gasification include air, pure oxygen, steam, CO_2, or a combination of these (Wang et al. 2008). The use of atmospheric air as gasifying agent is widely used though it results in lower heating value of syngas (3–6 MJ/m^3), due to the higher amount of nitrogen in air that will dilute the syngas produced. The use of pure oxygen and steam as gasification agent is commonly reported to results in increased heating value of syngas (10–15 MJ/Nm^3), but will lead to higher cost in terms of oxygen and steam production. The use of CO_2 as a gasification agent is reported to better facilitate the reduction of CO_2 and H_2O during the gasification process, hence improving the heating value of syngas. However, the use of CO_2 as gasification agent requires external heating of the reactor incurring additional investment cost (Wang et al. 2008).

4.1 *Types of Reactors*

Gasification processes could be carried out at ambient pressure or at increased pressure using various reactor designs. The major types of reactor designs for gasification include the fixed bed systems and fluidized bed gasifiers. Fixed bed gasifiers are generally described by their feature, where the gas stream passes through a bed packed with particles. Fuel will generally be fed through an opening at the top, while gasification agent will usually be supplied at the grate level. Fixed bed gasifiers are also sometimes referred to as moving bed reactors considering the movement of the fuel bulk filling inside the reactor. Yet in moving bed gasifiers, the feed particles are packed together and could be assumed not to detach from each

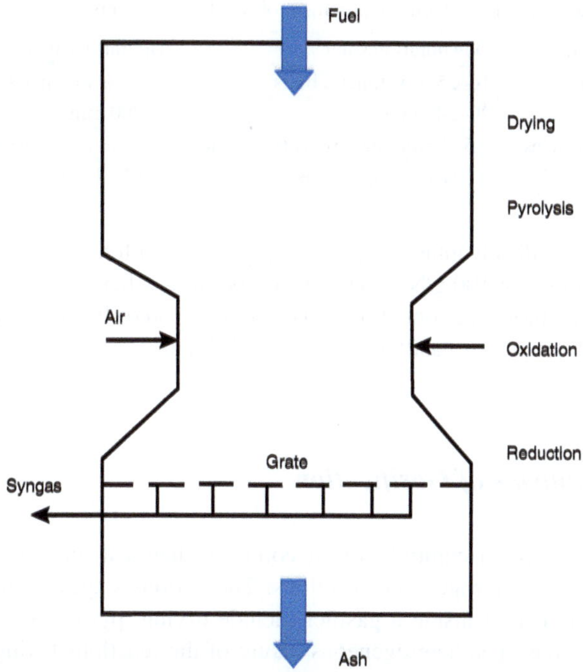

Fig. 2 Schematics of throated downdraft gasifiers

other while the fuel bed (bulk filling) moves down to the grate as the reaction progress. Depending on the direction of flow of gasification medium and the bulk filling inside the reactor, fixed bed gasifiers are further classified into cocurrent (downdraft) and countercurrent (updraft) gasifiers (McKendry 2002a; Mitta et al. 2006; Mohammed et al. 2011a). Fixed bed gasifiers generally have the advantage of simplicity in design and fuel quality requirements, as well as possibility of application in low- and medium-scale power generation.

In downdraft gasifiers, the movement of gasification medium and the sinking flow of the bulk filling take place in the same direction (cocurrently). In addition, the syngas produced in downdraft gasifiers typically passes through the glowing bed of fuel at the grate before it eventually leaves the gasifier as shown in the schematics in Fig. 2. This is a huge advantage in terms of cracking of tar condensates as they pass through the glowing fuel bed at high temperature, resulting in production of relatively cleaner syngas of tar content lower than 1 g/Nm^3. Whereas in the case of updraft gasifiers, the flow of the gasification medium and the sinking flow of the fuel bed follow opposite directions in a countercurrent manner. In addition to syngas, the products of decomposition released in the pyrolysis zone and the steam released in the drying zone will be discharged out of updraft reactor near the top of the gasifier. As a result, the tar contaminants produced will be discharged with syngas without passing through the high-temperature oxidation zone, and hence they are not suitably split up or oxidized. As a result updraft

Table 6 Comparison of fuel quality requirement of fixed bed gasifiers

Fuel characteristics	Downdraft gasifier	Updraft gasifier
Moisture content	10–25 %wet, mass basis	<40 % wet, mass, wet
Grain size	20–200 mm	5–200 mm
Share of fine particles	<5 % wet, mass basis (≤ 5 mm)	<30 % wet, mass basis (≤5 mm)
Ash content	<6 % dry, mass basis	<6 % dry, mass basis

gasifiers have the disadvantage of producing raw gas with higher tar concentration usually in the range of 100 g/Nm^3. However, updraft gasifiers have the advantage of tolerating higher moisture content fuel. The comparison of fuel quality requirement of updraft and downdraft gasifiers is given in Table 6.

4.2 *Mechanisms of Gasification*

Gasification involves a combination of complex chemical reactions that include homogeneous and heterogeneous reactions. The various stages of the gasification process of a solid fuel inside a gasifier include drying, pyrolysis, oxidation, and reduction reactions. The heterogeneous nature of the reactions taking place as the solid char interacts with gaseous volatile components involves complicated heat and mass transfer phenomena as the feedstock undergoes drying, pyrolysis, oxidation, and reduction reactions. Each of these stages of the process displays an array of physical and chemical changes, some of them taking place simultaneously. For example, considering gasification of a single feedstock particle, as gasification takes place, the heat transfer advances from the particle surface to its core causing drying and pyrolysis to take place inside the core, while the already charred surface of the particle could simultaneously be undergoing oxidation and reduction reactions. Such complicated reaction paths and overlap of the different gasification reactions in time create a major problem in understanding the detail reaction mechanism of gasification. Hence, for an easier investigation of the process, the different physical and chemical reactions taking place inside fixed bed gasifiers are theoretically classified into the four zones: drying, pyrolysis, oxidation, and reduction reaction zones.

4.2.1 Drying Zone

The initial moisture content of as produced biomass could be as high as 70 %; however, the moisture content of the feedstock used for gasification is typically in the range of 5–35 %. The drying zone of the gasifier is usually at lower temperature in the range of 70–200 °C where the moisture content of the biomass will be reduced or totally removed and released as vapor (Dogru et al. 2002).

4.2.2 Pyrolysis Zone

The second stage in the gasification process is pyrolysis, which is responsible for the release of volatile components of fuel. Pyrolysis is basically thermal breakdown of the fuel with no or limited oxygen. In addition to the volatile organic compounds and gaseous products released during pyrolysis, a solid char with increased porosity and resulting higher surface area will be produced. As the main oxidation and reduction processes that produce the major permanent gas products make use of these intermediate products of char and volatiles, pyrolysis is usually referred as a precursor to the gasification process. Therefore, it is crucial to understand the pyrolysis behavior of the biomass material and its relation with reactor temperature, heating rate, size of particles, and residence time. The temperature of the pyrolysis zone is typically reported to be in the range of 350–500 °C in literature (Dogru et al. 2002).

4.2.3 Oxidation Zone

Inside the oxidation zone of gasification partial combustion of the fuel is done with a limited air/oxygen supply. This partial combustion of part of the solid carbon and hydrogen generates the heat required for the endothermic reactions taking place in the pyrolysis and reduction zones. To make sure only partial oxidation of the char, an amount of air much lower than the stoichiometric amount needed for complete combustion will be used. Usually equivalence ratio in the range of 0.2–0.4 is reported in literature. In addition to the combustion of part of the solid carbon to CO_2, CO will also be formed due to the substoichiometric presence of air/oxygen. Due to the exothermic oxidation reaction, the oxidation zone temperature of downdraft gasifiers was typically reported to reach up to 1200 °C (Dogru et al. 2002; Hsi et al. 2008).

4.2.4 Reduction Zone

Reduction of solid carbon (char) inside the reduction zone produces a number of permanent gases like CO, H_2, and CH_4 in the presence of limited oxygen. Although the reduction zone is in close proximity with the oxidation zone, lower reduction zone temperature values are expected as a result of endothermic reactions of the reduction zone. The syngas leaving the reduction zone is reported to have a temperature of 200–300 °C (Dogru et al. 2002).

4.3 Chemistry of Biomass Gasification

The chemical composition is known to vary across the different parts of the solid particle. However, the relative atomic composition for large samples could be expressed using empirical relation, CH_xO_y, based on ultimate analysis results (Lee et al. 2007; Detournay et al. 2011). With the solid biomass represented with such empirical formula, the different chemical reactions taking place in the different zones of the gasifier could be summarized as follows:

Pyrolysis reaction

$$CH_xO_y + \text{Heat}\ (300 - 400^{\circ}C) \rightarrow \text{gases}\ (CO + CO_2 + H_2 + \text{light HC's}) + \text{pyrolyzed vapor} + \text{char} \quad (1)$$

Oxidation reactions

$$C + \tfrac{1}{2} O_2 \leftrightarrow CO + 123.1\ KJ/mol \quad (2)$$

$$C + O_2 \leftrightarrow CO_2 + 393.8\ KJ/mol \quad (3)$$

$$CO + \tfrac{1}{2} O_2 \leftrightarrow CO_2 + 283.9\ KJ/mol \quad (4)$$

Reduction reactions

$$C + CO_2 + 172.6\ KJ/mol \leftrightarrow 2CO \quad (5)$$

$$C + H_2O + 131.4\ KJ/mol \leftrightarrow CO + H_2 \quad (6)$$

$$C + 2H_2 \leftrightarrow CH_4 + 74.9\ KJ/mol \quad (7)$$

$$CO + H_2O \leftrightarrow CO_2 + H_2 + 41.2\ KJ/mol \quad (8)$$

$$CH_4 + H_2O + 206.3\ KJ/mol \leftrightarrow CO + 3H_2 \quad (9)$$

$$\text{Pyrolyzed vapors} + (\text{Heat and high residence time}) \rightarrow \text{gases}\ (CO, CO_2, H_2, \text{light hydrocarbons}) \quad (10)$$

The pyrolysis reaction shown in Eq. (1) mostly takes place within temperature range of 280–500 °C, and it produces gaseous products, volatile tar vapor, and solid char. The oxidation zone reactions in Eqs. (2)–(4) are exothermic reactions producing the heat required for endothermic reactions in the pyrolysis and reduction zones. The char partial oxidation reactions (Eqs. 2 and 3) are known to be the slowest and are the rate controlling factors of the gasification process. The principal reduction reactions which produce the most part of CO and H_2 concentration in the syngas are the Boudouard and water–gas reactions shown in Eqs. (5) and (6), respectively (Ratnadhariya and Channiwala 2009).

All the reduction zone reactions are endothermic in nature with the exception of the methanation and water–gas shift reactions of Eqs. (7) and (8), respectively. This endothermic nature of the reduction reactions, which take place at temperature higher than 900 °C, has a cooling effect on the final gas product. Most of the gasification reactions are heterogeneous reactions that involve the interaction of solid and gaseous states with the exception of the reactions in Eqs. 8 and 9. For downdraft gasification, thermal cracking of the tar and volatile vapor produced during pyrolysis to gaseous products (Eq. 10) will take place as the gas and condensate pass through the oxidation zone at temperature higher than 900 °C.

5 Studies on Biomass Gasification

5.1 *Experimental Studies*

A number of studies were reported in literature to study gasification of various solid fuels. Studies conducted in earlier times are found to mainly focus on gasification of coal and woody biomass feedstocks: coal (Guo et al. 1997; Li and Suzuki 2009), woody biomass (Baker et al. 1984; Mathieu and Dubuisson 2002; Chen et al. 2003; Hsi et al. 2008; Kumar et al. 2008), furniture wood (Chen et al. 2003; Lv et al. 2004), pine sawdust, and pine bark (Lv et al. 2004; Nikoo and Mahinpey 2008; Tinaut et al. 2008). In recent years, gasification of various agricultural residue biomass types had been investigated: rice husk (Bhat et al. 2001), hazelnut shell (Midilli and Dogru 2001; Dogru et al. 2002), peanut shell (Hanping et al. 2008), vine shoot (Ganan et al. 2006), and coconut shell (Bhattacharya and Hla 2001). The search for biomass energy sources has nowadays expanded to gasification and co-gasification of industrial and household disposals like tire (Larsen et al. 2006; Mitta et al. 2006; Donatelli et al. 2012), plastics (Aznar et al. 2006; Ponzio et al. 2006; Mastellone et al. 2010), and municipal solid waste (Dong et al. 2002; Folgueras et al. 2003). Though there are a number of studies on gasification of various biomass fuels, very little number of studies existed on plantation products like oil palm biomass waste, which is found abundantly in oil palm-producing countries like Malaysia. Even these few number of studies were focused mainly on pyrolysis studies of palm oil mill by-products like EFBs, shells, and fibers (Guo and Lua 2001; Lee et al. 2007; Akhtar et al. 2010), while there is no study and recorded data on gasification of OPF biomass. Only very recently in 2011, studies were reported on fluidized bed gasification of EFB (Lahijani and Zainal 2011; Mohammed et al. 2011b). The studies investigated that the performance of fluidized bed gasification of EFB in terms of calorific value, conversion of carbon, and the cold gas efficiency and the results showed that EFB gave results comparable with that of sawdust biomass.

The experimental studies in literature were carried out using different types of reactors. The studies on fixed bed gasification include those carried out in downdraft gasifiers and updraft gasifiers (Midilli and Dogru 2001; Dogru et al. 2002; Hsi et al. 2008). A number of studies were also conducted using fluidized bed gasifiers (Dong et al. 2002; de Jong et al. 2003; Hanping et al. 2008; Hannula and Kurkela 2012). The studies on downdraft gasification of biomass showed that downdraft gasifiers are simple in terms of design and fabrication and also have the advantage of medium- and low-scale applications as well as production of syngas with low tar concentration (McKendry 2002b).

The common interest in most of the gasification studies in literature was on syngas composition and the resulting heating values of the syngas. In addition, most of the researches focused on how to improve the gasification efficiency through control of operating parameters and feedstock characteristics. The operating conditions and feedstock characteristics considered in various literature include reactor temperature (Ruggiero and Manfrida 1999; Mathieu and Dubuisson 2002; Jand et al. 2006; Nikoo and Mahinpey 2008), pressure (Ruggiero and Manfrida 1999; Mathieu and Dubuisson 2002), equivalence ratio, ER (Mathieu and Dubuisson 2002; de Jong et al. 2003; Jand et al. 2006; Ratnadhariya and Channiwala 2009), particle size (Nikoo and Mahinpey 2008; Tinaut et al. 2008), air preheat temperature (Mathieu and Dubuisson 2002; Tinaut et al. 2008), fuel moisture content (Ratnadhariya and Channiwala 2009), and steam to biomass ratio (Mathieu and Dubuisson 2002; Nikoo and Mahinpey 2008). The results reported on effect of ER showed that it is the major operating condition controlling the rate of reaction, composition, and heating value of syngas as well as overall performance in terms of gasification efficiency.

5.1.1 Influence of Biomass Type

Comparison of experimental results on gasification studies carried out by various researchers on different biomass types showed that the composition of syngas produced and its heating value are significantly dependent on the type of biomass used. The study of Skoulou et al. (2008) on downdraft gasification of olive kernel biomass demonstrated that the amount of char produced from olive kernel biomass was higher than that of olive tree cuttings as expected from the higher fixed carbon content of the kernels (16.39 % dry weight basis fixed carbon of olive tree kernels as compared to only 8.47 % of olive tree cuttings). The olive kernels showed higher concentration of syngas (CO and H_2) as compared to olive tree cuttings which was explained in terms of the high cellulose and hemicellulose composition of the earlier as compared to the latter. In the work of Lahijani and Zainal (2011), comparison of gasification output of EFB and sawdust was experimentally investigated. Higher calorific values of 5.37 and 5.88 MJ/Nm^3, 2.04 and 2.0 Nm^3/kg of rate of production of gas, and gasification efficiency of 72 % and 71 % were reported on empty fruit bunches (EFB) and sawdust, respectively.

The sustainability of gasification of ten different biomass types mostly composed of various wood species and other materials in a circulating fluidized bed gasifier (CFBG) was experimentally investigated by Van der Drift et al. (2001). The study carried out comparison of gasification of the different biomass types at given moisture content for better comparability of data. In addition, the authors suggested that the moisture content needs to be defined as ash-free basis as the ash component of the different fuels widely varies. Syngas of higher heating value (HHV) varying between 4.41 MJ/Nm^3 and 5.13 MJ/Nm^3 was reported for the various fuel materials. The carbon conversion efficiency of the various biomass materials was reported to vary between a relatively close range of 85.2–97.0 %, while the cold gas efficiency was found to vary widely from the lowest of 37 % for inferior-quality woody biomass to highest value of 66 % for willow and park wood biomass materials. Generally, the study concluded that the circulating fluidized bed gasification process was found to be flexible to handle different fuel types so long as the necessary modification of the feeding system was done depending on type of fuel. The study was found to be useful in demonstrating the range of variation of the performance of gasification of various fuels; however, it did not further investigate the optimum operation parameters for gasification of the fuels.

Hanping et al. (2008) reported that lower heating value of syngas from fluidized bed gasification of sawdust was higher than those obtained from peanut shells and wheat straw. This was evidenced from the higher volatile matter content of sawdust as compared to the other biomass types (peanut shells and wheat straw). The study demonstrated that higher volatile matter content of sawdust resulted in higher H_2 and CO content, while that of wheat straw was the lowest. Furthermore, the study by Hanping et al. (2008) reported that the optimum equivalence ratio of gasification varies depending on the O/C ratio of the biomass type, and woody biomass materials like sawdust which have a high O/C ratio ($\geq$0.6) could be gasified at lower equivalence ratio value of 0.15.

Experimental results for gasification of forestry waste (*Pinus*), agricultural waste (grapevine and olive), and industrial waste (sawdust and grape) inside air-blown circulating gasifier were reported in the work of Lapuerta et al. (2008). The study was carried out with the objective of achieving the use of various fuels inside the same gasifier in order to avoid seasonal availability problems of a given fuel type. The results of the study showed that sawdust and agricultural wastes (grapevine and olive prunings) resulted in better gasification output (with high H_2 and CO concentration in the range of 25 %) as compared to the other biomass materials. The study further showed that the K_2O content of the ash from marc of grapes biomass played a positive role on its gasification performance though the specific biomass has the lowest volatile matter content (65.77 %) and relatively high ash content (7.83 %) as compared to the other fuels.

The results from literature showed that the composition and heating value of syngas, as well as the gasification performance, are highly dependent on the type and characteristics of the biomass type. In addition to the gasification output, the optimum operating conditions like temperature and equivalence ratio of gasification were also reported to vary with the biomass type (Lahijani and Zainal 2011). Hence, it is important to investigate gasification behavior of each biomass type.

5.1.2 Influence of Reactor Temperature

Most of the studies reported on biomass gasification have investigated the influence of temperature on gasification performance and showed that the concentration of component gases in syngas and its heating value and yield is highly dependent on temperature. The studies discussed the variation of syngas composition with temperature in terms of the equilibrium conditions of the various gasification reactions (Lahijani and Zainal 2011; Mohammed et al. 2011a). The results of the studies showed that high gas yield was achieved for high gasification temperature above 900 °C (Hsi et al. 2008; Lapuerta et al. 2008; Skoulou et al. 2008; Sharma 2009). A study on olive kernels and olive tree cuttings by Skoulou et al. (2008) reported that the percentage of main syngas components (CO and H_2) increased with temperature, while concentration of CO_2, CH_4, tar, and light hydrocarbons decreased at high temperature for externally heated reactor. The study further demonstrated that syngas with lower heating value (LHV) of 9.41 MJ/Nm^3 and 8.6 MJ/Nm^3 was obtained in the study of the tree cuttings and kernels of olive, respectively, for operation with optimum reactor temperature of 950 °C.

The concentration of CO in syngas was reported in various studies (Skoulou et al. 2008; Lahijani and Zainal 2011; Mohammed et al. 2011b) to increase sharply with rise in temperature. In the study of Lahijani and Zainal (2011), the concentration of CO showed a sharp rise (from 11.5 % to 20 %) for increase in temperature between 650 °C and 1050 °C for gasification of EFB in a fluidized bed gasifier. The concentration of H_2 in the product gas was reported to show only slight increase, while CH_4 and CO_2 concentrations decreased with increase in temperature (Skoulou et al. 2008). Lower oxidation zone temperature in the range of 650–750 °C was reported to give the highest CH_4 content (up to 3.6 %) in product gas and showed a decrease with increase in temperature. The work of Mohammed et al. (2011a) on fluidized bed gasification of EFB biomass showed that maximum concentrations of CO and H_2 (36.36 and 38.02 %, respectively) were obtained for high reactor temperature of 1000 °C. In the work of Dogru et al. (2002), maximum syngas higher heating value (HHV) of 5.15 MJ/Nm^3 and the lowest tar concentration of 0.023 kg/h were reported corresponding to optimum oxidation zone temperature of 1020 °C for downdraft gasification of hazelnut shells.

Zhao et al. (2010) studied gasification of sawdust making use of entrained flow reactor, in terms of four main parameters, namely, calorific value, syngas yield, conversion of carbon, and cold gas efficiency of gasification. The study concluded that concentration of CO decreases with increase in reactor temperature, while concentration of the gases CO_2 and H_2 increases with temperature. The study also reported that the concentration of light hydrocarbons (CH_4 and C_2H_4), heating value, carbon conversion, as well as cold gas efficiency of gasification were found to be maximum corresponding to reaction temperature value of 800 °C and optimal ER of 0.28. Similar trends of variation in relative concentration of the syngas components and its heating value had been reported in other studies (Cao et al. 2006; Hanping et al. 2008; Hsi et al. 2008; Lapuerta et al. 2008). In addition to the influence on the composition and heating value of syngas produced from gasification, the studies showed that high temperature has the advantage of reducing tar concentration in syngas and increasing the gas yield. Mohammed et al. (2011a) reported that gas yield up to 92 wt% (on the basis of total mass of feed) was reached for temperature above 700 °C in a fluidized bed gasifier.

5.1.3 Influence of Moisture Content

Results of various studies showed that excessive moisture content in the feed adversely affected the gasifier performance and heating value of produced syngas as a large part of the input energy will be consumed for moisture evaporation inside the drying zone. The range of initial moisture content of feedstock for downdraft gasification was reported to be in the range of 10–25 % (wet, mass basis), while updraft gasifiers could handle moisture content up to 40 % (wet, mass basis). In the study of Hsi et al. (2008) on downdraft gasification of woody biomass, high concentration of H_2 and CO in syngas was reported to be obtained for 18 % moisture content of the feed. Narvaez et al. (1996) studied fluidized bed gasification of sawdust, and the authors reported that increasing the moisture content of the feed from 10 % to 25 % improved the H_2 content in the product gas from about 5 % to 9 % for a constant reactor temperature of 800 °C.

5.1.4 Influence of Equivalence Ratio

Equivalence ratio is a major parameter to investigate the gasification process as it could represent the combined influence of most other operating conditions like airflow rate, rate of fuel consumption, and total operation time (Zainal et al. 2002). Zainal et al. (2002) studied effect of equivalence ratio (ER) on syngas composition and energy content and reported peak heating value of 5.34 MJ/Nm^3, for ER of 0.388 and corresponding cold gas efficiency as high as 86.8 % for downdraft gasification of furniture wood. In the work of Dogru et al. (2002), higher heating value (HHV) of 5 MJ/Nm^3 with 8–9 Nm^3/h of syngas is reported corresponding to optimum air–fuel ratios (AFR) of 1.44 and 1.47 Nm^3/kg for downdraft gasification of hazelnut shell feedstock. Also the work of Midilli and Dogru (2001) demonstrated that optimum production of H_2 gas is achieved at AFR of 1.44 and 1.52 Nm^3/kg with H_2 yield of 2.4 kg/h for downdraft gasification of hazelnut shell biomass.

A summary of the optimum ER for gasification of different fuels and reactors reported in literature is given in Table 7. These results clearly showed that the

Table 7 Optimum ER of gasification for different biomass fuels

Reactor	Feedstock	Optimum ER	Peak HV (MJ/Nm^3)	Gasification efficiency (%)	References
Downdraft	Furniture wood	0.38	5.34	86.60	Zainal et al. (2002)
Downdraft	Rice husk	0.20–0.55	4.50–6.00	60–70	Natarajan, et al. (1998)
Downdraft	Hazelnut shells	0.28	5.20	na	Dogru et al. (2002)
Fluidized bed	Pine sawdust	0.30	4.5–6.5	na	Gil et al. (1999)

optimum ER of gasification varied with the type of biomass. This showed the need for the study of gasification behavior of each biomass type.

5.2 *Theoretical Studies*

A significant number of theoretical studies of gasification process for different feedstock and reactor types were carried out with the major objective of improving the general understanding of the gasification process. In addition, the modeling and simulation studies facilitated performance prediction of the units. Most of the studies demonstrated the phenomenological modeling and simulation of the gasification process (Ruggiero and Manfrida 1999; Brown et al. 2006), while few studies employ nonmechanistic prediction tools like neural network applications (Guo (B) et al. 2001; Chen et al. 2004). Ratnadhariya and Channiwala (2009) presented a biomass gasifier model that does not involve both equilibrium and kinetic model approaches. It predicted gas composition and temperature profile. The first zone of the model simulates drying and pyrolysis processes combined, with the second zone simulating oxidation zone, and the third zone is the reduction zone. The model has been formulated with (1) chemical reaction equations, (2) conservation of elements, and (3) conservation of energy with assumptions—without the needs for kinetic or equilibrium considerations.

In most of the modeling studies, biomass was represented with the general formula CH_xO_y (Ruggiero and Manfrida 1999; Jand et al. 2006; Nikoo and Mahinpey 2008; Ratnadhariya and Channiwala 2009), where x and y represent H/C and O/C fractions, respectively. A few of the common assumptions made include: char is modeled as pure carbon or graphitic carbon (Ruggiero and Manfrida 1999; Jand et al. 2006; Mitta et al. 2006; Nikoo and Mahinpey 2008; Ratnadhariya and Channiwala 2009), 0 % char carries over (Ratnadhariya and Channiwala 2009), reactions proceed adiabatically (Ruggiero and Manfrida 1999), and particle geometry is considered as spherical (Nikoo and Mahinpey 2008; Tinaut et al. 2008; Ratnadhariya and Channiwala 2009). And the major products of the gasification process are considered to be CO, CH_4, H_2, CO_2, and H_2O (Ruggiero and Manfrida 1999; Mitta et al. 2006; Ratnadhariya and Channiwala 2009). The method of solutions used to determine the composition of syngas used by different authors involves application of equations of:

1. Species conservation (C, O, H, N, S) (Ruggiero and Manfrida 1999; Mitta et al. 2006; Nikoo and Mahinpey 2008; Tinaut et al. 2008; Ratnadhariya and Channiwala 2009)
2. Energy balance (Ruggiero and Manfrida 1999; Tinaut et al. 2008; Ratnadhariya and Channiwala 2009)
3. Minimization of Gibbs free energy (Li et al. 2001; Mathieu and Dubuisson 2002; Mitta et al. 2006)
4. Equilibrium relations (Rashidi 1997; Ruggiero and Manfrida 1999; Li et al. 2001; Brown et al. 2006)

5.2.1 Equilibrium Model of Gasification

The level of complexity of the models developed ranges from simpler ones to more complicated models that use reaction kinetic approaches. Equilibrium models which assume thermodynamic equilibrium and do not consider chemical kinetics and transport rate phenomena are used by a number of authors (Rashidi 1997; Ruggiero and Manfrida 1999; Li et al. 2001; Mathieu and Dubuisson 2002; Brown et al. 2006), to see the effect of operating conditions and feedstock characteristics on output syngas composition. Though the equilibrium model can only estimate the maximum yield which can be achieved ideally, without considering kinetic limitations, they are found to be a valuable tool in predicting thermodynamic limits. Knowledge of the thermodynamic limit of the process is a valuable tool in the design, evaluation, and improvement of the gasification process (Li et al. 2001). Moreover, studies showed that prediction of equilibrium models is in close agreement with experimental results and sufficiently accurate for gasification processes where the operating temperature is sufficiently high (Schuster et al. 2001).

The equilibrium modeling studies could generally be divided into stoichiometric and non-stoichiometric models, where the latter use minimization of the Gibbs free energy to solve the equilibrium problem without considering stoichiometry of the reactions (Li et al. 2001). The two equilibrium modeling approaches are basically similar with the stoichiometric models also known to use the Gibbs free energy data to determine the equilibrium constants of the various reactions (Li et al. 2001). Mathieu and Dubuisson (2002) developed an equilibrium model using Gibbs free energy with Aspen Plus process simulator to model a fluidized bed gasifier that uses wood biomass. Jand et al. (2006) proposed a method to improve the predicting capability of the equilibrium-based calculation, by considering two-stage conversions via pyrolysis and the slower process of converting methane and char. The input parameters were adjusted based on experimental data.

The equilibrium model formulation depends on whether stoichiometric or non-stoichiometric modeling approach was followed. The detailed account of the model formulation for non-stoichiometric equilibrium model that works based on Gibbs free energy is given in the work of Li et al. (2001). Equilibrium models developed by various authors based on stoichiometric modeling approach are usually formulated considering the overall reaction and the main reduction zone reaction (namely, the shift and methane formation reactions). The reaction for atmospheric air gasification of biomass can be represented as follows:

$$\begin{aligned} \mathrm{CH}_x\mathrm{O}_y + \mathrm{wH_2O} + \mathrm{mO_2} + 3.76\ \mathrm{mN_2} \leftrightarrow\ & x_1\mathrm{H_2} + x_2\mathrm{CO} + x_3\mathrm{CO_2} + x_4\mathrm{H_2O} \\ & + x_5\mathrm{CH_4} + 3.76\ \mathrm{mN_2} \end{aligned} \tag{11}$$

where the coefficient w is the water content per kmol of fuel and m stands for amount of oxygen per kmol of wood. The other five factors for the constituent products (x_1, x_2, x_3, x_4, and x_5) represent composition of syngas from gasification. The coefficient w which stands for the amount of water per kmol of fuel could be determined from a simple relation involving the moisture content of fuel and molecular weight of the fuel material calculated from the empirical formula CH_xO_y.

The major reduction reactions of gasification include the combined shift reaction and methane formation reactions given in Eqs. (12) and (13):

$$CO + H_2O \rightarrow CO_2 + H_2 \quad (12)$$

$$C + 2H_2 \rightarrow CH_4 \quad (13)$$

The global gasification reaction involves six unknowns including the five product species coefficients and the coefficient m for oxygen amount in the reactant. Therefore, six equations need to be formulated as follows:

Species balance for carbon, hydrogen, and oxygen gives Eqs. (14), (15), and (16), respectively:

$$1 = x_2 + x_3 + x_5 \quad (14)$$

$$2\mathrm{w} + x = 2x_1 + 2x_4 + 4x_5 \quad (15)$$

where x is the subscript for hydrogen in the empirical formula of the fuel CH_xO_y.

$$\mathrm{w} + \mathrm{y} + 2\mathrm{m} = x_2 + 2x_3 + x_4 \quad (16)$$

where y is the subscript for oxygen in the empirical formula of the fuel CH_xO_y. The constant for the methane formation reaction will be given as:

$$K_1 = \frac{x_5}{x_1^2} \quad (17)$$

And the constant for the shift reaction:

$$K_2 = \frac{x_1 x_3}{x_2 x_4} \quad (18)$$

The Gibbs free energy functions given in Eqs. (19) and (20) below will be used to calculate values of K_1 and K_2.

$$\ln K(T) = \frac{-\Delta G_T}{RT} \quad (19)$$

$$\Delta G_T = \sum_i x_i \Delta \bar{g}^o_{f,T,i} \quad (20)$$

The final equation was formulated considering enthalpy balance of the gasification process after assuming $H^0_{fH_2}$, $H^0_{fN_2}$, and $H^0_{fO_2}$ to be zero at ambient temperature:

$$H^0_{fFuel} + w\left(H^0_{fH_2O(1)} + H_{Vap}\right) = x_2 H^0_{fCO} + x_3 H^0_{fCO_2} + x_4 H^0_{fH_2O(Vap)} + x_5 H^0_{fCH_4}$$
$$+\Delta T(x_1 C_{PH_2} + x_2 C_{PCO} + x_3 C_{PCO_2} + x_4 C_{PH_2O} + x_5 C_{PCH_4} + 3.76mC_{PN_2}) \quad (21)$$

Mathematical computation tools and different algorithms could be used to solve the six unknowns (x_1–x_5 and m), from Eqs. (14)–(18) and (21).

5.2.2 Kinetic Models

Kinetic models of gasification account for the residence time and kinetics of the various homogeneous and heterogeneous reactions taking place inside the gasifier. Kinetic models are mainly used for design of the reactor (Sharma 2008). Tinaut et al. (2008) developed a kinetic model considering mass and energy conservation, energy exchange between the different phases (gas and solid), as well as radiation from solid particles, via the ash segregation (exposed-core model). The kinetic parameters of the gas–gas and solid–gas reactions were adopted from literature. Nikoo and Mahinpey (2008) and de Jong et al. (2003) developed a kinetic model of fluidized bed gasifiers, using Aspen Plus process simulator for pine sawdust and woody biomasses, respectively.

Sharma (2008) considered the fact that conversion of char is influenced by surface reaction kinetics, which is a major parameter in deciding the optimum reduction zone size, and developed a model to study the critical condition of temperature that allows the effective conversion of char to syngas. Moreover, the equilibrium and kinetic reactions and the relations for energy conservation were used to study the influence of char bed dimensions on the syngas, calorific value, conversion efficiency, and power output of a downdraft biomass gasifier.

The major limitation of kinetic modeling of gasification processes is its dependency on a vast number of parameters which are usually unique to each biomass type (according to its cellulose, hemicelluloses, and lignin constituents). The determination of such parameters is very difficult because of the complexity of the gasification process, which makes development of a phenomenological model that tries to imitate the actual reactions in a mechanistic approach very difficult. For example, if we consider pyrolysis, which is considered to be a precursor to gasification process, it involves all three phases interacting with each other (heterogeneous reactions) and complicated heat and mass transfers.

6 Syngas Composition and Heating Value

Syngas is a synthetic gaseous fuel produced usually from atmospheric air gasification of wood and other biomass materials, charcoal, and coal. It usually consists up to 40 % by volume of combustible gas mainly composed of CO, H_2, and trace amount of CH_4. The rest of the syngas composition is generally noncombustible and mainly consists of N_2 up to about 50 %, CO_2, and water vapor (Zainal et al. 2002). As the inert part of the syngas does not contribute to the energy content, the characterization of usefulness of the syngas is limited to the relative composition of CO, H_2, and CH_4 only.

The heating value of syngas produced from atmospheric air gasification of biomass is reported by various researchers to be in the range of 4–6 MJ/Nm^3 and characterized as lower heating value fuel (McKendry 2002b). A typical syngas composition and heating values for atmospheric air gasification of different biomass materials in a downdraft gasifier reported in literature are summarized in Table 8. As shown in Table 8, syngas composition varies widely with biomass type. The comparison of composition of syngas from gasification of woody biomass (of different gasifying agents) with the composition of biogas and natural gas is given in Table 9.

The heating value of syngas is low compared to that of methane with factors of 8–13 (McKendry 2002b). However, syngas requires a much lower amount of air–fuel ratio for stoichiometric combustion compared to natural gas. Higher air–fuel ratio requirement of methane causes significant thinning of the fuel–air mixture, considerably reducing the energy value of the mix. The stoichiometric syngas/air mixture is only 33 % lower than the stoichiometric methane/air mixture. Hence, the advantage of lower air–fuel ratio requirement of syngas allowed it to be suitably used as fuel in IC engines with some or no modifications needed.

Table 8 Typical results from downdraft gasification of different biomass types

Components	Typical syngas composition and heating value				
CO (vol. %)	23.0	18.0	18.0	16.5	16.1
H_2 (Vol %)	15.2	18.0	18.3	16.5	9.6
CH_4 (Vol %)	1.6	na	na	na	0.9
CO_2 (Vol %)	16.4	12.5	12.5	13.0	na
N_2 (Vol %)	42.3	na	na	na	na
LHV [MJ/Nm3]	5.4	4.5	5.8	5.3	3.3
Biomass type	FW	WS	CH	SC	RH
References	Zainal et al. (2002)	Dogru et al. (2002)	Dogru et al. (2002)	Dogru et al. (2002)	Dogru et al. (2002)

na not available, *FW* furniture wood, *WS* wheat straw, *CH* coconut husk, *SC* sugarcane, *RH* rice husk

Table 9 Comparison of composition of syngas, biogas, and natural gas

Components	Wood gas air blown	Biogas	Natural gas
CO	15–25	–	–
CO_2	12–25	15–25	~1.4
H_2	6–20	–	–
H_2O, vapor	2–6	2–6	~0
CH_4	1–5	40–75	96–97

All values in volume fraction

7 Syngas Quality Requirements

Various studies showed that syngas from biomass gasification contains small fractions of contaminants like tar, acids, and particulates (Knoef and Koele 2000). These highly oxidized condensates and particulate contaminants are known to create problems like corrosion and abnormal wear of component parts in end-use equipment like IC engines and turbines (Wang et al. 2008). Hence, a major challenge in gasification technology is to control the formation of tar and other contaminants during the process and to further clean the syngas before being used as fuel to run different equipment. Generally, the requirement for syngas characteristics varies with the specific end-use application intended. For example, applications as turbine fuel and as fuel for IC engines typically require a strict guideline in the cleanliness and energy density of the syngas, while applications for direct burner use are adaptable to lower-quality ranges. Various studies showed that syngas from biomass gasification contains small fractions of contaminants like tar, acids, and particulates (Knoef and Koele 2000). These highly oxidized condensates and particulate contaminants are known to create problems like corrosion and abnormal wear of components parts in end use equipment like IC engines and turbines (Wang et al. 2008). Hence, a major challenge in gasification technology is to control the formation of tar and other contaminants during the process, and to further clean the syngas before being used as fuel to run different equipment. Generally, the requirement for syngas characteristics vary with the specific end use application intended. For example applications as turbine fuel and as fuel for IC engines typically require a strict guideline in the cleanliness and energy density of the syngas, while applications for direct burner use are adaptable to lower quality ranges. The major points of consideration in quality of syngas include less fluctuation in chemical composition of syngas and its resulting energy density and less fluctuation in syngas yield from the gasifier plant. Syngas with proper energy density and consistent gas composition ensures more stable combustion properties of fuel gas (like firing point, flame speeds, heating value, etc.). In addition, concentration of contaminates and impurities like particulates and tars should be below the amount permitted for the application equipment used. A summary of pollutant and contaminant concentration for syngas application in IC engines and gas turbines is shown in Table 10.

Table 10 Syngas quality requirement for IC engine and turbine application

Application	Particulate content [mg/Nm^3]	Tar content [mg/Nm^3]	Alkali content [mg/Nm^3]
IC engine	<50	<50	na
Gas turbine	<30	na	<0.24

na not available

8 Conclusions

Biomass is mainly composed of organic volatile matter which reaches up to 80 % by mass of the dry fuel. This organic volatile matter content of biomass which contains hydrocarbon compounds with properties resembling conventional petroleum-based fuel derivatives makes it a highly reactive form of solid fuel. Studies showed that the physical and chemical characteristics as well as thermo-chemical conversion behavior of biomass materials widely vary with the type of fuel material. Downdraft gasifiers which have temperature stratification with clearly defined four zones of gasification were found to give a better look into the gasification process. Downdraft gasifiers have the advantage of producing relatively cleaner syngas with lower tar load. Moreover, they are found suitable for decentralized and low- and medium-scale power output plants which could be conveniently located near the biomass production site.

There are various modeling and simulation approaches for theoretical study of gasification processes, which include equilibrium models, kinetic models, and kinetic- and equilibrium-free models. The theoretical studies of gasification process mainly focused on prediction of syngas composition and investigating influence of various operating conditions on process output. The equilibrium modeling assumes conditions of the ideal, well-stirred reactor with sufficiently long residence time to allow the reactions to reach equilibrium. Kinetic models present detailed information on the residence time and kinetic limitations; however, equilibrium models were widely used as a valuable tool in predicting the thermodynamic limits of chemical processes. Knowledge of the thermodynamic limits of the gasification process is a valuable tool in the design evaluation and improvement of the gasification process. Studies showed that equilibrium models give sufficiently accurate prediction for processes taking place at higher temperature. In addition, the major limitation of kinetic modeling is its dependency on knowledge of many physical and chemical parameters which are unique to each biomass type.

Acknowledgments The authors wish to thank the support of Fundamental Research Grant Scheme (FRGS) of Malaysian Ministry of Education (MoE).

References

Akhtar J, Kuang SK, Amin NAS (2010) Liquefaction of empty palm fruit bunch (EPFB) in alkaline hot compressed water. Renew Energy 35:1220–1227. doi:10.1016/j.renene.2009.10.003

Atnaw SM, Sulaiman SA, Yusup S (2013) Syngas production from downdraft gasification of oil palm fronds. Energy 61:491–501. doi:10.1016/j.energy.2013.09.039

Aznar MP, Caballero MA, Sancho JA, Francés E (2006) Plastic waste elimination by co-gasification with coal and biomass in fluidized bed with air in pilot plant. Fuel Process Technol 87:409–420. doi:10.1016/j.fuproc.2005.09.006

Baker EG, Mudge LK, Mitchell DH (1984) Oxygen/steam gasification of wood in a fixed-bed gasifier. Ind Eng Chem Proc Des Dev 23:725–728. doi:10.1021/i200027a016

Bhat A, Ram Bheemarasetti JV, Rajeswara Rao T (2001) Kinetics of rice husk char gasification. Energy Convers Manage 42:2061–2069. doi:10.1016/S0196-8904(00)00173-4

Bhattacharya SC, Hla S (2001) A study on a multi-stage hybrid gasifier-engine system. Biomass Bioenergy 21:445–460. doi:10.1016/S0961-9534(01)00048-4

Brown D, Fuchino T, Maréchal F (2006) Solid fuel decomposition modelling for the design of biomass gasification systems. Comp Aided Chem Eng 21:1661. doi:10.1016/S1570-7946(06)80286-5

Cao Y, Wang Y, Riley JT, Pan WP (2006) A novel biomass air gasification process for producing tar-free higher heating value fuel gas. Fuel Process Technol 87:343–353. doi:10.1016/j.fuproc.2005.10.003

Chen G, Andries J, Luo Z, Spliethoff H (2003) Biomass pyrolysis/gasification for product gas production: the overall investigation of parametric effects. Energy Convers Manage 44:1875–1884. doi:10.1016/S0196-8904(02)00188-7

Chen L, Hontoir Y, Huang D, Zhang J, Morris AJ (2004) Combining first principles with black-box techniques for reaction systems. Control Eng Pract 12:819–826. doi:10.1016/j.conengprac.2003.09.006

de Jong W, Unal Ö, Andries J, Hein KRG, Spliethoff H (2003) Biomass and fossil fuel conversion by pressurised fluidised bed gasification using hot gas ceramic filters as gas cleaning. Biomass Bioenergy 25:59–83. doi:10.1016/S0961-9534(02)00186-1

Demirbas A (2009) Biofuels securing the planet's future energy needs. Energy Convers Manage 50:2239–2249. doi:10.1016/j.enconman.2009.05.010

Detournay M, Hemati M, Andreux R (2011) Biomass steam gasification in fluidized bed of inert or catalytic particles: comparison between experimental results and thermodynamic equilibrium predictions. Powder Technol 208:558–567. doi:10.1016/j.powtec.2010.08.059

Dogru M, Howarth CR, Akay G, Keskinler B, Malik AA (2002) Gasification of hazelnut shells in a downdraft gasifier. Energy 27:415–427. doi:10.1016/S0360-5442(01)00094-9

Donatelli A, Iovane P, Molino A (2012) High energy syngas production by waste tyres steam gasification in a rotary kiln pilot plant. Experimental and numerical investigations. Fuel 89:2721–2728. doi:10.1016/j.fuel.2010.03.040

Dong C, Jin B, Zhong Z, Lan J (2002) Tests on co-firing of municipal solid waste and coal in a circulating fluidized bed. Energy Convers Manage 43:2189–2199. doi:10.1016/S0196-8904(01)00157-1

Folgueras MB, Diaz RM, Xiberta J, Prieto I (2003) Thermogravimetric analysis of the co-combustion of coal and sewage sludge. Fuel 82:2051–2055. doi:10.1016/S0016-2361(03)00161-3

Ganan J, Al-Kassir Abdulla A, Cuerda Correa EM, Macías-García A (2006) Energetic exploitation of vine shoot by gasification processes a preliminary study. Fuel Process Technol 87:891–897. doi:10.1016/j.fuproc.2006.06.004

Gil J, Corella J, Aznar MP, Caballero MA (1999) Biomass gasification in atmospheric and bubbling fluidized bed: effect of the type of gasifying agent on the product distribution. Biomass Bioenergy 17:389–403. doi:10.1016/S0961-9534(99)00055-0

Guo B, Li D, Cheng C, Lü Z, Shen Y (2001) Simulation of biomass gasification with a hybrid neural network model. Bioresour Technol 76:77–83. doi:10.1016/S0960-8524(00)00106-1

Guo B, Shen Y, Li D, Zhao F (1997) Modelling coal gasification with a hybrid neural network. Fuel 76:1159–1164. doi:10.1016/S0016-2361(97)00122-1

Guo J, Lua AC (2001) Kinetic study on pyrolytic process of oil-palm solid waste using two-step consecutive reaction model. Biomass Bioenergy 20:223–233. doi:10.1016/S0961-9534(00)00080-5

Hanaoka T, Inoue S, Uno S, Ogi T, Minowa T (2005) Effect of woody biomass components on air-steam gasification. Biomass Bioenergy 28:69–76. doi:10.1016/j.biombioe.2004.03.008

Hannula I, Kurkela E (2012) A parametric modelling study for pressurised steam/O2- blown fluidised-bed gasification of wood with catalytic reforming. under publication. Biomass Bioenergy 38:58–67. doi:10.1016/j.biombioe.2011.02.045
Hanping C, Bin L, Haiping Y, Guolai Y, Shihong Z (2008) Experimental investigation of biomass gasification in a fluidized bed reactor. Energy Fuel 22:3493–3498. doi:10.1021/ef800180e
Hsi CL, Wang TY, Tsai CH, Chang CY, Liu CH, Chang YC, Kuo JT (2008) Characteristics of an air-blown fixed-bed downdraft biomass gasifier. Energy Fuel 22:4196–4205. doi:10.1021/ef800026x
Jand N, Brandani V, Foscolo PU (2006) Thermodynamic limits and actual product yields and compositions in biomass gasification processes. Ind Eng Chem Res 45:834–843. doi:10.1021/ie050824v
Jarana G, Sanchez-Oneto J, Portela JR, Nebot Sanz E, Martinez de la Ossa EJ (2008) Supercritical water gasification of industrial organic wastes. J Supercrit Fluids 46:329–334. doi:10.1016/j.supflu.2008.03.002
Kelly Yong TL, Lee KT, Mohamed AR, Bhatia S (2007) Potential of hydrogen from oil palm biomass as a source of renewable energy worldwide. Energy Policy 35:5692–5701. doi:10.1016/j.enpol.2007.06.017
Knoef HAM, Koele HJ (2000) Survey of tar measurement protocols. Biomass Bioenergy 18:55–59. doi:10.1016/S0961-9534(99)00067-7
Kumar SS, Pitchandi K, Natarajan ZE (2008) Modeling and simulation of down draft wood gasifier. J App Sci 8:271–279. doi:10.3923/jas.2008.271.279
Lahijani P, Zainal ZA (2011) Gasification of palm empty fruit bunch in a bubbling fluidized bed: a performance and agglomeration study. Bioresour Technol 102:2068–2076. doi:10.1016/j.biortech.2010.09.101
Lapuerta M, Hernández JJ, Pazo A, López J (2008) Gasification and co-gasification of biomass wastes: Effect of the biomass origin and the gasifier operating conditions. Fuel Process Technol 89:828–837. doi:10.1016/j.fuproc.2008.02.001
Larsen MB, Schultz L, Glarborg P, Skaarup-Jensen L, Dam-Johansen K, Frandsen F, Henriksen U (2006) Devolatilization characteristics of large particles of tyre rubber under combustion conditions. Fuel 85:1335–1345. doi:10.1016/j.fuel.2005.12.014
Lee DH, Yang H, Yan R, Liang DT (2007) Prediction of gaseous products from biomass pyrolysis through combined kinetic and thermodynamic simulations. Fuel 86:410–417. doi:10.1016/j.fuel.2006.07.020
Li C, Suzuki K (2009) Tar property, analysis, reforming mechanism and model for biomass gasification—an overview. Renew Sust Energy Rev 13:594–604. doi:10.1016/j.rser.2008.01.009
Li X, Grace JR, Watkinson AP, Lim CJ, Ergüdenler A (2001) Equilibrium modeling of gasification: a free energy minimization approach and its application to a circulating fluidized bed coal gasifier. Fuel 80:195–207. doi:10.1016/S0016-2361(00)00074-0
Lv PM, Xiong ZH, Chang J, Wu CZ, Chen Y, Zhu JX (2004) An experimental study on biomass air–steam gasification in a fluidized bed. Bioresour Technol 95:95–101. doi:10.1016/j.biortech.2004.02.003
Mastellone ML, Zaccariello L, Arena U (2010) Co-gasification of coal, plastic waste and wood in a bubbling fluidized bed reactor. Fuel 89:2991–3000. doi:10.1016/j.fuel.2010.05.019
Mathieu P, Dubuisson R (2002) Performance analysis of a biomass gasifier. Energy Convers Manage 43:1291–1299. doi:10.1016/S0196-8904(02)00015-8
McKendry P (2002a) Energy production from biomass (part 1): overview of biomass. Bioresour Technol 83:37–46. doi:10.1016/S0960-8524(01)00118-3
McKendry P (2002b) Energy production from biomass (part 3): gasification technologies. Bioresour Technol 83:55–63. doi:10.1016/S0960-8524(01)00119-5
Midilli A, Dogru M (2001) Hydrogen production from hazelnut shell by applying air-blown downdraft gasification technique. Int J Hydrogen Energy 26:29–37. doi:10.1016/S0360-3199(00)00049-5

Mitta NR, Ferrer-Nadal S, Lazovic AM, Perales JF, Velo E, Puigjaner L (2006) Modelling and simulation of a tyre gasification plant for synthesis gas production. Comput Aided Chem Eng 21:1771. doi:10.1016/S1570-7946(06)80304-4

Mohammed MAA, Salmiaton A, Wan Azlina W, Mohammad Amran MS, Fakhru'l-Razi A (2011a) Air gasification of empty fruit bunch for hydrogen-rich gas production in a fluidized-bed reactor. Energy Convers Manage 52:1555–1561. doi:10.1016/j.enconman.2010.10.023

Mohammed MAA, Salmiaton A, Wan Azlina W, Mohammad Amran MS, Fakhru'l-Razi A, Taufiq-Yap YH (2011b) Hydrogen rich gas from oil palm biomass as a potential source of renewable energy in Malaysia. Renew Sust Energy Rev 15:1258–1270. doi:10.1016/j.rser.2010.10.003

Narvaez I, Orio A, Aznar MP, Corella J (1996) Biomass gasification with air in an atmospheric bubbling fluidized bed. Effect of six operational variables on the quality of the produced raw gas. Ind Eng Chem Res 35:2110–2120. doi:10.1021/ie9507540

Natarajan E, Nordin A, Rao AN (1998) Overview of combustion and gasification of rice husk in fluidized bed reactors. Biomass Bioenergy 14:533–546. doi:10.1016/S0961-9534(97)10060-5

Nikoo MB, Mahinpey N (2008) Simulation of biomass gasification in fluidized bed reactor using ASPEN PLUS. Biomass Bioenergy 32:1245–1254. doi:10.1016/j.biombioe.2008.02.020

Ponzio A, Kalisz S, Blasiak W (2006) Effect of operating conditions on tar and gas composition in high temperature air/steam gasification (HTAG) of plastic containing waste. Fuel Process Technol 87:223–233. doi:10.1016/j.fuproc.2005.08.002

Rashidi M (1997) Calculation of equilibrium composition in combustion products. Chem Eng Technol 20:571–575. doi:10.1002/ceat.270200809

Ratnadhariya JK, Channiwala SA (2009) Three zone equilibrium and kinetic free modeling of biomass gasifier–a novel approach. Renew Energy 34:1050–1058. doi:10.1016/j.renene.2008.08.001

Ruggiero M, Manfrida G (1999) An equilibrium model for biomass gasification processes. Renew Energy 16:1106–1109. doi:10.1016/S0960-1481(98)00429-7

Saidur R, Abdelaziz EA, Demirbas A, Hossain MS, Mekhilef S (2011) A review on biomass as a fuel for boilers. Renew Sust Energy Rev 15:2262–2289. doi:10.1016/j.rser.2011.02.015

Schuster G, Loffler G, Weigl K, Hofbauer H (2001) Biomass steam gasification—an extensive parametric modeling study. Bioresour Technol 77:71–79. doi:10.1016/S0960-8524(00)00115-2

Sharma AK (2008) Equilibrium and kinetic modeling of char reduction reactions in a downdraft biomass gasifier: a comparison. Sol Energy 82:918–928. doi:10.1016/j.solener.2008.03.004

Sharma AK (2009) Experimental study on 75 kWth downdraft (biomass) gasifier system. Renew Energy 34:1726–1733. doi:10.1016/j.renene.2008.12.030

Skoulou V, Zabaniotou A, Stavropoulos G, Sakelaropoulos G (2008) Syngas production from olive tree cuttings and olive kernels in a downdraft fixed-bed gasifier. Int J Hydrogen Energy 33:1185–1194. doi:10.1016/j.ijhydene.2007.12.051

Tinaut FV, Melgar A, Pérez JF, Horrillo A (2008) Effect of biomass particle size and air superficial velocity on the gasification process in a downdraft fixed bed gasifier. An experimental and modelling study. Fuel Process Technol 89:1076–1089. doi:10.1016/j.fuproc.2008.04.010

Van der Drift A, Van Doorn J, Vermeulen JW (2001) Ten residual biomass fuels for circulating fluidized-bed gasification. Biomass Bioenergy 20:45–56. doi:10.1016/S0961-9534(00)00045-3

Wang L, Weller CL, Jones DD, Hanna MA (2008) Contemporary issues in thermal gasification of biomass and its application to electricity and fuel production. Biomass Bioenergy 32:573–581. doi:10.1016/j.biombioe.2007.12.007

Zainal ZA, Rifau A, Quadir GA, Seetharamu KN (2002) Experimental investigation of a downdraft biomass gasifier. Biomass Bioenergy 23:283–289. doi:10.1016/S0961-9534(02)00059-4

Zhao Y, Sun S, Zhou H, Sun R, Tian H, Luan J, Qian J (2010) Experimental study on sawdust air gasification in an entrained-flow reactor. Fuel Process Technol 91:910–914. doi:10.1016/j.fuproc.2010.01.012

Activated Carbon from Renewable Sources: Thermochemical Conversion and Activation of Biomass and Carbon Residues from Biomass Gasification

Davide Bergna, Henrik Romar, Sari Tuomikoski, Hanna Runtti, Teija Kangas, Pekka Tynjälä, and Ulla Lassi

Abstract Activated carbon is one of the most widely applied adsorbent. As a porous carbon, it is used for the purification of both gaseous and liquid emissions. Activated carbon is prepared from fossil resources, such as coal, or from biomass through (hydro)thermal processing followed by chemical and/or physical activation. Further, some biomass thermal treatment processes, such as biomass gasification, produce carbon residues that can be modified to activated carbon with physical or chemical activation methods. The desired properties of activated carbon, i.e. high specific surface area and porosity, high carbon content and excellent sorption capacity, can be modified and optimized during thermochemical treatment and activation. Those properties, which are shortly considered, are important in different applications for activated carbon.

Keywords Renewable sources • Activated carbon • Carbonization • Physical activation • Chemical activation • Characterization • Adsorbent • Novel applications

D. Bergna • H. Romar • P. Tynjälä • U. Lassi (✉)
Research Unit of Sustainable Chemistry, University of Oulu, P.O. Box 3000, Oulu 90014, Finland

Applied Chemistry Unit, University of Jyvaskyla, Kokkola University Consortium Chydenius, Talonpojankatu 2B, 67100 Kokkola, Finland
e-mail: ulla.lassi@oulu.fi

S. Tuomikoski • H. Runtti • T. Kangas
Research Unit of Sustainable Chemistry, University of Oulu, P.O. Box 3000, Oulu 90014, Finland

L. Singh, V.C. Kalia (eds.), *Waste Biomass Management – A Holistic Approach*,
DOI 10.1007/978-3-319-49595-8_9

1 Introduction

Commercial activated carbon is widely used as an (ad)sorbent in gas and water purification applications. It can be produced by carbonization and activating various materials, such as coal and cellulosic raw materials (Ahmadpour and Do 1996). Recently, more attention has been paid onto the use of biomass and different types of biowaste as source materials for activated carbon. The production process and quality of activated carbon, for example, affect the price of the material, but in general, commercial activated carbon is fairly expensive. The preparation of activated carbon is also an energy-consuming process. Therefore other substitutes, such as biomass-based adsorbents, must be investigated (Bart and von Gemmingen 2005; Marsh and Rodríquez-Reinoso 2006; Guo et al. 2009). Among biomasses, waste biomass would be a better alternative because they are abundant, renewable and cheap. The use of biowaste for preparing (ad)sorbents is also very attractive because it helps to decrease the costs of waste disposal and to promote environmental protection. In fact, agricultural wastes and residues (Ioannidou and Zabaniotou 2007), rice hulls (Teker et al. 1997), palm-tree cobs (Avom et al. 1997), walnut shells (Martínez et al. 2006), corn cobs (Chang et al. 2000), coconut husks (Tan et al. 2008), coconut coir pith (Namasivayam and Sangeetha 2008), macadamia nutshells (Ahmadpour and Do 1997), coffee endocarps (Nabais et al. 2008), lignin (Ragan and Megonnell 2011) and cherry stones (Lussier et al. 1994), for example, have been studied and used to prepare activated carbons with high carbon contents and low price. The Council Directive (1999/31/EC) obliges member states to reduce the amount of biodegradable municipal waste they landfill to 35 % of 1995 levels by 2016 and for some countries by 2020. In Finland, the landfilling of biodegradable and any other organic waste that contains more than 10 % organic carbon has been forbidden since the beginning of 2016 (Government decree 331/2013). One option is to use these kinds of waste materials as precursors for activated carbon production and to further utilize them in many applications. Due to existing legislation, and to avoid emissions caused by transportation, there is a growing need to study locally available waste materials as precursor materials.

Carbons can be classified as graphitizing and non-graphitizing (Franklin 1951). Activated carbon is derived from non-graphitizable carbons which cannot be converted to crystalline graphite at 2700 °C and above atmospheric or low pressures. Non-graphitizable carbons exhibit structural disorder, but they are never amorphous (International Committee for Characterization and Terminology of Carbon 1982). Activated carbon typically has a well-developed porous structure. In addition, high internal surface area and good mechanical properties are preferred with activated carbons (Guo et al. 2009). It has a unique porous structure of a three-dimensional network featuring imperfect sections of small-sized graphitic lamellas bonded together extremely closely. The porosity of carbon material is formed when lamellas create a three-dimensional network, and then spaces between those lamellas form pores. A majority of carbon atoms have a hexagonal arrangement, with some pentagons and other non-hexagonal rings (Marsh and Rodríquez-Reinoso 2006).

External and internal structure of activated carbon is useful to distinguish. The external surface of activated carbon is surrounded by the discrete particles, but the solid surface of carbon is rarely smooth at the atomic scale. The external surface includes all the prominences and also the surface of those cracks which are wider than they are deep. The internal surface consists of pores, cracks and cavity walls which are deeper than they are wide and which are accessible to the adsorptive (IUPAC 1985).

Basically, preparation of activated carbon is a two-stage process of carbonization and activation (Ahmadpour and Do 1996). Carbonization takes place under an anoxic atmosphere (usually molecular nitrogen) at temperatures of 450–600 °C. The raw material is decomposed to hydrocarbon compounds and carbonaceous products. In this step, the most volatile compounds are eliminated as gaseous products (Bansal and Meenakshi 2005). The porosity of carbon formed during the carbonization step should be developed further since high-enough porosity for most applications is not reached thus far (Marsh and Rodríquez-Reinoso 2006). Basically, activated carbons can be prepared using either physical or chemical activation. However, the preparation process can consist of variable combinations of these two. Physical activation is typically implemented by using carbon dioxide or steam (water vapour) at high temperatures of 700–900 °C. In chemical activation, the most common chemicals are dehydrating agents, such as zinc chloride or phosphoric acid. In addition, alkali metal salts, hydroxides and carbonates of potassium and sodium, for example, are widely used (Srinivasakannan and Zailani 2004).

Different carbonization methods of biomass and physical and chemical activation are considered. Also, case studies of carbonization and physical/chemical activation of biomass and the direct activation of the previous carbonized carbon residue of biomass gasification are considered. The advantages and drawbacks of each method are given and compared. Finally, the physicochemical properties of activated carbon are considered from the end-user viewpoint, and different utilization applications are presented.

2 Carbonization of Biomass

Biomass carbonization can be performed by pyrolysis or hydrothermal treatment. After carbonization, activation can be done physically or chemically. Physical activation is typically done in a separate step, but chemical activation can be performed in the same process with carbonization. Different carbonization and activation steps for biomass are presented in Fig. 1. In Table 1, reaction conditions and yields for different thermochemical processes are compared. Pyrolysis can be performed at different speeds, slow, medium or fast, and the reaction condition varies a lot among those. Reaction conditions for hydrothermal carbonization (HTC) and gasification are also presented, and the composition of the product is variable with different thermochemical treatment methods. In this chapter, HTC and dry pyrolysis will be discussed. The chemical or physical activation of carbonized biomass

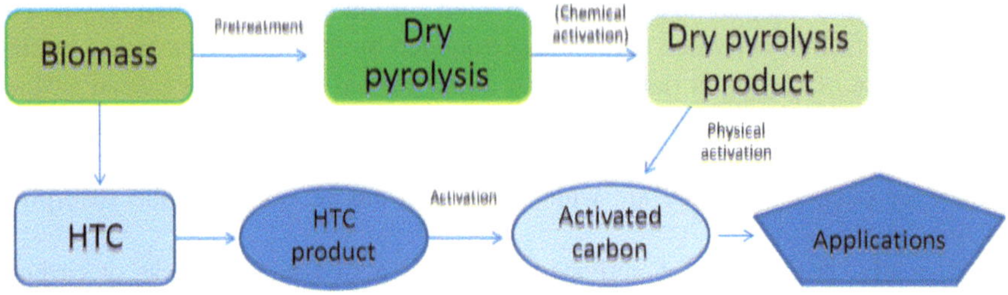

Fig. 1 Possibilities for carbonization and activation of biomass

Table 1 Comparison of reaction conditions and product yields for different thermochemical conversion processes (Funke and Ziegler 2010; Libra et al. 2011; Brewer 2012)

Process	Reaction temperature (°C)	Steam delay time	Product	Composition of product (m-%)		
				Solid	Liquid	Gas
Pyrolysis, slow	ca 400	1 h–week	Biochar	35	30	35
Pyrolysis, medium	ca 500	10–20 s	Biochar	20	50	30
Pyrolysis, fast	ca 500	ca 1 s	Bio oil	12	75	13
Gasification	ca 800	10–20 s	Synthesis gas	10	5	85
HTC	180–250	No steam delay, process time 1–12 h	HTC carbon	50–80	5–20	2–5

and carbon residue from the biomass gasification process will be discussed in Chap. 3.

2.1 *Hydrothermal Carbonization*

Hydrothermal carbonization (HTC) is a combination of the dehydration and decarboxylation of a biomass to increase its carbon content. The process of converting biomass to HTC is carried out at elevated temperatures (180–220 °C) in a water suspension at saturated pressures for several hours. With this HTC conversion process, a lignite-like material with the well-defined properties can be prepared from biomass residues. Even high-moisture-content raw material can be utilized in the HTC process. Thus, the HTC process may contribute to a wider range of applications of biomass, not just for energetic purposes. The HTC process has been known for nearly a century, but it still has received only little attention in the current biomass conversion research. In this paper, the HTC process is not given much more attention.

2.2 Dry Pyrolysis

Dry pyrolysis is the most well-known thermochemical process that produces carbon with a high yield. In addition, dry pyrolysis can produce pyrolysis oil and gas. Dry pyrolysis can be slow or fast, and it is performed without oxygen at temperatures below 500 °C. Biomass used for dry pyrolysis must have quite low water content. Suitable moisture content in slow pyrolysis is 15–20 %. Slow pyrolysis produces carbon, distillation products and gases, such as carbon monoxide, carbon dioxide and hydrogen. In fast pyrolysis, biomass is very fine due to very fast heating and the heat transfer condition. It is used mainly for pyrolysis oil production (Bridgwater 1999; Antal et al. 2000; Bridgwater and Peacocke 2000; Antal and Grønli 2003; Libra et al. 2011; Fagernäs et al. 2014).

2.3 Comparison of HTC Treatment and Dry Pyrolysis

HTC treatment and dry pyrolysis have several advantages and disadvantages. Temperatures are higher in dry pyrolysis as compared with hydrothermal treatment, and raw material has to be dried before pyrolysis, which is not the situation in HTC treatment. Harmful gases, such as PAH compounds and aromatic hydrocarbons, are formed in dry pyrolysis. In addition, tar compounds formation in dry pyrolysis is a problem that does not exist in HTC process (Fischer and Bieńkowski 1999; Antal and Grønli 2003; Sadaka and Negi 2009; Sevilla et al. 2011).

The disadvantages of HTC treatment are the energy consumption that takes place when the saturated steam (up to 220 °C) is being warmed up in the reactor, and the relative high pressure required (25 bars) (Ramke et al. 2009). In dry pyrolysis, energy can be obtained by burning formed synthesis gas. In addition, the separation process for HTC carbon and process water treatment is needed (Funke and Ziegler 2010; Libra et al. 2011).

3 Carbonization and Activation of Raw Materials

Biomass carbonized by hydrothermal treatment or dry pyrolysis can be activated physically or chemically to produce activated carbon (Ahmadpour and Do 1996). Typically, the carbonization step is performed first and after that activation can be done. In the case of chemical activation, carbonization and activation can be done in the same step. The porous structure is one of the most important properties of activated carbon. The porosity of carbon material formed during the carbonization step is typically not high enough for most applications which is the reason it is necessary to develop the porosity of the carbon surface further through physical or chemical activation methods (Marsh and Rodríquez-Reinoso 2006). During the activation process, other specific properties of the carbon surface, such as cationic, anionic or neutral surface functional groups, can be developed. The biomass raw material, activation method and conditions play a key role in the characteristics of activated carbon including, for example, the porosity and shapes of pores (Guo et al. 2009).

In addition to biomass, various carbonaceous by-products and wastes from industry can be used as precursors in activated carbon production. In the future, it is evaluated that the commercial applications for the thermochemical or hydrothermal conversion processes of biomass will increase, and the suitable utilization applications for formed by-products and wastes should be developed. Waste biomass residue or some other organic residues with high carbon content are formed during these processes. Therefore, some reuse would be needed. By utilizing carbonaceous waste materials as precursors for activated carbon production, only the activation step is needed. In the case study presented in Sects. 3.1.1 and 3.2.1, studied carbon residue was obtained from a biomass gasification pilot plant in which carbon residue is formed as a waste. Carbon residue was used as a raw material for the preparation of adsorbents through chemical and physical activation. Wood chips (pine and spruce) were used as a fuel to the downdraft gasifier (150 kW). Gasifier was operated at temperature of 1000 °C, and a feeding rate for raw material was 50 kg/h (Tuomikoski 2014).

3.1 Physical Activation

In the activation process, the removal of the carbon atoms from the nanostructure of the precursor accessing and interconnecting the inherent structural porosity occurs (Ragan and Megonnell 2011). In the physical activation, the carbonized raw material is activated by the physical activating agent. Typically used activating agents are water vapour (steam) or carbon dioxide. Additionally, a mixture of water vapour (steam) and carbon dioxide can be used. The activation temperature varies between 600 °C and 900 °C (Guo et al. 2009). Typically the choice of the activation agent and activation time depends on the type of raw material (Guo et al. 2009). Two to six hours of activation time has been applied for coconut shells, for example.

Physical activation including the carbonization step is typically a two-stage process, and therefore, activation is performed after the carbonization process. A physical activating agent, such as steam, is fed to the precursor material typically at temperatures of between 600 °C and 800 °C. During this phase, the carbon changes physically, and in addition to physical changes, its properties change chemically by opening the carbon matrix and changing its functional groups (Fu et al. 2013). The quantity of water fed depends on the quantity and the material that is processed. Generally, the efficiency of this activation is influenced by the contact between the activating agent and the mass of carbon. Therefore, rotating activation reactors are preferable during this stage. The entire process has to be realized under an inert atmosphere, typically nitrogen, because the water molecules include hydrogen and

Table 2 Typical reactions that took place under steam activation (Chun et al. 2012)

Reaction		
Water-gas reaction	$C + H_2O \rightarrow CO + H_2$	$\Delta H = 131.3$ kJ/mol
Boudouard reaction	$C + CO_2 \rightarrow 2CO$	$\Delta H = 171.7$ kJ/mol
Hydrogasification	$C + 2H_2 \rightarrow CH_4$	$\Delta H = -74.9$ kJ/mol
Decomposition reaction of tar		
Tar pyrolysis	$Tar \rightarrow wH_2 + xCO + yCO_2 + zC_nH_m$	
Tar steam gasification	$Tar + vH_2O \rightarrow xCO + yH_2$	
Reaction of light gas		
Methanation reaction	$CO + 3H_2 \leftrightarrow CH_4 + H_2O$	$\Delta H = -206.2$ kJ/mol
Water-gas shift reaction	$CO + H_2O \leftrightarrow CO_2 + H_2$	$\Delta H = -41.1$ kJ/mol
Steam reforming	$C_nH_m + nH_2O = nCO + (n + m/2)H_2$ C_nH_m is the light hydrocarbon	

oxygen, which generates a potential explosive atmosphere under particular temperature conditions. Steam activation time is also a parameter that has to be optimized because it can influence the structure and the yield of the carbon mass. Generally, higher activation times correspond to higher yield loss but also higher porosity, creating micropores (Chun et al. 2012).

From a chemical point of view, during steam activation, the following compound exists: O_2, CO_2, H_2 and H_2O. Typical reactions for steam activation are listed in Table 2.

3.1.1 Physical Activation of Carbon Residue From Biomass Gasification

In the case study of Tuomikoski (2014), the precursor material for physical activation was carbon residue, which is formed as a waste material in the biomass gasification process. Carbonization step was not done in the presented case study because it was assumed to have already taken place during the thermal gasification process of biomass, in which precursor material was formed as a waste material. Physical activation was carried out by using carbon monoxide (CO) and carbon dioxide (CO_2) as physical activating agents. Additionally, reference samples were prepared by thermal treatment without any activating agent under a nitrogen atmosphere. The temperature and duration of activation were the two parameters investigated in addition to the activating agent. Temperatures of 600 °C and 800 °C were used. The duration of activation was one or three hours. The activation conditions during physicochemical activation are presented in detail in Kilpimaa et al. (2015). The success of the activation was evaluated by characterization of the produced material. The specific surface area and pore volume, for example, from the produced adsorbents were measured. FESEM image of physically activated (by carbon dioxide) carbon residue is presented in Fig. 2.

Fig. 2 FESEM image of the carbon residue sample which was activated with carbon dioxide (CO_2), magnification x8000

3.1.2 Carbonization and Physical Activation of Virgin Lignocellulosic Biomass

In the other case study, biomass was carbonized and activated physically (Bergna et al. 2016). In this study, three sawdust samples originating from birch, spruce and pine were used as precursors for activated carbon production. Prior to carbonization/activation, the precursors were dried at 105 °C overnight. The sawdusts were carbonized and activated in a one step process consisting of carbonization with a temperature ramp from an ambient temperature to 800 °C. During this step, the reactor, a rotating quartz tube fit into a tubular oven, was flushed with nitrogen. At 800 °C, the samples were activated physically by steam for a fixed period of time by passing steam through the reactor using nitrogen as the carrier gas. The specific surface area was increased notably during the carbonization and activation steps. FESEM images of the precursors and the activated carbons are presented in Fig. 3, in which a pore structure can be observed.

3.2 Chemical Activation

Chemical activation can be considered a single-step process in which the carbonization of raw material and activation are done simultaneously. Substances which contain alkali and alkaline earth metal and some general acids are typically used as a chemical activating agent such as KOH, NaOH, $ZnCl_2$ and H_3PO_4 (Ahmadpour and Do 1996). Chemicals during activation step can act as a dehydrating agent (e.g. $ZnCl_2$ and H_3PO_4) and those types of chemicals are commonly used for the activation of especially lignocellulosic materials. Alkali hydroxides are typically

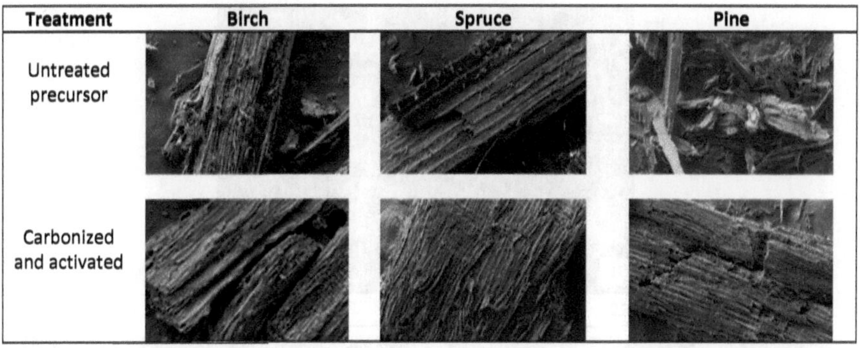

Fig. 3 FESEM image obtained before and after carbonization and physical activation

used as chemical activating agents for the activation of coal or chars (Srinivasakannan and Zailani 2004).

3.2.1 Chemical Activation of the Carbon Residue Formed in Biomass Gasification

In the case study of Tuomikoski (2014), the carbon residue from the biomass gasification process was chemically activated to produce suitable water purification material. In the case study, a wet impregnation method for the chemical activation was used. Carbon residue pretreatment is described in detail in Kilpimaa et al. (2014). After the pretreatment, the sample and the chemical were mixed, and the used contact time was five minutes. Chemicals used for the chemical activation of the carbon residue were 0.1 M HCl, 0.1 M H_2SO_4, 5 M $ZnCl_2$, 5 M KOH, HCl or HNO_3. The used liquid-to-solid ratio (L/S) between chemical activating agent and carbon residue was 10 (w/w). After that, the mixture of the carbon residue and chemical activating agent was dried at 110 °C overnight. Activation was done at 500 °C, and the duration of activation was 1 h. The products were washed sequentially with 0.5 M HCl, and used time for that was 10 min by using L/S (liquid-to-solid ratio) of 10. Additionally, hot distilled water and finally cold distilled water were used in the washing stage. This step was done to remove any remaining organic and inorganic residues. Before use, products were dried at 110 °C overnight, crushed and sieved to obtain particle size below 150 μm.

The surface area and porosity of the carbonaceous material are known to depend on the variables of the preparation process. For that reason, the L/S ratio of the carbon residue and chemical activating agent, the concentration of the impregnation solution and the activation time were optimized. In the case study (Tuomikoski 2014), the highest specific surface area for the carbon residue sample was obtained by using a 5 M zinc chloride solution with L/S ratio of 10 and activation time of 1 h in the chemical activation at 500 °C. In Fig. 4, FESEM images for chemically activated ($ZnCl_2$) carbon residue, commercial activated carbon and carbon residue from biomass gasification are presented as a reference. These images illustrate the

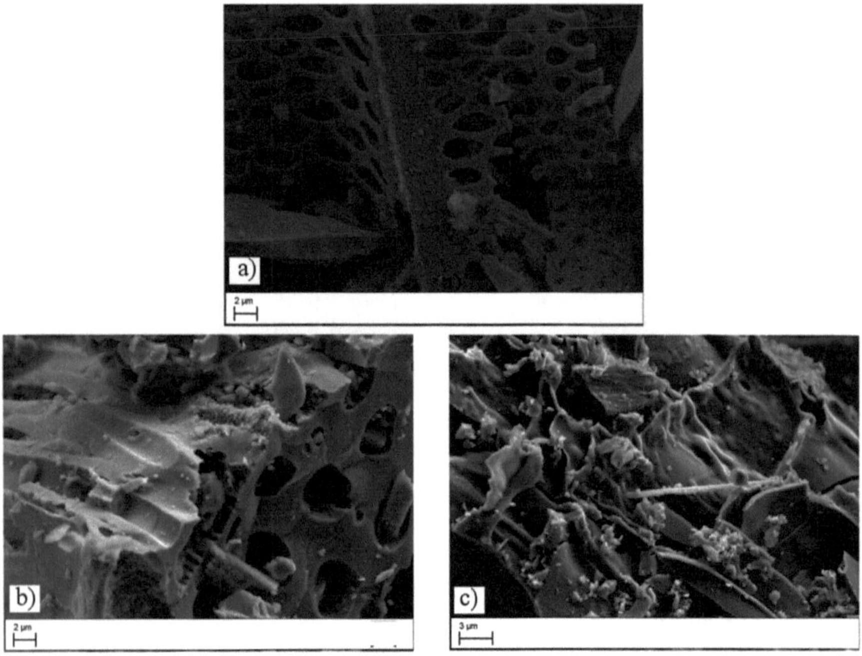

Fig. 4 FESEM images for (**a**) chemically ($ZnCl_2$) activated carbon residue, (**b**) commercial activated carbon and (**c**) carbon residue without any activation, magnification ×7000

crystals on the surfaces of the adsorbents and the porous structure of the adsorbents. Activated carbon residue and activated carbon seem to be clearly more porous materials than carbon residue. The measured values of specific surface areas presented in Table 4 agree with this argument (Kilpimaa et al. 2012, 2014).

3.3 Comparison of Physical and Chemical Activation

There are several advantages and disadvantages in the physical and chemical activation methods. When comparing chemical activation to physical one, the main advantages in the chemical activation comparing are lower activation temperature, shorter activation time, and, typically, the higher porosity of produced carbons are obtained. The main disadvantage of the chemical activation is the secondary pollution which is formed from the activation chemicals. Additionally, chemical consumption and price must be taken into account in the discussion of advantages and disadvantages. A washing stage is required in the chemical activation method, and, therefore, physical activation could be a more environmentally friendly and effective method to produce an adsorbent from biowaste material (Marsh and Rodríquez-Reinoso 2006).

4 Physicochemical Properties of Activated Carbon

The properties of activated carbons depend on a number of factors, including the biomass used, carbonization conditions, activation conditions and the physical or chemical activating agents used. According to the methods published by the European Council of Chemical Manufacturers' Federations *Test Methods for Activated Carbon*, there are some suggestions which parameters should be determined from the activated carbons. These parameters include bulk properties measurement such as density (kg/m^3), water content and mechanical strength and in addition, adsorption properties should be determined. The widely used application for activated carbon is use as an adsorbent, and therefore, adsorption tests are typically done. Those are tests that include both liquid and gas phase adsorption (European Council of Chemical Manufacturers' Federations 1986).

Density measurements are divided into bulk density, absolute density and particle density. Bulk density is defined as "mass of a unit volume of the sample in air, including both the pore system and voids between the particles". The bulk density depends on parameters such as the shapes, sizes and densities of the individual particles. The determination is performed by transferring a representative portion of the sample into a graduated measuring cylinder (about 100 mL), recording the volume and measuring the weight of the sample. Samples can be used as received or dried prior to measurement. Bulk density (kg/m^3) is calculated from Eq. (1):

$$D_b = 1000^*M/V \tag{1}$$

where M is the mass of the sample in grams and V is the volume of the sample in mL.

Absolute density (or helium density) is defined as "the mass of a unit volume of the carbon skeleton that is inaccessible to helium". Measurements are made with commercial helium pycnometers, according to the instructions of the manufacturer. Prior to analysis, the samples are dried at 150 °C to a constant weight (European Council of Chemical Manufacturers' Federations 1986).

Particle density or Hg-density is defined as "the mass of a unit volume of the carbon particles, including its pore system". Measurements are made using a pycnometer constructed to operate under vacuum using mercury as medium (European Council of Chemical Manufacturers' Federations 1986).

Water content or moisture content tests can be performed if water is known to be the only volatile material which presents in the carbon. Xylene distillation is typically used for heat-sensitive carbons, or if water-insoluble material is present in the carbon. This test is performed by drying a sample (1–2 g for powdered carbon or 5–10 g for granular carbon) at 145–155 °C to a constant weight (European Council of Chemical Manufacturers' Federations 1986).

In addition, the success of physical or chemical activation can be evaluated by characterization of the produced material, i.e. the specific surface area, the pore

Table 3 Porosity in typical activated carbon (Ruthven 1984; Marsh and Rodríquez-Reinoso 2006)

Parameter	Micropore	Mesopore	Macropore
Pore diameter (nm)	< 2	2–50	> 50
Pore volume (cm^3/g)	0.15–0.5	0.02–0.1	0.2–0.5
Surface area (m^2/g)	100–1500	10–100	0.5–2

Table 4 Summary of the properties for studied adsorbents (Tuomikoski 2014)

Parameter	CR	PACR	CACR	AC
Carbon content (%)	69.5	52.1	61.8	91.9
Surface area (m^2/g)	52.4	590	285	786
Pore diameter (nm)	8.16	3.44	3.94	2.21
Pore volume (cm^3/g)	0.107	0.335	0.26	0.435

CR carbon residue, *PACR* physically (by carbon dioxide) activated carbon residue, *CACR* chemically ($ZnCl_2$) activated carbon residue, *AC* activated carbon

volume and the pore size were measured. By measuring the carbon content, the burn-off of the material can be measured, and the success of the carbonization can be evaluated. Adsorptive capacity results from the porous network that exists within the grains of activated carbon. The internal surface area in addition to the pore volume is notably greater comparing with the external surface area and the volume of the grains themselves (Ragan and Megonnell 2011). Commercial activated carbon exhibits a high internal surface area of 1000 m^2/g or higher and large internal pore volume (0.3–3 cm^3/g). Porosity within adsorbents has been classified by International Union of Pure and Applied Chemistry (IUPAC) (1972) according to the widths of the pores as follows: macropores (access pores, 1000 nm down to 50 nm in width), mesopores (transport pores, between 50 nm and 2 nm in width) and micropores (smallest pores with highest adsorption energy, < 2 nm wide). The porosity of typical activated carbon is presented in Table 3. A summary of properties for adsorbents in the case study of Tuomikoski (2014) is presented in Table 4.

5 Utilization Application as an Adsorbent

Produced activated carbon has been applied in water purification applications for the removal of several impurities such as organics, metal cations and anions from wastewaters. The high adsorption capacity of activated carbon is based on the strong interaction, i.e. the attractive forces exist between the atoms in the walls of the nanopores and the contaminant molecules which are to be removed at the atomic level. The adsorptive capacity of activated carbon results from the complex, three-dimensional, interconnected porous network that exists within the activated

Table 5 Measured point of zero charge values for the adsorbents studied in Tuomikoski (2014)

Sample	pH_{ZPC}
CACR	4
PACR	3
CR	2.1
AC	2.1

CACR chemically ($ZnCl_2$) activated carbon residue, *PACR* physically (CO_2) activated carbon residue, *CR* carbon residue, *AC* activated carbon

carbon grains (Ragan and Megonnell 2011). Although the adsorption capacity is mainly determined by the microporous structure, the surface chemistry of the carbon adsorbent also plays an important role in determining its adsorption properties and behaviour. Typically, surface of activated carbon is nonpolar, and therefore, it favours the adsorption of nonpolar molecules by non-specific mechanisms (Moreno-Castilla 2004; Marsh and Rodríquez-Reinoso 2006). The polarity of the surface can be altered, and the adsorption of polar molecules increased by an adequate modification. The surface chemistry therefore depends on the heteroatom contents. On the surface, chemisorbed oxygen, hydrogen, etc. may exist, and oxygen-containing functional groups have great importance for the surface characteristics because they determine the surface charge and the hydrophobicity of activated carbon. The surface charge can be determined by titration or by using electrokinetic methods. The first method provides a measure of the total surface charge, whereas the latter primarily measures the surface charge at the more external surface of the particles. The point of zero charge (pH_{ZPC}) is the pH in which the total surface charge is zero, and the isoelectric point (pH_{IEP}) describes pH in which the pH of the external surface is zero (Rodríquez-Reinoso and Molina-Sabio 1998; Marsh and Rodríquez-Reinoso 2006). The pH_{ZPC} values for adsorbents in case study of Tuomikoski (2014) are presented in Table 5. In the pH values above pH_{ZPC}, the adsorbent is favoured to remove anions, and in the pH values measured below pH_{ZPC}, the adsorbent removes cations.

When evaluating utilization potential as an adsorbent, adsorption isotherms are vital to calculate. An isotherm is constructed by plotting the residual concentration of the test substance against the mass of carbon. Adsorption isotherms are used for designing an adsorption system since they represent the amount of species adsorbed versus the amount of species left in the solution phase at equilibrium. Isotherms give information about the nature of the adsorption, and adsorption capacity (q_m) can be calculated. Capacity makes different adsorbents and their adsorption capacities comparable because it is not dependent on conditions in which adsorption experiments have been done. Adsorption isotherms furthermore give information about a larger scale adsorption process.

The most commonly used adsorption isotherm is the Langmuir model (Langmuir 1918). The following assumptions are made in the Langmuir isotherm model: (1) there is only one kind of adsorption site on the surface, (2) the energy of the adsorption is independent of how many of the surrounding sites are occupied, (3) maximum adsorbate coverage is one monolayer and (4) all adsorption occurs through the same mechanism. The linear form of the Langmuir isotherm model is given by Eq. (2), as follows:

$$\frac{1}{q_e} = \frac{1}{q_m} + \frac{1}{q_m b C_e} \tag{2}$$

where q_e (mg/g) is the adsorption amount, C_e (mg/L) is the equilibrium concentration and q_m (mg/g) and b (L/mg) are the Langmuir constants which are related to the adsorption capacity and adsorption energy, respectively. Furthermore, the constants of Langmuir model were obtained from the linear plots of $1/q_e$ versus $1/C_e$. In addition to the Langmuir model, there are several other isotherm models, such as the bi-Langmuir model, Temkin model, Freundlich model, D-R model and Sips model, and every model has its own characteristics and hypotheses (Freundlich 1906, Temkin and Pyzhev 1940; Karthikeyan et al. 2004; Repo et al. 2009; Zheng et al. 2009; Repo et al. 2011; Roosta et al. 2014a, b, c).

In addition to the adsorption isotherm model, the rate of adsorption is important to know. The kinetic studies where the progress of the adsorption processes is followed in the function of time provide useful information about the efficiency of adsorption. Kinetic results are also crucial when the feasibility of the process for large scale applications is considered. Both pseudo-first-order and pseudo-second-order kinetic models are most commonly utilized to identify the kinetics of the adsorption process (Lagergren 1898; Ho and McKay 1999). The equation of Lagergren (1898) is widely used to describe the adsorption of liquid phase compounds to the solid adsorbate surface. Most often, the modelling has been accomplished using the linear form of pseudo-first-order rate expression which is given as follows (Bhatnagar et al. 2010b):

$$\log(q_e - q_t) = \log q_e - \frac{k_f}{2.303} t \tag{3}$$

where q_e and q_t are the amounts of impurities adsorbed (mg/g) at equilibrium and at time t, respectively. The rate constants in the pseudo-first-order equation are marked as k_f [min^{-1}].

Pseudo-second-order process can be expressed in a linear form (Bhatnagar et al. 2010b):

$$\frac{t}{q_t} = \frac{1}{k_s q_e^2} + \frac{1}{q_e} t \tag{4}$$

where k_s [g mg^{-1} min^{-1}] is the rate constant of pseudo-second-order kinetics.

5.1 Adsorption Experiments over Carbon-Based Adsorbents

5.1.1 Standard Method

In the adsorption tests for liquids, the activated carbon comes into contact with a solution containing a specific test substance. The contact time has to be long enough for equilibrium formation. A number of test substances can be used, for instance, phenol, iodine and methylene blue are typically used model substances. If the activated carbon is to be used for pharmaceutical purposes, phenazone adsorption is used. In addition to standard methods, specific test solutions for the applications of customers can be used. A summary of different test substances is presented in Table 6 (European Council of Chemical Manufacturers' Federations 1986).

In addition to liquid phase experiments, activated carbon works as an adsorbent also in gas phase adsorption. The adsorption properties of gaseous compounds are tested by feeding the gas at a known concentration and velocity through a bed of activated carbon and registering the breakthrough volume, i.e. the gas volume needed to saturate all binding places of the carbon (European Council of Chemical Manufacturers' Federations 1986).

5.1.2 Case Study

In the studies of Runtti et al. (2014) and Kilpimaa et al. (2014, 2015), laboratory experiments were carried out to examine the adsorption capacity for the produced adsorbents. During experiments, the effects of different variables such as the effect of initial pH, initial concentration of adsorbate solution and adsorption time on the adsorption efficiency of phosphates, nitrates, iron(II), copper(II) and nickel(II) were studied. Used adsorbents were carbon residue from the wood gasification process, carbon residue which was activated physically or chemically and commercial activated carbon provided by Norit. The properties of adsorbents were previously presented in Tables 4 and 5. The experiments were done using a batch type of process, and the effect of the initial pH was studied in the pH range of 4–8. The effect of the initial solution concentration was studied with a concentration range of

Table 6 Summary of test substances for evaluating of adsorption capacity in liquid phase and gas phase (European Council of Chemical Manufacturers' Federations 1986)

Liquid/gas phase	Property tested	Adsorbate
Liquid phase	Mesopores	Methylene blue
Liquid phase	Micropores	Iodine
Liquid phase	Pharmaceutical use	Phenazone
Liquid phase		Application-specific compounds
Gas phase	Gas-binding capacity	Carbon tetrachloride (g)
Gas phase	Gas-binding capacity	Application specific compounds

25–125 mg/L (chemically activated samples) or concentrations of 10–140 mg/L (physically activated samples), and experiments were done at optimum pH determined in the previous step. The model solution was prepared by adding solid KH_2PO_4 or $NaNO_3$ to distilled water in the case of phosphate and nitrate, respectively. Model metal solutions were prepared from iron sulphate ($FeSO_4 \cdot 7H_2O$), copper sulphate ($CuSO_4 \cdot 5H_2O$) and nickel sulphate ($NiSO_4 \cdot 6H_2O$). In the all adsorption experiments, the adsorbent dose used was 5 g/L.

In the studies in which the optimum initial pH and effect of initial concentrations were determined, a solution of adsorbate and adsorbent was mixed. After that, the pH of the mixture was adjusted by HCl or NaOH. pH was adjusted after adding the adsorbent to adsorbate solution to ensure the correct pH of the adsorbent-adsorbate mixture due to the fact that the carbon residue from the gasification process is clearly alkaline (pH $\approx$ 9) which increases the pH. In addition, the carbon residue has some buffer capacity which means that material is able to retain an almost constant pH when a small amount of acid is added to the solution, and therefore, it possesses ions that have the ability to neutralize the added hydrogen ions. These ions are similar to hydrogen carbonates, carbonates and hydroxides (Kilpimaa et al. 2011, 2013). The solution was shaken at room temperature for 24 h.

Further in the studies of Kilpimaa et al. (2014, 2015) and Runtti et al. (2014), the effect of the adsorption time was studied. These experiments were done by using optimum initial pH and initial concentration of adsorbate. The rate of adsorption was determined with different time intervals of 1 min–24 h. Before analysis, all samples, including the initial samples, were filtered through a 0.45 μm filter. The phosphate and nitrate concentrations were analyzed by ion chromatography (Metrohm 761 Compact IC). Heavy metals were determined by an atomic absorption spectrometer (AAS) (PerkinElmer AAnalyst 200) whose absorption wavelengths were found to be Fe^{2+} (248.3 nm), Cu^{2+} (324.7 nm) and Ni^{2+} (232.0 nm), respectively.

The amount of ions adsorbed (q_e in mg/g) was determined as follows:

$$q_e = \frac{C_0 - C_e}{m} V \tag{5}$$

where C_0 and C_e are the initial and equilibrium concentrations of impurity ion in a solution (mg/L), respectively, V is the volume of solution (L) and m is the mass of the adsorbent (g) (Bhatnagar et al. 2010a, b).

5.1.3 Literature Survey

Commercial activated carbon and carbon from biomass have been widely applied as adsorbent for water purification. Table 7 lists some of the known references for experimental adsorption capacities ($q_{m,\ \mathrm{exp}}$) and Langmuir adsorption capacities

Table 7 Experimental sorption capacities ($q_{m,\ exp}$) and Langmuir adsorption capacity for different adsorbents and adsorbates

Adsorbate	Adsorbent	$q_{m,\ exp}$(mg/g)	q_m(mg/g)	References
Phosphate	Chemically activated carbon residue	20.5	14.3	Kilpimaa et al. (2014)
	Iron oxide tailings	–	8.21	Zeng et al. (2004)
	Physically (CO_2) activated carbon residue		30.2	Kilpimaa et al. (2015)
	Goethite	–	6.42	Borggaard et al. (2005)
	Commercial activated carbon	8.7	6.44	Kilpimaa et al. (2014)
	Iron-hydroxide modified eggshell	–	14.5	Mezenner and Bensmaili (2009)
	Natural zeolite with Na	–	2.19	Wu et al. (2006)
Nitrate	Chemically activated carbon residue	7.97	3.06	Kilpimaa et al. (2014)
	Physically (CO_2) activated carbon residue		11.2	Kilpimaa et al. (2015)
	Commercial activated carbon	14.6	11.4	Kilpimaa et al. (2014)
	Chemically modified sugar beet bagasse		9.14–27.55	Demiral and Gündüzoğlu (2010)
	Zr(IV)-loaded sugar beet pulp	63		Hassan et al. (2010)
	Sepiolite activated by HCl		38.16	Öztürk and Bekta (2004)
	Impregnated almond shell activated carbon		16–17	Rezaee et al. (2008)
	$ZnCl_2$ treated coconut granular activated carbon		10.2	Bhatnagar et al. (2008)
Iron(II)	Chemically activated carbon residue	20.5	21.4	Runtti et al. (2014)
	Carbon residue	24.1	25.4	Runtti et al. (2014)
	Commercial activated carbon	13.9	12.1	Runtti et al. (2014)
	Cow bone charcoal	–	31.43	Moreno et al. (2010)
	Commercial activated carbon	–	31	Kouakou et al. (2013)
	Hazelnut hull	–	13.59	Sheibani et al. (2012)

(continued)

Table 7 (continued)

Adsorbate	Adsorbent	$q_{m,\ \mathrm{exp}}$(mg/g)	q_m(mg/g)	References
Copper(II)	Chemically activated carbon residue	23.1	23.3	Runtti et al. (2014)
	Carbon residue	11.1	10.1	Runtti et al. (2014)
	Commercial activated carbon	5.1	4.4	Runtti et al. (2014)
	Granular biomass	55.0	–	Hawari and Mulligan (2006)
	Activated slag	–	30.0	Gupta (1998)
	Activated carbon produced from pomegranate peel	–	18.1–22.0	El-Ashtoukhy et al. (2008)
	Coal fly ash pellets	–	20.9	Papandreou et al. (2007)
	Red mud	–	19.7	Lopez et al. (1998)
	Na-bentonite		17.9	Liu and Zhou (2010)
	Rubber wood sawdust-activated carbon	–	5.7	Kalavathy et al. (2005)
	Bagasse fly ash	–	2.3	Gupta and Ali (2000)
	Olive stone waste	–	1.87	Fiol et al. (2006)
Nickel(II)	Carbon residue	5.6	5.7	Runtti et al. (2014)
	Chemically activated carbon residue	18.2	62.9	Runtti et al. (2014)
	Irradiation-grafted activated carbon	–	55.7	Ewecharoen et al. (2009)
	Activated slag	–	29.4	Gupta (1998)
	Granular biomass	26.0	–	Hawari and Mulligan (2006)
	Activated carbon from *Hevea brasiliensis*	–	16.9	Kalavathy et al. (2010)
	Commercial activated carbon	2.9	3.1	Runtti et al. (2014)
	Coir pith waste	–	16.0	Parab et al. (2006)
	Na-bentonite	–	14.0	Liu and Zhou (2010)
	Activated carbon prepared from almond husk	–	30.8–37.2	Hasar (2003)
	Red mud	–	11.0	Lopez et al. (1998)
	Olive stone waste	–	2.3	Fiol et al. (2006)

(q_m) for different adsorbents and adsorbates. Results indicate that adsorption capacities vary a lot between activated carbons produced from different raw materials with different activation methods.

6 Novel Utilization Applications for Activated Carbon

Utilization applications of thermochemically produced carbon have been studied widely, and several potential applications have been reported in the literature. In this chapter, some utilization applications for activated carbon have been presented. Utilization applications as adsorbents are discussed in Chap. 5.

6.1 Activated Carbon in Catalytic Applications

In the production process of a heterogeneous catalyst, active material is added to the surface of the support material. The support material should have specific properties, such as a large surface area. Activated carbons possess a number of properties making them very suitable support materials for heterogeneous catalysts. Carbons have good chemical stability under both acidic and alkaline conditions but also good mechanical stability. The large surface areas and the porosity make carbons ideal supports. The recovery of active metals is rather easy, because the carbon support can be removed by burning the coal and recovering the metal, a property that is most important when using rare and expensive metals as catalysts. The surface properties of activated carbons can easily be modified by adding different functional groups. Activated carbon has been used as a catalyst support, for example, in the Fischer-Tropsch production of biodiesel starting from synthesis gas using cobalt as the active metal (Tingjun and Li 2015).

Another type of biodiesel produced by the transmethylation of fatty acids can be catalysed by activated-carbon supported catalysts (Konwar et al. 2014). The activated carbon itself can in some applications act as a catalyst. Actually, it has been used as a catalyst for the hydrolysis of cellulose in a mechano-catalytic application using ball-mill grinding (Yabushita et al. 2014).

Activated carbon has been used as a support material in some catalytic processes (Lam and Luong 2014):

- Hydrogenation of nitro aromatics on Pt-V/AC
- Reductive alkylation with Pt/C (fine and specialty chemicals)
- Hydrogenation of dinitrotoluene to toluenediamine on Pd/carbon black
- Butanediol synthesis (hydrogenation of maleic acid) on Pd-Ag-Re/AC
- Terephthalic acid purification on Pt/AC

6.2 Activated Carbon as Precursor for Graphene in Electrochemical Applications

One of the most recent discoveries regarding the use of biomass-based carbon is to produce graphene-like nanosheets. Studies by Shams et al. (2015) show that it is possible to obtain few-layer graphene (FLG) from biomass. From the mechanical and electrical point of view, the potential properties of graphene are well known (Al-Saleh 2015). Specific applications in that research area are made, for example, in the use of manganese dioxide nanoplates, which are anchored to the biomass-derived cross-linked carbon nanosheets for high-performance asymmetric supercapacitors (Li et al. 2015). These supercapacitors showed impressive electrical performance because of the specific properties of graphene, such as high specific surface and conductivity.

The syntheses of porous graphene and activated-carbon composite have been studied. In the studies of Zheng et al. (2014), the high packing density and a large specific surface area for supercapacitor electrode material have been used (Zheng et al. 2014) with promising results. Particularly, the interest of researchers is to develop carbon as a precursor for graphene production from different biomass matrices. It is already known that different biomasses can generate carbons with different properties. For instance, pomelo peel carbon was used as an anode for lithium-based batteries (Sun et al. 2013). The use of biomass-derived material in the battery industry is still under investigation, and according to Zhang et al. (2015), biomass-based graphene could bring an almost totally renewable material (separator, binder and electrode) to battery production. This possibility of using porous activated carbon from biomass to the production of graphene and then to electrochemistry is still at a basic research level, and further applications have to be made and studies done (Konwar et al. 2014).

6.3 Other Utilization Applications

Biomass-based dry pyrolysis carbon or HTC carbon have been used in addition to the applications mentioned earlier, for example, soil improvement, adsorption in the purification of gases, electronic applications, catalyst support material for producing sustainable catalysts, novel anode material in production of batteries, a bio-based reducer in the steelmaking industries and an environmental application to reduce carbon dioxide. The utilization of an adsorbent for removing impurities in an aqueous solution has been presented in Chap. 5 (Cakan et al. 2008; Makowski et al. 2008; Sevilla et al. 2011; Sevilla and Fuertes 2011; Suopajärvi et al. 2014). In Table 8, some utilization applications based on literature and production conditions for the utilized materials are presented.

Table 8 Utilization applications for dry pyrolysis and HTC products

Raw material	Carbonization	Activation	Utilization application	Reference
α-D-glucose, potato starch, cellulose, sawdust from eucalyptus, furfural	HTC (230 °C, 250 °C)	Chemical (KOH, 600–800 °C)	Hydrogen storage	Sevilla et al. (2011)
Glucose + Si-nanoparticles	HTC & pyrolysis (900 °C)	No activation	C/Si-nanocomposites as anode material for lithium ion batteries	Cakan et al. (2008)
Furfural + palladium	HTC (190 °C)	No activation	As Pd-catalyst support in selective hydrogenation to cyclohexanone	Makowski et al. (2008)
Starch, cellulose, sawdust	HTC (230–250 °C)	Chemical (KOH, 600–800 °C)	Adsorption of carbon dioxide	Sevilla et al. (2011)

7 Conclusions

Biomass can be used as a precursor in activated carbon production. Biomass carbonization can be done by using thermochemical or hydrothermal treatments. After the carbonization step, activation can be performed physically or chemically. Biomass-based carbon can also be produced from carbonaceous waste materials. In this chapter, carbon residue from the wood gasification process was used as a raw material for activated-carbon production, and it was presented as a case study. In chemical activation studies, the highest specific surface area for carbon residue was obtained using $ZnCl_2$ as a chemical activating agent. In the physical activation procedure, the highest surface area was obtained using carbon dioxide as a physical activating agent. The widely used application for activated carbon is use as an adsorbent for the purification of gases and liquids. In the case study, physically and chemically activated carbon residue was used in the removal of phosphates, nitrates, iron(II), copper(II) and nickel(II) in wastewater treatment. Based on the results presented in the case study, activated carbon residue had a notably higher adsorption capacity compared with commercial activated carbon in the case of phosphate and heavy metal removal. Adsorption capacity is also substantially better for activated carbon residue compared with carbon residue, and therefore, the adsorption capacity can clearly be enhanced by activation. The use of gasification carbon residue is not only an effective and inexpensive adsorbent, but also it has the potential to significantly reduce waste streams. Other potential utilization applications for activated carbon are, for example, used as a catalyst support and as a precursor in graphene production in battery production.

8 Opinion

The carbonization of biomass and the physical and/or chemical activation of carbon have been considered. The properties of obtained activated carbon were considered from an adsorption viewpoint. In addition to adsorption, other applications for carbon are available. It can be used as a catalyst support, as a carbon dioxide storage material and as an energy storage material in lithium ion batteries, for example.

Acknowledgements The authors wish to thank the EU Interreg Nord programme within the project no 304-8455-10 (HighBio2-Biomass to fuels and chemicals) and accompanying companies for their financial and technical support. The authors also thank project SULKA (A32164, 524/2012) and Maa- ja vesitekniikan tuki ry. Davide Bergna acknowledges the Central Ostrobothnia Regional Fund for its personal grant.

References

Ahmadpour A, Do DD (1996) The preparation of active carbons from coal by chemical and physical activation. Carbon 34:471–479. doi:10.1016/0008-6223(95)00204-9

Ahmadpour A, Do DD (1997) The preparation of activated carbon from macadamia nutshell by chemical activation. Carbon 35:1723–1732. doi:10.1016/S0008-6223(97)00127-9

Al-Saleh MH (2015) Electrical and mechanical properties of graphene/carbon nanotube hybrid nanocomposites. Synth Met 209:41–46. doi:10.1016/j.synthmet.2015.06.023

Antal MJ, Grønli M (2003) The art, science, and technology of charcoal production. Ind Eng Chem Res 42:1619–1640. doi:10.1021/ie0207919

Antal MJ, Allen SG, Dai X, Shimizu B, Tam MS, Grønli M (2000) Attainment of the theoretical yield of carbon from biomass. Ind Eng Chem Res 39:4024–4031. doi:10.1021/ie000511u

Avom J, KetchaMbadcam J, Noubactep C, Germain P (1997) Adsorption of methylene blue from an aqueous solution on to activated carbons from palm-tree cobs. Carbon 35:365–369. doi:10.1016/S0008-6223(96)00158-3

Bansal RC, Meenakshi G (2005) Activated carbon adsorption. Taylor & Francis, Boca Raton, FL, p 495

Bart H-J, von Gemmingen U (2005) Ullmann's encyclopedia of industrial chemistry–adsorption. Wiley, Weinheim, pp 549–1620

Bergna D, Romar H, Tuomikoski S, Lassi U (2016) A demineralization process for lower ash content in activated carbons, manuscript (unpublished data)

Bhatnagar A, Ji M, Choi Y-H, Jung W, Lee S-H, Kim S-J, Lee G, Suk H, Kim H-S, Min B, Kim S-H, Jeon B-H, Kang J-W (2008) Removal of nitrate from water by adsorption onto zinc chloride treated activated carbon. Sep Sci Technol 43:886–907. doi:10.1080/01496390701787461

Bhatnagar A, Kumar E, Sillanpää M (2010a) Nitrate removal from water by nano-alumina: characterization and sorption studies. Chem Eng J 163:317–323. doi:10.1016/j.cej.2010.08.008

Bhatnagar A, Minocha AK, Sillanpää M (2010b) Adsorptive removal of cobalt from aqueous solution by utilizing lemon peel as biosorbent. Biochem Eng J 48:181–186. doi:10.1016/j.bej.2009.10.005

Borggaard OK, Raben-Lange B, Gimsing AL, Strobel BW (2005) Influence of humic substances on phosphate adsorption by aluminium and iron oxides. Geoderma 127:270–279. doi:10.1016/j.geoderma.2004.12.011

Brewer C.E (2012) Biochar characterization and engineering. Dissertation, Iowa State University, p 182

Bridgwater AV, Peacocke GVC (2000) Fast pyrolysis processes for biomass. Renew Sustain Energy Rev 4:1–73. doi:10.1016/S1364-0321(99)00007-6

Bridgwater AV (1999) Principles and practice of biomass fast pyrolysis processes for liquids. J Anal Appl Pyrolysis 51:3–22. doi:10.1016/S0165-2370(99)00005-4

Cakan R, Titirici M-M, Antinietti M, Cui G, Maier J, Hu Y-S (2008) Hydrothermal carbon spheres containing silicon nanoparticles: synthesis and lithium storage performance. Chem Commun:3759–3761. doi:10.1039/B805671B

Chang C-F, Chang C-Y, Tsai W-T (2000) Effects of burn-off and activation temperature on preparation of activated carbon from corn cob agrowaste by CO_2 and steam. J Colloid Int Sci 232: 45–49. doi:10.1006/jcis.2000.7171

Chun YN, Lim MS, Yoshikawa K (2012) Characteristics of the product from steam activation of sewage sludge. J Ind Eng Chem 18:839–847. doi:10.1016/j.jiec.2011.11.144

Council Directive (1999) 1999/31/EC Council Directive on the Landfill of Waste. http://eur-lex.europa.eu/legal-content/EN/TXT/?uri=CELEX:31999L0031. Accessed 12.11.2015

Demiral H, Gündüzoğlu G (2010) Removal of nitrate from aqueous solutions by activated carbon prepared from sugar beet bagasse. Bioresour Technol 101:1675–1680. doi:10.1016/j.biortech.2009.09.087

El-Ashtoukhy E-SZ, Amin NK, Abdelwahab O (2008) Removal of lead (II) and copper (II) from aqueous solution using pomegranate peel as a new adsorbent. Desalination 223:162–173. doi:10.1016/j.desal.2007.01.206

European Council of Chemical Manufacturers' Federations (1986) Test methods for activated carbon. http://www.cefic.org/Documents/Other/Test-method-for-Activated-Carbon_86.pdf. Accessed 20.10.2015

Ewecharoen A, Thiravetyan P, Wendel E, Bertagnolli H (2009) Nickel adsorption by sodium polyacrylate-grafted activated carbon. J Hazard Mater 171:335–339. doi:10.1016/j.jhazmat.2009.06.008

Fagernäs L, Kuoppala E, Ranta J, Arpiainen V, Tiilikkala K, Kemppainen R, Hagner M, Setälä H (2014) Hidaspyrolyysituotteiden hyödyntäminen ja tuotannon kannattavuus: Biohiili ja tisle Espoo. VTT Publications, Finland, 182. ISBN:978-951-38-8277-8

Fischer Z, Bieńkowski P (1999) Some remarks about the effect of smoke from charcoal kilns on soil degradation. Environ Monit Assess 58:349–358. doi:10.1023/A:1006058512644

Fiol N, Villaescusa I, Martínez M, Miralles N, Poch J, Serarols J (2006) Sorption of Pb (II), Ni(II), Cu(II) and Cd(II) from aqueous solution by olive stone waste. Sep Purif Technol 50:132–140. doi:10.1016/j.seppur.2005.11.016

Franklin RE (1951) Crystallite growth in graphitizing and non-graphitizing carbons. Proc R Soc Lond 209:196–218

Freundlich HMF (1906) Over the adsorption in solution. J Phys Chem 57:385–470

Fu K, Yue Q, Gao B, Sun Y, Zhu L (2013) Preparation, characterization and application of lignin-based activated carbon from black liquor lignin by steam activation. Chem Eng J 228: 1074–1082. doi:10.1016/j.cej.2013.05.028

Funke A, Ziegler F (2010) Hydrothermal carbonization of biomass: A summary and discussion of chemical mechanisms for process engineering. Biofuels Bioprod Biorefining 4:160–177. doi:10.1002/bbb.198

Government Decree (2013) 331/2013 Valtioneuvoston asetus kaatopaikoista (in Finnish). https://www.finlex.fi/fi/laki/alkup/2013/20130331. Accessed 23.10.2015

Guo S, Peng J, Li W, Yang K, Zhang L, Zhang S, Xia H (2009) Effects of CO_2 activation on porous structures of coconut shell-based activated carbons. Appl Surf Sci 255:443–8449. doi:10.1016/j.apsusc.2009.05.150

Gupta VK (1998) Equilibrium uptake, sorption dynamics, process optimization, and column operations for the removal of copper and nickel from aqueous solution and wastewater using activated slag, a low-cost adsorbent. Ind Eng Chem Res 37:192–202. doi:10.1021/ie9703898

Gupta VK, Ali I (2000) Utilisation of bagasse fly ah (a sugar industry waste) for the removal of copper and zinc from wastewater. Sep Purif Technol 18:131–140. doi:10.1016/S1383-5866(99)00058-1

Hasar H (2003) Adsorption of nickel(II) from aqueous solution onto activated carbon prepared from almond husk. J Hazard Mater B97:49–57. doi:10.1016/S0304-3894(02)00237-6
Hassan ML, Kassem NF, El-Lader AHA (2010) Novel Zr(IV)/sugar beet pulp composite for removal of sulfate and nitrate anions. J Appl Polym Sci 117:2205–2212. doi:10.1002/app.32063
Hawari AH, Mulligan CN (2006) Biosorption of lead(II), cadmium(II), copper(II) and nickel(II) by anaerobic granular biomass. Bioresour Technol 97:692–700. doi:10.1016/j.biortech.2005.03.033
Ho YS, McKay G (1999) Pseudo-second-order model for sorption processes. Process Biochem 34: 451–465. doi:10.1016/S0032-9592(98)00112-5
International Committee for Characterization and Terminology of Carbon (1982) First publication of 30 tentative definitions. Carbon 20:445–449
Ioannidou O, Zabaniotou A (2007) Agricultural residues as precursors for activated carbon production—a review. Renew Sustain Energy Rev 11:1966–2005. doi:10.1016/j.rser.2006.03.013
IUPAC Commision on Colloid and Surface Chemistry Including Catalysis (1985) Reporting physisorption Data of gas/solid systems with special reference to the determination of surface area and porosity. Pure Appl Chem 57:603
IUPAC Manual Of Symbols And Terminology, Appendix 2, Pt.1 (1972) J Colloid Interface Chem Pure Appl Chem 31:578
Kalavathy MH, Karthikeyan T, Rajgopal S, Miranda LR (2005) Kinetic and isotherm studies of Cu (II) adsorption onto H3PO4- activated rubber wood sawdust. J Colloid Interf Sci 292:354–362. doi:10.1016/j.jcis.2005.05.087
Kalavathy H, Karthik B, Miranda LR (2010) Removal and recovery of Ni and Zn from aqueous solution using activated carbon from Heveabrasiliensis: batch and column studies. Colloids Surf B Biointerfaces 78:291–302. doi:10.1016/j.colsurfb.2010.03.014
Karthikeyan KG, Mandla A, Wang TD, Kalbasi M (2004) Solution chemistry effects on orthophosphate adsorption by cationized solid wood residues. Environ Sci Technol 38:904–911. doi:10.1021/es034819z
Kilpimaa S, Kuokkanen T, Lassi U (2011) Physical and chemical properties of wood ash from burning and gasification processes. The Journal of Solid Waste Technology and Management, Proceedings in ICSW 2011, pp 879–887
Kilpimaa S, Runtti H, Lassi U, Kuokkanen T (2012) Chemical activation of gasification carbon residue for phosphate removal. Porous Media and its applications in Science, Engineering and Industry, AIP Conference proceedings 2012, pp 293–298
Kilpimaa S, Kuokkanen T, Lassi U (2013) Characterization and utilization potential of wood ash from combustion process and carbon residue from gasification process. Bioresources 8: 1011–1027
Kilpimaa S, Runtti H, Kangas T, Lassi U, Kuokkanen T (2014) Removal of phosphate and nitrate over a modified carbon residue from biomass gasification. Chem Eng Res Des 92:1923–1933. doi:10.1016/j.cherd.2014.03.019
Kilpimaa S, Runtti H, Kangas T, Lassi U, Kuokkanen T (2015) Physical activation of carbon residue from biomass gasification: novel sorbent for the removal of phosphates and nitrates from aqueous solution. J Ind Eng Chem 21:1354–1364. doi:10.1016/j.jiec.2014.06.006
Konwar LJ, Boro J, Deka D (2014) Review on latest developments in biodiesel production using carbon-based catalysts. Renew Sustain Energy Rev 29:546–564. doi:10.1016/j.rser.2013.09.003
Kouakou U, Ello AS, Yapo JA, Trokourey A (2013) Adsorption of iron and zinc on commercial activated carbon. J Environ Chem Ecotoxic 5:168–171. doi:10.5897/JECE2013.0264
Lagergren S (1898) About the theory of so-called adsorption of soluble substances. K Svenska Vetenskapsakad Handl 24:1–39
Lam E, Luong JHT (2014) Carbon materials as catalyst supports and catalysts in the transformation of biomass to fuels and chemicals. ACS Catal 4:3393–3410. doi:10.1021/cs5008393
Langmuir I (1918) The adsorption of gases on plane surfaces of glass, mica and platinum. J Am Chem Soc 40:1361–1403

Libra JA, Ro KS, Kammann C, Funke A, Berge ND, Neubauer Y, Titirici M-M, Führer C, Bens O, Kern J (2011) Hydrothermal carbonization of biomass residuals: a comparative review of the chemistry, process and applications of wet and dry pyrolysis. Biofuels 2:71–106. doi:10.4155/bfs.10.81

Li Y, Yu N, Yan P, Li Y, Zhou X, Chen S, Fan Z (2015) Fabrication of manganese dioxide nanoplates anchoring on biomass-derived cross-linked carbon nanosheets for high-performance asymmetric supercapacitors. J Power Sources 300:309–317. doi:10.1016/j.jpowsour.2015.09.077

Liu Z-R, Zhou S-Q (2010) Adsorption of copper and nickel on Na-bentonite. Process Saf Environ Prot 88:62–66. doi:10.1016/j.psep.2009.09.001

Lopez E, Soto B, Arias M, Nun´ez A, Rubinos D, Barral MT (1998) Adsorbent properties of red mud and its use for wastewater treatment. Water Res 32:1314–1322. doi:10.1016/S0043-1354(97)00326-6

Lussier MG, Shull JC, Miller DJ (1994) Activated carbon from cherry stones. Carbon 32:1493–1498. doi:10.1016/0008-6223(94)90144-9

Makowski P, Cakan RD, Antonietti M, Goettmann F, Titirici M-M (2008) Selective partial hydrogenation of hydroxy aromatic derivatives with palladium nanoparticles supported on hydrophilic carbon. Chem Commun 8:999–1001. doi:10.1039/B717928F

Marsh H, Rodríquez-Reinoso F (2006) Activated carbon. Elsevier, Amsterdam, p 536

Martínez ML, Torres MM, Guzmán CA, Maestri DM (2006) Preparation and characteristics of activated carbon from olive stones and walnut shells. Ind Crops Prod 23:23–28. doi:10.1016/j.indcrop.2005.03.001

Mezenner NY, Bensmaili A (2009) Kinetics and thermodynamic study of phosphate adsorption on iron hydroxide-eggshell waste. Chem Eng J 147:87–96. doi:10.1016/j.cej.2008.06.024

Moreno JC, Gómez R, Giraldo L (2010) Removal of Mn, Fe, Ni and Cu Ions from wastewater using cow bone charcoal. Materials 3:452–466. doi:10.3390/ma3010452

Moreno-Castilla C (2004) Adsorption of organic molecules from aqueous solutions on carbon materials. Carbon 42:83–94. doi:10.1016/j.carbon.2003.09.022

Nabais JV, Carrott P, Ribeiro Carrott MM, Luz V, Ortiz AL (2008) Influence of preparation conditions in the textural and chemical properties of activated carbons from a novel biomass precursor: the coffee endocarp. Bioresour Technol 99:7224–7231. doi:10.1016/j.biortech.2007.12.068

Namasivayam C, Sangeetha D (2008) Application of coconut coir pith for the removal of sulfate and other anions from water. Desalination 219:1–13. doi:10.1016/j.desal.2007.03.008

Öztürk N, Bekta TE (2004) Nitrate removal from aqueous solution by adsorption onto various materials. J Hazard Mater 112:155–162. doi:10.1016/j.jhazmat.2004.05.001

Papandreou A, Stournaras CJ, Panias D (2007) Copper and cadmium adsorption on pellets made from fired coal fly ash. J Hazard Mater 148:538–547. doi:10.1016/j.jhazmat.2007.03.020

Parab H, Joshi S, Shenoy N, Lali A, Sarma US, Sudersanan M (2006) Determination of kinetic and equilibrium parameters of the batch adsorption of Co(II), Cr(III) and Ni(II) onto coir pith. Process Biochem 41:609–615. doi:10.1016/j.procbio.2005.08.006

Ragan S, Megonnell N (2011) Activated carbon from renewable resources—lignin. Cellul Chem Technol 45:527–531

Ramke H-G, Blöhse D, Lehmann H -J, Fettig J (2009) Hydrothermal carbonization of organic waste. In: Cossu R, Diaz LF, Stegmann R (eds) Sardinia 2009: twelfth international waste management and landfill symposium, Sardinia, Italy, 05–09 October 2009, Proceedings, CISA Publisher

Repo E, Petrus R, Sillanpää M, Warchoł JK (2011) Equilibrium studies on the adsorption of Co(II) and Ni(II) by modified silica gels: one-component and binary systems. Chem Eng J 172:376–385. doi:10.1016/j.cej.2011.06.019

Repo E, Kurniawan TA, Warchol JK, Sillanpää MET (2009) Removal of Co(II) and Ni(II) ions from contaminated water using silica gel functionalized with EDTA and/or DTPA as chelating agents. J Hazard Mater 171:1071–1080. doi:10.1016/j.jhazmat.2009.06.111

Rezaee A, Godini H, Dehestani S, Khavanin A (2008) Application of impregnate almond shell activated carbon b y zinc and zinc sulfate for nitrate removal from water. Iran J Environ Health Sci Eng 5:125–130

Rodríquez-Reinoso F, Molina-Sabio M (1998) Textural and chemical characterization of microporous carbons. Adv Colloid Interf Sci 76–77:271–294. doi:10.1016/S0001-8686(98)00049-9

Roosta M, Ghaedi M, Daneshfar A, Sahraei R, Asghari A (2014a) Optimization of the ultrasonic assisted removal of methylene blue by gold nanoparticles loaded on activated carbon using experimental design methodology. Ultrason Sonochem 21:242–252. doi:10.1016/j.ultsonch.2013.05.014

Roosta M, Ghaedi M, Daneshfar A, Sahraei R (2014b) Experimental design based response surface methodology optimization of ultrasonic assisted adsorption of safaranin O by tin sulfide nanoparticle loaded on activated carbon. Spectrochim Acta A Mol Biomol Spectrosc 122:223–231. doi:10.1016/j.saa.2013.10.116

Roosta M, Ghaedi M, Shokri N, Daneshfar A, Sahraei R, Asghari A (2014c) Optimization of the combined ultrasonic assisted/adsorption methods for the removal of malachite green by gold nanoparticles loaded on activated carbon: experimental design. Spectrochim Acta A Mol Biomol Spectrosc 118:55–65. doi:10.1016/j.saa.2013.08.082

Runtti H, Tuomikoski S, Kangas T, Lassi U, Kuokkanen T, Rämö J (2014) Chemically activated carbon residue from biomass gasification as a sorbent for iron(II), copper(II) and nickel(II) ions. J Water Process Eng 4:12–24. doi:10.1016/j.jwpe.2014.08.009

Ruthven DM (1984) Principles of adsorption and adsorption processes. Wiley, New York, NY, 433 p

Sadaka S, Negi S (2009) Improvements of biomass physical and thermochemical characteristics via torrefaction process. Environ Prog Sustain Energy 28:427–434. doi:10.1002/ep.10392

Sevilla M, Maciá-Agulló JA, Fuertes AB (2011) Hydrothermal carbonization of biomass as a route for the sequestration of CO_2: chemical and structural properties of the carbonized products. Biomass Bioenergy 35:3152–3159. doi:10.1016/j.biombioe.2011.04.032

Sevilla M, Fuertes AB (2011) Sustainable porous carbons with a superior performance for CO_2 capture. Energy Environ 4:1765–1771. doi:10.1039/C0EE00784F

Shams SS, Zhang LS, Hu R, Zhang R, Zhu J (2015) Synthesis of graphene from biomass: a green chemistry approach. Mater Lett 161:476–479. doi:10.1016/j.matlet.2015.09.022

Sheibani A, Shishehbor MR, Alaei H (2012) Removal of Fe(III) ions from aqueous solutionby-hazelnut hull as an adsorbent. Int J Ind Chem 3:1–4. doi:10.1186/2228-5547-3-4

Srinivasakannan C, Zailani ABM (2004) Production of activated carbon from rubber wood sawdust. Biomass Bioenergy 27:89–96. doi:10.1016/j.biombioe.2003.11.002

Sun X, Wang X, Feng N, Qiao L, Li X, He D (2013) A new carbonaceous material derived from biomass source peels as an improved anode for lithium ion batteries. J Anal Appl Pyrolysis 100:181–185. doi:10.1016/j.jaap.2012.12.016

Suopajärvi H, Pongrácz E, Fabritius T (2014) Bioreducer use in Finnish blast furnace ironmaking—analysis of CO_2 emission reduction potential and mitigation cost. Appl Energy 124:82–93. doi:10.1016/j.apenergy.2014.03.008

Tan IAW, Ahmad AL, Hameed BH (2008) Preparation of activated carbon from coconut husk: optimization study on removal of 2,4,6-trichlorophenol using response surface methodology. J Hazard Mater 153:709–717. doi:10.1016/j.jhazmat.2007.09.014

Teker M, Saltabas Ö, İmamoğlu M (1997) Adsorption of cobalt by activated carbon from the rice hulls. J Environ Sci Health 32:2077–2086. doi:10.1080/10934529709376668

Temkin MJ, Pyzhev V (1940) Recent modifications to Langmuir isotherms. Acta Physiochim URSS 12:217–222

Tingjun F, Li Z (2015) Review of recent development in Co-based catalysts supported on carbon materials for Fischer–Tropsch synthesis. Chem Eng Sci 135:3–20. doi:10.1016/j.ces.2015.03.007

Tuomikoski S (2014) Utilisation of gasification carbon residues—activation, characterisation and use as an adsorbent. D.Sc. thesis, University of Oulu, Department of Physics and Chemistry

Wu D, Zhang B, Li C, Zhang Z, Kong H (2006) Simultaneous removal of ammonium and phosphate by zeolite synthesized from fly ash as influenced by salt treatment. J Colloid Int Sci 304:300–360. doi:10.1016/j.jcis.2006.09.011

Yabushita M, Kobayashi H, Haraa K, Fukuoka A (2014) Quantitative evaluation of ball-milling effects on the hydrolysis of cellulose catalysed by activated carbon. Catal Sci Technol 4:2312. doi:10.1039/C4CY00175C

Zeng L, Li X, Liu J (2004) Adsorptive removal of phosphate from aqueous solutions using iron oxide tailings. Water Res 38:1318–1326. doi:10.1016/j.watres.2003.12.009

Zhang L, Liu Z, Cui G, Chen L (2015) Biomass-derived materials for electrochemical energy storages. Prog Polym Sci 43:136–164. doi:10.1016/j.progpolymsci.2014.09.003

Zheng H, Liu D, Zheng Y, Liang S, Liu Z (2009) Sorption isotherm and kinetic modeling of aniline on Cr-bentonite. J Hazard Mater 167:141–147. doi:10.1016/j.jhazmat.2008.12.093

Zheng C, Zhou X, Cao H, Wang G, Liu Z (2014) Synthesis of porous graphene/activated carbon composite with high packing density and large specific surface area for supercapacitor electrode material. J Power Sources 258:290–296. doi:10.1016/j.jpowsour.2014.01.056

Pyrolysis of Biomass

Dooshyantsingh Oochit, Anurita Selvarajoo, and Senthil Kumar Arumugasamy

Abstract The energy crisis is a major issue of concern which is mainly associated with rapid population and economic growth. With the increasing trend in annual energy demand, there is a high dependence on fossil fuels to meet this expected increase in energy demand. However, the use of fossil fuels results in a substantial amount of environmental pollution, and this is why nowadays, a lot of researches are geared towards the generation of green and renewable energy. As such, thermochemical processes, particularly the pyrolysis process, are getting a lot of recognition lately. In this chapter, slow, fast and flash pyrolysis is reviewed. For the production of biochar, a low heating rate and temperature are employed. Bio-oil production is favoured at a high temperature together with a high heating rate. A short vapour residence time is also employed for bio-oil production. While for gas production, a long vapour residence time is favoured. This is employed together with a high temperature and a high heating rate. The operating conditions to be employed will depend on the final product desired. Pyrolysis effect on biochar properties and characterisation of biomass and biochar are included here. As thermochemical processes claim to be carbon neutral processes, investing in those technologies for biofuel production will help achieve a sustainable and green future.

Keywords Pyrolysis • Biomass • Agricultural wastes • Renewable energy • Biochar

1 Introduction

Current energy provision is almost totally dependent on fossil fuel, which is non-renewable. According to the International Energy Agency (2015), 81.4 % of the world energy needs in year 2013 were covered by fossil resources such as petroleum, natural gas and coal. Rapid industrialisation and highly increasing

D. Oochit • A. Selvarajoo • S.K. Arumugasamy (✉)
Department of Chemical and Environmental Engineering, Faculty of Engineering, University of Nottingham Malaysia Campus, Jalan Broga, Semenyih 43500, Selangor Darul Ehsan, Malaysia
e-mail: SenthilKumar.Arumugasamy@nottingham.edu.my

L. Singh, V.C. Kalia (eds.), *Waste Biomass Management – A Holistic Approach*,
DOI 10.1007/978-3-319-49595-8_10

world's population are the major causes for the high demand in energy. Fossil fuel combustion products are causing global problems. Some of the major global problems are global warming, acid rain, ozone layer depletion and urban air pollution (Veziroglu and Sahin 2008). It was stated by Lancaster (2002) that the main driver for moving away from fossil resources, other than long-term supply issues, is pollution. Eventually, global problems adversely affect our environment, which leads to negative impacts on the living creatures on earth. The availability of fossil energy is predicted to last only for a few more decades. In recent years, interest in alternative energy sources has increased tremendously. The awareness of the need for a sustainable future, and in recent years, is being fostered due to the rapid exhaustion of fossil fuel. Thus, it is important to consider alternative fuels that can sustain rapid development and also substitute fossil fuel when its reserves are exhausted.

Biomass is considered to be among the most viable renewable sources of energy due to its abundance (Sumathi et al. 2008). However, biomass is usually not suitable for direct energy production in its initial state due to its high moisture content and low bulk energy density (Chin et al. 2013). As such, energy is derived from biomass through various conversion technologies. The two main categories of biomass conversion technologies are thermochemical conversion and biological conversion. Pyrolysis is one of the several thermochemical methods to degrade the biomass. Without the presence of oxygen, the biomass is degraded by heat. The products of this biomass thermochemical conversion are char, bio-oil and gases (Hosseini et al. 2015). Char is the solid product while bio-oil is the liquid product. Another thermochemical method which is used to derive gaseous products from solid or liquid fuel is called gasification. High yield of gaseous products which are rich in carbon monoxide, carbon dioxide and methane can further be processed to produce hydrogen. Liquefaction is also a thermochemical process converting lignocellulose material, woodchips, organic wastes and other renewable fuels into liquid and solid fractions. The liquid fraction undergoes neutralisation, fermentation and distillation processes for the production of ethanol. Lignin components resulting from fermentation process are used for the generation of process heat and electricity (Lee 2007).

Biological conversion includes two process options, digestion and fermentation. Digestion process converts biomass using bacteria. Biogases which consist of a mixture of methane and carbon dioxide are produced in an anaerobic environment. Fermentation process produces ethanol from starch crops and sugar. The starch is converted to sugars by enzymes with yeast once the biomass is ground. Then the sugars are converted to ethanol with the help of yeast. This process has been commercialised in various countries at a very large scale (McKendry 2002).

2 Biomass as Alternative Energy

Energy from biomass is considered renewable as plants obtain their energy through photosynthesis from the sun, an inexhaustible source. Biomass is considered important since it can be cultivated almost anywhere and, therefore, provides social and economic development (Özçimen and Ersoy-Meriçboyu 2008). Currently biomass is the fourth largest energy source. Biomass represents 10 % of the world's total primary energy supply and 54.6 % of the global renewable energy supply (International Energy Agency 2014). Tremendously abundant crop residues are readily available in the agriculture-based countries. However, the residues from agriculture are not well utilised for alternative fuel production.

The energy derived from biomass is gaining importance. Depleting supply of fossil fuels is the main reason driving the interest on biomass energy. Together with rising environmental concerns, there is a need to shift away from the conventional fossil fuel (Liu et al. 2013). Generally, the conversions of biomass into fuels are done through thermochemical processes. Gasification, pyrolysis and liquefaction are mainly the processes involved. Biochar is a by-product in all of these conversion processes (Ghani et al. 2013). According to the International Biochar Initiative Organization, about 80 % of all crop and forestry residues may be converted to biochar and energy, an estimation by the year 2050 (Kołodyńska et al. 2012). Most companies are repeatedly searching for fuels that are less expensive than coal, oil and gas. Biochar can be produced from biomass during the thermochemical conversion process. When biomass is burnt or biochar being used, the carbon in the parent biomass or the biochar reacts with oxygen in the air to form carbon dioxide (CO_2) which is released into the atmosphere. The amount of CO_2 which is taken from the atmosphere during growing stage is equal to amount of carbon dioxide produced if fully combusted. So biomass can be considered as a carbon sink as there is no net addition of CO_2 (Saidur et al. 2011).

3 Pyrolysis

Pyrolysis is a thermal degradation process conducted without the presence of oxygen or any other oxidising agent. The thermal reactions proceed at temperatures from a few hundred up to 1500 °C (Lancaster 2002). The products are char, a carbon-rich solid residue and volatile fraction, that is, inclusive gaseous, vaporous and tar components (Demirbas 2002). Pyrolysis, a basic thermochemical process, is being used to convert biomass to a more useful fuel (Demirbas 2004a, b, c). Char that is produced from biomass, which is also known as biochar, can be used as fuel. Usually it is used in the form of briquettes, because it has a high heating value to sustain combustion (Apaydin-Varol and Pütün 2007). It can also be used as a feedstock to prepare activated carbon, used for soil remediation purposes (Suárez-Garcia et al. 2002; Daifullah et al. 2004; Karagöz et al. 2008; Chandra et al. 2009;

Tay et al. 2009; Denyes et al. 2013). The activated carbon produced can also be used in wastewater treatment (Liu et al. 2013; Muhammad et al. 2013; Krishnan et al. 2011; Rattier et al. 2014; Qing et al. 2015).

Reaction mechanisms of biomass pyrolysis are complex. It can be defined in three main steps as shown below (Demirbas 2004a, b, c):

$$\text{Biomass} \rightarrow \text{Water} + \text{Unreacted Residue} \tag{Eq.1}$$

$$\text{Unreacted Residue} \rightarrow (\text{Volatile} + \text{Gases}) + \text{Char} \tag{Eq.2}$$

$$\text{Char} \rightarrow (\text{Volatile} + \text{Gases}) + \text{Char} \tag{Eq.3}$$

The first step can be called pre-pyrolysis (Eq. 1). Internal rearrangements such as water elimination, bond breakage and appearance of free radicals occur in this stage. Formation of functional groups such as carbonyl, carboxyl and hydroperoxide groups also takes place in this stage. In the second step (Eq. 2), formation of volatiles and gases and primary biochar occurs (Demirbas 2004a, b, c). In the last step (Eq. 3), biochar undergoes chemical rearrangement and decomposes at a very slow rate. In this phase, carbon-rich residual solids (char) are formed.

There are three types of pyrolysis process. They are slow or conventional pyrolysis, fast pyrolysis and flash pyrolysis (Naik et al. 2010). Each of the processes has different operating conditions, and the product yield differs from one process to another. Conventional pyrolysis is mainly used for the generation of biochar. If the aim is to produce biochar, a low temperature and heating rate are employed (Tsai et al. 2007). Fast and flash pyrolysis are commonly used for bio-oil production. A high temperature, high heating rate and a short vapour residence time are employed to favour liquid yield. While gas production is favoured with a high temperature, high heating rate and long vapour residence time. The operating conditions are described in more detail in the section below.

3.1 Slow/Conventional Pyrolysis

This is one of the most commonly employed pyrolysis process since it has the highest biomass to biochar conversion ratio, which is around 35 %wt (Mohan et al. 2006). The temperature range for slow/conventional pyrolysis is around 300–650 °C with a residence time of a few minutes to a few hours. The heating rate is usually set at 10–30 °C/min but a lower heating rate of 5 °C/min is not unusual (Onay and Kockar 2003; Wang et al. 2015). Slow pyrolysis is mainly used when the main objective is to favour biochar yield.

Table 1 Typical operating parameters for pyrolysis process

Pyrolysis process	Solid residence time (s)	Heating rate (K/s)	Particle size (mm)	Temp. (K)
Slow	450–550	0.1–1	5–50	550–950
Fast	0.5–10	10–200	<1	850–1250
Flash	<0.5	>1000	<0.2	1050–1300

Source: Balat et al. (2009)

3.2 *Fast Pyrolysis*

When bio-oil yield is desirable, high operating temperatures and heating rate along with short vapour residence time are usually employed (Demirbas 2001a, b). The reaction temperature is in the range of 577–977 °C (Naik et al. 2010) with a heating rate of more than 100 °C/min. Preferred particle size of biomass for fast pyrolysis is less than 1 mm (Tsai et al. 2007).

Production of gaseous products, also known as syngas, is usually favoured when high heating rates are involved to achieve temperatures above 650 °C along with long residence time (Fornasiero and Graziani 2011).

3.3 *Flash Pyrolysis*

With a heating rate of more than 12 °C/min, flash pyrolysis occurs in a temperature range of 777–1027 °C. The residence time employed for flash pyrolysis is less than 0.5 s. Particle size of the sample is less than 0.2 mm. Flash pyrolysis is mostly used for the production of bio-oil, which has high conversion efficiencies of up to 70 % wt (Naik et al. 2010). For pyrolysis processes, the typical operating parameters are shown in Table 1.

The various pyrolysis studies reported in the literature are summarised in Table 2.

4 Pyrolysis Effects on Biochar Properties

A lot of researches (Yuan et al. 2015; Kim et al. 2012) have been done on the conversion of different biomass during pyrolysis. It can be concluded that the temperature during pyrolysis is one of the main key parameters that governs the chemical and physical parameters of the biochar. At high pyrolysis temperatures, Asadullah et al. (2010) reported that a more ordered char structure is achieved, and this was confirmed by Tan et al. (2014) who observed that the char microstructure effectively develops better at higher temperatures. A more ordered structure would result in a char having a higher surface area. For char that will be subsequently used

Table 2 Summary of previous pyrolysis studies

Sample	Sample size	Operating conditions	Results obtained	Reference
Waste wood, cardboard wastes, textile wastes	20 × 20 × 20 mm (150–200 g)	T: 350–700 °C HR: 10 °C/min RT: 2 h N_2 flowrate: 2L/min	• At 700 °C, the yield of char was 31–34 % of the original mass. At 350 °C, the yield of char was 21–25 % at 700 °C • At increasing nitrogen flowrate, residence time of pyrolysis reduced and re-polymerisation prevented • Char, oil and gas yields affected by nitrogen flowrate	Yang et al. (2007)
Rice straw, sawdust	2 × 10 mm 0.25 mm	P: 1 atm T: 800 °C	• Higher product gas yield achieved at higher heating rate • Temperature of gasifier strongly influenced gas yield • For rice straw: – Gas yield was 39 vol % at 600 °C and increased to 49 % at 700 °C due to tarry component cracking – Gas yield increased by 12.7 % when there is reduction in the particle size (original to 0.132 mm) • For sawdust: – Gas yield increased relatively up to 14 % when the diameter was reduced to 0.25 mm – As temperature increased, gas yield increased, but solid fraction decreased, while liquid fraction yield reached a maximum at about 500 °C • For rice straw and sawdust, increasing gas yield found when increase in residence time of volatile phase Due to better cracking reaction at a longer residence time	Chen et al. (2003)

Beech trunk bark	0.25–0.425 mm	T: 277, 327, 377, 427, 477, 527 °C HR: 4.5, 7.5, 12.5, 22.5, 42.5, 102.5 °C/min	• Heating value of products increased with increasing heating rate and temperature • Char yield decreased when temperature and heating rate increased • Char yield decreased from 59 to 42 % for 2 °C/s run and from 43 to 29 % for 100 °C/s run when temperature was raised from 278 to 527 °C • A fast depolymerisation of the solid material to primary volatiles occurs when there is a rapid heating	Demirbas (2004a, b, c)
Rice husk	< 0.5 mm	T: 400–800 °C HR: 100–500 °C/min RT: 1–8 min N_2 flowrate: 500–1500 cm^3/min	• Char yield decreased when temperature increased • When residence time increased, oil yield increased and then decreased slightly. This was possibly due to thermal cracking of the pyrolysis products • When particle size reduced from 0.5 to 0.125 mm, char yield decreased • Higher nitrogen flowrate did not affect pyrolysis product yield	Tsai et al. (2007)
Pistachio shell	1.82 mm	T: 300, 400, 500, 550, 700 °C RT: 30 min	• Char yield about 28 % attained at 300 °C, and it decreased as temperature was increased, reaching 23 % at 700 °C • At 300 °C, gas yield about 25 % was achieved, and it increased with temperature, reaching 30 % at 700 °C	Apaydin-Varol and Pütün (2007)
Cotton cocoon shell, tea factory waste and olive husk	0.063–0.015 mm	T: 500, 650, 700 and 750 °C HR: 4.5 °C/min Catalyst: $ZnCl_2$, Na_2CO_3 and K_2CO_3	• Gas yield increased with increasing temperature	Demirbas (2001a, b)

(continued)

Table 2 (continued)

Sample	Sample size	Operating conditions	Results obtained	Reference
Olive husk, corncob, tea waste	0.5–2.2 mm	T: 675–975 °C	• Char yield decreased as particle size reduced (from 2.2 mm to 0.5 mm) • Char produced with smaller biomass particle has higher reactivity when reacted with CO_2 and H_2O	Demirbas (2004a, b, c)
Rice straw, rice husk, corncob	Less than 74 μm	Sensitive thermobalance was used with: HR: 10 °C/min T: 600 °C Helium flowrate of 50 ml/min	• At 600 °C, char yield for rice straw was 30 %, rice husk was 47 % and corncob was 19 %	Worasuwannarak et al. (2007)
Rice husk	0.1 m × 0.1 m × 0.2 m	Fixed-bed reactor T: 200–650 °C HR: 10 °C/min RT: 60 min	• Drastic weight loss occurred from 350 to 400 °C was due to destruction of cellulose and hemicelluloses in biomass • Volatile matter dropped rapidly from 300–400 °C and remained constant from 500 °C	Maiti et al. (2006)
Beech wood	0.35–0.80 mm	Solar reactor T: 800, 1200 and 2000 °C HR: 5, 50, 150, 450 °C/s T: 600, 1600 °C. HR: 50 °C/s Argon flowrate: 6 L/min RT: 6 min	• More ordered char structure with increased temperature and heating rate (> 50 °C/s) • Surface area and pore volume increased with increasing pyrolysis temperature and reached maximum at 1200 °C. Then surface area and pore volume reduced significantly at 2000 °C • Char reactivity correlated with the evolution of char surface area and pore volume with temperature and heating rate	Zeng et al. (2015)

Empty fruit bunch	710–1000 μm	Fixed-bed reactor Temperature: 442.15 °C Heating rate: 50 °C/min N_2 flowrate: 150, 200, 300, 400, 500 cm^3/min RT: 6, 8, 10, 12 min	• The highest bio-oil yield of 45.72 % was obtained at 10 min holding time With biochar yield of 25.88 % and 28.40 % gaseous yield • At N_2 flowrate of 200 cm^3/min, bio-oil yield of 46.02 %, maximum yield • Biochar and gas production were 26.69 % and 27.29 %, respectively	Mohamed et al. (2013)
Oil palm trunk (OPT) and empty fruit bunch (EFB)	$dp_1 < 0.25$ mm $0.25 \leq dp_2 \leq 0.30$ mm	Pyrolysis in thermogravimetric analyser T: 700 °C N_2 flowrate: 150 ml/min HR: 50 °C/min and 80 °C/min	• Smaller OPT produced 0.04–0.86 % more volatile product when degraded at both heating rates • EFB showed a 3.81–9.81 % increment in volatile product • As heating rate increased, peak degradation was sped up by 1.42–1.56 mg/s at dp_1 and 3.13–4.92 mg/s at dp_2	Alias et al. (2015)
Empty fruit bunch	0.2–0.5 mm	Fixed-bed reactor Catalysts: K_2CO_3, $Ca(OH)_2$, MgO T: 400, 500, 600 °C HR: 10 °C/min N_2 flowrate: 200 mL/min Catalyst: 10 wt. % RT: 10 min	• Maximum percentage yield of 42.64 % of bio-oil was obtained from the $Ca(OH)_2$ catalysed EFB pyrolysis at 600 °C • $Ca(OH)_2$ catalysed pyrolysis showed an increase of 10.68 % in the desirable phenolic content and a reduction of 35.21 % in the undesirable acidic components of the bio-oils as compared to a phenolic content of 16.74 % and acidic components of 42.88 % of the non-catalysed EFB pyrolysis bio-oil.	Auta et al. (2014).

T temperature, *HR* heating rate, *RT* residence time, *P* pressure for gasification, higher surface area exhibits better reactivity when reacted with gasifying agent than biochar produced at low temperatures (Zeng et al. 2015). Furthermore, it was observed that as temperature increases, biochar yield decreases (Mohan et al. 2006; Zhang et al. 2015a, b). Employing high temperature also enhances the carbon conversion efficiency, and it is an important factor to consider if the biochar is going to be used for carbon dioxide reduction (Kambo and Dutta 2015; Zhang et al. 2015a, b). Also, char produced at high temperature results in a biochar which is more resistant to biological decay, and hence the release of CO_2 to the atmosphere is minimal as compared to a biochar which is produced at lower pyrolysis temperature (Al-Wabel et al. 2013).

Other important parameters which are affected by the temperature are the surface area. Porosity of the biochar is also affected by the temperature. Méndez et al. (2013) reported that porosity and surface area of the char increased with increasing temperature. Knowing how pyrolysis temperature influences both the parameters is important because the porosity and surface area will dictate the final use of the biochar. For example, when preparing decontamination adsorbent, the surface area and porosity have to be studied well as they affect the adsorption capacity (Demirbas et al. 2006). Burhenne and Aicher (2014) experimented the effects of pyrolysis temperature on the structure and surface area of produced biochar from dry spruce woodchips. It was noted that when the temperature increased from 500 to 800 °C, the surface area along with the pore volume of the char increased. Park et al. (2013) observed that biochar from debarked loblolly pine at 300 and at 350 °C had very low surface area as compared to biochar from the same biomass produced at 500 and 700 °C, respectively.

5 Characterisation of Biochar from Biomass

The quality of biomass fuel for usage in the pyrolysis unit can be determined from the proximate analysis, which gives information about the products evolved as a result of controlled heating. Proximate analysis includes moisture content, volatile matter, fixed carbon and ash content (Miskam et al. 2009). The volatile and fixed carbon content indicates their importance on how the biomass will be utilised as a fuel source and how easily biomass can be ignited and subsequently combusted depending on the amount of volatile and fixed carbon content. Fuel with high volatile matter content produces more tar. A high volatile matter content is favourable if the aim is to produce bio-oil or gaseous products. In thermal conversion it is necessary to reduce the moisture content as moisture content contributes to a low heating value of the feedstock. It reduces the thermal efficiency since heat will be used for removing the moisture content (Miskam et al. 2009). McKendry (2002) reported that moisture content above 30 % is difficult to ignite and reduces the heating value of the char due to the need to evaporate the additional moisture before pyrolysis takes place. Smooth running of furnace is affected by ash content,

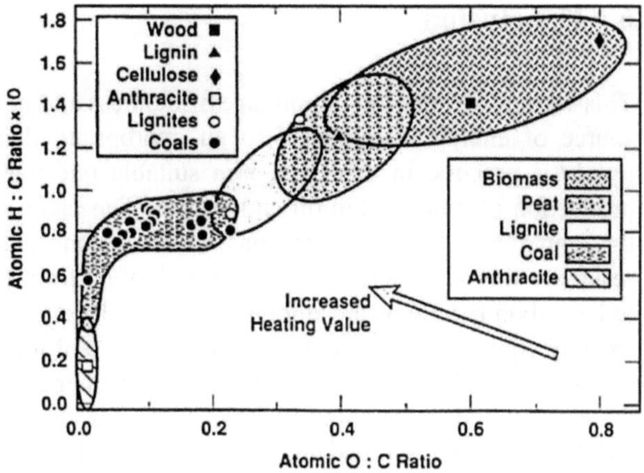

Fig. 1 Van Krevelen diagram (McKendry 2002)

and high mineral content composition makes thermal conversion impossible (Miskam et al. 2009). Furthermore, low ash content is usually desired if the biomass will be used as a source of fuel since ash will affect the fusion temperatures and will induce clinker formation.

Ultimate analysis gives the elemental composition of carbon (C), hydrogen (H), oxygen (O), nitrogen (N) and sulphur (S) in the biomass in weight percentage. According to Miskam et al. (2009), biomass has substantially more oxygen and less carbon than other fuels and also has relatively low sulphur content. The main elements of biomass are carbon, oxygen and hydrogen. Nitrogen may be present in certain biomass types. In comparison with coal, biomass contains less carbon and more oxygen. Atomic ratio of H:C and O:C for biomass, peat, lignite, coal and anthracite region is shown in Fig. 1, Van Krevelen diagram (McKendry 2002). Energy value of a biofuel reduces with higher proportion of oxygen and hydrogen. While compared with fossil fuels, such as coal which has high carbon content, the energy value of fuel increases. In carbon–carbon bonds, the energy is higher. Lower energy values are found in carbon–oxygen and carbon–hydrogen bonds. Higher H:C atomic ratio with lower O:C atomic ratio results in increased heating value of solid fuels.

Another important parameter to consider when biomass is used as a source of biofuel is the heating value. The heating value gives an overall indication of the energy contained in the chemical bonds present in the biomass. However, the heating value is usually expressed as the higher heating value (HHV) which can be determined by using a bomb calorimeter. The HHV or gross calorific value can be described as the amount of heat released from combustion, including the latent heat of vaporisation. Another familiar term used is the lower heating value (LHV) which also represents the energy released during combustion but does not include the latent heat of vaporisation. The term HHV is mostly used by natural gas and coal industries, while the term LHV is mostly used for engine and turbine companies (Crocker 2010).

6 Conclusion

This section outlines the importance of biomass and how it can be used as a viable source of energy if subjected to an appropriate thermochemical process. The pyrolysis process, in this case, is a suitable one that can be employed for the production of biochar, bio-oil and syngas. The favoured products will depend on the employed operating conditions which include temperature, heating rate and residence time. Furthermore, generating biofuel from the pyrolysis process is said to be carbon neutral as the amount of carbon dioxide generated during pyrolysis is roughly equal to the amount of carbon dioxide absorbed by plants for photosynthesis (Zhu et al. 2008). Thus, generating energy from biomass can be beneficial to the environment as opposed to fossil fuels, for example, coal, which is the most carbon-intensive fossil fuel and contributes largely to the worldwide carbon dioxide emission (Reaver and Khare 2014).

7 Opinion

Using biomass for the production of biofuel through thermochemical processes is a good initiative that needs to get more recognition in a way such that investors will be willing to invest in the technology. By employing thermochemical process, energy can be produced out of a raw material or biomass that would have normally been disposed of in a landfill or burned in an incinerator. The pyrolysis process seems to be a viable choice in an environmental and an energy-wise point of view. Biomass seems to be a good feedstock for pyrolysis process for the production of biochar. The produced biochar can be used as an alternative solid fuel for energy source. It can also be used for soil remediation and as adsorbent in wastewater treatment.

References

Alias NB, Ibrahim N, Hamid MKA, Hasbullah H, Ali RR, Kasmani RM (2015) Investigation of oil palm wastes pyrolysis by thermogravimetric analyser for potential biofuel production. Energy Procedia 75:78–83. doi:10.1016/j.egypro.2015.07.141

Al-Wabel M, Al-Omran A, El-Naggar AH, Nadeem M, Usman ARA (2013) Pyrolysis temperature induced changes in characteristics and chemical composition of biochar produced from conocarpus wastes. Bioresour Technol 131:374–379. doi:10.1016/j.biortech.2012.12.165

Apaydin-Varol E, Pütün AE (2007) Slow pyrolysis of pistachio shell. Fuel 86:1892–1899. doi:10.1016/j.fuel.2006.11.041

Asadullah M, Zhang S, Li CZ (2010) Evaluation of structural features of chars from pyrolysis of biomass of different particle sizes. Fuel Process Technol 91:877–881. doi:10.1016/j.fuproc.2009.08.008

Auta M, Ern LM, Hameed BH (2014) Fixed-bed catalytic and non-catalytic empty fruit bunch biomass pyrolysis. J Anal Appl Pyrol 107:67–72. doi:10.1016/j.jaap.2014.02.004

Balat M, Kirtay E, Balat H (2009) Main routes for the thermo-conversion of biomass into fuels and chemicals. Part 1: pyrolysis systems. Energy Convers Manage 50:3147–3157. doi:10.1016/j.enconman.2009.08.014

Burhenne L, Aicher T (2014) Benzene removal over a fixed bed of wood char: the effect of pyrolysis temperature and activation with CO_2 on the char reactivity. Fuel Process Technol 127:40–148. doi:10.1016/j.fuproc.2014.05.034

Chandra TC, Mirna MM, Sunarso J, Sudaryanto Y, Ismadji S (2009) Activated carbon from durian shell: preparation and characterization. J Taiwan Inst Chem E 40:457–462. doi:10.1016/j.jtice.2008.10.002

Chen G, Andries J, Luo Z, Spliethoff H (2003) Biomass pyrolysis/gasification for product gas production: the overall investigation of parametric effects. Energy Convers Manage 44:1875–1884. doi:10.1016/S0196-8904(02)00188-7

Chin KL, H'ng PS, Go WZ, Wong WZ, Lim TW, Maminski M, Paridah MT, Luqman AC (2013) Optimization of torrefaction conditions for high energy density solid biofuel from oil palm biomass and fast growing species available in Malaysia. Ind Crop Prod 49:768–774. doi:10.1016/j.indcrop.2013.06.007

Crocker M (2010) Thermochemical conversion of biomass to liquid fuels and chemicals. Royal Society of Chemistry, London. doi:10.1039/9781849732260

Daifullah AAM, Girgis BS, Gad HMH (2004) A study of the factors affecting the removal of humic acid by activated carbon prepared from biomass material. Colloid Surface A 235 (1–3):1–10. doi:10.1016/j.colsurfa.2003.12.020

Demirbas A (2001a) Biomass resource facilities and biomass conversion processing for fuels and chemicals. Energy Convers Manage 42:1357–1378. doi:10.1016/S0196-8904(00)00137-0

Demirbas A (2001b) Yields of hydrogen-rich gaseous products via pyrolysis from selected biomass samples. Fuel 80:1885–1891. doi:10.1016/S0016-2361(01)00070-9

Demirbas A (2002) Gaseous products from biomass by pyrolysis and gasification: effects of catalyst on hydrogen yield. Energy Convers Manage 43:897–909. doi:10.1016/S0196-8904(01)00080-2

Demirbas A (2004a) Combustion characteristics of different biomass fuels. Prog Energy Combust 30:219–230. doi:10.1016/j.pecs.2003.10.004

Demirbas A (2004b) Determination of calorific values of bio-chars and pyro-oils from pyrolysis of beech trunkbarks. J Anal Appl Pyrol 72:215–219. doi:10.1016/j.jaap.2004.06.005

Demirbas A (2004c) Effects of temperature and particle size on bio-char yield from pyrolysis of agricultural residues. J Anal Appl Pyrol 72:243–248. doi:10.1016/j.jaap.2004.07.003

Demirbas A, Pehlivan E, Altun T (2006) Potential evolution of Turkish agricultural residues as bio-gas, bio-char and bio-oil sources. Int J Hydrogen Energy 31:613–620. doi:10.1016/j.ijhydene.2005.06.003

Denyes MJ, Rutter A, Zeeb BA (2013) In Situ application of activated carbon and biochar to PCB-contaminated soil and the effects of mixing regime. Environ Pollut 182:201–208. doi:10.1016/j.envpol.2013.07.016

Fornasiero P, Graziani M (2011) Renewable resources and renewable energy: a global challenge. CRC Press, New York, NY

Ghani WAWAK, Mohd A, da Silva G, Bachmann RT, Taufiq-Yap YH, Rashid U, AAH A-M (2013) Biochar production from waste rubberwood-sawdust and its potential use in C sequestration: chemical and physical characterization. Ind Crop Prod 44:18–24. doi:10.1016/j.indcrop.2012.10.017

Hosseini SE, Wahid MA, Ganjehkaviri A (2015) An overview of renewable hydrogen production from thermochemical process of oil palm solid waste in Malaysia. Energy Convers Manage 94:415–429. doi:10.1016/j.enconman.2015.02.012

International Energy Agency (2014) Key world energy statistics. Accessed 28 Oct 2015. https://www.iea.org/publications/freepublications/publication/KeyWorld2014.pdf
International Energy Agency (2015) Key world energy statistics. Accessed 23 Dec 2015. https://www.iea.org/publications/freepublications/publication/KeyWorld_Statistics_2015.pdf
Kambo HS, Dutta A (2015) A comparative review of biochar and hydrochar in terms of production, physico-chemical properties and applications. Renew Sust Energy Rev 45:359–378. doi:10.1016/j.rser.2015.01.050
Karagöz S, Tay T, Ucar S, Erdem M (2008) Activated carbons from waste biomass by sulfuric acid activation and their use on methylene blue adsorption. Bioresour Technol 99:6214–6222. doi:10.1016/j.biortech.2007.12.019
Kim KH, Kim JY, Cho TS, Choi JW (2012) Influence of pyrolysis temperature on physicochemical properties of biochar obtained from the fast pyrolysis of pitch pine (Pinusrigida). Bioresour Technol 118:158–162. doi:10.1016/j.biortech.2012.04.094
Kołodyńska D, Wnętrzak R, Leahy JJ, Hayes MHB, Kwapiński W, Hukicki Z (2012) Kinetic and adsorptive characterization of biochar in metal ions removal. Chem Eng J 197:295–305. doi:10.1016/j.cej.2012.05.025
Krishnan KA, Sreejalekshmi KG, Baiju RS (2011) Nickel(II) adsorption onto biomass based activated carbon obtained from sugarcane bagasse pith. Bioresour Technol 102:10239–10247. doi:10.1016/j.biortech.2011.08.069
Lancaster M (2002) Green chemistry: an introductory text. RSC Publishing, Cambridge
Lee S (2007) Energy from biomass conversion. In: Lee S, Speight JG, Loyalka SK (eds) Handbook of alternative fuel technologies. CRC Press Taylor's & Francis Group, Boca Raton, FL, pp 377–394
Liu Z, Quek A, Kent Hoekman S, Balasubramanian R (2013) Production of solid biochar fuel from waste biomass by hydrothermal carbonization. Fuel 103:943–949. doi:10.1016/j.fuel.2012.07.069
Maiti S, Dey S, Purakayastha S, Ghosh B (2006) Physical and thermochemical characterization of rice husk char as a potential biomass energy source. Bioresour Technol 97:2065–2070. doi:10.1016/j.biortech.2005.10.005
McKendry P (2002) Energy production from biomass (part 1): overview of biomass. Bioresour Technol 83:37–46. doi:10.1016/S0960-8524(01)00118-3
Méndez A, Terradillos M, Gasco G (2013) Physicochemical and agronomic properties of biochar from sewage sludge pyrolysed at different temperatures. J Anal Appl Pyrol 102:124–130. doi:10.1016/j.jaap.2013.03.006
Miskam A, Zainal ZA, Yusof IM (2009) Characterization of sawdust residues for cyclone gasifier. J Eng Appl Sci 9:2294–2300. doi:10.3923/jas.2009.2294.2300
Mohamed AR, Hamzah Z, Daud MZM, Zakaria Z (2013) The effects of holding time and the sweeping nitrogen gas flowrates on the pyrolysis of EFB using a fixed-bed reactor. Procedia Eng 53:185–191. doi:10.1016/j.proeng.2013.02.024
Mohan D, Pittman CU, Steele PH (2006) Pyrolysis of wood/biomass for bio-oil: a critical review. Energy Fuel 20:848–889. doi:10.1021/ef0502397
Muhammad B, Jehanzeb AS, Tayyab A, Syed MHG, Adnan AT, Arshid P, Hajira H, Qaisar M (2013) Waste biomass adsorbents for copper removal from industrial wastewater—a review. J Hazard Mater 263:322–333. doi:10.1016/j.jhazmat.2013.07.071
Naik SN, Goud VV, Rout PK, Dalai AK (2010) Production of first and second generation biofuels: a comprehensive review. Renew Sust Energy Rev 14:578–597. doi:10.1016/j.rser.2009.10.003
Onay O, Kockar OM (2003) Slow, fast and flash pyrolysis of rapeseed. Renew Energy 28:2417–2433. doi:10.1016/S0960-1481(03)00137-X
Özçimen D, Ersoy-Meriçboyu A (2008) A study on the carbonization of grapeseed and chestnut shell. Fuel Process Technol 89:1041–1046. doi:10.1016/j.fuproc.2008.04.006
Park J, Hung I, Gan Z, Rojas OJ, Lim KH, Park S (2013) Activated carbon from biochar: Influence of its physicochemical properties on the sorption characteristics of phenanthrene. Bioresour Technol 149:383–389. doi:10.1016/j.biortech.2013.09.085

Qing YH, Meng L, Can W, Min J (2015) Influence of powdered activated carbon addition on water quality, sludge properties, and microbial characteristics in the biological treatment of commingled industrial wastewater. J Hazard Mater 295:1–8. doi:10.1016/j.jhazmat.2015.03.070

Rattier M, Reungoat J, Keller J, Gernjak W (2014) Removal of micropollutants during tertiary wastewater treatment by biofiltration: role of nitrifiers and removal mechanisms. Water Res 54:89–99. doi:10.1016/j.watres.2014.01.030

Reaver NGF, Khare SV (2014) Imminence of peak in US coal production and overestimation of reserves. Int J Coal Geol 131:90–105. doi:10.1016/j.coal.2014.05.013

Saidur R, Abdelaziz EA, Demirbas A, Hossain MS, Mekhilef S (2011) A review on biomass as a fuel for boilers. Renew Sust Energy Rev 15:2262–2289. doi:10.1016/j.rser.2011.02.015

Suárez-García F, Martínez-Alonso A, Tascón JMD (2002) Pyrolysis of apple pulp: chemical activation with phosphoric acid. J Anal Appl Pyrol 63:283–301. doi:10.1016/S0165-2370(01)00160-7

Sumathi S, Chai SP, Mohamed AR (2008) Utilization of oil palm as a source of renewable energy in Malaysia. Renew Sust Energy Rev 12:2404–2421. doi:10.1016/j.rser.2007.06.006

Tan C, Yaxin Z, Hongtao W, Wenjing L, Zeyu Z, Yuancheng Z, Lulu R (2014) Influence of pyrolysis temperature on characteristics and heavy metal adsorptive performance of biochar derived from municipal sewage sludge. Bioresour Technol 164:47–54. doi:10.1016/j.biortech.2014.04.048

Tay T, Ucar S, Karagöz S (2009) Preparation and characterization of activated carbon from waste biomass. J Hazard Mater 165:481–485. doi:10.1016/j.jhazmat.2008.10.011

Tsai WT, Lee MK, Chang YM (2007) Fast pyrolysis of rice husk: product yields and compositions. Bioresour Technol 98:22–28. doi:10.1016/j.biortech.2005.12.005

Veziroglu TN, Sahin S (2008) 21st century's energy: hydrogen energy system. Energ Convers Manage 49:1820–1831. doi:10.1016/j.enconman.2007.08.015

Wang Y, Li J, Gu J, Zhou Z, Wang Z (2015) Artificial neural networks for infectious diarrhea prediction using meteorological factors in Shangai (China). Appl Soft Comput 35:280–290. doi:10.1016/j.asoc.2015.05.047

Worasuwannarak N, Sonobe T, Tanthapanichakoon W (2007) Pyrolysis behaviors of rice straw, rice husk, and corncob by TG-MS technique. J Anal Appl Pyrol 78:265–271. doi:10.1016/j.jaap.2006.08.002

Yang YB, Phan AN, Ryu C, Sharifi V, Swithenbank J (2007) Mathematical modelling of slow pyrolysis of segregated solid wastes in a packed-bed pyrolyser. Fuel 86:169–180. doi:10.1016/j.fuel.2006.07.012

Yuan H, Lu T, Huang H, Zhao D, Kobayashi N, Chen Y (2015) Influence of pyrolysis temperature on physical and chemical properties of biochar made from sewage sludge. J Anal Appl Pyrol 112:282–289. doi:10.1016/j.jaap.2015.01.010

Zeng K, Minh DP, Gauthier D, Weiss-Horsala E, Nzihou A, Flamant G (2015) The effect of temperature and heating rate on char properties obtained from solar pyrolysis of beech wood. Bioresour Technol 182:114–119. doi:10.1016/j.biortech.2015.01.112

Zhang J, Liu J, Liu R (2015a) Effects of pyrolysis temperature and heating time on biochar obtained from the pyrolysis of straw and lignosulfonate. Bioresour Technol 176:288–291. doi:10.1016/j.biortech.2014.11.011

Zhang H, Voroney RP, Price GW (2015b) Effects of temperature and processing conditions on biochar chemical properties and their influence on soil C and N transformations. Soil Biol Biochem 83:19–28. doi:10.1016/j.soilbio.2015.01.006

Zhu W, Song W, Lin W (2008) Catalytic gasification of char from co-pyrolysis of coal and biomass. Fuel Process Technol 89:890–896. doi:10.1016/j.fuproc.2008.03.001

Liquefaction of Biomass for Bio-oil Products

Hua-jun Huang, Xing-zhong Yuan, and Guo-qiang Wu

Abstract Biomass is considered to be one of the most important renewable energy sources and will become an important part of future sustainable energy system. Liquefaction is a promising thermochemical conversion process that is applied to convert biomass into bio-oil (target product), biochar, and gases. The process is usually carried out in water or another suitable solvent at 250–400 °C under pressures of 5–25 MPa. This chapter firstly collects and sorts out the available review literatures on the liquefaction of biomass. Next, the effects of processing parameters on the liquefaction of biomass are briefly summarized. Then, the characterization and application studies of biochar (solid-phase product) are discussed. Finally, the main upgradation methods of bio-oil (liquid-phase product) are introduced. The ultimate aim of this chapter is to offer some reference for the study of biomass liquefaction.

Keywords Biomass • Liquefaction • Biochar • Bio-oil

1 Introduction

The application of biomass-derived compounds is increasingly sought as a good alternative to replace petroleum as a result of the depletion of fossil resources and the increasing environmental concerns (Faba et al. 2015). Biomass is a matter derived directly or indirectly from plants and can be utilized for energy generation or in various materials. Biomass is often considered a bio-resource or a bio-derived

H.-j. Huang (✉)
School of Land Resources and Environment, Jiangxi Agricultural University, Zhimin Road 1101, Nanchang 330045, Jiangxi Province, PR China
e-mail: huanghuajun2004@126.com

X.-z. Yuan
College of Environmental Science and Engineering, Hunan University, Changsha 410082, PR China

G.-q. Wu
College of Food Science and Engineering, Jiangxi Agricultural University, Nanchang 330045, PR China

L. Singh, V.C. Kalia (eds.), *Waste Biomass Management – A Holistic Approach*,
DOI 10.1007/978-3-319-49595-8_11

resource. Its resource base contains hundreds of thousands of terrestrial and aquatic plant species; various forestry, agricultural and industrial residues, and process wastes; animal wastes; and sewage sludge (Ibrahim et al. 2014). Biofuel from biomass can be produced by biochemical/biological and thermochemical processes. As listed in Fig. 1, thermochemical processes consist of pyrolysis, direct combustion, liquefaction, and gasification. The most widely used technology for biomass conversion is the direct combustion. The biomass is usually directly burnt with enough air existing, and the chemical energy stored in biomass is converted into mechanical power, heat, electricity, etc. Decomposition through pyrolysis process takes place in the absence of oxygen at elevated temperatures. The biomass is converted into solid charcoal, gases, and liquid (bio-oil). Biomass gasification is used to produce combustible gases (e.g., H_2, CH_4, CO, and CO_2) from carbonaceous biomass. This process is operated at specific heating values with a supply of partial oxygen (O_2) or suitable oxidants, for example, steam and CO_2. The liquefaction of biomass is often carried out at low temperature and high pressure in water or other suitable solvents. And biomass is converted into three products, i.e., a bio-oil fraction (target product), a solid residue fraction (biochar), and a gas fraction (Goyal et al. 2008; Zhang et al. 2010; Ma et al. 2012; Panwar et al. 2012).

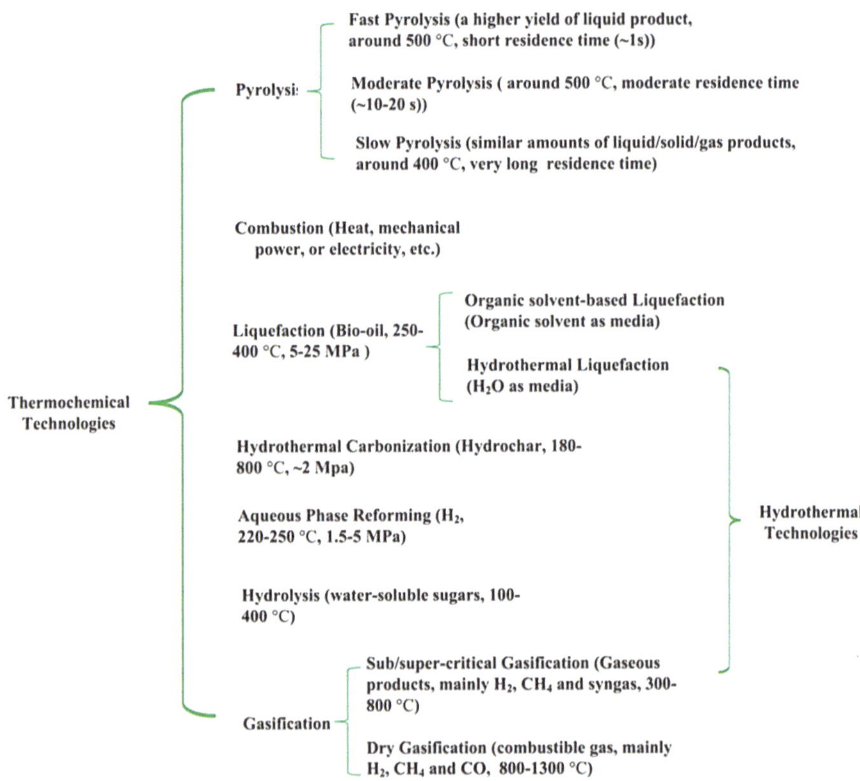

Fig. 1 Thermochemical processes for bioenergy production from biomass

The gasification/liquefaction of biomass in water also belongs to the hydrothermal technology.

Hydrothermal technology could be described as various processes of treating/disposing biomass with water as media at certain temperatures and pressures, namely, in sub- and supercritical water. The main principle of this technology is to oxidize the organic compounds in biomass on the basis of the capacity of sub- and supercritical water. Various products, such as bio-oils, smaller liquid compounds, biochars, and gases, were formed. According to the types (or phases) of products, hydrothermal technology could be subdivided into the following processes: hydrothermal carbonization, hydrothermal liquefaction, aqueous-phase reforming, and sub-/supercritical water gasification (Fig. 1) (Pavlovič et al. 2013). Regarding the reaction temperature, hydrothermal carbonization can be classified as two kinds: high temperatures (300–800 °C) or low temperatures (180–250 °C). In both processes, hydrochar (biochar), a carbonaceous material, is produced. Aqueous-phase reforming is performed at relatively mild reaction conditions (220–250 °C, 1.5–5.0 MPa), which is viewed as a very promising route to covert biomass into H_2 and alkanes. In particular, biomass carbohydrates or biomass-derived alcohols (ethylene glycol, methanol, sorbitol, and glycerol) are converted with the addition of appropriate heterogeneous catalysts, primarily leading to the formation of H_2 and CO_2. Hydrothermal liquefaction is performed at middling temperatures (250–400 °C) and high pressures (5–25 MPa), in which biomass is hydrolyzed and decomposed to unstable small components, which are further re-polymerized, resulting in various products, including water-insoluble bio-oil, biochar, water-soluble substances, and light gases. Hydrothermal gasification is used to produce gaseous products, mainly H_2, CH_4, CO_2, CO, and C_1–C_4 carbon gases, by reacting biomass with water at high temperatures and pressures. When this process is performed at temperatures of 300–500 °C (in sub- and near-critical water), it is also called as catalytic wet gasification. The dominant gaseous product is CH_4. When at temperatures of 500–800 °C, it is also named supercritical water gasification. And H_2 is the dominant gaseous product (Knez et al. 2015). The hydrolysis of biomass is a process which can be applied to convert biomass into the water-soluble sugars with sub-/supercritical water as medium. These sugars can be further manufactured into bioethanol and other biofuels (Guo et al. 2012; Loppinet-Serani and Aymonier 2014).

The key difference between the liquefaction technology and other three thermochemical conversion processes (combustion, pyrolysis, and gasification) is that water or other suitable solvents must be adopted as the reaction medium during the liquefaction process. The mechanism of biomass liquefaction is roughly described as follows: (1) biomass is firstly depolymerized into biomass monomers; (2) biomass monomers are further decomposed by cleavage, deamination, dehydration, and decarboxylation into light fragments of small molecules, which are unstable and active; and (3) light fragments are rearranged through cyclization, condensation, and polymerization, resulting in the production of new compounds (Huang and Yuan 2015). As depicted in Fig. 2, the liquefaction of biomass includes seven basic steps: (1) preparation of feedstock, (2) mixing the feedstock with proper

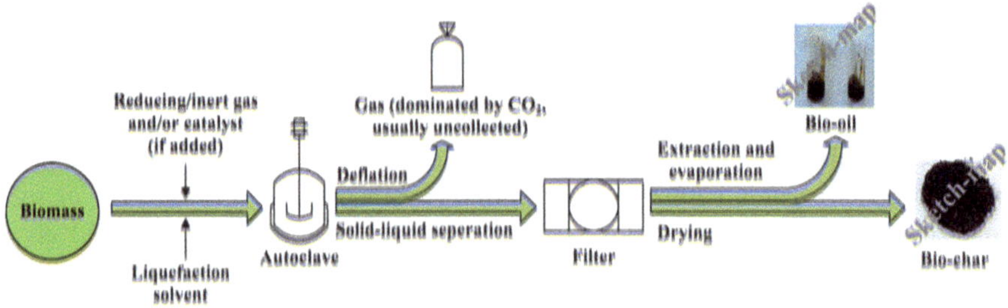

Fig. 2 The basic steps of the liquefaction of biomass

liquefaction solvent, (3) addition of reducing/inert gas and/or catalyst (optional), (4) heating the mixtures to reaction conditions, (5) main reaction, (6) product separation, and (7) recovery of solvent (optional) (Behrendt et al. 2008). The type of liquefaction solvents can be generally classified into two categories, water and organic solvents (methanol, ethanol, acetone, phenol, etc.). When organic solvents were adopted as liquefaction solvent, the recycling of organic solvents should be considered in order to avoid raising some environmental concerns.

In this chapter, we firstly collect and sort out the available review literatures on the liquefaction of biomass (Sect. 2). Next, Section 3 briefly summarizes the effects of processing parameters on the liquefaction of biomass. Section 4 focuses on the quality and application studies of the solid-phase product (biochar) obtained from the liquefaction of biomass. Finally, the main upgradation methods of the liquid-phase product (bio-oil) derived from the liquefaction of biomass are discussed in Sect. 5.

2 Available Review Literatures on the Liquefaction of Biomass in Recent Years

Liquefaction is a relatively novel method of converting biomass to bio-oil products. In recent years, many review works have been published, focusing on the liquefaction of biomass (Table 1). Pavlovič et al. (2013), Ruiz et al. (2013), and Knez et al. (2015) introduced the hydrothermal liquefaction of agricultural and food processing wastes, agriculture residues and marine biomass, organic wastes and by-products, respectively. The liquefaction of algal biomass can refer to the works of López Barreiro et al. (2013), Tian et al. (2014), and Guo et al. (2015). The effects of processing parameters on the liquefaction of biomass were widely discussed by Behrendt et al. (2008) and Akhtar and Admin (2011). The catalytic liquefaction of biomass is summarized by Tekin and Karagöz (2013) and Yeh et al. (2013). Elliott et al. (2015) described the recent application results of continuous flow processing systems in hydrothermal liquefaction of biomass. A review written by Pan (2011) was focused on the synthesis of resin products from the liquefaction of biomass in

Table 1 Review papers focused on the liquefaction of biomass in recent years

S. no.	Main contributions	References
1	Hydrothermal liquefaction of agricultural and food processing wastes	Pavlovič et al. (2013)
2	Hydrothermal liquefaction of organic wastes and by-product	Knez et al. (2015)
3	Hydrothermal liquefaction of agriculture residues and marine biomass	Ruiz et al. (2013)
4	Liquefaction of algal biomass	López Barreiro et al. (2013), Guo et al. (2015), Tian et al. (2014)
5	The effect of process parameters	Akhtar and Admin (2011), Behrendt et al. (2008)
6	Catalytic liquefaction of biomass	Tekin and Karagöz, (2013), Yeh et al. (2013)
7	Liquefaction of biomass in continuous flow processing systems	Elliot et al. (2015)
8	The synthesis of resin products from the liquefaction of biomass in organic solvents	Pan (2011)
9	Liquefaction of biomass in water	Toor et al. (2011), Peterson et al. (2008), Kang et al. (2013), Tekin et al. (2014), Toor et al. (2014)
10	Liquefaction of biomass in alcohol-water cosolvents	Xu et al. (2014)
11	The effect of liquefaction solvent type and biomass type; the liquefaction of biomass in ethanol and organic solvent-water mixed solvent; the liquefaction of sewage sludge	Huang and Yuan (2015)

organic solvents, such as polyesters, phenolic, epoxy, polyurethane, etc. The current status and related knowledge of the liquefaction of biomass in water were systematically introduced by Peterson et al. (2008), Toor et al. (2011, 2014), Kang et al. (2013), and Tekin et al. (2014). Xu et al. (2014) discussed the liquefaction of biomass in alcohol-water cosolvents. Huang and Yuan (2015) presented the recent progress in the liquefaction of biomass from the following five aspects of work: first, the effects of solvent type on the liquefaction behaviors of biomass; second, the effects of biomass type on the liquefaction behaviors of biomass; third, the liquefaction of biomass in sub- and supercritical ethanol; fourth, the liquefaction of biomass in organic solvent-water mixed solvents; and fifth, the liquefaction of sewage sludge.

3 The Influence of Liquefaction Parameters

The yield and quality of the solid, gas, and liquid products obtained from the liquefaction of biomass are determined by the specific operating parameters. These process parameters include biomass type, solvent type, heating conditions (reaction temperature, reaction time, and heating rate), the mass ratio of biomass to solvent, atmosphere, catalyst, etc. (Fig. 3). The liquefaction solvent type and biomass type are seen as the two primary factors that determine the distributions and properties of the products. The others can be seen as secondary factors.

3.1 *The Effects of Biomass Type*

As the main components in biomass such as lipid, protein, lignin, and carbohydrate behave differently during the liquefaction process, heterogeneity in biomass feedstock will lead to the modification in the distributions and compositions of liquefaction products.

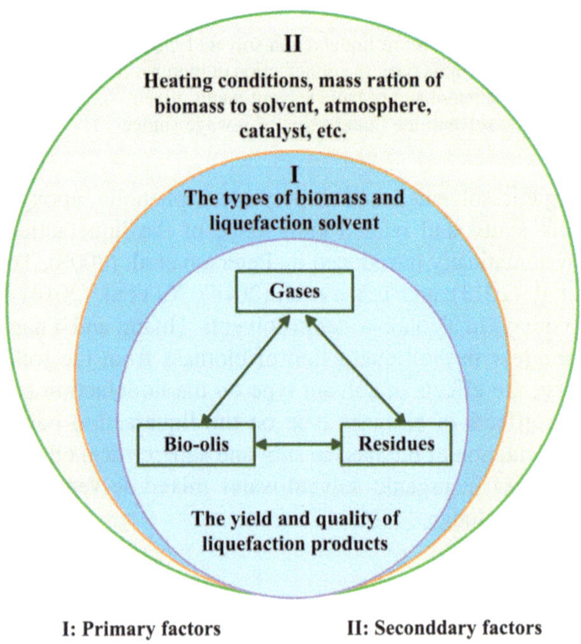

Fig. 3 The effect of processing parameters on the liquefaction of biomass

3.1.1 Lignocellulosic Biomass

Lignocellulose is the most common feedstock for producing bio-oils by liquefaction process (Zhang et al. 2010). The major components of lignocellulosic biomass include lignin, cellulose, and hemicellulose (Cheng and Timilsina 2011).

The distribution of liquefaction products obtained from the liquefaction of lignocellulose is mainly determined by the content of lignin in the biomass feedstock. In general, the higher the lignin content in the feedstock, the lower the conversion rate and bio-oil yield. The degradation of lignin is difficult and so it mostly appears in solid residue fraction (biochar). The content of holocellulose in lignocellulosic biomass could also alter the distribution of liquefaction products. The decomposition of holocellulose is favorable to the production of water-soluble hydrocarbons. However, it is disadvantageous to the formation of oily compounds (water insoluble). And thus, a low yield of bio-oils will be obtained. The differences in the components of lignocellulosic biomass also have significant influence on the compositions of bio-oils. The degradation of lignin mainly forms the phenolic compounds, whereas the decomposition of cellulose yields a large amount of furan derivatives (Behrendt et al. 2008; Akhtar and Amin 2011; Huang and Yuan 2015).

3.1.2 Algal Biomass

Algae biomass belongs to aquatic photosynthetic microorganisms. It could grow rapidly in various environments, including municipal wastewater, coastal seawater, saline water, and land which is not suitable for farming and agriculture. The major components of algal biomass are lipids, proteins, and carbohydrates. Researchers generally recognize that microalgal biomass is the promising feedstock for the third-generation biofuel (Nigam and Singh 2011; Ho et al. 2014).

Different algal biomasses have distinct liquefaction behaviors. Some significant and referable results can be summarized as follows (Chen et al. 2015; Huang and Yuan 2015):

1. The higher the content of lipid, the higher the yield of bio-oil.
2. The contribution rate of different components to the formation of bio-oil complies with the following trend: lipids > proteins > carbohydrates, 55–80 % of bio-oil yields from lipids, 11–18 % of bio-oil yields from proteins, and 6–15 % of bio-oil yields from carbohydrates.
3. Proteins mainly form a lot of nitrogen heterocycles, indoles, and pyrroles. Carbohydrates are mainly converted into cyclic ketones as well as phenols. The degradation of lipids mainly produces fatty acids.
4. The conversion of algae biomass with high lipid content seems more suitable in water or in a reducing atmosphere (the addition of formic acid).
5. During the processing of high-protein algae, the use of catalyst is usually redundant.

6. Alkali catalyst is the most suitable option for the liquefaction of high-carbohydrate algae.

3.2 The Effects of Solvent Type

The liquefaction solvents mainly contained two categories: water and organic solvents (methanol, ethanol, acetone, phenol, etc.). Liquefaction of biomass in different solvents usually results in distinct liquefaction behaviors. The specific reason may be that the interaction between biomass and different solvents presents different forms and intensities. Generally speaking, the adopted liquefaction solvent should be able to strongly react with biomass. Thus, those solvents derived from the liquefaction of biomass itself, such as phenol and its derivatives, simple alcohol, and polyalcohol, should be first used (Behrendt et al. 2008).

The influences of solvents on the liquefaction behaviors of different kinds of biomass are also different. *Lignocellulosic biomass*: On the whole, three solvents, acetone, phenol, and 1,4-dioxane, are found to be more effective in the liquefaction of lignocelluloses. In addition, the liquefaction of lignocelluloses in high-polarity or high-hydrogenation ability solvents will also have higher efficiency. *Algal biomass*: (1) The solvent with strong polarity usually results in higher conversion rate. (2) The polarity of the solvent doesn't present obvious trend with the yield of bio-oil. Almost certainly, when the solvents with strong polarity are adopted as the liquefaction medium, the yield of bio-oil is always higher. (3) Ethanol is considered to be the most suitable solvent. *Sewage sludge*: Ethanol usually performs better (Huang and Yuan 2015).

The use of water or organic solvents as liquefaction solvent has its own advantages and disadvantages. In view of this, the liquefaction of biomass in organic solvent-water mixed solvents is proposed to combine the advantages of water and organic solvents. In general, isometric organic solvent-water mixed solvents will present the highest efficiency. Among all the organic solvent-water mixed solvents tested, alcohol-water mixed solvents have been more commonly employed. One of the most important advantages of selecting alcohols for the liquefaction of biomass is that they are renewable chemical reagents. Alcohols can be produced from biomass. Furthermore, after the liquefaction process, they are easy to be recycled and reused (Xu et al. 2014; Huang and Yuan 2015).

3.3 The Effects of Heating Conditions

3.3.1 Reaction Temperature

The reaction temperature imparts a sequential influence on the quantities of liquefaction products. In the beginning, the yield of bio-oil will quickly increase to a maxima with the rise in reaction temperature. However, the yield of bio-oil will decrease with the further increase in reaction temperature. The liquid bio-oil is the target product of biomass liquefaction. Thus, too high reaction temperature is unsuitable from the views of operational cost and the yield of liquid bio-oil. In general, researchers give two mechanisms for explaining this behavior. Firstly, the formation of gaseous products will be enhanced at high reaction temperatures because the secondary decompositions and Bourdard gas reactions come to be active. Secondly, at high reaction temperatures, the concentrations of free radical are high and the recombination of free radical reactions become active, which promotes the formation of biochar. At high reaction temperatures, above two mechanisms become dominant, which results in the decrease of the yield of bio-oil. Furthermore, at temperatures below 280 °C, the degradation of individual biomass components is usually incomplete, which suppresses the yield of bio-oil. Thus, it is widely accepted that the reaction temperatures ranging from 300 to 350 °C would be effective for the conversion of biomass by liquefaction technology (Akhtar and Amin 2011).

Behrendt et al. (2008) proposed a similar view on the effect of reaction temperature on the bio-oil yield. In general, the yield of bio-oil increases to a maximum in an intermediate reaction temperature range. The higher the reaction temperature, the more active is the defragmentation of the biomass components into bio-oil products. The degradation of these fragments into gaseous species will be enhanced when the reaction temperature is further increased. The bio-oil-forming process is dominant below a critical reaction temperature. However, the gases-forming process is chief above this critical reaction temperature.

3.3.2 Reaction Time

The reaction time is sometimes called the residence time, which represents the continuous time of liquefaction reaction at a set reaction temperature. The duration of reaction times may determine the compositions of liquefaction products and overall conversion rate of biomass. The influence of residence time on the bio-oil yield obeys a volcano-type behavior. The yield of bio-oil reaches a maximum at a critical residence time. Too long residence time usually means that the liquid products will be further decomposed and re-polymerized, leading to the formation of gaseous products and solid residue, respectively. During the liquefaction process, the chances for the secondary and tertiary reactions always exist. The heavy intermediates are converted into liquid, gas, or residue species by above reactions. Once the saturation point of the conversion of biomass reaches, it is more probable for the chances of the decrease in bio-oil yields (Behrendt et al. 2008; Akhtar and Amin 2011).

3.3.3 Heating Rate

The higher the heating rates, the easier the bulk fragmentation of biomass. Higher heating rates also could inhibit the formation of biochar. At slow heating rates, secondary reactions usually result in the formation of biochar. Interestingly, at very high heating rates, secondary reactions will also become dominant, which leads to high yields of gaseous products as in case of supercritical liquefaction. Furthermore, the large variations in high heating rates have a weak influence on the yield of liquid bio-oil. It is significant to select a suitable heating rate to obtain a large amount of fragmentation but with a minimum number of secondary reactions. On the whole, to overcome the limitations of heat transfer, moderate heating rates is enough. Meanwhile, at moderate heating rates, higher yields of liquid bio-oil will be obtained (Akhtar and Amin 2011).

3.4 The Effects of the Mass Ratio of Biomass to Solvent

During the liquefaction process, the liquefaction solvent promotes the dissolution of biomass fragments. The higher the biomass to solvent ratio, the less influential is the relative interactions among molecules of solvent and that of biomass. And thus the dissolution of biomass components is suppressed. Moreover, at very high biomass to solvent ratios, the liquefaction process is inclined to behave like pyrolysis process. Compared with the liquefaction process, the pyrolysis of biomass usually yields more gases. Hence, low biomass to solvent ratios will enhance the conversion of biomass and produce high yield of bio-oil. However, too low biomass to solvent ratios will lower the yield of bio-oil, which may be due to that the formation of light fractions is enhanced at high solvent to biomass ratios. These light fractions in liquid products are difficult to recover and will be lost during the separation process of liquefaction products (Karagöz et al. 2006; Behrendt et al. 2008; Akhtar and Amin 2011).

3.5 The Effects of Atmosphere

The liquefaction of biomass is usually carried out with the presence of a reducing gas (H_2 or CO) or an inert gas (N_2 or Ar) to obtain higher yields of bio-oil. The reducing/inert gases have two important roles: (1) the stabilization of the fragmented products of liquefaction and (2) the inhibition of the cyclization, condensation, or re-polymerization of free radicals. And thus, the formation of biochar will be reduced. The reducing gases possess higher reactivity than inert gases. H_2 and CO usually have more affinity toward biomass radicals. And so the biomass radicals get stabilized easily under reducing atmosphere. In practice, H_2

gas or mixture of CO/H_2 gases is often selected as reducing gases. Remarkably, researchers usually ignored the selection of CO as reducing gas for its hazardous nature (Xu and Etcheverry 2008; Akhtar and Amin 2011).

3.6 *The Effects of the Type of Catalysts*

The common catalysts include two types: heterogenous catalysts and homogeneous catalysts. Heterogenous catalysts are usually applied to hydrogenation, for example, Raney nickel. There is no obvious catalytic activity for heterogenous catalysts toward the liquefaction of biomass. However, the homogeneous catalysts exhibit higher activity than heterogenous catalysts. The main research results are summarized as follows:

1. Compared with inorganic acids, the usage of organic acids as catalysts could yield lower solid residue content.
2. At typical liquefaction temperatures, NaOH shows higher catalytic activity toward biomass liquefaction than salts (chlorides, phosphates, carbonates, acetates, sulfates).
3. Acids and bases both have catalytic activity toward the liquefaction of biomass. The production of water-soluble products, e.g., furfural, sugars, 5-hydroxymethyl furfural, and carboxylic acids, can be enhanced by acids. However, the yield of bio-oil can be enhanced by alkali salts at subcritical conditions. Meanwhile, the formation of solid residue will be reduced.
4. The effect of the concentration of the homogeneous catalyst: below a critical value, the yield of bio-oil increases with the rise in catalyst concentration, and the content of solid residue correspondingly decreases. When the catalyst concentration further increases above this critical value, the yield of bio-oil will decrease and the content of solid residue will increase.

It is recognized that the addition of catalyst not only promotes de-polymerization reactions but also re-polymerization reactions. At high catalyst concentrations, the latter process (re-polymerization reactions) may become dominant (Behrendt et al. 2008; Tekin and Karagöz 2013; Yeh et al. 2013; Knez et al. 2015).

4 The Characterization and Application of Liquefaction Biochars

Currently, liquefaction has become a promising biomass-to-oil technology. One of the major by-products is solid residue, namely, biochar. As more and more attention is focused on the production of bio-oil products, large amounts of biochar products need to be disposed and reused in the near future. During the study of biomass liquefaction, researchers usually pay most attention on the liquid product (bio-oil). However, less attention has been focused on the solid product (biochar) (Liu and Zhang 2009). Notably, the biochar derived from the liquefaction of biomass is quite different from that produced from the pyrolysis of biomass. Liu et al. (2010) reported that the pinewood biochar produced by hydrothermal liquefaction performed better on the removal of copper than pinewood-pyrolyzed biochar, due to that the surface of liquefaction biochar contained abundant functional groups. Researchers usually attribute the prominent performances of liquefaction biochar to the insightful advantageous surface chemistry rather than the porosity. The surface area and pore volume of liquefaction biochar is often very small, but the surface functional groups are always rich. Surface functional groups can provide possible affinities for the target pollutants and the favorable surface chemistry could be of greater importance than a high porosity. Liquefaction biochar was found to be abundant in oxygen-containing functional groups, for example, carboxyl and hydroxyl groups (Leng et al. 2015c).

There are many applications of biochars, such as soil remediation, carbon sequestration, and modification-free adsorbents for the removal of pollutants. So far, the liquefaction biochars are mainly used as adsorbents for the removal of phenol, heavy metals, and dye. Liu and Zhang (2009) prepared two liquefaction biochars from pinewood (300 °C, P300) and rice husk (300 °C, R300), respectively. And it was found that the surface of two biochars both contained a lot of oxygen-containing groups. And the biochars were quite effective for the removal of lead from aqueous solution. The removal capacities of P300 and R300 reached 4.25 and 2.40 mg/g, respectively. Liu and Zhang (2011) further produced porous carbons from pinewood/rice husk liquefaction biochars through carbon dioxide activation and applied these porous carbons to the removal of copper and phenol. Leng et al. (2015c) tested the adsorption capacity of rich husk biochars for the removal of cationic dye. The biochars with ethanol as liquefaction solvent (adsorption capacities of 32.5–67.6 mg/g) were more effective that those with water or isometric water-ethanol as liquefaction solvent on the removal of cationic dye. Leng et al. (2015a) further reported that the maximum adsorption capacities of liquefaction biochars of sewage sludge, *Spirulina*, and rice straw on the removal of methylene blue were found to be 144.2, 128.6, and 160.5 mg/g, respectively, exceeding that of pyrolysis biochars reported in literatures (12–130 mg/g). The biochars derived from the liquefaction of sewage sludge in different solvents (methanol, ethanol, or acetone) at 260–380 °C were effective on the removal of malachite green and methylene blue from aqueous solution. Additionally, the liquefaction of sewage sludge with acetone at 280 °C, i.e., at lower temperature, favors biochar yield and the adsorption capacities of malachite green and methylene blue (Leng et al. 2015b). Marx et al. (2014) carried out the conversion of sunflower husks into biochar by liquefaction under different reaction atmospheres and solvents. It was suggested that the sunflower husk biochars might have a wide application potential in co-gasification or combustion.

In the future, the application of biochars derived from the liquefaction of biomass in the field of soil remediation and carbon sequestration should also be investigated.

5 The Upgradation Methods of Liquefaction Bio-oils

The bio-oils produced by the liquefaction of biomass usually have some undesired properties compared to heavy petroleum fuel oil. For example, the contents of water, ash, and oxygen in liquefaction bio-oils are all too high. And liquefaction bio-oils have high viscosity and corrosiveness (acidity) but low heating value. The application range of liquefaction bio-oils has been limited by these undesired properties to an extent. Therefore, it is necessary to carry out the upgradation of liquefaction bio-oils so that the properties of liquefaction bio-oils meet the criteria for direct use as liquid fuel. Current methods applied to the upgradation of liquefaction bio-oils are summarized in Table 2 (Butler et al. 2011; Bridgwater 2012; Xiu and Shahbazi 2012; Zhang et al. 2013).

5.1 Hydrotreating

Hydrotreating (HDT) is usually used to improve the quality of bio-oil without appreciable change of the boiling range. It is seen as a simple or nondestructive hydrogenation process. The content of oxygen in bio-oils can be reduced via hydrotreating. Hydrotreating is carried out at mild conditions. However, the process produces relatively low yield of bio-oil. Meanwhile, a lot of char, coke, and tar are also produced, which will lead to the deactivation of catalyst and the clogging of reactor (Xiu and Shahbazi 2012). The hydrotreatment of bio-oil for the production of value-added chemicals and fuels is briefly reviewed by Graça et al. (2013) and Jacobson et al. (2013).

5.2 Hydro-cracking

Hydro-cracking process accompanies hydrogenation with cracking, which is performed at relatively high temperature (> 350 °C) and pressure (0.6–13.8 Mpa). There is often appreciable alteration in the quality or character of the end products. Hydro-cracking process usually leads to a wide range of products, resulting from the combination of catalytic cracking reactions and hydrogenation (the multiplicity of reactions may occur). Although hydro-cracking is an effective method of producing a lot of light product, it needs more rigorous conditions to deal with acids, for example, higher temperature and hydrogen

Table 2 Brief description, treatment condition, and technical feasibility of the current techniques used for upgrading bio-oil

Upgrading methods	Treating condition/ requirement	Reaction mechanism/ process description	Technique feasibility	
			Pros.	Cons.
Hydrotreating/ hydrofining	Mild conditions (~500 °C/low pressure), chemical needed: H_2/ CO, catalyst (e.g., CoMo, HDS, NiMo, HZSM-5)	Hydrogenation without simultaneous cracking (eliminating N, O, and S as NH_3, H_2O, and H_2S)	Cheaper route, commercialized already	High coking (8–25 %) and poor quality of fuels obtained
Hydro-cracking/ hydrogenolysis/ catalytic cracking	Severe conditions (>350 °C, 0.6–13.8 MPa), chemical needed: H_2/CO or H_2 donor solvents, catalyst (e.g., Ni/Al_2O_3-TiO_2)	Hydrogenation with simultaneous cracking Destructive (resulting in low molecular product)	Makes large quantities of light products	Need complicated equipment, excess cost, catalyst deactivation, reactor clogging
Sub-/supercritical fluid	Mild conditions, organic solvents needed such as alcohol, acetone, ethyl acetate, and glycerol	Promotes the reaction by its unique transport properties: gas-like diffusivity and liquid-like density, thus dissolved materials not soluble in either liquid or gaseous phase of solvent	Higher oil yield, better fuel quality (lower oxygen content, lower viscosity)	Solvent is expensive
Solvent addition (direct add solvent or esterification of the oil with alcohol and acid catalysts)	Mild conditions, polar solvents needed such as water, methanol, ethanol, and furfural	Reduces oil viscosity by three mechanisms: (1) physical dilution; (2) molecular dilution or by changing the oil microstructure, and (3) chemical reactions like esterification and acetalization	The most practical approach (simplicity, the low cost of some solvents and their beneficial effects on the oil properties)	Mechanisms involved in adding solvent are not quite understand yet
Emulsification/ emulsions	Mild conditions, need surfactant (e.g., CANMET)	Combines with diesel directly. Bio-oil is miscible with diesel fuels with the aid of surfactants	Simple, less corrosive	Requires high energy for production

(continued)

Table 2 (continued)

Upgrading methods	Treating condition/ requirement	Reaction mechanism/ process description	Technique feasibility	
			Pros.	Cons.
Steam reforming	High temperature (800–900 °C), need catalyst (e.g., Ni)	Catalytic steam reforming + water-gas shift	Produces H_2 as a clean energy resource	Complicated, requires steady, dependable, fully developed reactors
Chemical extracted from the bio-oils	Mild conditions	Solvent extraction, distillation, or chemical modification	Extract valuable chemicals	Low-cost separation and refining techniques still needed

pressure. In general, it is considered to be not economical and energy efficient (Xiu and Shahbazi 2012). Al-Sabawi and Chen (2012) systematically summarized the hydro-cracking and hydrotreating of bio-oils and their blends with petroleum feedstocks. For more information on the catalytic cracking of bio-oil and their blends with petroleum feedstocks, the reader can refer to the review contributed by Al-Sabawi et al. (2012).

5.3 Supercritical Fluids (SCFs)

The materials, normally insoluble in liquid or gaseous phase of one solvent, can be dissolved in SCFs. Hence, the gasification/liquefaction reactions of biomass in SCFs have higher efficiency. Recently, SCFs have been applied to improve the yield and quality of bio-oil. In general, the bio-oils obtained with SCFs as medium have much lower viscosity and higher caloric values. Supercritical water is the most common supercritical fluid. To further improve the yield and quality of bio-oil, organic solvents, such as methanol, ethanol, 1,4-dioxane, acetone, etc., are adopted. It is environmentally friendly to upgrade bio-oil by using SCFs, and the upgradation process can be performed at a relatively lower temperature. However, in consideration of the high cost of organic solvents, it is economically unfeasible to carry out this upgradation process on a large scale (Xiu and Shahbazi 2012; Zhang et al. 2013).

5.4 Solvent Addition/Esterification

Adding polar solvents, for example, methanol, furfural, and ethanol, into biomass-derived bio-oils is a traditional upgradation method. This upgradation process has been used for many years. The viscosity of bio-oils can be immediately decreased with the addition of these polar solvents, and the heating value can be significantly increased. Now that these polar solvents usually have higher calorific value than that of most bio-oils, the mixtures of bio-oils and some polar solvents have higher heating value. The decrease in the viscosity of bio-oils can be explained by the following three aspects: (1) physical dilution effect, and no influence on the chemical reaction rates; (2) molecular dilution effect or the changes of the micro-structure of bio-oil, and the reaction rates are reduced; (3) chemical reactions between the bio-oil components and the solvent, and the further growth of chain is prevented. Compared with simple solvent addition, esterification treatment is a more effective upgradation process, in which the bio-oil components are reacted with alcohols (e.g., methanol and ethanol). The esterification process is usually performed at moderate conditions by using reactive distillation, and some acid catalysts (e.g., acetic acid) are adopted. The obtained bio-oils have a better quality. The viscosity, acidity, and aging rate are obviously decreased. The volatility and heating value are increased. In addition, the miscibility of bio-oils with diesel fuels is also improved (Bridgwater 2012; Xiu and Shahbazi 2012; Zhang et al. 2013).

5.5 Emulsification (Emulsions)

The formation of emulsions by mixing bio-oils with other fuel sources is a practical method of utilizing bio-oils as a combustion fuels in boilers or transportation. In general, the liquefaction bio-oils have little miscibility in hydrocarbon fuels. But, liquefaction bio-oils can be solubilized in diesel oil by emulsification with the aid of surfactants. On the whole, it is a relatively simple method of upgrading bio-oil through the formation of emulsions with diesel oil. In other words, this upgradation process could offer a short-term way to the utilization of bio-oils in diesel engines. The ignition characteristics of emulsions are promising, but other properties are still unsatisfied, including cetane, heating value, and corrosiveness. Furthermore, the implementation of this process required high energy input. Special injectors and fuel pumps should be designed, produced, and tested. Stainless steel or other better materials are also required (Xiu and Shahbazi 2012; Zhang et al. 2013).

5.6 Steam Reforming

Steam reforming process is used to convert bio-oil to synthesis gas, consisting of hydrogen and carbon monoxide. However, the formation of coke during this upgradation process will lead to the deactivation of catalyst, which is a large obstacle to producing sustainable hydrogen through the steam reforming of bio-oil (Xiu and Shahbazi 2012; Zhang et al. 2013).

5.7 Chemicals Extracted from the Bio-oils

The biomass-derived bio-oils contain a lot of valuable substances, for example, volatile organic acids, phenols, n-alkanes, and nitrogen heterocycles. These substances can be recovered from bio-oils by some extraction procedures. The recovered phenols are important feed stocks used in the resins industry. More attention should be paid to the development of dependable separation and refining techniques with low cost for the commercialization of special chemicals extracted from bio-oils (Xiu and Shahbazi 2012).

Other methods for upgrading of bio-oil, such as hot vapor filtration and gasification for synfuels, are described in detail in review literature (Bridgwater 2012). The integrated upgrading approaches and coprocessing of bio-oils are discussed by Butler et al. (2011).

In recent years, there are many technologies developed for the upgradation of bio-oils, such as chemical extraction, hydrotreating, emulsification, hydrocracking, supercritical fluids, steam reforming, and solvent addition/esterification. The upgraded bio-oils can be applied to replace conventional fossil energy and used in boiler and internal combustion engine. However, the restrictions in costs and technologies make it impossible to completely replace the fossil fuel. It is a formidable task to realize the industrial application of biomass-derived bio-oil by dint of upgrading technologies. The development and utilization of upgrading technologies require more devotion. It is necessary to develop novel integrated refinery processes for the systematic upgradation of bio-oils into transportation fuels. The upgraded bio-oils have desirable qualities. Meanwhile, to make the upgradation process economically feasible, the exploration of other value-added coproducts should also be considered (Xiu and Shahbazi 2012; Zhang et al. 2013).

6 Conclusions

The liquefaction of biomass for bio-oil products has received more and more attention due to their potential application in replacing conventional fossil energy as fuels and resources for the production of chemical materials. The selection of proper processing conditions, including biomass type, solvent type, heating conditions (reaction temperature, reaction time, and heating rate), the mass ratio of biomass to solvent, atmosphere, catalyst, etc., is significant for improving the yield and quality of bio-oils. A fact that cannot be ignored is that the biomass-derived bio-oils are complicated and chemically unstable mixtures. Therefore, more efforts should be made to stabilize and upgrade bio-oils. Appropriate upgradation technologies should be developed and used to improve the properties of biomass-derived bio-oils to the transport biofuel grade. Thus, the bio-oils can be miscible in the existing fuel stream. And a gradual but complete replacement of fossil fuels can be realized. In future applications, the liquefaction of biomass to produce bio-oil will take place in industrial practice. Furthermore, attention should be also paid to the characterization and utilization of liquefaction biochars, which could in turn promote the development of biomass liquefaction in industry.

Acknowledgments The authors gratefully acknowledge the financial support provided by the Natural Science Foundation of Jiangxi, China (No. 20151BAB213024), the Scientific Research Fund of Jiangxi Provincial Education Department (GJJ14302), and the Hunan Province Innovation Foundation for Postgraduate (CX2012B139).

References

Akhtar J, Amin NAS (2011) A review on process conditions for optimum bio-oil yield in hydrothermal liquefaction of biomass. Renew Sustain Energy Rev 15:1615–1624. doi:10.1016/j.rser.2010.11.054

Al-Sabawi M, Chen JW (2012) Hydroprocessing of biomass-derived oils and their blends with petroleum feedstocks: a review. Energy Fuel 26:5373–5399. doi:10.1021/ef3006405

Al-Sabawi M, Chen JW, Ng S (2012) Fluid catalytic cracking of biomass-derived oils and their blends with petroleum feedstocks: a review. Energy Fuel 26:5355–5372. doi:10.1021/ef3006417

Behrendt F, Neubauer Y, Oevermann M, Wilmes B, Zobel N (2008) Direct liquefaction of biomass. Chem Eng Technol 31:667–677. doi:10.1002/ceat.200800077

Bridgwater AV (2012) Review of fast pyrolysis of biomass and product upgrading. Biomass Bioenergy 38:68–94. doi:10.1016/j.biombioe.2011.01.048

Butler E, Devlin G, Meier D, McDonnell K (2011) A review of recent laboratory research and commercial developments in fast pyrolysis and upgrading. Renew Sustain Energy Rev 15:4171–4186. doi:10.1016/j.rser.2011.07.035

Chen WH, Lin BJ, Huang MY, Chang JS (2015) Thermochemical conversion of microalgal biomass into biofuels: a review. Bioresour Technol 184:314–327. doi:10.1016/j.biortech.2014.11.050

Cheng JJ, Timilsina GR (2011) Status and barriers of advanced biofuel technologies: a review. Renew Energy 36:3541–3549. doi:10.1016/j.renene.2011.04.031

Elliott DC, Biller P, Ross AB, Schmidt AJ, Jones SB (2015) Hydrothermal liquefaction of biomass: developments from batch to continuous process. Bioresour Technol 178:147–156. doi:10.1016/j.biortech.2014.09.132

Faba L, Díaz E, Ordóñez S (2015) Recent developments on the catalytic technologies for the transformation of biomass into biofuels: a patent survey. Renew Sustain Energy Rev 51:273–287. doi:10.1016/j.rser.2015.06.020

Goyal HB, Seal D, Saxena RC (2008) Bio-fuels from thermochemical conversion of renewable resources: a review. Renew Sustain Energy Rev 12:504–517. doi:10.1016/j.rser.2006.07.014

Graça I, Lopes JM, Cerqueira HS, Ribeiro MF (2013) Bio-oils upgrading for second generation biofuels. Ind Eng Chem Res 52:275–287. doi:10.1021/ie301714x

Guo F, Fang Z, Xu CC, Smith RL Jr (2012) Solid acid mediated hydrolysis of biomass for producing biofuels. Prog Energy Combust 38:672–690. doi:10.1016/j.pecs.2012.04.001

Guo Y, Yeh T, Song WH, Xu DH, Wang SZ (2015) A review of bio-oil production from hydrothermal liquefaction of algae. Renew Sustain Energy Rev 48:776–790. doi:10.1016/j.rser.2015.04.049

Ho DP, Ngo HH, Guo WS (2014) A mini review on renewable sources for biofuel. Bioresour Technol 169:742–749. doi:10.1016/j.biortech.2014.07.022

Huang HJ, Yuan XZ (2015) Recent progress in the direct liquefaction of typical biomass. Prog Energy Combust 49:59–80. doi:10.1016/j.pecs.2015.01.003

Ibrahim N, Kamarudin SK, Minggu LJ (2014) Biofuel from biomass via photo-electrochemical reactions: an overview. J Power Sources 259:33–42. doi:10.1016/j.jpowsour.2014.02.017

Jacobson K, Maheria KC, Dalai AK (2013) Bio-oil valorization: a review. Renew Sustain Energy Rev 23:91–106. doi:10.1016/j.rser.2013.02.036

Kang SM, Li XL, Fan J, Chang J (2013) Hydrothermal conversion of lignin: a review. Renew Sustain Energy Rev 27:546–558. doi:10.1016/j.rser.2013.07.013

Karagöz S, Bhaskar T, Muto A, Sakata Y (2006) Hydrothermal upgrading of biomass: effect of K_2CO_3 concentration and biomass/water ratio on products distribution. Bioresour Technol 97:90–98. doi:10.1016/j.biortech.2005.02.051

Knez Ž, Markočič E, Hrnčič MK, Ravber M, Škerget M (2015) High pressure water reforming of biomass for energy and chemicals: a short review. J Supercrit Fluid 96:46–52. doi:10.1016/j.supflu.2014.06.008

Leng LJ, Yuan XZ, Huang HJ, Wang H, Wu ZB, Fu LH, Peng X, Chen XH, Zeng GM (2015a) Characterization and application of bio-chars from liquefaction of microalgae, lignocellulosic biomass and sewage sludge. Fuel Process Technol 129:8–14. doi:10.1016/j.fuproc.2014.08.016

Leng LJ, Yuan XZ, Huang HJ, Shao JG, Wang H, Chen XH, Zeng GM (2015b) Bio-char derived from sewage sludge by liquefaction: characterization and application for dye adsorption. Appl Surf Sci 346:223–231. doi:10.1016/j.apsusc.2015.04.014

Leng LJ, Yuan XZ, Zeng GM, Shao JG, Chen XH, Wu ZB, Wang H, Peng X (2015c) Surface characterization of rice husk bio-char produced by liquefaction and application for cationic dye (Malachite green) adsorption. Fuel 155:77–85. doi:10.1016/j.fuel.2015.04.019

Liu ZG, Zhang FS (2009) Removal of lead from water using biochars prepared from hydrothermal liquefaction of biomass. J Hazard Mater 167:933–939. doi:10.1016/j.jhazmat.2009.01.085

Liu ZG, Zhang FS (2011) Removal of copper (II) and phenol from aqueous solution using porous carbons derived from hydrothermal chars. Desalination 267:101–106. doi:10.1016/j.desal.2010.09.013

Liu ZG, Zhang FS, Wu JZ (2010) Characterization and application of chars produced from pinewood pyrolysis and hydrothermal treatment. Fuel 89:510–514. doi:10.1016/j.fuel.2009.08.042

López Barreiro D, Prins W, Ronsse F, Brilman W (2013) Hydrothermal liquefaction (HTL) of microalgae for biofuel production: state of the art review and future prospects. Biomass Bioenergy 53:113–127. doi:10.1016/j.biombioe.2012.12.029

Loppinet-Serani A, Aymonier C (2014) Chapter 7-Hydrolysis in near- and supercritical water for biomass conversion and material recycling. In: Fan VA (ed) Supercritical fluid technology for energy and environmental applications. Elsevier, Boston, MA, pp 139–156. doi:10.1016/B978-0-444-62696-7.00007-1

Ma LL, Wang TJ, Liu QY, Zhang XH, Ma WC, Zhang Q (2012) A review of thermal-chemical conversion of lignocellulosic biomass in China. Biotechnol Adv 30:859–873. doi:10.1016/j.biotechadv.2012.01.016

Marx S, Chiyanzu I, Piyo N (2014) Influence of reaction atmosphere and solvent on biochar yield and characteristics. Bioresour Technol 164:177–183. doi:10.1016/j.biortech.2014.04.067

Nigam PS, Singh A (2011) Production of liquid biofuels from renewable resources. Prog Energy Combust 37:52–68. doi:10.1016/j.pecs.2010.01.003

Pan H (2011) Synthesis of polymers from organic solvent liquefied biomass: a review. Renew Sustain Energy Rev 15:3454–3463. doi:10.1016/j.rser.2011.05.002

Panwar NL, Kothari R, Tyagi VV (2012) Thermo-chemical conversion of biomass—eco friendly energy routes. Renew Sustain Energy Rev 16:1801–1816. doi:10.1016/j.rser.2012.01.024

Pavlovič I, Knez Ž, Škerget M (2013) Hydrothermal reactions of agricultural and food processing wastes in sub- and supercritical water: a review of fundamentals, mechanisms, and state of Research. J Agr Food Chem 61:8003–8025. doi:10.1021/jf401008a

Peterson AA, Vogel F, Lachance RP, Froling M, Antal JMJ, Tester JW (2008) Thermochemical biofuel production in hydrothermal media: a review of sub- and supercritical water technologies. Energy Environ Sci 1:32–65. doi:10.1039/b810100k

Ruiz HA, Rodríguez-Jasso RM, Fernandes BD, Vicente AA, Teixeira JA (2013) Hydrothermal processing, as an alternative for upgrading agriculture residues and marine biomass according to the biorefinery concept: a review. Renew Sustain Energy Rev 21:35–51. doi:10.1016/j.rser.2012.11.069

Tekin K, Karagöz S (2013) Non-catalytic and catalytical hydrothermal liquefaction of biomass. Res Chem Intermediat 39:485–498

Tekin K, Karagöz S, Bektaş S (2014) A review of hydrothermal biomass processing. Renew Sustain Energy Rev 40:673–687. doi:10.1016/j.rser.2014.07.216

Tian CY, Li BM, Liu ZD, Zhang YH, Lu HF (2014) Hydrothermal liquefaction for algal biorefinery: a critical review. Renew Sustain Energy Rev 38:933–950. doi:10.1016/j.rser.2014.07.030

Toor SS, Rosendahl L, Rudolf A (2011) Hydrothermal liquefaction of biomass: a review of subcritical water technologies. Energy 36:2328–2342. doi:10.1016/j.energy.2011.03.013

Toor SS, Rosendahl LA, Hoffmann J, Pedersen TH, Nielsen RP, Søgaard EG (2014) Hydrothermal liquefaction of biomass. In: Jing FM (ed) Application of hydrothermal reactions to biomass conversion. Springer, Berlin, pp 189–218. doi:10.1007/978-3-642-54458-3_9

Xiu SG, Shahbazi A (2012) Bio-oil production and upgrading research: a review. Renew Sustain Energy Rev 16:4406–4414. doi:10.1016/j.rser.2012.04.028

Xu CB, Etcheverry T (2008) Hydro-liquefaction of woody biomass in sub- and super-critical ethanol with iron-based catalysts. Fuel 87:335–345. doi:10.1016/j.fuel.2007.05.013

Xu CC, Shao YY, Yuan ZS, Cheng SN, Feng SH, Nazari L, Tymchyshyn M (2014) Chapter 8-Hydrothermal liquefaction of biomass in hot-compressed water, alcohols, and alcohol-water co-solvents for biocrude production. In: Jing FM (ed) Application of hydrothermal reactions to biomass conversion. Springer, Berlin, pp 171–188. doi:10.1007/978-3-642-54458-3_8

Yeh TM, Dickinson JG, Franck A, Linic S, Thompson LT Jr, Savage PE (2013) Hydrothermal catalytic production of fuels and chemicals from aquatic biomass. J Chem Technol Biotechnol 88:13–24. doi:10.1002/jctb.3933

Zhang LH, Xu CB, Champagne P (2010) Overview of recent advances in thermochemical conversion of biomass. Energy Convers Manage 51:969–982. doi:10.1016/j.enconman.2009.11.038

Zhang L, Liu RH, Yin RZ, Mei YF (2013) Upgrading of bio-oil from biomass fast pyrolysis in China: a review. Renew Sustain Energy Rev 24:66–72. doi:10.1016/j.rser.2013.03.027

Intensified Synthesis of Bioethanol from Sustainable Biomass

Saurabh M. Joshi and Parag R. Gogate

Abstract The need for alternate energy resources has significantly increased in recent years due to the rapid depletion of currently available resources and strain on economic and environmental conditions. Production of biofuels from the sustainable lignocellulosic feedstocks can be one of the promising approaches to meet the energy requirements in an renewable fashion also minimizing the impact on environment. The production process is usually comprised of three processing steps as pretreatment for the removal of lignin, hydrolysis to yield fermentable sugars followed by actual fermentation giving biofuels like bioethanol, bio-butanol, bio-hydrogen, and biogas. The current chapter provides an overview of potential sustainable feedstock with subsequent discussion on different processes for pretreatment with an objective of intensifying the biofuel production. Dilute acid pretreatment, alkaline hydrolysis, and steam explosion have been established to be the generally applicable processes for pretreatment. Also the use of ultrasound, microwave, and hydrodynamic cavitation has been demonstrated to give process intensification in different processing steps including pretreatment, hydrolysis and fermentation overviewed in the work with specific reference to bioethanol production. Employment of these intensified processes can help in efficient processing of sustainable lignocellulosic feedstock like sugarcane waste, corn waste, food waste, paper waste, etc. which can lead to energy generation from renewable resources in economically efficient manner also giving the advantage of waste utilization.

Keywords Bioethanol • Lignocellulosic feedstock • Hydrolysis • Sustainable feedstock • Pretreatment

S.M. Joshi • P.R. Gogate (✉)
Chemical Engineering Department, Institute of Chemical Technology, Matunga, Mumbai 400 019, India
e-mail: pr.gogate@ictmumbai.edu.in

L. Singh, V.C. Kalia (eds.), *Waste Biomass Management – A Holistic Approach*,
DOI 10.1007/978-3-319-49595-8_12

1 Introduction

Energy crisis and environmental pollution are the major issues faced globally with direct impact on the quality of life. The detrimental outcome of global warming, primarily coming from the combustion of fossil fuels and greenhouse gas emissions, is being felt at significantly increasing levels in recent years. Also the fossil fuel resources are getting depleted, and continuous fluctuations in the prices have led to a considerable strain on the economic conditions. Typically, 60 % of the world's total oil consumption is by the transportation sector and is also responsible for half of the total global CO_2 emissions. These problems associated with the more commonly applied fossil fuels have led to renewed interest in search of alternative sources and most importantly the renewable sources of fuels (Cinelli et al. 2015). Among the renewable sources, biofuels account for a significant contribution and can be produced from natural sources such as biomass available in different forms. Production of renewable carbon by ecosystem is in significant quantity, out of which 77×10^9 tons is fixed via photosynthesis producing 100 billion tons of biomass per year (Bozell 2001). The renewable resources take up CO_2 during the growth and give it up at the end of life, which completes the cycle. Photosynthesis is a process by which solar energy is captured and stored in the form of polymers such as cellulose, hemicellulose, and lignin in the biomass, and these polymers can be utilized to produce biofuels using various processes (Khare et al. 2015). Lignocellulosic biomasses can be a clean and renewable feedstock as they are available abundantly, and biofuels can be generated from them, which will further help in the reduction of emissions and improve the energy security. Typically 86 % of the greenhouse gases can be reduced based on the use of ethanol produced from cellulosic biomass and other sources (Kumar et al. 2009).

Typically biofuels are classified into solid, liquid, and gaseous fuels. Wood and charcoal are in the category of solid fuels, whereas bio-alcohol and biodiesel are recognized as liquid fuels. Bio-hydrogen and methane (also described as biogas) come under the category of gaseous fuels (Katiyar et al. 2015). The present chapter focuses on bio-alcohols which are generally produced with the help of microbial fermentation process. Bioethanol is the most widely synthesized alcohol, at times on larger scale, from variety of substrates, and mostly by action of microorganisms and enzymes on biological sources. The microorganisms utilize the sugars which are obtained from the hydrolysis of starch, hemicellulose, and cellulose present in feedstock (Kumar et al. 2009).

Bioethanol is an oxygenated fuel containing 35 % oxygen, has higher octane number (102), and possesses properties which make it theoretically more efficient than gasoline in an internal combustion (IC) engine (Balat and Balat 2009). Bioethanol with high octane number and low cetane number makes it also a suitable fuel for possible mixing with gasoline. Bioethanol can be produced with the help of different microbes or yeasts, and among them *Saccharomyces cerevisiae* (also called as baker's yeast or brewer's yeast) is the predominantly used yeast for alcohol fermentation. Theoretically, 51.4 g of bioethanol and 48.8 g of carbon dioxide will be produced from 100 g of glucose. In practice, the actual yield obtained is less than 100 % as the glucose is also utilized by microorganisms for their growth (Ergun et al. 1997). Bioethanol can be further classified as follows based on the source of feedstock and the approach used for the production:

1. First generation: ethanol produced from starch and sugar (mostly edible source of biomass such as corn and sugarcane)
2. Second generation: ethanol produced from nonfood crops (waste biomass, special energy or biomass crops that are nonedible)
3. Third generation: ethanol production from algae

Second- and third-generation ethanol is categorized as advanced biofuels, and currently more focus is given on their process development especially considering the scale-up aspects that can yield a sustainable solution. Bioethanol consumption leads to no net CO_2 and sulfur emissions and also has lower particulate and toxic emissions as compared to the fossil fuels. CO_2 released during fermentation and combustion is reabsorbed by plants via photosynthesis only to form more vegetation giving a shorter carbon cycle. To reduce oil imports and improve quality of air, many countries have focused on the commercialization of bioethanol production. World bioethanol production is 89 billion liters as per the 2013 data (Table 1), and of the total ethanol production, on an average 73 % is currently used as fuel, 17 % for beverage production, and 10 % directly consumed as industrial grade ethanol (Gupta and Verma 2015, Srivastava et al. 2014). Bioethanol production from sources like corn, sugar beet, wheat, maize, barley, etc. leads to food versus fuel concerns. To avoid these issues, lignocellulosic feedstocks are being increasingly looked upon as sustainable source, and considerable efforts for developing effective processes for commercial-scale utilization have been directed currently. Lignocellulosic feedstock consists of cellulose, hemicellulose, and lignin in complex form and requires multistep processing

Table 1 Total production of ethanol in year 2013

Country	Feedstock	Ethanol production (billion liters)
USA	Corn/maize	50.3
Brazil	Sugarcane	25.5
Germany	Wheat	0.8
France	Sugar beet, wheat	1.0
China	Corn, sugarcane, maize, cassava	2.0
Argentina	Sugarcane	0.5
Spain	Barley, wheat	0.4
India	Sugarcane, wheat	1.7
Canada	Wheat/cereal	1.8
Poland	Rye	0.2
Colombia	Sugarcane	0.4
Thailand	Sugarcane, cassava	1.0
Australia	Sugarcane	0.3
Belgium	Wheat	0.4
EU	Various plants/cereal and sugar beet	4.5
World total		88.9

starting with the breakage of the complex feedstock using some pretreatment that can lead to the better utilization of the feedstock and higher yields. We now analyze the different available sources of lignocellulosic feedstock followed by highlighting the required processing so as to obtain biofuels in an efficient manner.

2 Lignocellulosic Feedstock

Ethanol production is generally achieved from materials classified into three groups, i.e., sugars (cane or sweet sorghum), starch (corn, cassava, or potato), and cellulose. Availability of cellulosic materials is abundant, and these materials are considered as nonfood competitive resources, also many times available as waste. These waste materials are generally described as the lignocellulosic feedstocks and are typically the wastes generated from agricultural, forestry, municipal, and other activities (Saini et al. 2015). Typically, in a lignocellulosic complex, cellulose is the core of complex with crystalline fibrous structure, hemicellulose is a heteropolymer found between macrofibrils and microfibrils of cellulose, and structural support to the matrix is provided by lignin (Rubin 2008).

2.1 Cellulose

The most abundant polysaccharide found on the earth is cellulose, which is also the main integral component of plant cell walls. Cellulose consists of repetitive units of cellobiose (glucose disaccharide) linked by β-1,4 bonds, and $(C_6H_{10}O_5)n$ is the chemical formula representing cellulose. The degree of polymerization (DP) is the main factor on which the properties of cellulose are dependent. Generally, 8000–10,000 units of D-glucose are linked by β-1,4 linkages in a cellulose polymer which also consist of hydroxide groups evenly distributed on both sides (Costa et al. 2013). Hydroxyl groups lead to formation of hydrogen bonds between the same chains or with the neighbouring chains, and this further leads to development of microfibrils of cellulose with high tensile strength. Different levels of crystallinity (tending toward amorphous or highly crystalline) are observed based on the different orientations present throughout structure. Hemicelluloses and/or pectin covered with lignin also helps in the attachment of cellulose, making it complex form and protected from chemical and biological attacks (Zheng et al. 2014).

2.2 Hemicellulose

Hemicellulose is a heteropolymer of various hexoses (glucose, mannose, galactose, and/or rhamnose), pentoses (xylose and arabinose), and acids (galacturonic acid, glucuronic acid, and methyl glucuronic acid). Polysaccharide chains are short branched and amorphous, which assist in the built-up of network with cellulose microfibrils, and interact with lignin as well as render hemicellulose susceptible to biological, chemical, and thermal hydrolysis. Xylose and mannose are the monomers found abundantly in hemicellulose (Xu and Huang 2014).

2.3 Lignin

Lignin is a heterogeneous, amorphous, and cross-linked aromatic polymer which consists of three-dimensional polyphenolic network made up of sinapyl alcohol and coniferyl alcohol, with methoxyl, hydroxyl, and carbonyl functional groups (Zheng et al. 2014). Content and composition of lignin differs based on the plant groups, and it acts as a binding material between cellulose and hemicellulose to form a rigid cell wall. Lignin plays a vital role in providing the recalcitrant property to lignocellulosic feedstock. Softwood is the most recalcitrant material to bioconversion and has a very high content of lignin (Xu and Huang 2014).

Wood residues, food industry residues, grasses, municipal solid wastes, agricultural residues (straw, nutshells, stover, seeds, peelings, cobs, stalks, bagasse), and domestic wastes (lignocellulosic garbage and sewage) are the wastes which come under the category of lignocellulose waste. Nonfood crops (i.e., *Jatropha*, poplar, switch grass, and *Eucalyptus* sp.) can also be utilized as sustainable lignocellulosic feedstock for energy production (Katiyar et al. 2015). Lignocellulosic feedstock production in terms of the growing of the nonfood crops for energy can be carried out on marginal quantum of land without affecting the land required for food crops (Rubin 2008).

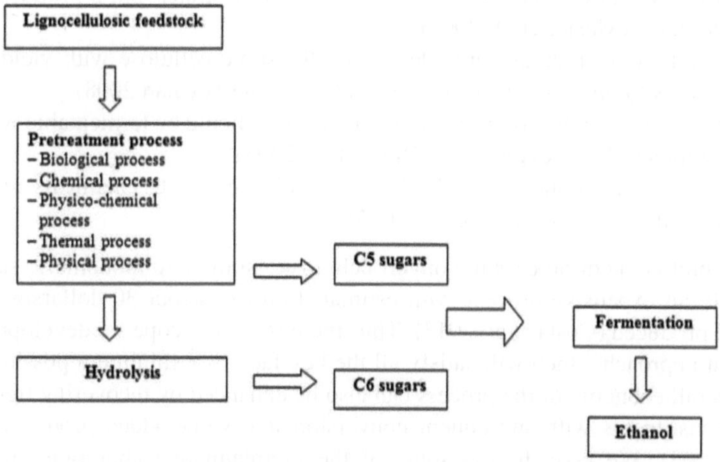

Fig. 1 Processing of lignocellulosic feedstocks

Lignocellulosic feedstock can be utilized for the biofuel production with the help of three essential steps (Fig. 1) to be performed in sequential manner as follows:

1. Pretreatment step consisting of disruption of lignocellulosic interactions to make hemicelluloses and cellulose accessible based on the removal of maximum lignin content
2. Saccharification step that converts the accessible polysaccharides into monosaccharides
3. Fermentation step where the generated monosaccharides are converted into desired biofuels or other platform chemicals (Khare et al. 2015)

3 Pretreatment Processing

The process of utilizing lignocellulosic feedstock requires intensive labor and high processing cost coupled with significant reaction times due to its recalcitrant nature. The porosity of substrate can be increased using the pretreatment operations with redistribution or removal of lignin. The pretreatment processes can be physical (i.e., biomass size reduction) or thermochemical processes, i.e., disruption of recalcitrant material present in the biomass (Limayem and Ricke 2012). The selection of the appropriate pretreatment approach for the processing depends on the proportion of constituents present in the feedstock, i.e., cellulose, hemicellulose, and lignin. The key requirements of an efficient treatment process are as follows:

- The pretreatment technology should be able to treat various feedstocks with the same efficiency which is an important criteria considering the fact that some of the pretreatment techniques used are effective on specific substrates only. At the same time, the approach should be environmentally friendly. For example, acid hydrolysis treatment is effective on most of the substrates, but it produces toxic by-products (Mosier et al. 2005).
- The pretreatment should provide highly digestible cellulose with yield above 90% so as to give favorable economics (Yang and Wyman 2008).
- Hydrolysate obtained after the pretreatment step should be fermentable with low requirement of processing cost (Oliva et al. 2003).
- The pretreatment should not lead to generation of waste chemicals requiring additional disposal (Shafiei et al. 2015).

The pretreatment process to convert cellulosic biomass to monomeric sugars is typically an expensive process with estimated cost of about 30 dollars/gallon of ethanol produced (Costa et al. 2013). Thus there is ample scope for development of efficient approach which will satisfy all the key factors at minimum possible cost. The overall economy of the process can also be enhanced by recovering lignin and other constituents with subsequent conversion into value-added products (Oliva et al. 2003). We now discuss some of the pretreatment technologies used for lignocellulosic feedstocks with discussion on relative merits and demerits.

3.1 *Biological Pretreatment*

Biological pretreatment is typically performed with microorganisms like brown, soft-rot, and white fungi, which acts on lignin and hemicellulose at the same time and more importantly has very less effect on cellulose, which is more resistant than other components. In the case of pretreatment using white-rot fungi, the degradation of lignin is based on the peroxidase and laccase enzymes (Limayem and Ricke 2012). The efficacy of degradation of lignocellulosic materials is dependent on the C/N (carbon to nitrogen) ratio present in the material, and a definite portion of nitrogen is always required to degrade single carbon molecule (Costa et al. 2013). Several white-rot fungi with high delignification capacity such as *Phanerochaete chrysosporium*, *Ceriporia lacerata*, *Cyathus stercoreus*, *Ceriporiopsis subvermispora*, *Pycnoporus cinnabarinus*, and *Pleurotus ostreatus* (Kumar et al. 2009) have been reported to be effectively applied for a variety of sustainable feedstock. Microbial consortium development based on screening of the microbes from natural environment which use rotten lignocellulosic material as substrate is one of the viable options. The consortiums can be combined with pure yeast and cellulolytic cultures to increase the overall efficiency of treatment (Sanchez 2009, Zhang et al. 2011).

In general, requirement of low capital cost, low energy, and mild environmental conditions along with no chemical usage are the distinct advantages of biological pretreatment, though these processes are many times inefficient and require significantly higher retention time for hydrolysis. For commercialization of an effective biological pretreatment process, the issues like cost, efficiency, and selectivity need to be addressed (Zheng et al. 2014).

3.2 *Physical Pretreatment*

The different physical pretreatment processes reported include milling (ball milling, hammer milling, two-roll milling), high-pressure steaming, extrusion, pyrolysis, hydrothermal process, and microwave irradiation. An optimized combination of these processes typically can give higher overall efficiency of process. The selected pretreatment should not only consider the operating cost but also the compatibility with subsequent downstream processes, i.e., hydrolysis and fermentation. A brief overview of the different physical processes has now been presented.

3.2.1 Milling

Milling is generally used to grind biomass with an objective of reducing the polymerization and increasing the available surface area of biomass based on the reduction in size. Treatment of lignocellulosic biomass with milling can lead to efficient utilization in the subsequent enzymatic hydrolysis process (Taherzadeh and Karimi 2008). Different types of milling include hammer milling, ball milling, two-roll milling, vibro energy milling, and colloid milling. Selection of the type of milling equipment depends on the moisture content of the material. Materials with moisture content up to 10–15 % wet basis are usually treated with two-roll, hammer, attrition, and knife mills, whereas colloid mills and extruders are used for materials with moisture content more then 10–15 % (wet basis) (Kratky and Jirout 2011). Though there is an increase in the surface area, lignin, which is responsible for restriction of cellulose accessibility to enzymes, cannot be removed with the help of milling process. Also the power requirement of this process depends on the moisture content and desired particle size and is typically very high which makes the pretreatment an economically non-feasible process.

3.2.2 Extrusion

An extruder consists of single screw with a rotating barrel and subjects material to heating, mixing, and shearing, which leads to physical and chemical modifications. The extent of disintegration of lignocellulosic feedstock depends on the barrel temperature and screw speed. The changes in physical properties occur when the material is forced through the narrow clearance between screws and barrel. Enzyme application during extrusion process can lead to efficient hydrolysis (Alvira et al. 2010). A major disadvantage is that lignocellulosic feedstock cannot be treated efficiently and the flowability of the feedstock is very low, which further results in the separation of solids from liquid, affecting the overall shearing stress on the biomass.

3.2.3 Hydrodynamic Cavitation

Hydrodynamic cavitation is one of the upcoming technologies which can be used for pretreatment of lignocellulosic waste. In this process, cavities are generated under controlled conditions with the help of Venturi or orifice plates as the cavitation chamber. Cavities are generated at the vena contracta and undergo subsequent oscillations in the size due to the pressure fluctuations and finally collapse downstream of constriction. This collapse of cavities leads to formation of shock waves which causes disruption or hydrolysis of material present nearby (Gogate and Pandit 2000). Iskalieva et al. (2012) have overviewed the different aspects related to the delignification of wheat straw using hydrodynamic cavitation. Kim et al. (2015) have also employed hydrodynamic cavitation pretreatment on reed for the production of ethanol and reported significant improvement based on the use of hydrodynamic cavitation. However, apart from these two studies, not much work has been reported on the application of hydrodynamic cavitation for the pretreatment of feedstock. More research needs to be performed in order to employ this process for processing of lignocellulosic feedstocks in an efficient and cost-effective manner.

3.2.4 Ultrasound Pretreatment

The mechanism of ultrasound-induced pretreatment is similar to the hydrodynamic cavitation with only difference being cavitation is induced by the passage of ultrasound (>20 kHz) through a liquid medium. Conditions of very high temperature and pressures (500−15000 K and 100–5000 atm, respectively) are generated based on the ultrasound-induced cavity implosion (Gogate and Kabadi 2009). The collapse of cavitational bubbles in liquid also leads to generation of high-energy intermediates such as H• (hydrogen radical), OH• (hydroxyl), and HO_2 (superoxide) as well as H_2O_2 (hydrogen peroxide) and H_2 (molecular hydrogen). Hydrodynamic shear force is also generated by cavitation which leads to disruption of the structure of the lignocellulosic materials (Subhedar and Gogate 2013). Some of the successful case studies for the disintegration of biomass based on the use of ultrasound include the modifications in the soy flakes biomass due to microfractures generated using ultrasound (Karki et al. 2010) as well as reported reduction in the size of sorghum biomass particles by 50 % (Shewale and Pandit 2009). The cavitational intensity and active cavitational volume are the two key factors in deciding the efficacy of cavitation for pretreatment as well as the applicability at large scale of operation. These two attributes of cavitation are strongly influenced by the operating parameters such as temperature, duration of sonication, frequency of ultrasound, amplitude and power dissipation, transfer efficiency of the transducer, properties of the suspension, as well as the type of biomass (Gogate 2008). The application of ultrasound has been also successfully reported to increase the effectiveness of other pretreatment processes (alkaline, dilute acid, organosolv, ionic liquid, and enzymatic pretreatments) as per the details mentioned in Table 2. It has been established that the application of ultrasound reduces the particle size, lowers the requirement of chemicals and process time, and increases efficiency of hemicellulose and lignin removal. These factors contribute to overall increase in hydrolysis efficiency and ethanol production. Problems related to the application of ultrasound pretreatment on large scale are significant energy requirements, lack of suitable scale-up strategies, and nonuniform distribution of the cavitational activity (Gogate and Kabadi 2009; Luo et al. 2014).

3.2.5 Microwave Pretreatment

The action of the microwave pretreatment is decided by the various thermal and nonthermal effects induced by the passage of microwaves, which are a form of electromagnetic energy in frequency range of 300–300,000 MHz. Direct interaction between the object to be heated and electromagnetic field applied results in heating within significantly quick time intervals. Polar molecules interact with this form of energy with higher efficiency resulting in rapid heating, which gives faster reaction rate and better yields. In the case of biomass pretreatment, generally combination of microwave with the chemical methods has been found to be very efficient. For

Table 2 Effect of ultrasound (US) on pretreatment process

Pretreatment	Biomass	Liquid medium	Result	References
Alkaline pretreatment	Olive tree pruning residues	Acetic acid, sodium hydroxide, water	Hemicellulose and lignin were removed. Only Alkali treatment required a longer exposure period	García et al. (2011)
	Sugarcane bagasse	Sodium hydroxide	About 75 % of lignin was removed. 79 % hemicellulose and 99 % cellulose were recovered	Velmurugan and Muthukumar (2011)
	Sugarcane bagasse	Sodium hydroxide	Reducing sugar yield predicted and achieved was 97 %	Velmurugan and Muthukumar (2012)
	Wheat straw	Potassium hydroxide	Lignin removed was more than 50 %	Sun and Tomkinson (2002)
	Corn stover	Sodium hydroxide	Alkali requirement reduced by 3 % and processing time was shortened by twofold relative to control. Lignin removal, 46 %	Zhang et al. (2008)
	Newspaper	Potassium hydroxide	The delignification with US increased delignification by twofold	Subhedar and Gogate (2014)
Dilute acid hydrolysis	Corncob	Sulfuric acid	39 % of xylose was extracted and 97 % reduction in processing time was achieved	Yang et al. (2009)
	Oil palm empty fruit bunch	Sulfuric acid	Xylose yield was increased by threefold	Yunus et al. (2010)
Organosolv	Wheat straw	Sodium hydroxide, methanol, and water	9 % increase in hemicellulose extraction as compared to control	Sun et al. (2002)
	Poplar wood	Ethanol, methanol, dioxane, dimethyl sulfoxide, sodium hydroxide	96 % of lignin and 75.5 % of hemicelluloses extraction was achieved with the help of sonication	Yuan et al. (2010)
Ionic liquid (IL)	Kenaf core fiber	Various ionic liquids	60–95 % increase in glucose yield in all ionic liquids was obtained as compared to heat pretreatment	Ninomiya et al. (2012)
	Bamboo powder	Cholinium	92 % increase in cellulose saccharification ratio obtained using combined ultrasound and ionic liquid as compared to 55 % obtained using only the thermal pretreated sample in the presence of cholinium (IL)	Ninomiya et al. (2013)

Enzymatic hydrolysis	Carboxymethyl cellulose	Acetate buffer	Enzymatic hydrolysis rate was improved	Imai et al. (2004)
	Corn slurry	Acetate buffer and water	20-fold reduction in particle size and threefold increase in total sugar release, with the help of sonication applied at high amplitude for 120 s	Khanal et al. (2007)
	Corn	Acetate buffer	300 % increase in sugar release	Montalbo et al. (2010)
	Cassava chips	Acetate buffer	2.7-fold increase in ethanol yield relative to control and fermentation time reduced by 24 h	Nitayavardhana et al. (2008)
	Sorghum flour	Acetate buffer and $CaCl_2$	50 % reduction of particle size and 8 % increase in saccharification	Shewale and Pandit (2009)
	Paper pulp	Acetate buffer	Complete conversion of pulp to sugars	Li et al. (2004)
	Waste paper	Acetate buffer	Waste paper was effectively hydrolyzed	Li et al. (2005)
	Sugarcane bagasse	N-Methylmorpholine-N-oxide	96.5 % hydrolysis of biomass was achieved	Li et al. (2012)
	Waste paper	Citrate buffer and water	Bioethanol yield was increased by 20 %. Ethanol production found to be decreased with continuous US exposure, an important observation highlighting the need of optimization of ultrasonic parameters	Wood et al. (1997)
Alkaline pretreatment+ enzymatic hydrolysis	Newspaper	Potassium hydroxide, citrate buffer	Optimization of process increased delignification by twofold and lignin was recovered. Further enzymatic hydrolysis of delignified waste newspaper with US gave 2.4 times increase in reducing sugar release	Subhedar et al. (2015)

example, it was reported that the pretreatment of the biomass slurry using dilute chemical agents is significantly improved by using microwave exposure for 5–20 min (Keshwani and Cheng 2009). Also combination of alkali (sodium hydroxide) and microwave has been reported to be effective (Merino et al. 2014). The microwave-induced pretreatment can also be combined with ultrasound or ionic liquids giving enhanced effectiveness for the combination approach (Zhu et al. 2005). The major drawback of the microwave-based process, especially as individual operation, is the economic feasibility and restricted treatment efficacy at large-scale operations.

3.3 *Chemical and Physicochemical Treatment*

The structure of lignocellulose can be modified with the help of chemicals based on the reduction in the cellulose crystallinity and removal of the lignin and hemicellulose. We now present overview of different chemical-based methods used for pretreatment of biomass.

3.3.1 Steam Explosion

Steam explosion is one of the extensively used techniques for pretreatment of lignocellulosic feedstock with low requirement of chemicals and limited consumption of energy. In this approach, reactor is filled with biomass, saturated steam with high pressure is dosed into reactor, and pressure is reduced suddenly which causes material to undergo explosive decompression giving degradation of hemicellulose and disruption of lignin matrix. The steam explosion is carried out over the temperature range of 160–260 °C (corresponding pressure range as 0.69–4.83 MPa) for few seconds to minutes. The mechanical decompression effects are generated as there is a sudden decrease in the pressure, which causes the fibers to separate (Xu and Huang 2014). During the process, lignin gets redistributed along with partial hemicellulose hydrolysis and solubilization. Also the cellulose gets exposed due to the removal of hemicelluloses offering increased accessibility to the enzymes. Residence time, particle size, and temperature are the factors deciding the effectiveness of the pretreatment (Alvira et al. 2010). The process of steam explosion can be operated in batch or continuous mode.

Significant increase in the removal of hemicellulose can be achieved by the addition of acid catalyst preferably H_2SO_4 during the steam explosion. Hydrolysis of the hemicellulose is intensified due to the external acids which are added as catalysts or by the acetic acid which is released from acetylated hemicellulose. Water also contributes to the hydrolysis, as it possesses certain acidic properties at high temperatures though at lower rates as compared to the externally added acids (Weil et al. 1997). The degree of polymerization and degree of crystallinity determine the structural configuration of cellulose. Steam explosion pretreatment causes selective hydrolysis of amorphous portions and converts them into crystalline constituents which further increases the crystallinity of cellulose (Corredor et al. 2009; Cherian et al. 2010). Steam explosion technology has been reported as an effective pretreatment approach giving enhanced production of ethanol from a wide range of raw materials, viz., poplar (Oliva et al. 2003), herbaceous residues as corn stover (Varga et al. 2004), olive residues (Cara et al. 2006), and wheat straw (Ballesteros et al. 2006).

Steam explosion pretreatment needs to be optimized according to material to be treated because treating with harsh conditions can lead to formation of inhibitory compounds which have negative effect on the further hydrolysis and fermentation steps. Inhibitors mainly produced are furan derivatives, weak acids, and phenolic compounds. Pentose and hexose degradation leads to formation of furfural and 5-hydroxyl methyl furfural (HMF), respectively. Weak acids typically generated are acetic, formic, and levulinic acid. Higher levels of acetic acid are produced from acetic groups of hemicelluloses, whereas levulinic and formic acids are generated from degradation of HMF and furfural (Taherzadeh and Karimi 2008). Lignin breakdown leads to formation of phenolic compounds depending on the lignin content of material. It is important to note here that detoxification methods have been developed to reduce the effects of these compounds, and also there have been some interests in developing genetically engineered microbes, which can function effectively even in the presence of inhibitors. Still, with the current level of knowledge, it can be said that it is important to control the intensity of steam explosion so that the inhibitory products are not produced in significant proportions rather than thinking about additional processing in terms of detoxification.

3.3.2 Liquid Hot Water Treatment

Liquid hot water (LHW) treatment is typically performed at high temperatures over the range of 150–230 °C under pressure which keeps water in liquid form (Shafiei et al. 2015). Different terms that have been used for this treatment based on hot water are hydrothermolysis, aqueous or steam/aqueous fractionation (Bouchard et al. 1991), uncatalyzed solvolysis (Mok and Antal 1992), and aquasolv (Allen et al. 1996). In LHW process, hydronium ion is released by water which acts as a weak acid in the system, affecting the polymerization of hemicellulose with hydrolysis of glycosidic linkages. *O*-acetyl group and other acid moieties are released, which further form acetic and uronic acids, which also help in the hydrolysis process. Liquid hot water mainly solubilizes hemicellulose and makes the cellulose more accessible without formation of inhibitors. After pretreatment, the slurry generated consists of solid fraction with cellulose and liquid fraction with solubilized hemicellulose. It is important to note that 90 % of the recovered hemicellulose is in the form of monomers (Alvira et al. 2010). The important factors that decide the efficiency of pretreatment are pH, temperature, time, and mode of operation. During the pretreatment process, pH is maintained over the range of 4–7, as it helps in retaining hemicellulose in oligomeric form. 40–60 % of the total biomass is typically dissolved in the process with contributions from 4 % to 22 % of cellulose, 35–60 % of lignin, and also most of the hemicellulose content is dissolved (Potumarthi et al. 2013). Cocurrent, counter-current, and flow-through are the three different configurations used for the LHW treatment. In cocurrent process, slurry of biomass and water are held at controlled conditions for desired residence time and then cooled. In counter-current process, reactor is designed to pass lignocellulosic material and water in opposite directions, and in the flow-through process, reactor consists of stationary bed made up of lignocellulosic material, and the hot water is passed over it, which results in the hydrolysis of material present in the bed with dissolution of specific components. Flow-through system removes more hemicellulose and lignin as compared to other systems and hence is recommended for better effectiveness (Mosier et al. 2005; Alvira et al. 2010).

No additive such as acid catalyst is required in the operation which minimizes the formation of inhibitory products and there is no requirement of final washing step, as water is the only solvent used in the processing. The requirement of high pressure and large amount of water makes it a highly demanding process (Xu and Huang 2014).

3.3.3 Acid Hydrolysis

Acid hydrolysis treatment makes cellulose more accessible by solubilizing the hemicellulosic fraction. Dilute acid treatment is generally preferred over concentrated acid as the higher loading of acid leads to production of inhibitory compounds as well as can result in equipment corrosion. Also the acid recovery is difficult, which coupled with higher maintenance and operational cost of process makes use of concentrated acid pretreatment non-feasible. On the other hand, dilute acid pretreatment is found out to be inexpensive and most effective. Hemicellulose sugars can be recovered easily and inhibitory compounds like furfural, hydroxyl methyl furfural (HMF), and other aromatic lignin degradation products are also produced in lesser amount in the case of dilute acids. Acid hydrolysis has been reported effective for pretreatment of various lignocellulosic feedstocks using different types of reactors such as percolation, shrinking-bed, plug flow, batch, and counter-current reactors (Taherzadeh and Karimi 2008). The actual process is generally performed using two approaches of high temperature (e.g., 180 °C) with lower retention time or lower temperature (e.g., 120 °C) with higher retention time (30–90 min). The recommended levels of acid concentrations are typically less than 4 wt %. The most preferred acid is the dilute sulfuric acid (H_2SO_4) as it generates lesser inhibition products as compared to other acids such as hydrochloric acid, phosphoric acid, and nitric acid. Organic acids like maleic and fumaric acids have also been reported to give negligible production of inhibitory compounds. It has been reported that, at 150 °C with substrate loading of 20–30 % w/v, organic acids can hydrolyze the feedstock with same efficiency as that of sulfuric acid with negligible production of inhibitory compounds (Kootstra et al. 2009).

3.3.4 Alkaline Hydrolysis

Alkaline hydrolysis is based on the use of alkali such as sodium, calcium, potassium, and ammonium hydroxides. Effectiveness of this process is dependent on the lignin content in the biomass, as alkali results in solubilization of only lignin and has very little effect on hemicellulose and cellulose content. Processing can be performed at room temperature and the required treatment times range from few minutes to days depending on the feedstock used. Production of degradation compounds, which can give inhibition, is lower as compared to acid pretreatment, and the process has been found to be suitable for agricultural residues. Calcium hydroxide (lime) is widely used as the treatment agent as it helps in the removal of amorphous substances like lignin which are responsible for high crystallinity index (Taherzadeh and Karimi 2008). The acetyl groups from hemicellulose which are responsible for the reduction in the steric hindrance for enzymes are also removed by addition of lime. The effectiveness of the process can be improved by addition of the oxidant (H_2O_2/oxygen) in combination with lime. As compared to potassium and sodium hydroxide, lime pretreatment is less costly with lower safety concerns and easy recovery is possible based on reaction with CO_2 (Mosier et al. 2005).

3.3.5 Ammonia Fiber Explosion (AFEX)

In ammonia fiber explosion (AFEX) pretreatment, lignocellulosic biomass is subjected to anhydrous ammonia at high pressures and under moderate temperature (60–100 °C) with subsequent rapid pressure release which leads to disruption of material. Rapid release of pressure causes expansion of ammonia gas resulting into swelling and physical disruption of material with partial decrystallization of cellulose. Ammonia loading, temperature, water loading, blow-down pressure, time, and number of treatment cycles are the key parameters in deciding the efficacy of the process. AFEX is typically performed in batch reactor, but some flow-through configurations are also possible. In the case of AFEX process, lignin typically gets degraded leaving hemicellulose and cellulose intact (Potumarthi et al. 2013). Extended AFEX leads to disruption of lignocarbohydrate linkage, hydrolysis of hemicellulose, and ammonolysis of glucuronic cross-linked bonds with partial decrystallization of cellulose (Xu and Huang 2014). AFEX pretreatment is effective on feedstock with lower lignin content, i.e., agricultural residues and herbaceous crops, and has limitations when subjected on woody biomass and other high-lignin-containing feedstocks. No inhibitors are generated in the AFEX process, and recovery of the ammonia from the process is 95 % with residual small amount of remaining ammonia acting as a nitrogen source to microbes during the fermentation.

Ammonia recycle percolation (ARP) is another pretreatment process in which ammonia is used. Aqueous ammonia (5–15 wt %) is made to pass over a column packed with biomass. The typical parameters for this process are temperature over the range of 140–210 °C, reaction time up to 90 min, and percolation rate of about 5 mL/min (Sun and Cheng 2002; Kim et al. 2008a,b). ARP process causes solubilization of hemicellulose but leaves short chains of cellulose intact with high content of glucan. Energy cost of this process can be reduced based on optimization to use lesser liquid loading or lower process temperature.

3.3.6 Organosolv Pretreatment

In this type of pretreatment, aqueous or organic solvent mixtures are used, which include ethanol, methanol, acetone, ethylene glycol, and tetrahydrofurfuryl alcohol. The used chemicals are responsible for solubilization of lignin and help in increasing accessibility to the cellulose (Zhao et al. 2009). Major advantage of this process is recovery of lignin in its pure form as a by-product. Organosolv pretreatment can be coupled with acid catalysts (HCl, H_2SO_4, oxalic acid, or salicylic acid) to efficiently break the hemicellulose bonds as well, and this leads to the effective separation of lignocellulosic constituents, i.e., cellulose, hemicellulose, and lignin. High yield of xylose can be obtained with addition of acids like HCl, H_2SO_4, or oxalic and salicylic acids to the solvents. Organosolv process is typically carried out at 150–200 °C with solvent and water. Organosolv approach can be used with acid hydrolysis to separate lignin and hemicellulose by two-stage fractionation. In the first stage, feedstock is subjected to dilute aqueous acid (0.5–2.5 N sulfuric acid) at 100 °C for 10–60 min which leads to solubilization of hemicellulose, and in the subsequent second stage, pretreated lignocellulose is delignified under acidic conditions at about 80 °C for 90 min (Papatheofanous et al. 1995). Araque et al. (2008) used the organosolv process based on acetone water mixture for pretreatment, and the highest ethanol yield of 99.5 % was reported after pretreatment at 195 °C for 5 min with pH of 2 and 1:1 ratio of acetone water. The most commonly used solvents are ethanol and methanol (Alvira et al. 2010) though there is a requirement that these should be removed before fermentation, as it acts as inhibitor for fermentation and enzymatic hydrolysis. The solvent recovery can be based on the evaporation and condensation.

3.3.7 Wet Oxidation

Wet oxidation pretreatment is the treatment of material with air or oxygen, which acts as a oxidizing agent in process. Temperature, oxygen pressure, and reaction time are the key parameters, and the typical recommended temperature is around 170–200 °C with treatment time of 5–15 min and oxygen pressure as 10 to 12 bar. The process can be further improved by the use of a catalyst, but this can also add significant cost especially at larger scales of operation. In wet oxidation process, acid formation occurs due to hydrolytic processes and oxidative reactions. Introduction of oxygen at temperature above 170 °C makes process exothermic and decreases the overall energy demand (Alvira et al. 2010). In this process, monomeric sugars are produced from hemicelluloses, whereas lignin undergoes cleavage and oxidation. End products do not consist of phenolic compounds as they are degraded into carboxylic acids, and also the production of HMF and furfural is lower as compared to steam explosion and liquid hot water pretreatment. Na_2CO_3 addition during the process helps in maintaining pH in alkaline or neutral range which further helps in the reduction of inhibitory product formation (Ahring et al. 1996). Wet oxidation has been reported effective for number of biomass sources such as spruce and corn stover giving high yields of monomeric sugar (Palonen et al. 2004) as well as wheat straw giving 70 % hemicellulose yield and 96 % cellulose yield in the presence of Na_2CO_3 (Klinke et al. 2002). Lower generation of inhibitory compounds and greater lignin removal are achieved with this type of pretreatment, but cost of oxygen is the main hindrance in the application of this process at commercial scale. One of the other disadvantages of the process is that lignin produced by this method cannot be used as fuel, as it is oxidized or undergoes cleavage during process.

3.3.8 CO_2 Explosion

CO_2 explosion is a process similar to the steam explosion and ammonia fiber explosion process. CO_2 is used in the pretreatment under super critical conditions with better efficacy as compared to steam- or ammonia-based process due to the lower size of carbon dioxide molecules as compared to ammonia and water. Lower size gives higher penetration into the small pores, which improves the hydrolysis rate. Carbon dioxide when used with aqueous solution forms carbonic acid which also helps in polymer hydrolysis. The explosive release of CO_2 pressure during the treatment leads to disruption of cellulose and hemicellulose (Taherzadeh and Karimi (2008)). The operating temperature is lower as compared to other processes resulting in lesser generation of inhibitory compounds (Zheng et al. 1998). The process can also be used as supplement to enzymatic hydrolysis with improved efficacy as reported for the case of aspen and southern yellow pine (Kim and Hong 2001). Utilization of CO_2 which is abundantly available and co-produced during ethanol production is a viable option. Requirement of high pressures in the process is bit of concern, and for industrial-scale applications, the operation might be too costly. Also, the production of sugars is less compared to the other methods.

3.3.9 Ionic Liquids Pretreatment

Ionic liquids (IL) are the salts which are in liquid state at relative low temperatures and consist of large organic cations and inorganic anions. Ionic liquids are called green solvents, as no toxic or explosive gases are released by them. By adjusting the anion and alkyl constituents of cation, their solvent properties can be varied. The important properties that drive its effective use include nonflammability, low vapor pressures, thermal and chemical stability, and attribute of remaining in the liquid state over wide range of temperatures (Hayes 2009). Formation of hydrogen bonds occurs between the non-hydrated chloride ions of IL and sugar hydroxyl protons which cause simultaneous dissolution of carbohydrates and lignin. The complex network of non-covalent interactions in biopolymer is disrupted without generation of degradation products (Alvira et al. 2010). Wheat straw and wood have been reported to be effectively treated using IL (Li et al. 2009; Lee et al. 2009). Currently, limited information is available about different IL and their effects on variety of feedstocks, and more research needs to be performed to establish the economic feasibility and applicability for pretreatment of sustainable feedstocks. Also, energy-efficient recycling methods should be developed for the application of IL on larger scale.

3.3.10 Ozonolysis

Ozone is a strong oxidizing agent and can be applied to delignify the biomass. Ozonolysis does not produce any inhibitory compounds and can be carried out at room temperature and normal pressure. Typical recommended flow rate of ozone is 0.5–0.8 L/min with ozone consumption of 2–7 % (% dry weight of biomass). Compounds with conjugated double bonds and functional groups with high electron densities are highly reactive with ozone. Lignin with presence of high content of C=C bonds is most susceptible to oxidation by ozone (Xu and Huang 2014). Soluble compounds of less molecular weight are released during the action of ozone on lignin. The residual ozone can be decomposed by increasing the temperature or by using a catalytic bed. Higher requirement of ozone is major disadvantage of this process (Taherzadeh and Karimi 2008).

Table 3 gives a comparative analysis of different pretreatment processes that can be applied for lignocellulosic feedstocks. It can be established that steam explosion, dilute acid pretreatment, and alkaline hydrolysis are the processes which have been employed on larger scale and have been reported to give enhanced ethanol production with economic viability. Organosolv, wet oxidation, ionic liquids, CO_2 explosion, and AFEX pretreatment processes are currently at developmental stage though quite energy intensive, and hence more research needs to be carried out to employ these processes at commercial scale. Processes like milling, ultrasound, and microwave are better suited as supportive approaches which can be employed to increase the efficiency of other pretreatment processes. Hydrodynamic cavitation pretreatment is still under development stage, and more investigations are needed to utilize it as an effective pretreatment process particularly based on the fact that it can be a useful alternative to ultrasound-based treatment.

An effective pretreatment approach should open the structures as well as remove refractory materials so as to increase the efficiency of hydrolysis giving higher extent of production of the monomeric sugars in the subsequent enzymatic hydrolysis approach. Enzymes in general are the proteins which act as a catalyst in the process. In ethanol production process, enzymes are employed in hydrolysis/saccharification step to produce monomeric sugars from pretreated feedstock. There

Table 3 Different effects of pretreatment technologies on lignocellulosic feedstock

Pretreatment	Increase of accessible surface area	Decrystallization of cellulose	Solubilization of hemicellulose	Solubilization of lignin	Alteration of lignin structure	Formation of furfural/ hydroxyl-methylfurfural (HMF)
Biological	√√	ND	√√	√√	√√	–
Mechanical	√√	√√	–	–	–	–
Steam explosion	√√	√	√√	√	√√	√√
Liquid hot water (LHW)	√√	ND	√√	√	√	√
Acid	√√	√	√√	√	√√	√√
Alkaline	√√	√	√	√√	√	√√
Ammonia fiber explosion (AFEX)	√√	√√	√	√√	√√	√
Organosolv	√√	√	√√	√√	√√	√√
Wet oxidation	√√	√	√√	√√	–	√
CO_2 explosion	√√	–	√√	–	–	–
Ionic liquids	√√	√√	√	√	√√	–
Microwave	√√	√	√	√	√	√
Ultrasound	√√	√	–	–	–	√
Hydrodynamic cavitation	√	ND	ND	ND	ND	ND

√√ high effect, √ medium effect, *ND* not determined, blank (-) no effect are a variety of enzymes which can be employed in the process of hydrolysis depending on the nature of the substrate. Cellulase and hemicellulases (xylanase, glucanase, and amylase) are the enzymes mostly used to hydrolyze lignocellulosic feedstocks. Monomeric sugars are further subjected to microbial fermentation to produce ethanol. Now we focus the discussion on potential lignocellulosic feedstocks which can be utilized for ethanol production.

4 Potential Lignocellulosic Feedstocks

Lignocellulosic feedstocks are abundantly available with the types varying according to different regions based on geographical conditions. In this section, the discussion about potential feedstocks has been presented with some guidelines in terms of modifications in the processing steps to maximize the ethanol production.

4.1 Sugarcane Waste

The top three producers of sugarcane are Brazil (40 %), India (19 %), and China (6 %). Four hundred ninety-five billion tons is the total global production of sugarcane per year. In India, total land under sugarcane cultivation is 4.94 million hectares with Maharashtra state being the highest producer with 80.1 t/ha as per the data for 2011–2012 (Macedo 1998; Yarnal and Puranik 2009). Twelve million tons of sugar is produced annually by 500 sugar mills across country, and there are over 300 molasses-based alcohol distilleries (Tewari et al. 2007).

Sugar manufacturing process generates different types of solid wastes including sugarcane trash, bagasse, press mud (PM), and bagasse fly ash (Fig. 2). Sugarcane trash consists of leaves, stems, etc. that are obtained on sugarcane harvesting. Sugarcane bagasse is a fibrous residue obtained after juice extraction and consists of 50 % cellulose, 25 % hemicellulose, and 25 % lignin (Balakrishnan and Batra 2011). Generation of bagasse from one ton of sugarcane is approximately 0.2–0.3 ton (Pessoa et al. 1997), which is a significant quantum considering the overall

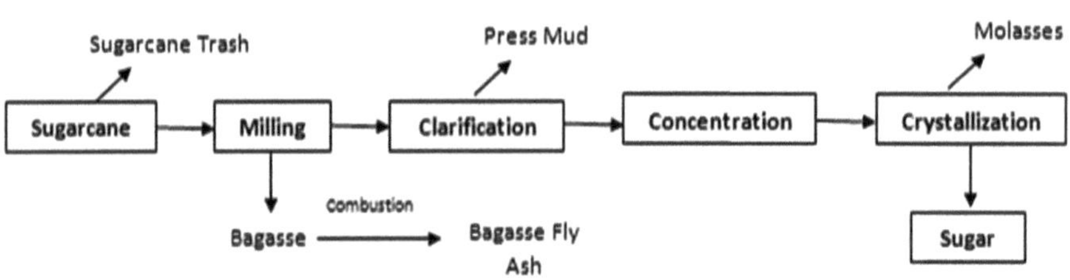

Fig. 2 Solid waste generation from cane sugar production process

production of sugar. Press mud is obtained during the sugarcane juice clarification process. One ton of sugarcane processing generates around 0.03 ton of press mud and consists of crude wax (5–14 %), fiber (15–30 %), crude protein (5–15 %), sugar (5–15 %), SiO (4–10 %), CaO (1–4 %), PO (1–3 %), MgO (0.5–1.5 %), and ash (9–10 %) (Yadav and Solomon 2006). The combustion of bagasse leads to production of bagasse fly ash which mainly consists of silica as main component with other metal oxides and unburnt carbon. Crushing of one ton sugarcane produces around 0.005–0.066 tons of fly ash (Umamaheswaran and Batra 2008). As these wastes are abundantly available, they can be utilized to produce value-added products. It is important to develop effective processing and perform studies related to the optimization of operating parameters to give maximum yields of any value added product.

Da Silva et al. (2010) performed a study on treatment of sugarcane bagasse with ball milling and wet disk milling with subsequent use for ethanol production. It was reported that bagasse processed with ball mill gave 78.7 % and 72.1 % of glucose and xylose yield, respectively. Further the obtained hydrolysate subjected to fermentation by C6 fermenting strain resulted in ethanol production yield as 89.8 %. De Albuquerque et al. (2013) employed steam explosion as a pretreatment approach for sugarcane bagasse followed by enzymatic hydrolysis in batch and fed batch mode with 8 % w/v consistency in citrate buffer (pH-4.8, 50 mM). It was reported that 53 g/l of glucose was produced and ethanol production of 0.23–0.39 g ethanol/g bagasse was obtained. Velmurugan and Muthukumar (2011) also reported effective conversion of sugarcane bagasse to produce sugar monomers based on the use of ultrasound-assisted acid hydrolysis pretreatment. Ninety-nine percent and 78.95 % recovery of cellulose and hemicelluloses, respectively, was reported to be achieved with 75.44 % lignin removal. It was also reported that optimized acid concentration was 2 % giving glucose yield of 41.02–69.09 % and pentose yield of 55.05–81.35 %. Cao and Aita (2013) performed a study to treat sugarcane bagasse with alkali in the presence of surfactants (Tween 80, Tween 20, PEG 4000, or PEG 6000). The ratios of the various components used were 1:0.5:20 for sugarcane bagasse, ammonium hydroxide (28 % v/v), and water. The reaction mixture was heated to 160 °C for 1 h, and it was reported that high cellulose digestibility (62 % and 66 %) is obtained with ethanol yield of 45.3 % and 52.8 % when treated with PEG 4000 and Tween 80, respectively. Cheng et al. (2008) also investigated the use of sugarcane bagasse as the feedstock and reported that reducing sugar production was increased from 28 to 63.5 g/l with two cycles of acidic treatment, and electrodialysis process led to 90 % removal of acetic acid and 88 % recovery of sulfuric acid. It was thus established that acid used in the acid hydrolysis process can be recovered via electrodialysis, which leads to detoxification of the medium and recycle of acid. The final ethanol production from the process was reported to be 19 g/l with productivity of 0.57 g/l/h.

Press mud (PM) is generally used as organic manure after composting, but recently processes are developed to generate biofuels from this waste product. Kuruti et al. (2015) investigated the hydrolysis of press mud slurry with 10 % solid concentration under acidic thermal conditions. Mixed microbial consortium was subsequently used for the digestion of obtained pretreated press mud. Ethanol produced was reported to be 0.04 g/g of PM and volatile fatty acids (VFA) generated were 0.27 g/g of PM. It was established that one ton of press mud can produce approximately 40 kg of bioethanol and 270 kg of VFA. The abundant availability of sugarcane waste makes it a potential candidate for development of processes which can produce ethanol in cost-effective way.

4.2 Corn Waste

Corn stover is an abundantly available cellulosic (40–50 wt % cellulose) biomass resource which can also be utilized for bioethanol production (Menon and Rao 2012). It is a nonfood biomass resource consisting of stalk, cob, leaf, and husk fractions. We now overview some of the studies related to production of different biofuels from corn waste with different processing methods. Saha et al. (2013) investigated bioethanol production from corn stover based on pretreatment using the hydrothermal approach followed by enzymatic hydrolysis. The optimal conditions reported for hydrothermal pretreatment were 10 % w/v of substrate loading, temperature of 200 °C, and time of 5 min. The optimum conditions for enzymatic saccharification were temperature of 45 °C, pH of 5.0, and reaction time of 72 h giving 550 ± 5 mg of reducing sugars per gm of corn stover and 20.9 ± 0.5 g ethanol from 42.8 ± 1.7 g sugars/lit loading. Steam explosion pretreatment with and without sulfuric acid as catalyst was also investigated to process corn stover at different residence times (5–10 min) and temperatures (190–210 °C) for maximizing glucose and xylose recovery. It was reported that the maximum yield of glucose as 86 % was obtained at 210 °C and time of treatment as 10 min without catalyst. After the pretreatment, simultaneous saccharification and fermentation (SSF) was performed to produce 22.6 g/l of ethanol (Bondesson et al. 2013).

Zhao and Xia (2010) investigated the use of three hydrolysates of corn stover (two hydrolysates obtained were from acid and alkaline pretreatment, whereas the third hydrolysate obtained was hemicellulosic hydrolysate produced by acid pretreatment only) for ethanol production using enzymatic hydrolysis. It was reported that the alkaline-pretreated corn stover gave highest ethanol production of 41.2 g/l as compared to other processes. Hybrid two-stage fractionation process was also reported based on the use of zinc chloride to solubilize the hemicellulose in the first step with further treatment based on the use of cellulase enzymes for ethanol production. 93.8 % of glucan, 89.7 % of xylan, and 71.1 % of arabinan were reported to be recovered with optimized loading of zinc chloride (5 % acidified $ZnCl_2$, 7.5 ml/min, 150 °C for 10 min, and 170 °C for 10 min) and 74.9 % of lignin was also recovered. Maximum ethanol yield of 69–98 % was reported from SSF-treated solids (Yoo et al. 2012). Xu et al. (2009) used milled corn stover pretreated with organic acids (acetic and lactic acid) for hydrolysate generation. The pretreatment without acid catalyst gave highest xylan recovery, whereas pretreatment with acids gave highest glucan recovery (95.66 %). Inhibition effect was not reported in the process, and optimized parameters obtained were 195 °C temperature for 15 min treatment with acetic acid. Further SSF of water-insoluble solids gave ethanol production of 0.51 g ethanol/g sugar from C5 and C6 sugars, and total estimated ethanol production was 241.1 kg/ton.

Ma et al. (2015) investigated the pretreatment of corncobs with alkaline potassium permanganate (APP) solution. The reported optimum parameters were 2 % (w/v) loading of potassium permanganate with solid to liquid ratio of 1:10, treatment time of 6 h, and temperature of 50 °C. It was also reported that 94.56 % cellulose and 81.47 % hemicellulose were recovered, and 46.79 % lignin was removed with simple one-step pretreatment. Zheng et al. (2015) investigated the effect of type of screw elements (conveying screw elements, kneading screw elements, and reverse screw elements) on the treatment efficiency for the case of corncobs as feedstock. The extruded corncobs were further treated with sodium hydroxide, and it was reported that 22 % increase in glucan conversion was obtained due to the pretreatment using extrusion process. Thirty-one percent, 30 %, and 37 % lignin were removed from corncobs extruded with conveying, kneading, and reverse screw elements, respectively. Delignified extruded corncobs with conveying, kneading, and reverse screw elements gave 56 %, 64 %, and 68 % glucan conversion when further treated with alkali. It was established that the use of screw elements increased the efficiency of enzymatic hydrolysis. Thus it has been demonstrated that corn residues can be effectively used for the bioethanol production, and different pretreatments can be used to improve the productivity.

4.3 Rice Waste

India and China are the main producers of rice and it is staple crop in many countries in Asia. The agricultural wastes generated from rice milling are rice husk and rice straw. The generation of rice straw is 0.7 kg of per kg paddy harvested and rice husk generated is 20–33 % of paddy weight (Khoo 2015). The global production of rice straw per year is 731 million tons out of which 667.6 million tons is produced by Asia, which is indeed a significant amount. Rice straw consists of 37 % cellulose, 24 % hemicelluloses, and 14 % lignin (Karimi et al. 2006), whereas rice husk consists of 36 % cellulose, 12 % hemicellulose, and 16 % lignin (Kim and Dale 2004, Saha and Cotta 2007).

Lin et al. (2012) investigated the use of rice straw as a feedstock for ethanol production at pilot scale. Twin screw extruder equipped with a hot water washing system and acid-catalyzed steam explosion were employed as pretreatment. 38.8 % glucan, 16.0 % xylan, 3.3 % arabinose, and 19.5 % lignin were obtained after the pretreatment. Hemicellulosic hydrolysate was generated from twin screw extruder, and highest ethanol yield from whole process was 0.44 g/g sugars. Patel et al. (2007) employed microbial pretreatment for intensifying the ethanol production from rice straw. The different fungi used in the investigation for pretreatment were *Aspergillus niger*, *Aspergillus awamori*, *Trichoderma reesei*, *Phanerochaete chrysosporium*, and *Pleurotus sajor-caju*. Fermentation was performed using *Saccharomyces cerevisiae* (NCIM 3095). It was reported that pretreatment with *A. niger* and *A. awamori* gave maximum yield of ethanol as 2.2 g/L.

Wi et al. (2013) investigated the influence of popping pretreatment on rice straw with subsequent optimization of the enzyme loading for ethanol production. Popping treatment combines mechanical effects of explosion with that of chemical effects of hydrolysis, and equipment used in process consists of direct burner and rotary reactor without steam generator. The sugar recovery of 0.567 g/g with glucose of 0.394 g/g biomass was reported after using the optimized enzyme loading and popping pretreatment for substrate loading (w/v) as 15 % and treatment time of 48 h. Popping pretreatment was reported to increase the surface area twofold as compared to non-treated biomass. Jamshid et al. (2005) also reported efficient conversion of rice straw into ethanol based on pulping using diethylene glycol, mixture of diethylene glycol, and ethylene glycol at atmospheric pressure. Kun et al. (2009) reported that pretreatment of rice straw with alkali-assisted photocatalysis resulted in change in the physical properties and microstructure of rice straw. Lignin content was found to be decreased which increased the enzyme accessibility. Alkali pretreatment led to solubilization of smaller molecules from hemicelluloses, and same process in the presence of H_2O_2 led to solubilization of larger molecules of hemicelluloses as well. Sheikh et al. (2013) reported the use of torrefaction process (thermal processing at 200–300 °C and inert atmosphere) to pretreat rice straw. The optimized parameters obtained were temperature of 220 °C and treatment time of 40 min which led to 60.68 % increase in the reducing sugar yield as compared to control sample. Saha and Cotta (2007) studied the effect of lime pretreatment on enzymatic hydrolysis and fermentation of rice hulls for ethanol production. It was reported that lime pretreatment did not produce any inhibitory compounds, and ethanol yield obtained was 0.49 g/g. Singh et al. (2014) investigated the application of crude unprocessed hydrolytic enzymes in hydrolysis process. The rice husk treated with microwave alkali pretreatment was subjected to crude enzymes. Reducing sugars produced were used for ethanol production using a coculture of *Saccharomyces cerevisiae* and *Scheffersomyces stipitis*. It was reported that 32 % more ethanol was produced from coculture as compared to the use of *S. cerevisiae* alone. Use of microwave was also reported to intensify the ethanol production. Blandino et al. (2014) reported that alkaline hydrogen peroxide (AHP) pretreatment at high pressures (10–30 bar) gave intensified processing with lower overall time and concentration of H_2O_2 as well as significant increase in reducing sugar concentration. The maximum hydrolysis yield obtained was 98.5 % with requirement of peroxide concentration as 3 % w/v and time as 30 min.

4.4 Energy Crops

Energy crops (perennial grasses) are also potential lignocellulosic feedstock for biofuel production. These crops have higher lignin and cellulose content than annual crops and have an efficient photosynthetic pathway which helps them to grow faster with lower water requirement as compared to other crops. These crops also possess a rhizome system which recycles nutrient and hence requires lesser external nutrient input. The rhizomatous grasses (perennial grass), namely, miscanthus (*Miscanthus* spp.), switch grass (*Panicum virgatum*), giant reed (*Arundo donax*), and reed canary grass (*Phalaris arundinacea*), have been reported to be effective feedstock for producing biofuels (Lewandowski et al. 2003).

Switch grass (*Panicum virgatum)* is perennial C4 grass, native to North America, and highly adaptable which can be grown in different regions. Annual yield of switch grass can be up to 16 t DM/ha (Keshwani and Cheng 2009). Switch grass when compared as alternative to other annual row crops leads to the reduction in the soil erosion by 95 % and pesticide usage by 90 % (David and Ragauskas 2010). Cellulosic content is found to be higher in these grasses, and many studies have been reported for production of ethanol from switch grass. Tao et al. (2011) investigated the effect of six different pretreatment on ethanol production from switch grass as (1) liquid hot water (LHW), (2) lime, (3) dilute acid (DA), (4) ammonia fiber expansion (AFEX), (5) sulfur dioxide-impregnated steam explosion (SO_2), and (6) soaking in aqueous ammonia (SAA). SAA pretreatment was reported to give the lowest monomer production, whereas SO_2 gave the highest production of monomer sugars. Ewanick and Bura (2011) reported a study to predict effect of moisture on steam pretreatment impregnated with SO_2. It was observed from the study that moisture had a great effect on the pretreatment process and increased the ethanol yield by 18–28 %. Zhang et al. (2011) proposed an integrated mathematical model to determine the optimal comprehensive supply chain/logistics decisions for a switch grass-to-ethanol facility. It was reported that 313 l of ethanol can be produced from one ton of switch grass. The results demonstrated that efficient utilization of 61 % of the available marginal land will lead to ethanol production equivalent to 100 % of current gasoline consumption.

Miscanthus giganteus is an energy crop with low maintenance requirement and high yield/energy content which makes it a suitable candidate for ethanol production. The crop consists of 40–60 % wt of cellulose, 20–40 % wt of hemicelluloses, and 10–30 % wt of lignin. Ten to thirty ton of dry matter/hectare is the harvest yield of miscanthus depending on the strain and growth area (Guo et al. 2008). Different pretreatment studies (Table 4) along with optimized parameters of one-step processes (AFEX, dilute acid, and organosolv) and two-step processes (dilute acid + wet explosion and enzyme + organosolv) have been reported for efficient conversion of *Miscanthus giganteus* into monomer sugars, which can be utilized for ethanol production.

4.5 Wheat Straw

Wheat (*Triticum aestivum*) is one of the most popularly grown food crops in the world with 21 % of food requirement being based on wheat. Wheat straw is a lignocellulosic waste obtained from wheat cultivation, and it amounts to 589.670 to

Table 4 Overview of pretreatment of *Miscanthus*

Pretreatment	Conditions/Results	References
One step		
Soda	145 °C, 30 min, 1.5 M NaOH Higher yield of cellulose obtained than other processes employed	Serrano et al. (2010)
AFEX	160 °C, 5 min, 2:1 (w/v) ammonia to biomass. 90–95 and 80–85 % glucan and xylan conversions were obtained on further hydrolysis	Murnen et al. (2007)
Dilute acid	130 °C, 15 min, 1–4 % H_2SO_4 70–75 % xylose yield and lesser acetic acid formation with higher ethanol yield	Guo et al. (2008)
Organosolv	Step 1. Ethanol-water 170–190 °C, 60 min Step 2. H_2SO_4 0.5–1.2 % ethanol-water 180 °C, 90 min 95 % recovery of glucan and 71 % of lignin recovery	Brosse et al. (2009)
Two step		
Dilute acid and wet explosion	1. 80–100 °C, 3–25 h, 0.5–1.5 % H_2SO_4 2. 170 °C, 5 min, 18 bars, O_2, H_2O_2 94 % xylose yield and 65 % glucose yield	Sorensen et al. (2008)
Enzyme and ethanol organosolv	1.Cellulyve® enzyme 2. 150–170 °C, 30–60 min, H_2SO_4, 0.5–1 % 75 % of cellulose to glucose conversion and 93 % delignification	Obama et al. (2012)

635.029 g/g of wheat grain (Montane et al. 1998). Wheat straw produced is mostly burnt or dumped which leads to environmental issues. Wheat straw consists of cellulose (33–40 %), hemicellulose (20–25 %), and lignin (15–20 %).

Saha et al. (2005) investigated the effect of dilute acid pretreatment on wheat straw at different temperatures with an objective of intensification of production of ethanol. Highest yield of monomers (565 mg/g) was reported when wheat straw was pretreated with dilute sulfuric acid (0.75 % v/v, 121 °C, 1 h) and hydrolyzed enzymatically (45 °C, pH 5.0, 72 h) with cellulase, β-glucosidase, xylanase, and esterase. It was also reported that detoxification of the hydrolysate with lime treatment as well as coupling hydrolysis and fermentation process significantly reduced the required fermentation time. Ballesteros et al. (2006) studied the application of steam explosion pretreatment on wheat straw. Steam explosion was carried out with diluted acid [H_2SO_4 0.9 % (w/w)] and water as pre-impregnation agent. The highest ethanol yield of 140 L/ton of wheat straw was obtained when steam explosion was carried out at 180 °C for 10 min. Saha and Cotta (2007) investigated the effect of lime pretreatment on ethanol production and reported that lime pretreatment can efficiently help in enhancing the production of reducing sugars for ethanol production as compared to dilute acid pretreatment, and detoxification step was also not needed. Kootstra et al. (2009) performed a study to compare the efficiency of acid pretreatment based on the use of fumaric and maleic acid with that of sulfuric acid using wheat straw as feedstock. It was reported that with substrate loading of 30 % w/w, organic acid pretreatment gave lower production of degradation products as compared to sulfuric acid. The study established that fumaric and maleic acid can be used as a replacement to sulfuric acid in pretreatment process. Han et al. (2015) investigated the use of auto-hydrolysis process to pretreat wheat straw with subsequent enzymatic hydrolysis to estimate the total sugar recovery. 25.8 g/100 g reducing sugar were reported to be obtained based on the use of auto-hydrolysis at 180 °C for 40 min, followed by mechanical refining. Economic evaluation of the process revealed that process of producing ethanol from wheat straw using auto-hydrolysis was economical with 28 % internal rate of return based on current ethanol wholesale price.

4.6 Food Waste

The wastage of food throughout food supply chain is nearly 1.3 billion tons and the various sources include households, restaurants, and cafeterias (Kiran and Liu 2015). Landfill disposal is a common way to dispose food waste which leads to environmental problems and shortage of disposal sites. Waste food is a rich organic source and can be utilized for production of many value-added products. Kim et al. (2008a,b) studied statistical optimization of enzymatic saccharification and production of ethanol from food waste. Based on response surface methodology, the optimum conditions were established giving a maximum ethanol production of 57.6 g/L. Kiran and Liu (2015) also reported a study to utilize food waste to produce ethanol with the help of fungal mash, which is a cocktail of various crude enzymes that can be employed to hydrolyze food waste. Ethanol production was observed to be 58 g/L with yield of 0.5 g/g glucose in 32 h of treatment. Matsakas et al. (2014) reported that use of separate liquefaction or saccharification step gives rapid reduction in the viscosity giving proper mixing of microorganisms resulting in significant increase in production of ethanol from food waste. Yang et al. (2014) investigated the use of instant noodle waste for biofuel production. In the processing, used oil was first isolated from waste and converted to biodiesel with the help of chemical catalyst (KOH, H_2SO_4), and leftover starch residues were utilized for ethanol production by subjecting them to enzymatic hydrolysis. It was reported that the maximum ethanol concentration was 61.1 g/l.

Fruit processing industries also produce ample amount of waste rich in organic content. The typical types of wastes obtained are apple pomace, banana waste, orange peel waste, pineapple waste, coffee waste, and many more. Apple pomace is the waste generated from pressing of apples for juice, which consists of pulp and peels as well as cores, and accounts for 25–35 % dry weight of processed apple (Joshi and Sandhu 1996). Orange peel waste is made up of peel, rags (segment membranes and core), and seeds left after juice extraction process. Composition of orange peel (Table 5) indicates that it is a potential feedstock for biofuel production. Indeed, there have been many reports related to biofuel production from orange peel waste (Pourbafrani et al. 2010; Joshi et al. 2015). Banana wastes mostly consist of peels and stalks, whereas pineapple wastes consist of skin, seeds, and remaining

Table 5 Proximate composition of fruit waste

Constituents (%)	Composition of apple pomace (dry weight basis)	Orange peel (dry weight basis)
Total carbohydrates	45–62	28–40
Glucose	22.7	8.2
Fructose	24	11.0
Sucrose	2.1	2.6
Protein	4.5–6	6
Pectin	3–14	25
Ash	1.6	3.7

Table 6 Characteristics of banana and pineapple peel

Constituents (%)	Banana peel (percent total weight)	Pineapple peel (percent total weight)
Total carbohydrates	24	33
Cellulose	11.1	20.5
Hemicellulose	5.5	10.6
Ash	13	12
Total nitrogen	1	1.2

parts after juice extraction process. The physicochemical compositions of these waste mentioned in Table 6 also clearly confirm that they can be utilized for ethanol production.

Manufacturing process of soluble coffee results in generation of waste which consists of solid residues left after extraction process. Waste produced is a potential feedstock for biofuel production. The oil extracted from waste can be utilized for biodiesel production, whereas residues which consist of 35–50 % dry weight of carbohydrates can be used for ethanol production (Mussatto et al. 2011, Mussatto et al. 2012).

4.7 *Palm Oil Mill Waste*

Substantial amount of agricultural waste is generated in the palm oil processing consisting of empty fruit bunch, mesocarp fiber, fronds, trunk, and shell. It has been reported that 182 million dry tons of solid palm biomass waste is generated per year (Wiloso et al. 2015), and these wastes are rich in carbohydrate content. Cheng et al. (2008) investigated the production of ethanol from empty palm oil fruit bunch using two-stage approach of pretreatment with sodium hydroxide followed by hydrolysis with sulfuric acid. The maximum reported yield of ethanol was 13.8 % w/w. Piarpuzán et al. (2011) studied the effect of alkaline pretreatment on ethanol production from empty palm oil fruit bunch. It was reported that alkaline pretreatment followed by enzymatic hydrolysis using acetate buffer gave ethanol production of 65 dm^3 per ton empty fruit bunch.

4.8 Paper Waste

The world annual consumption of paper is around 400 million tons as per the survey of 2010, (Yuan et al. 2012) and majority of this consumption finally appears as a waste material. Municipal solid wastes typically consist of 50 % paper waste which can be a rich source of fermentable sugars. Cellulose content of waste paper ranges from 35 % to 87.4 % depending on the type of paper; office paper contains around 55 % cellulose and newspaper contains around 49 % cellulose (Guerfali et al. 2015; Subhedar et al. 2015). Yuan et al. (2012) performed a study on the determination of economically feasible technology for production of ethanol from different waste papers. Four different types of paper, i.e., office paper, newspaper, magazines, and cardboard, were used in the study, and it was reported that enzymatic hydrolysis is the most efficient method for ethanol production, and cardboard is the most economical feedstock. The two-step approach of dilute acid pretreatment followed by hydrolysis/fermentation resulted in 444 L/ton production of ethanol, and it was established that waste paper can be an economically viable feedstock for ethanol production. Pretreatment of waste newspaper for intensified production of ethanol using ultrasound has been recently reported by Subhedar and Gogate (2014) with about 80 % delignification under optimized conditions. The use of ultrasound in the enzymatic hydrolysis also increased the release of reducing sugars by 2.8 times as compared to untreated sample. Subhedar et al. (2015) also investigated the effect of ultrasound on fermentation of the hydrolysate for ethanol production and reported 1.8 times increase in ethanol production as compared to the untreated waste.

It can be established that processes like steam explosion, dilute acid pretreatment, and alkaline pretreatment have been employed extensively for treatment of different feedstocks, and the discussion presented earlier gives the degree of improvement that can be obtained due to the pretreatments. Also a variety of feedstocks are available including sustainable sources which can allow dealing with the environmental concerns. There is great scope of development of technologies to process lignocellulosic material which can help in the reduction of process cost for ethanol production.

5 Conclusions

Bioethanol production from various lignocellulosic sources including the sustainable sources from the waste streams can be an effective approach to green environment and to decrease the dependency on fossil fuels. Biofuels are one of the potential renewable source of fuel which can provide energy security and pollution control. Optimized pretreatment processes can lead to production of ethanol at lower cost and help to address the environmental concerns. Various new pretreatment processes like torrefaction, popping, ultrasound, microwave, hydrodynamic cavitation, etc. can also be employed in combination with the more commonly used methods like alkali or acid induced treatments leading to much better productivity at lower costs. Utilization of sustainable feedstocks such as sugarcane bagasse, corn stover, and rice straw with optimization of the processing parameters can lead to efficient ethanol production. Wastes generated from food/fruit processing industries can also be a potential source for ethanol production as they are rich in organic content, though the processes for this conversion need to be optimized. The major hindrance in utilization of these wastes is efficient conversion of carbohydrates into monomeric sugars. There are several chemical, thermal, and enzymatic processes which are currently employed, but economic feasibility of these processes is the main setback for employment of these processes at commercial scale. There is a need to develop and modify these processes based on the use of process-intensifying approaches, which will help in achieving a green solution for energy generation. Overall, the efficient production of the biofuels from sustainable sources is the need of the hour, and the use of different process-intensifying approaches can allow us to establish the dream of large-scale manufacture.

References

Ahring BK, Jensen K, Nielsen P, Bjerre AB, Schmidt AS (1996) Pretreatment of wheat strawand conversion of xylose and xylan to ethanol by thermophilic anaerobic bacteria. Bioresour Technol 58:107–113. doi:10.1016/S0960-8524(96)00090-9

Allen SG, Kam LC, Zemann AJ, Antal MJ (1996) Fractionation of sugar cane with hot, compressed, liquid water. Ind Eng Chem Res 35:2709–2715. doi:10.1021/ie950594s

Alvira P, Tomás-Pejó E, Ballesteros M, Negro MJ (2010) Pretreatment technologies for an efficient bioethanol production process based on enzymatic hydrolysis: a review. Bioresour Technol 101:4851–4861. doi:10.1016/j.biortech.2009.11.093

Araque E, Parra C, Freer J, Contreras D, Rodríguez J, Mendonça R, Baeza J (2008) Evaluation of organosolv pretreatment for the conversion of *Pinus radiata* D. Don to ethanol. Enzyme Microb Technol 43:214–219. doi:10.1016/j.enzmictec.2007.08.006

Balakrishnan M, Batra VS (2011) Valorization of solid waste in sugar factories with possible applications in India: a review. J Environ Manag 92:2886–2891. doi:10.1016/j.jenvman.2011.06.039

Balat M, Balat H (2009) Recent trends in global production and utilization of bio-ethanol fuel. Appl Energy 86:2273–2282. doi:10.1016/j.apenergy.2009.03.015

Ballesteros I, Negro M, Oliva JM, Cabañas A, Manzanares P, Ballesteros M (2006) Ethanol production from steam-explosion pretreated wheat straw. Appl Biochem Biotechnol 130:496–508. doi:10.1007/978-1-59745-268-7_41

Blandino A, Belleli C, Caro I (2014) An effective process for pretreating rice husk to enhance enzyme hydrolysis. Ind Eng Chem Res 53:10870–10875. doi:10.1021/ie501354r

Bondesson PM, Galbe M, Zacchi G (2013) Ethanol and biogas production after steam pretreatment of corn stover with or without the addition of sulphuric acid. Biotechnol Biofuels 6:11. doi:10.1186/1754-6834-6-11

Bouchard J, Nguyen TS, Chornet E, Overend RP (1991) Analytical methodology for biomass pretreatment. Part 2: characterization of the filtrates and cumulative product distribution as a function of treatment severity. Bioresour Technol 36:121–131. doi:10.1016/0960-8524(91)90169-K

Bozell JJ (2001) Chemicals and materials from renewable resources, Chemicals and materials from renewable resources, vol 784, pp. 1–9. doi:10.1021/bk-2001-0784.ch001

Brosse N, Sannigrahi P, Ragauskas AJ (2009) Pre-treatment of *Miscanthus* × *giganteus* using the ethanol organosolv process for ethanol production. Ind Eng Chem Res 48:8328–8334. doi:10.1021/ie9006672

Cao S, Aita GM (2013) Enzymatic hydrolysis and ethanol yields of combined surfactant and dilute ammonia treated sugarcane bagasse. Bioresour Technol 131:357–364. doi:10.1016/j.biortech.2012.12.170

Cara C, Ruiz E, Ballesteros I, Negro MJ, Castro E (2006) Enhanced enzymatic hydrolysis of olive tree wood by steam explosion and alkaline peroxide delignification. Process Biochem 41:423–429. doi:10.1016/j.procbio.2005.07.007

Cheng KK, Cai BY, Zhang JA, Ling HZ, Zhou YZ, Ge JP (2008) Sugarcane bagasse hemicellulose hydrolysate for ethanol production by acid recovery process. Biochem Eng J 38:105–109. doi:10.1016/j.bej.2007.07.012

Cherian BM, de Leão AL, Souza SF, Thomas S, Pothan LA, Kottaisamy M (2010) Isolation of nanocellulose from pineapple leaf fibres by steam explosion. Carbohydr Polym 81:720–725. doi:10.1016/j.carbpol.2010.03.046

Cinelli BA, Castilho LR, Freire DMG, Castro AM (2015) A brief review on the emerging technology of ethanol production by cold hydrolysis of raw starch. Fuel 150:721–729. doi:10.1016/j.fuel.2015.02.063

Corredor DY, Salazar JM, Hohn KL, Bean S, Bean B, Wang D (2009) Evaluation and characterization of forage sorghum as feedstock for fermentable sugar production. Appl Biochem Biotech 158:164–179. doi:10.1007/s12010-008-8340-y

Costa JC, Sousa DZ, Pereira MA, Stams JM (2013) Biomethanation potential of biological and other wastes. In: Gupta VK, Tuohy MG (eds) Biofuel technologies. Springer, Berlin, pp 369–396. doi:10.1007/978-3-642-34519-7

da Silva ASS, Inoue H, Endo T, Yano S, Bon EPS (2010) Milling pretreatment of sugarcane bagasse and straw for enzymatic hydrolysis and ethanol fermentation. Bioresour Technol 101:7402–7409. doi:10.1016/j.biortech.2010.05.008

David K, Ragauskas AJ (2010) Switchgrass as an energy crop for biofuel production: a review of its ligno-cellulosic chemical properties. Energy Environ Sci 3:1182–1190. doi:10.1039/b926617h

de Albuquerque WMC, Martín C, deMoraes RGJ, Gouveia ER (2013) Increase in ethanol production from sugarcane bagasse based on combined pretreatments and fedbatch enzymatic hydrolysis. Bioresour Technol 128:448–453. doi:10.1016/j.biortech.2012.10.131

Ergun M, Mutlu SF, Gurel O (1997) Improved ethanol production by *Saccharomyces cerevisiae* with EDTA, ferrocyanide and zeolite X addition to sugar beet molasses. J Chem Technol Biot 68:147–150. doi:10.1002/(SICI)1097-4660(199702)68:2<147::AID-JCTB606>3.0.CO;2-3

Ewanick S, Bura R (2011) Bioresource technology the effect of biomass moisture content on bioethanol yields from steam pretreated switchgrass and sugarcane bagasse. Bioresour Technol 102:2651–2658. doi:10.1016/j.biortech.2010.10.117

García A, Alriols MG, Llano-Ponte R, Labidi J (2011) Ultrasound-assisted fractionation of the lignocellulosic material. Bioresour Technol 102:6326–6330. doi:10.1016/j.biortech.2011.02.045

Gogate PR (2008) Cavitational reactors for process intensification of chemical processing applications: a critical review. Chem Eng Process 47:515–527. doi:10.1016/j.cep.2007.09.014

Gogate PR, Kabadi AM (2009) A review of applications of cavitation in biochemical engineering/biotechnology. Biochem Eng J 44:60–72. doi:10.1016/j.bej.2008.10.006

Gogate PR, Pandit AB (2000) Engineering design methods for cavitation reactors II: hydrodynamic cavitation. AIChE J 46:1641–1649. doi:10.1002/aic.690460815

Guerfali M, Saidi A, Gargouri A, Belghith H (2015) Enhanced enzymatic hydrolysis of waste paper for ethanol production using separate saccharification and fermentation. Appl Biochem Biotechnol 175:25–42. doi:10.1007/s12010-014-1243-1

Guo G, Chen W, Chen W (2008) Characterization of dilute acid pretreatment of silvergrass for ethanol production. Bioresour Technol 99:6046–6053. doi:10.1016/j.biortech.2007.12.047

Gupta A, Verma JP (2015) Sustainable bio-ethanol production from agro-residues: a review. Renew Sustain Energy Rev 41:550–567. doi:10.1016/j.rser.2014.08.032

Han Q, Jin Y, Jameel H (2015) Autohydrolysis pretreatment of waste wheat straw for cellulosic ethanol production in a co-located straw pulp mill. Appl Biochem Biotechnol 175:1193–1210. doi:10.1007/s12010-014-1349-5

Hayes DJ (2009) An examination of biorefining processes, catalysts and challenges. Catal Today 145:138–151. doi:10.1016/j.cattod.2008.04.017

Imai M, Ikari K, Suzuki I (2004) High-performance hydrolysis of cellulose using mixed cellulase species and ultrasonication pretreatment. Biochem Eng J 17:79–83. doi:10.1016/S1369-703X(03)00141-4

Iskalieva A, Yimmou BM, Gogate PR, Horvath M, Horvath PG, Csoka L (2012) Cavitation assisted delignification of wheat straw: a review. Ultrason Sonochem 19:984–993. doi:10.1016/j.ultsonch.2012.02.007

Jamshid MR, Alireza T, Pejman RC (2005) Pulping of rice straw by high boiling solvents in atmospheric pressure. Iran Polym J 14:223–227

Joshi VK, Sandhu DK (1996) Preparation and evaluation of animal feed using solid state fermentation of apple pomace. Bioresour Technol 56:251–255. doi:10.1016/0960-8524(96)00040-5

Joshi SM, Waghmare JS, Sonawane KD, Waghmare SR (2015) Bio-ethanol and bio-butanol production from orange peel waste. Biofuels 6:55–61. doi:10.1080/17597269.2015.1045276

Karimi K, Emtiazi G, Taherzadeh MJ (2006) Ethanol production from dilute–acid pretreated rice straw by simultaneous saccharification and fermentation with *Mucor indicus*, *Rhizopus oryzae*, and *Saccharomyces cerevisiae*. Enzyme Microb Technol 40:138–144. doi:10.1016/j.enzmictec.2005.10.046

Karki B, Lamsal BP, Jung S, van Leeuwen J, Pometto AL, Grewell D, Khanal SK (2010) Enhancing protein and sugar release from defatted soy flakes using ultrasound technology. J Food Eng 96:270–278. doi:10.1016/j.jfoodeng.2009.07.023

Katiyar P, Srivastava SK, Tyagi VK (2015) A current scenario and novel approaches to degrade the lignocellulosic biomass for the production of biodiesel. J Fundam Renewable Energy Appl 5:161. doi:10.4172/20904541.1000161

Keshwani DR, Cheng JJ (2009) Switchgrass for bioethanol and other value-added applications: a review. Bioresour Technol 100:1515–1523. doi:10.1016/j.biortech.2008.09.035

Khanal SK, Montalbo M, van Leeuwen J, Srinivasan G, Grewell D (2007) Ultrasound enhanced glucose release from corn in ethanol plants. Biotechnol Bioeng 98:978–985. doi:10.1002/bit.21497

Khare SK, Pandey A, Larroche C (2015) Current perspectives in enzymatic saccharification of lignocellulosic biomass. Biochem Eng J 102:38–44. doi:10.1016/j.bej.2015.02.033

Khoo HH (2015) Review of bio-conversion pathways of lignocellulose-to-ethanol: sustainability assessment based on land footprint projections. Renew Sustain Energy Rev 46:100–119. doi:10.1016/j.rser.2015.02.027

Kim S, Dale BE (2004) Global potential bioethanol production from wasted crops and crop residues. Biomass Bioenergy 26:361–375. doi:10.1016/j.biombioe.2003.08.002

Kim K, Hong J (2001) Supercritical CO_2 pretreatment of lignocellulose enhances enzymatic cellulose hydrolysis. Bioresour Technol 77:139–144. doi:10.1016/S0960-8524(00)00147-4

Kim JS, Kim H, Lee JS, Lee JP, Park SC (2008a) Pretreatment characteristics of waste oak wood by ammonia percolation. Appl Biochem Biotechnol 148:15–22. doi:10.1007/s12010-008-8179-2

Kim JK, Oh BR, Shin HJ, Eom CY, Kim SW (2008b) Statistical optimization of enzymatic saccharification and ethanol fermentation using food waste. Process Biochem 43:1308–1312. doi:10.1016/j.procbio.2008.07.007

Kim I, Lee I, Jeon SH et al (2015) Hydrodynamic cavitation as a novel pretreatment approach for bioethanol production from reed. Bioresour Technol 192:335–339. doi:10.1016/j.biortech.2015.05.038

Kiran EU, Liu Y (2015) Bioethanol production from mixed food waste by an effective enzymatic pretreatment. Fuel 159:463–469. doi:10.1016/j.fuel.2015.06.101

Klinke HB, Ahring BK, Schmidt AS, Thomsen AB (2002) Characterization of degradation products from alkaline wet oxidation of wheat straw. Bioresour Technol 82:15–26. doi:10.1016/S0960-8524(01)00152-3

Kootstra AMJ, Hendrik HB, Scott EL, Sanders JPM (2009) Comparison of dilute mineral and organic acid pretreatment for enzymatic hydrolysis of wheat straw. Biochem Eng J 46:126–131. doi:10.1016/j.bej.2009.04.020

Kratky L, Jirout T (2011) Biomass size reduction machines for enhancing biogas production. Chem Eng Technol 34:391–399. doi:10.1002/ceat.201000357

Kumar P, Barrett DM, Delwiche MJ, Stroeve P (2009) Methods for pretreatment of lignocellulosic biomass for efficient hydrolysis and biofuel production. Ind Eng Chem Res 48:3713–3729. doi:10.1021/ie801542g

Kun N, Peng C, Xu Z, Wen-Song T (2009) Enhanced enzymatic hydrolysis of rice straw pretreated by alkali assisted with photocatalysis technology. J Chem Technol Biotechnol 84:1240–1245. doi:10.1002/jctb.2185

Kuruti K, Gangagni Rao A, Gandu B, Kiran G, Mohammad S, Sailaja S, Swamy YV (2015) Generation of bioethanol and VFA through anaerobic acidogenic fermentation route with press mud obtained from sugar mill as a feedstock. Bioresour Technol 192:646–653. doi:10.1016/j.biortech.2015.05.104

Lee SH, Doherty TV, Linhardt RJ, Dordick JS (2009) Ionic liquid-mediated selective extraction of lignin from wood leading to enhanced enzymatic cellulose hydrolysis. Biotechnol Bioeng 102:1368–1376. doi:10.1002/bit.22179

Lewandowski I, Scurlock JMO, Lindvall E, Christou M (2003) The development and current status of perennial rhizomatous grasses as energy crops in the US and Europe. Biomass Bioenergy 25:335–361. doi:10.1016/S0961-9534(03)00030-8

Li C, Yoshimoto M, Tsukuda N, Fukunaga K, Nakao K (2004) A kinetic study on enzymatic hydrolysis of a variety of pulps for its enhancement with continuous ultrasonic irradiation. Biochem Eng J 19:155–164. doi:10.1016/j.bej.2003.12.010

Li CZ, Yoshimoto M, Ogata H, Tsukuda N, Fukunaga K, Nakao K (2005) Effects of ultrasonic intensity and reactor scale on kinetics of enzymatic saccharification of various waste papers in continuously irradiated stirred tanks. Ultrason Sonochem 12:373–384. doi:10.1016/j.ultsonch.2004.02.004

Li Q, He YC, Xian M, Jun G, Xu X, Yang JM, Li LZ (2009) Improving enzymatichydrolysis of wheat straw using ionic liquid 1-ethyl-3-methyl imidazolium diethyl phosphate pretreatment. Bioresour Technol 100:3570–3575. doi:10.1016/j.biortech.2009.02.040

Li Q, Ji GS, Tang YB, Gu XD, Fei JJ, Jiang HQ (2012) Ultrasound-assisted compatible in situ hydrolysis of sugarcane bagasse in cellulase-aqueous–N-methylmorpholine-N-oxide system for improved saccharification. Bioresour Technol 107:251–257. doi:10.1016/j.biortech.2011.12.068

Limayem A, Ricke SC (2012) Lignocellulosic biomass for bioethanol production: current perspectives, potential issues and future prospects. Prog Energy Combust Sci 38:449–467. doi:10.1016/j.pecs.2012.03.002

Lin TH, Huang CF, Guo GL, Hwang WS, Huang SL (2012) Pilot-scale ethanol production from rice straw hydrolysates using xylose-fermenting *Pichia stipitis*. Bioresour Technol 116:314–319. doi:10.1016/j.biortech.2012.03.089

Luo J, Fang Z, Smith RL Jr (2014) Ultrasound-enhanced conversion of biomass to biofuels. Prog Energy Combust Sci 41:56–93. doi:10.1016/j.pecs.2013.11.001

Ma L, Cui Y, Cai R, Liu X, Zhang C, Xiao D (2015) Optimization and evaluation of alkaline potassium permanganate pretreatment of corncob. Bioresour Technol 180:1–6. doi:10.1016/j.biortech.2014.12.078

Macedo I (1998) Greenhouse gas emissions and energy balances in bio-ethanol production and utilization in Brazil. Biomass Bioenergy 14:77–81. doi:10.1016/S0961-9534(97)00038-X

Matsakas L, Kekos D, Loizidou M, Christakopoulos P (2014) Utilization of household food waste for the production of ethanol at high dry material content. Biotechnol Biofuels 7:4. doi:10.1186/1754-6834-7-4

Menon V, Rao M (2012) Trends in bioconversion of lignocellulose: biofuels, platform chemicals & biorefinery concept. Prog Energy Combust Sci 38:522–550. doi:10.1016/j.pecs.2012.02.002

Merino PO, Martínez PR, Labidi J, Luque R (2014) Microwave-assisted pretreatment of lignocellulosic biomass to produce biofuels and value-added products. Biofuels Biorefineries 3:197–224. doi:10.1007/978-94-017-9612-5_10

Mok WSL, Antal MJ (1992) Uncatalyzed solvolysis of whole biomass hemicellulose by hot compressed liquid water. Ind Eng Chem Res 31:1157–1161. doi:10.1021/ie00004a026

Montalbo-Lomboy M, Johnson L, Khanal SK, van Leeuwene J, Grewell D (2010) Sonication of sugary-2 corn: a potential pretreatment to enhance sugar release. Bioresour Technol 101:351–358. doi:10.1016/j.biortech.2009.07.075

Montane D, Farriol X, Salvado J, Jollez P, Chornet E (1998) Application of stream explosion to the fractionation and rapid vapourphase alkaline pulping of wheat straw. Biomass Bioenergy 14:261–276. doi:10.1016/S0961-9534(97)10045-9

Mosier N, Wyman CE, Dale BD, Elander RT, Lee YY, Holtzapple M, Ladisch CM (2005) Features of promising technologies for pretreatment of lignocellulosic biomass. Bioresour Technol 96:673–686. doi:10.1016/j.biortech.2004.06.025

Murnen HK, Balan V, Chundawat SPS (2007) Optimization of Ammonia Fiber Expansion (AFEX) pretreatment and enzymatic hydrolysis of Miscanthus x giganteus to fermentable sugars. Biotechnol Prog 23:846–850. doi:10.1021/bp070098m

Mussatto SI, Machado EMS, Martins S, Teixeira JA (2011) Production, composition, and application of coffee and its industrial residues. Food Bioprocess Technol 4:661–672. doi:10.1007/s11947-011-0565-z

Mussatto SI, Machado EMS, Carneiro LM, Teixeira JA (2012) Sugars metabolism and ethanol production by different yeast strains from coffee industry wastes hydrolysates. Appl Energy 92:763–768. doi:10.1016/j.apenergy.2011.08.020

Ninomiya K, Kamide K, Takahashi K, Shimizu N (2012) Enhanced enzymatic saccharification of kenaf power after ultrasonic pretreatment in ionic liquids at room temperature. Bioresour Technol 103:259–265. doi:10.1016/j.biortech.2011.10.019

Ninomiya K, Ohta A, Omote S, Ogino C, Takahashi K, Shimizu N (2013) Combined use of completely bio-derived cholinium ionic liquids and ultrasound irradiation for the pretreatment of lignocellulosic material to enhance enzymatic saccharification. Chem Eng J 215–216:811–818. doi:10.1016/j.cej.2012.11.020

Nitayavardhana S, Rakshit SK, Grewell D, van Leeuwen J, Khanal SK (2008) Ultrasound pretreatment of cassava chip slurry to enhance sugar release for subsequent ethanol production. Biotechnol Bioeng 101:487–496. doi:10.1002/bit.21922

Obama P, Ricochon G, Muniglia L, Brosse N (2012) Combination enzymatic hydrolysis and ethanol organosolv pre-treatments : effect on lignin structures, delignification yields and cellulose-to-glucose conversion. Bioresour Technol 112:156–163. doi:10.1016/j.biortech.2012.02.080

Oliva JM, Sáez F, Ballesteros I, Gónzalez A, Negro MJ, Manzanares P, Ballesteros M (2003) Effect of lignocellulosic degradation compounds from steam explosion pretreatment on ethanol fermentation by thermotolerant yeast *Kluyveromyces marxianus*. Appl Microbiol Biotechnol 105:141–154. doi:10.1385/ABAB:105:1-3:141

Palonen H, Thomsen AB, Tenkanen M, Schmidt AS, Viikari L (2004) Evaluation of wet oxidation pretreatment for enzymatic hydrolysis of softwood. Appl Biochem Biotechnol 117:1–17. doi:10.1385/ABAB:117:1:01

Papatheofanous MG, Billa E, Koullas DP, Monties B, Koukios EG (1995) Two-stage acidcatalyzed fractionation of lignocellulosic biomass in aqueous ethanol systems at low temperatures. Bioresour Technol 54:305–310. doi:10.1016/0960-8524(95)00152-2

Patel SJ, Onkarappa R, Shobha KS (2007) Study of ethanol production from fungal pretreated wheat and rice straw. Int J Microbiol 4

Pessoa JA, Manchilha de IM, Sato S (1997) Evaluation of sugar cane hemicelluloses hydrolyzate for cultivation of yeasts and filamentous fungi. J Ind Microbiol Biotechnol 18:360–363. doi:10.1038/sj.jim.2900403

Piarpuzán D, Quintero JA, Cardona CA (2011) Empty fruit bunches from oil palm as a potential raw material for fuel ethanol production. Biomass Bioenergy 35:1130–1137. doi:10.1016/j.biombioe.2010.11.038

Potumarthi R, Baadhe R, Bhattacharya S (2013) Fermentable sugars from lignocellulosic biomass: technical challenges. In: Gupta VK, MG T (eds) Biofuel technologies. Springer, Berlin, pp 3–26. doi:10.1007/978-3-642-34519-7_1

Pourbafrani M, Forgács G, Horváth IS, Niklasson C, Taherzadeh MJ (2010) Production of biofuels, limonene and pectin from citrus wastes. Bioresour Technol 101:4246–4250. doi:10.1016/j.biortech.2010.01.077

Rubin EM (2008) Genomics of cellulosic biofuels. Nature 454:841–845. doi:10.1038/nature07190

Saha BC, Cotta MA (2007) Enzymatic saccharification and fermentation of alkaline peroxide pretreated rice hulls to ethanol. Enzyme Microb Technol 41:528–532. doi:10.1016/j.enzmictec.2007.04.006

Saha BC, Iten LB, Cotta MA, Wu YV (2005) Dilute acid pretreatment, enzymatic saccharification and fermentation of wheat straw to ethanol. Process Biochem 40:3693–3700. doi:10.1016/j.procbio.2005.04.006

Saha BC, Yoshida T, Cotta MA, Sonomoto K (2013) Hydrothermal pretreatment and enzymatic saccharification of corn stover for efficient ethanol production. Ind Crops Prod 44:367–372. doi:10.1016/j.indcrop.2012.11.025

Saini JK, Saini R, Tewari L (2015) Lignocellulosic agriculture wastes as biomass feedstocks for second-generation bioethanol production: concepts and recent developments. 3. Biotech 5:337–353. doi:10.1007/s13205-014-0246-5

Sánchez C (2009) Lignocellulosic residues: biodegradation and bioconversion by fungi. Biotechnol Adv 27:185–194. doi:10.1016/j.biotechadv.2008.11.001

Serrano L, Egües I, Alriols MG (2010) Miscanthus sinensis fractionation by different reagents. Chem Eng J 156:49–55. doi:10.1016/j.cej.2009.09.032

Shafiei M, Kumar R, Karimi K (2015) Pretreatment of lignocelluosic biomass. In: Karimi K (ed) Lignocellulose-based bioproducts. Biofuel and biorefinery technologies, vol 1. Springer, Berlin, pp 85–154. doi:10.1007/978-3-319-14033-9

Sheikh MMI, Kim CH, Park HJ, Kim SH, Kim GC, Lee JY, Kim JW (2013) Effect of torrefaction for the pretreatment of rice straw for ethanol production. J Sci Food Agric 93:3198–3204. doi:10.1002/jsfa.6155

Shewale SD, Pandit AB (2009) Enzymatic production of glucose from different qualities of grain sorghum and application of ultrasound to enhance the yield. Carbohyd Res 344:52–60. doi:10.1016/j.carres.2008.10.006

Singh A, Bajar S, Bishnoi NR (2014) Enzymatic hydrolysis of microwave alkali pretreated rice husk for ethanol production by *Saccharomyces cerevisiae*, *Scheffersomyces stipitis* and their co-culture. Fuel 116:699–702. doi:10.1016/j.fuel.2013.08.072

Sorensen A, Teller PJ, Hilstrøm T, Ahring BK (2008) Hydrolysis of Miscanthus for bioethanol production using dilute acid pre-soaking combined with wet explosion pre-treatment and enzymatic treatment. Bioresour Technol 99:6602–6607. doi:10.1016/j.biortech.2007.09.091

Srivastava N, Rawat R, Singh Oberoi H, Ramteke PW (2014) A review on fuel ethanol production from lignocellulosic biomass. Int J Green Energy 12:949–960. doi:10.1080/15435075.2014.890104

Subhedar PB, Gogate PR (2013) Intensification of enzymatic hydrolysis of lignocellulose using ultrasound for efficient bioethanol production: a review. Ind Eng Chem Res 52:11816–11828. doi:10.1021/ie401286z

Subhedar PB, Gogate PR (2014) Alkaline and ultrasound assisted alkaline pretreatment for intensification of delignification process from sustainable raw-material. Ultrason Sonochem 21:216–225. doi:10.1016/j.ultsonch.2013.08.001

Subhedar PB, Babu NR, Gogate PR (2015) Intensification of enzymatic hydrolysis of waste newspaper using ultrasound for fermentable sugar production. Ultrason Sonochem 22:326–332. doi:10.1016/j.ultsonch.2014.07.005

Sun Y, Cheng J (2002) Hydrolysis of lignocellulosic materials for ethanol production: a review. Bioresour Technol 83:1–11. doi:10.1016/S0960-8524(01)00212-7

Sun RC, Tomkinson J (2002) Comparative study of lignins isolated by alkali and ultrasound-assisted alkali extractions from wheat straw. Ultrason Sonochem 9:85–93. doi:10.1016/S1350-4177(01)00106-7

Sun RC, Sun XF, Ma XH (2002) Effect of ultrasound on the structural and physiochemical properties of organosolv soluble hemicelluloses from wheat straw. Ultrason Sonochem 9:95–101. doi:10.1016/S1350-4177(01)00102-X

Taherzadeh MJ, Karimi K (2008) Pretreatment of lignocellulosic wastes to improve ethanol and biogas production: a review. Int J Mol Sci 9:1621–1651. doi:10.3390/ijms9091621

Tao L, Aden A, Elander RT (2011) Process and technoeconomic analysis of leading pretreatment technologies for lignocellulosic ethanol production using switchgrass. Bioresour Technol 102:11105–11114. doi:10.1016/j.biortech.2011.07.051

Tewari PK, Batra VS, Balakrishnan M (2007) Water management initiatives in sugarcane molasses based distilleries in India. Resour Conserv Recycl 52:351–367. doi:10.1016/j.resconrec.2007.05.003

Umamaheswaran K, Batra VS (2008) Physico-chemical characterization of Indian biomass ashes. Fuel 87:628–638. doi:10.1016/j.fuel.2007.05.045

Varga E, Reczey K, Zacchi G (2004) Optimization of steam pretreatment of corn stover to enhance enzymatic digestibility. Appl Biochem Biotechnol 113–116:509–523. doi:10.1385/ABAB:114:1-3:509

Velmurugan R, Muthukumar K (2011) Utilization of sugarcane bagasse for bioethanol production: sono-assisted acid hydrolysis approach. Bioresour Technol 102:7119–7123. doi:10.1016/j.biortech.2011.04.045

Velmurugan R, Muthukumar K (2012) Ultrasound-assisted alkaline pretreatment of sugarcane bagasse for fermentable sugar production: Optimization through response surface methodology. Bioresour Technol 112:293–299. doi:10.1016/j.biortech.2012.01.168

Weil J, Sarikaya A, Rau SL, Goetz J, Ladisch CM, Brewer M (1997) Pretreatment of yellow poplar sawdust by pressure cooking in water. Appl Biochem Biotechnol 68:21–40. doi:10.1007/BF02785978

Wi SG, Choi IS, Kim KH, Kim HM, Bae HJ (2013) Bioethanol production from rice straw by popping pretreatment. Biotechnol Biofuels 6:166. doi:10.1186/1754-6834-6-166

Wiloso EI, Bessou C, Heijungs R (2015) Methodological issues in comparative life cycle assessment: treatment options for empty fruit bunches in a palm oil system. Int J Life Cycle Assess 20:204–216. doi:10.1007/s11367-014-0815-1

Wood BE, Aldrich HC, Ingram LO (1997) Ultrasound stimulates ethanol production during the simultaneous saccharification and fermentation of mixed waste office paper. Biotechnol Prog 13:232–237. doi:10.1021/bp970027v

Xu Z, Huang F (2014) Pretreatment methods for bioethanol production. Appl Biochem Biotechnol 174:43–62. doi:10.1007/s12010-014-1015-y

Xu J, Thomsen MH, Thomsen AB (2009) Enzymatic hydrolysis and fermentability of corn stover pretreated by lactic acid and/or acetic acid. J Biotechnol 139:300–305. doi:10.1016/j.jbiotec.2008.12.017
Yadav RL, Solomon S (2006) Potential of developing sugarcane by-product based industries in India. Sugar Technol 8:104–111. doi:10.1007/BF02943642
Yang B, Wyman CE (2008) Pretreatment: the key to unlocking low-cost cellulosic ethanol. Biofuels Bioprod Bior 2:26–40. doi:10.1002/bbb.49
Yang W, Ajapur VK, Krishnamurthy K, Feng H, Yang R, Rababah TM (2009) Expedited extraction of xylan from corncob by power ultrasound. Int J Agric Biol Eng 2:76–83. doi:10.3965/j.issn.1934-6344.2009.04.076-083
Yang X, Lee SJ, Yoo HY, Choi HS, Park C, Kim SW (2014) Biorefinery of instant noodle waste to biofuels. Bioresour Technol 159:17–23. doi:10.1016/j.biortech.2014.02.068
Yarnal GS, Puranik VS (2009) Energy management in cogeneration system of sugar industry using system dynamics modeling. Cogener Distrib Gener J 24:7–22. doi:10.1080/15453660909509158
Yoo CG, Kuo M, Kim TH (2012) Ethanol and furfural production from corn stover using a hybrid fractionation process with zinc chloride and simultaneous saccharification and fermentation (SSF). Process Biochem 47:319–326. doi:10.1016/j.procbio.2011.11.018
Yuan TQ, Xu F, He J, Sun RC (2010) Structural and physico-chemical characterization of hemicelluloses from ultrasoundassisted extractions of partially delignified fast-growing poplar wood through organic solvent and alkaline solutions. Biotechnol Adv 28:583–593. doi:10.1016/j.biotechadv.2010.05.016
Yuan X, Cao Y, Li J, Wen B, Zhu W, Wang X, Cui Z (2012) Effect of pretreatment by a microbial consortium on methane production of waste paper and cardboard. Bioresour Technol 118:281–288. doi:10.1016/j.biortech.2012.05.058
Yunus R, Salleh SF, Abdullah N, Biak DRA (2010) Effect of ultrasonic pre-treatment on low temperature acid hydrolysis of oil palm empty fruit bunch. Bioresour Technol 101:9792–9796. doi:10.1016/j.biortech.2010.07.074.
Zhang Y, Fu E, Liang J (2008) Effect of ultrasonic waves on the saccharification processes of lignocellulose. Chem Eng Technol 31:1510–1515. doi:10.1002/ceat.200700407
Zhang Q, He J, Tian M, Mao Z, Tang L, Zhang J (2011) Enhancement of methane production from cassava residues by biological pretreatment using a constructed microbial consortium. Bioresour Technol 102:8899–8906. doi:10.1016/j.biortech.2011.06.061
Zhao J, Xia L (2010) Bioconversion of corn stover hydrolysate to ethanol by a recombinant yeast strain. Fuel Process Technol 91:1807–1811. doi:10.1016/j.fuproc.2010.08.002
Zhao X, Cheng K, Liu D (2009) Organosolvpretreatment of lignocellulosic biomass for enzymatic hydrolysis. Appl Microbiol Biotechnol 82:815–827. doi:10.1007/s00253-009-1883-1
Zheng Y, Lin HM, Tsao GT (1998) Pretreatment for cellulose hydrolysis by carbondioxide explosion. Biotechnol Prog 14:890–896. doi:10.1021/bp980087g
Zheng Y, Zhao J, Xu F, Li Y (2014) Pretreatment of lignocellulosic biomass for enhanced biogas production. Prog Energy Combust Sci 42:35–53. doi:10.1016/j.pecs.2014.01.001
Zheng J, Choo K, Rehmann L (2015) The effects of screw elements on enzymatic digestibility of corncobs after pretreatment in a twin-screw extruder. Biomass Bioenergy 74:224–232. doi:10.1016/j.biombioe.2015.01.022
Zhu S, Wu Y, Yu Z, Liao J, Zhang Y (2005) Pretreatment by microwave/alkali of rice straw and its enzymatic hydrolysis. Proc Biochem 40:3082–3086. doi:10.1016/j.procbio.2005.03.016

Current Approaches in Producing Oil and Biodiesel from Microalgal Biomass

Marcondes M. Pacheco, Michele Hoeltz, Diego de Souza, Lisianne B. Benitez, Rosana C.S. Schneider, and Maria V.G. Müller

Abstract Alternative biofuel sources, such as biodiesel, produced from nontoxic, biodegradable, and renewable materials are currently attracting great interest given the growing global energy demand. Microalgae are an exceptional biofuel source due to their potential for generating significant quantities of biomass and oil and for combining their production with environmental technologies such as wastewater treatment. Technologies for producing biodiesel from microalgae have been widely studied, especially increasing the lipid content in the cell, strain selection, lipid extraction, and transesterification methods and innovations that use wet biomass for extraction, which reduces the environmental impact and production cost. These aspects are explored in this chapter to identify the different production methods and technologies under development for expanding biodiesel production from microalgae.

Keywords Microalgae • Oil • Biodiesel • Production technologies

1 Introduction

Fossil fuels currently meet 27 % of the world energy demand. However, the scarce reserves, high prices, and environmental problems associated with using these fuels have driven the development of alternative energy sources to ensure economic and environmental sustainability of the energy supply (Stephens et al. 2010; Lam and Lee 2012; Hong et al. 2014).

Biodiesel is a fuel composed of alkyl esters obtained from long-chain fatty acid transformation. It can be produced by animal or plant triglyceride transesterification with short-chain alcohols, such as methanol and ethanol (Elshahed 2010; Velasquez-Orta et al. 2012; Likozar and Levec 2014).

M.M. Pacheco • M. Hoeltz (✉) • D. de Souza • L.B. Benitez • R.C.S. Schneider • M.V.G. Müller
Environmental Technology Program (PPGTA), University of Santa Cruz do Sul, Av. Independência, 2293, CEP 96815-900 Santa Cruz do Sul, Rio Grande do Sul, Brazil
e-mail: hoeltz@unisc.br

L. Singh, V.C. Kalia (eds.), *Waste Biomass Management – A Holistic Approach*,
DOI 10.1007/978-3-319-49595-8_13

Biodiesel is an exceptional renewable energy source because it releases less carbon dioxide, sulfur (SO_x), and carbon monoxide (CO) into the atmosphere. Moreover, it does not contain aromatic compounds and other chemical substances that are harmful to the environment and human health; it is also more biodegradable than diesel (Kirrolia et al. 2013). The important properties of biodiesel are that it is a clean energy source with adequate viscosity and its inherent lubricity, high flash point, and elevated cetane number for use as a fuel (Knothe and Steidley 2011; Hong et al. 2014; Huang and Su 2014).

Multiple raw material sources are used for biodiesel production, including microalgae. These microorganisms contain lipids and fatty acids as cell membrane components, reserve substances, metabolites, and energy sources. Certain species are excessively rich in oils and may have an oil content greater than 80 %, which is efficiently converted into methyl esters (biodiesel) (Demirbas 2010).

Lipids from microalgae are mostly neutral lipids and typically include a high level of saturation, which renders the microalgae even more attractive for biofuel research (Rawat et al. 2013). In this context, using mixed cultures is also important because it may be advantageous to use biomass generated from wastewater treatment lagoons in biodiesel production (Wahlen et al. 2011).

In addition to using microalgae with a high oil content and improving the culture conditions to maximize biomass and oil production, the limiting step in producing biodiesel from microalgae is extracting the lipid content from the cells. This step can be performed separately or concurrently with transesterification (in situ) in the wet or dry biomass.

The aspects that should be explored to facilitate biodiesel production from microalgae include culture conditions that yield greater lipid accumulation in cells, oil extraction techniques, combination methods to avoid certain production stages, and optimizing the transesterification process.

2 An Overview of Microalgae Production for Obtaining Oil

Microalgae are among the most promising sources of biofuels due to their ability to rapidly multiply, even in low-quality water; absorb large quantities of CO_2; and accumulate substantial quantities of reserve substances (Ashokkumar et al. 2014). Depending on the culture conditions and cultured species, large quantities of polysaccharides (sugars) and triglycerides (fats), which are raw materials for producing bioethanol and biodiesel, respectively, can be generated (Dębowski et al. 2013; Slade and Bauen 2013).

Maximizing lipid productivity is key for a successful biotechnological process using microalgae for biodiesel production, and one of the greatest challenges is scaling up from a laboratory to a commercial scale while maintaining high lipid productivity (Table 1) (Xu and Boeing 2014). Increased lipid extraction from microalgae depends on optimizing a series of nutritional conditions in the growth medium (Gupta et al. 2013; Huang and Su 2014).

Table 1 Lipid concentrations extracted from the biomass of certain microalgal species

Espécie	Teor de lipídios (%)	Referência
Haematococcus pluvialis	24.3	Wu et al. (2013)
Chlorella vulgaris	27.6	Wu et al. (2013)
Chlorella vulgaris	27.38	Heredia-Arroyo et al. (2011)
Chlorella pyrenoidosa	17.4	Wu et al. (2013)
Chlorella sorokiniana	32	Wu et al. (2013)
Chlorella sp.	26.43–27.96	Ahmad et al. (2014)
Scenedesmus obliquus	12–14	Dragone et al. (2010)
Scenedesmus dimorphus	16–40	Dragone et al. (2010)
Chlorella protothecoides	14.6–57.8	Maity et al. (2014)
Dunaliella tertiolecta	16.7–71	Maity et al. (2014)
Prymnesium parvum	22–38	Demirbas (2011)
Euglena gracilis	14–20	Demirbas (2011)
Chlamydomonas sp. TAI-2	18.4	Wu et al. (2012)
Chlorella emersonii	25–63	Mata et al. (2010)
Nannochloropsis sp.	12–53	Mata et al. (2010)
Neochloris oleoabundans	29–65	Mata et al. (2010)
Botryococcus braunii	25–75	Mata et al. (2010)
Synechocystis aquatilis	15.85	Kaiwan-Arporn et al. (2012)

Lipid accumulation in microalgae may result from a lack of nutrients, especially nitrogen, and excess organic carbon (Chandra et al. 2014). To abundantly produce valuable bioproducts from microalgae, particularly biofuels, it is important that the conditions offered to the microorganisms transition them toward a heterotrophic metabolism, where organic carbons, such as in sugars and organic acids, act as carbon and energy sources (Liang 2013).

Temperature, luminosity, pH, salinity, and salt minerals are other variables that, when controlled and applied as stress, can improve growth performance and energy reserve accumulation in microalgae (Venkata Mohan and Devi 2014).

Nitrogen, sulfur, and phosphorus are important components of algal cells and DNA and are essential for cell growth; these elements become limiting factors at low concentrations. Information on nitrogen deficiency in the algal metabolism is scarce, but certain reports indicate that a nitrogen deficit can lead to oxidative stress (Hockin et al. 2012). Zhang et al. (2013) observed substantial neutral lipid accumulation by *Chlorella sorokiniana C*3 in response to oxidative stress induced by a hydrogen peroxide treatment.

Nutrient limitation, particularly nitrogen, increases neutral lipid production and accumulation in microalgae. However, this limitation yields severe losses in biomass productivity. To reduce the negative effect of subjecting microalgae cultures to nutrient-limiting conditions, biomass production must first be maximized, and later, a neutral lipid accumulation-stimulating factor must be introduced (Bertozzini et al. 2014).

To maximize algal biomass production using *Chlorella vulgaris* and *Scenedesmus* sp., thereby increasing lipid accumulation, different concentrations of calcium, magnesium, and sodium chloride were examined. The results showed that a short-term reduction in the magnesium concentration induced lipid production (Chen et al. 2014).

Among the micronutrients, iron, potassium, and inorganic salts are the most important metabolic activators and are used in algal culture media formulations (Zeng et al. 2011; Cai et al. 2013).

Yang et al. (2014) applied a response surface method to define the optimal conditions for promoting greater lipid accumulation in *Scenedesmus* sp. The results showed that adding $NaHCO_3$, i.e., a source of inorganic carbon, as well as NaH_2PO_4, $2H_2O$ and $NaNO_3$ significantly enhanced lipid accumulation in microalgae without influencing the fatty acid composition of *Scenedesmus* sp.

To synthesize cyanocobalamin (vitamin B12), algae require cobalt (Co), which should be used at low concentrations. According to Li et al. (2007), such care is necessary due to both the potential for toxicity to the algae and the researcher in addition to introducing this toxic element into the food chain.

Another important factor that merits study is supplementing the culture medium with glycerol, which is a by-product of biodiesel production that increases lipid production in microalgae. In studies conducted by Cerón-Garcıa et al. (2005), glycerol, fructose, glucose, mannose, and lactose were used as carbon sources to stimulate growth and fatty acid accumulation in *Phaeodactylum tricornutum*, which enabled effective gains in biomass yield and intracellular oil accumulation.

High glycerol concentrations in microalgal cultures subjected to mixotrophic growth conditions tend to change the thylakoids in chloroplasts that favor glycerol assimilation, which can be considered an adaptive response to stress imposed by mixotrophic conditions (Cerón-García et al. 2013).

Sun et al. (2014) studied the effects of culture media with different glycerol concentrations (1, 5, 10, 15, and 30 g L^{-1}) and glucose (10 g L^{-1}) as supplementary carbon sources on growth and lipid accumulation in *Chlorella vulgaris*. The presence of glycerol in the culture medium enhanced lipid production and accumulation, whereas the presence of glucose favored cell growth. The positive effect of glycerol supplementation on lipid accumulation in microalgae is limited over a given concentration (10 g L^{-1}), after which inhibitory effects on biomass growth and lipid production were observed.

Biofuel production using current algal cultivation technology is not economical (Pittman et al. 2011). A more attractive option for reducing greenhouse gas emissions and the demand for freshwater resources and fertilizers is algal growth using wastewater. The high productivity of microalgal biomass grown in sewage and wastewater suggests that this culture method is a viable means for generating biofuels and will likely be one of the main technologies used for producing sustainable and renewable energy in the future.

Wastewaters contain organic carbon, nitrogen, phosphorus, and other compounds (Liang 2013; Sahu et al. 2013), which are used for microalgal growth and form a readily available substrate for biomass production that is easily assimilated and is inexpensive (Rawat et al. 2011; Ummalyma and Sukumaran 2014; Zhu et al. 2014).

Culturing microalgal using wastewater is influenced by multiple factors, and growth efficiency depends on controlling the variables considered critical, such as pH, temperature, light availability, CO_2, O_2, and especially nutrient concentrations, which can impede algal biomass development. Municipal wastewater typically contains approximately 350 mg L^{-1} chemical oxygen demand (COD), 50 mg L^{-1} of N—$NH4^{+}$, and 10 mg L^{-1} of P—PO_4^{-3} as well as a considerable organic load and high nutrient concentrations (nitrogen and phosphorus) (Boelee et al. 2014).

Treated wastewater effluents also contain trace elements (K, Ca, Mg, Fe, Cu, and Mn) that are essential for microalgal metabolism and growth. Therefore, secondary and tertiary effluents may be broadly applied to microalgal culture media (Schneider et al. 2012; Aravantinou et al. 2013). However, the levels of cadmium, mercury, toxic organic compounds, and biotic contaminants, such as pathogenic bacteria and predators (zooplankton), in the culture medium should be considered because they can inhibit microalgal growth (Pittman et al. 2011; Van Den Hende et al. 2014).

Luminosity is a critical factor for microalgal growth when using urban wastewater as a culture medium due to the levels of suspended solids and turbidity in this type of effluent (Zhou et al. 2014). Microalgal cultures may overcome light limitations by consuming high levels of total organic carbon (TOC) as acetic acid, propionic acid, butyric acid, or ethanol from municipal wastewater for rapid growth under photoheterotrophic or mixotrophic conditions (Ji et al. 2014).

Likely one of the main requirements for large-scale biofuel production from microalgae is using microalgal strains adapted to local environmental conditions. Thus, there is a need for effective and rapid isolation of microalgal strains with a high potential for growth and biomass production as well as sufficiently high lipid and/or polysaccharide content for economic exploitation (Abdelaziz et al. 2014).

Microorganism communities overcome the limitations inherent to isolation through symbiosis. These interactions involve subtle signals and horizontal gene transfer, and they create competitive or cooperative scenarios where microorganisms can compete or provide resources. Cultures with multiple microbial species contain a broader range of genes and better metabolic capacity compared with monocultures (Hays et al. 2015).

The symbiotic relationship between microalgae and bacteria has been well characterized with regard to the oxygen supplied by microalgae for the aerobic bacteria to biodegrade organic pollutants and, in turn, consume the CO_2 released during bacterial respiration. However, the interactions between these two microbial groups are not limited to CO_2/O_2 exchange. When properly characterized, the microalgae-bacteria consortium benefits both species; however, this association should be carefully evaluated because it can also produce antagonistic relationships (Oswald 2003; Schumacher et al. 2003).

Microbial consortia generally perform more complex tasks and can perform functions that are difficult or even impossible for individual species or strains (Brenner et al. 2008). Bacteria can increase the microalgal cell growth capacity by releasing certain growth factors, and microalgae assist bacteria by releasing biosurfactants and extracellular compounds in the medium, which improves bacteria-pollutant degradation activity (Tang et al. 2010).

Increased cell growth in the microalgae *Chlorella minutissima* was observed when it was cocultured with *Escherichia coli* under mixotrophic conditions, in which both light and organic compounds are used for energy and carbon production. This association demonstrates the potential for increasing microalgal biomass production with 1 % substrate consumption compared with isolated *Chlorella minutissima* cultures (Higgins et al. 2015).

Microalgae and bacteria co-selection in a consortium system is important when the objective is to increase biomass production, and the effects of these interactions may affect the system stability; thus, the system should be thoroughly investigated (Muñoz and Guieysse 2006).

Microalgal biomass can be converted into solid, liquid, or gaseous biofuels and may supply 30 % of the global fuel demand without affecting food production (López Barreiro et al. 2013) via various processes, including thermochemical (liquefaction, pyrolysis, and gasification) (Amin 2009) and biochemical processes (fermentation, transesterification, and anaerobic digestion) (Demirbas 2011).

When processed via chemical or biological reactions, in addition to biodiesel and bioethanol, microalgal biomass can provide different types of renewable biofuels, such as bio-hydrogen, bio-methane, and butanol, as well as light hydrocarbons, such as ethane (C_2H_6) and ethylene (C_2H_4), for direct use as energy sources or to generate electricity (Mussgnug et al. 2010; Efremenko et al. 2012; Nasir Uddin et al. 2013; Lakaniemi et al. 2013; Nayak et al. 2014). For microalgae-based fuels, the main focus is biodiesel production, whereas bioethanol and biomethane are considered a part of integrated processes (Oncel 2013) that generate residual biomass during oil extraction, which can be aerobically fermented into ethanol and anaerobically fermented into biomethane (Chinnasamy et al. 2010). The relevant factors for microalgal production and their potential applications in biofuel production are illustrated in Fig. 1.

3 Biodiesel Production

3.1 Oil Extraction

Several methods can be used to extract lipids from microalgae, including supercritical extraction, ultrasonic extraction, microwave extraction, high-pressure homogenizer extraction, hydrothermal liquefaction, and solvent extraction (Bligh and Dyer, 1959; Friedrich and Pryde 1984; Chao et al. 1993; Converti et al. 2009; Bucy et al. 2012; Iqbal and Theegala 2013; Toor et al. 2013; Reddy et al. 2014). However, high quantities of solvent are necessary, which may lead to

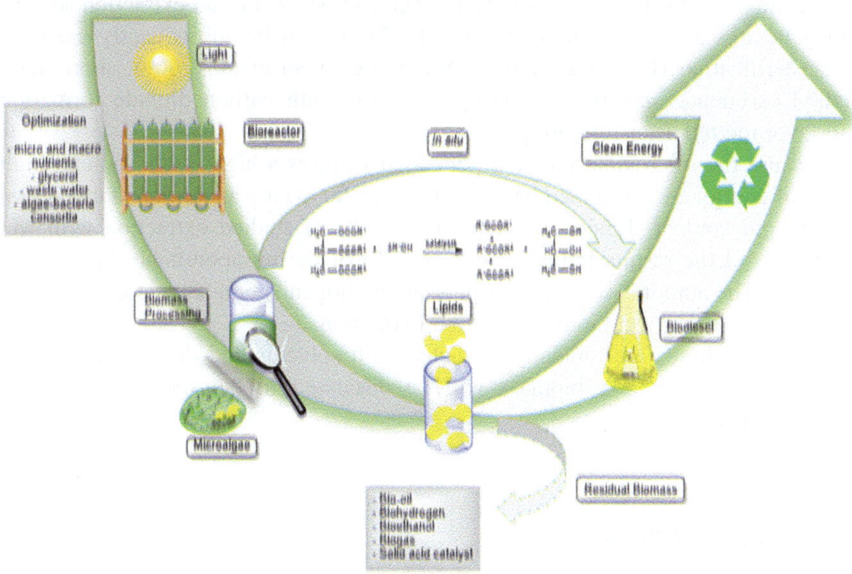

Fig. 1 Diagram illustrating the relevant factors for microalgal production and their potential applications in biofuel production

environmental pollution and increase costs as well as energy consumption during the process (Xu et al. 2006).

Since the introduction of the Folch method, researchers have investigated various techniques to effectively extract lipids using different ratios of chloroform and methanol. These techniques include the Bligh-Dyer method, which is considered a conventional lipid extraction method that successfully uses a mixture of chloroform and methanol and is currently adjusted to the conditions of several laboratories (Bligh and Dyer 1959).

Halim et al. (2012) compared lipid extraction in different solvents. Although chloroform can dissolve the lipid content of cells, including acilglicerois e ácidos graxos livres as well as polar lipids, fosfolipídios e glicolipídios, researchers are reluctant to use chloroform in lipid extraction due to the toxicity of chlorinated solvents. Hexane is often suggested as a hydrophobic alternative because it selectively extracts neutral lipids and is less toxic. Another option is mixing hexane/isopropanol (3:2 v/v), which features good lipid extraction yield.

Ethanol is an inexpensive solvent and has a strong affinity for the lipid complex; thus, lipids can be efficiently extracted. Moreover, ethanol is a safe and eco-friendly solvent, which suggests that the residual biomass can be used in the food industry (Reddy et al. 2014). Several authors (Ibáñez González et al. 1998; Fajardo et al. 2007) have efficiently used ethanol to extract lipids from microalgae, such as *Phaeodactylum tricornutum* and *P. tricornutum* UTEX 640. However, these studies use dehydration or high temperatures for extraction.

Dimethyl carbonate (DMC)/methanol can also be used as an extraction solvent and generates a 38.9 % yield. In fact, the DMC can be subsequently used for transesterification (Lee et al. 2013). According to Su et al. (2009), short-chain dialkyl carbonates can be acyl group acceptors with carbon dioxide formation and favor methyl ester formation.

Because the microalgal biomass generated features a high water content, high levels of energy are consumed before extraction for drying and efficient extraction of the lipid content. Jiménez Callejón et al. (2014) studied extraction with wet biomass, and the results indicated that 69.1 % of the unsaponifiable lipids were extracted. The authors used the high-pressure homogenization technique and added hexane when using wet biomass for lipid extraction.

Therefore, lipids are typically extracted with solvents and dry biomass; other processes for using wet biomass will be discussed in the section on in situ transesterification.

3.2 *Transesterification*

Several processes can be used for microalgal lipid content transesterification, which are combined through transesterification of extracted oil or biomass directly using in situ methods. During optimization, researchers consider the necessary physical treatments associated with whether or not to use catalysts or biocatalysts. Certain biodiesel production initiatives are presented in Table 2 and are the most investigated methods for assessing potential biodiesel production from microalgae.

Catalysts should also be considered an important aspect for converting microalgal lipids into biodiesel. Two catalyst groups are considered, chemical and biocatalysts (enzyme catalysts), with their specific modes of action.

3.3 *Chemical Catalysts*

Methods that involve chemical catalysts during transesterification are the most attractive and widely used due to high yield and increased reaction speed. Because this reaction is reversible, excess alcohol is typically used to shift the equilibrium toward product formation (Arias-Peñaranda et al. 2013; Kiran et al. 2014).

Homogeneous alkaline catalysis has been the most widely used pathway for biodiesel production because the reaction is catalyzed at a low temperature and atmospheric pressure, and it yields high levels of conversion in less time. The most frequently used alkaline catalysts are homogeneous catalysts, such as potassium hydroxide, sodium hydroxide, sodium methoxide, and potassium methoxide (Puna et al. 2010; Macario and Giordano 2013).

Often, an alkaline catalyst is not effective for converting microalgal lipids through transesterification due to the high free fatty acid content. Soaps may form

Table 2 Conditions for biodiesel production from microalgae using different methods and microalgal strains

Microalgae	Conditions	Yield (%)	References
Chlorella pyrenoidosa	120 °C/180 min/H_2SO_4/hexane/methanol Acid catalyst and cosolvent	> 90	Cao et al. (2013)
	60 °C/40 s/H_2SO_4/chloroform/methanol Microwave	10.5[a]	Cheng et al. (2013)
Chlamydomonas sp.	45 °C/15 min/ NaOH/hexane/methanol In situ and cosolvent	>97	Chen et al. (2015)
Chlorella vulgaris	40–45 °C/48 h/immobilized *Burkholderia* lipase/methanol/hexane/ in situ/from ultrasonic pretreated wet biomass	>90	Tran et al. (2013b)
	60 °C/75 min/NaOH/methanol/preheated in the oven at 100 °C for 1 h alkali (NaOH)	77.6	Velasquez-Orta et al. (2012)
	45 °C/48 h/enzyme/methanol/10 min sonication Enzymatic catalysis after sonication	> 95	Tran et al. (2013a)
Mixed cultures	60 °C/100 min/H_2SO_4/metanol/chloroform In situ microwave	74	Wahlen et al. (2011)
	65 °C/7 h/H_2SO_4/methanol Cosolvent chloroform and hexane	82.1 and 55.3	Soydemir et al. (2015)
Schizochytrium limacinum	211.6 °C/120 min/methanol Subcritical and supercritical methanol	37.5	Bi et al. (2015)
Nannochloropsis sp.	255 °C/25 min/methanol Supercritical methanol	>85	Patil et al. (2011)
Nannochloropsis salina	105 °C/30 min/H_2SO_4/methanol or ethanol In situ/wet biomass	>90	Kim et al. (2015a, b)
Nannochloropsis oceanica	95 °C/90 min/H_2SO_4/methanol/chloroform Acid catalyst in situ	> 90	Im et al. (2014)
Scenedesmus obliquus	35 °C/36 h/*Aspergillus niger* whole cell/methanol Enzymatic catalysis	53.7	Guldhe et al. (2016)

[a]Relative to dry biomass

under this condition, and they will consume the catalyst and impede separation of the esters and glycerol formed (Gog et al. 2012).

More reports describe biodiesel production from microalgae via acid catalysis than alkali metals because acid catalysts esterify free fatty acids and transesterify triglycerides (Suwannakarn et al. 2009; Leung et al. 2010; Nigam and Singh 2011; Lee et al. 2014). Acid catalysts are typically employed for lipids with free fatty acid content greater than 2 % (Velasquez-Orta et al. 2012).

The most commonly used acids for homogeneous catalysts in biodiesel production are H_2SO_4 and sulfonic acid; HCl, BF_3, and H_3PO_4 are also used (Bharathiraja et al. 2014). The use of this type of catalyst requires a greater response time and higher temperatures than alkaline catalysts (Hidalgo et al. 2013; Vonortas and Papayannakos 2014).

To overcome the problems with using homogeneous catalysts, certain authors report using both in the same transesterification process, which consists of two reaction steps. An acid catalyst may initially be used to convert free fatty acids to methyl esters by esterification such that the free fatty acid content in the oils is reduced to less than 1 %; thereafter, an alkaline catalyst is used (Park et al. 2015).

Reactions using homogeneous catalysts can generate many environmental and corrosion problems. Therefore, transesterification reactions with heterogeneous catalysts have emerged as an alternative for biodiesel production using different raw materials. Thus, researchers expect that heterogeneous catalysts will be easier to recover and reuse. Umdu et al. (2009) reported transesterification of oils from *Nannochloropsis oculata* using CaO and MgO supported on alumina with a 97.5 % yield. Using metal oxides composed of ZrO, TiO, and Al_2O_3, lipids from the microalgae *Dunaliella tertiolecta* and *Nannochloropsis oculata* were converted (85 %) into biodiesel with free fatty acid esterification and triglyceride transesterification simultaneously under supercritical conditions (Krohn et al. 2011).

3.4 Biocatalysts

Lipases (EC 3.1.1.3) are tools that catalyze various synthetic reactions, such as hydrolysis, esterification, transesterification, and aminolysis. Lipases are used as biocatalysts for transforming oil/lipids into biodiesel due to low operating costs and, particularly, high product purity under mild conditions (20–50 °C) (Gog et al. 2012; Tran et al. 2012; Christopher et al. 2014; Huang et al. 2015).

Furthermore, comparing the enzyme and chemical catalyst technologies, researchers have observed that enzymes catalyze free fatty acid (FFA) and triglyceride (TG) esterification in a reaction step without the need for a wash step and prevent by-product formation, which ensures an easily recovered product (Kulkarni and Dalai 2006; Robles-Medina et al. 2009).

Certain lipases synthesized by *Mucor miehei*, *Rhizopus oryzae*, *Pseudomonas cepacia*, *Candida antarctica* (Taher et al. 2011), *Candida rugosa*, *Candida cylindracea*, *Pseudomonas fluorescens*, *Rhizomucor miehei* (Teo et al. 2014), and *Burkholderia* sp. (Tran et al. 2012) are used as enzyme biocatalysts for biofuel production.

The lipases may exhibit regiospecificity, specificity for fatty acids or the nature of the alcohol, and stereospecificity, which includes *sn*-1,3 specific (hydrolyzable ester linkage at position sR1 or R3 of the triglyceride), *sn*-2 specific (hydrolyzable ester linkage in the R2 position of the TAG), and nonspecific (no distinction between the ester positions) (Kapoor and Gupta 2012).

In general, enzymatic transesterification for oils features certain disadvantages, including the reaction time and the potential enzyme inactivation by methanol

(Persson et al. 2002; Adamczak and Bednarski 2004; Jung et al. 2006). These disadvantages limit the industrial applications of biodiesel production processes due to the high cost of enzymes associated with this process (Bajaj et al. 2010). Using the whole cell or immobilized enzyme yields higher conversion and less activity loss but allows for enzyme reuse (Teo et al. 2014; Huang et al. 2015).

The lipids are more viscous compared with the organic solvents, and dissolving lipids in this class of solvents may improve reactant diffusion in the liquid phase and hence transesterification performance (Lam and Lee 2013).

Studies on biocatalysis for transesterification are only performed in media with the alcohol used for the conversion or with cosolvents, especially hexane, which solubilizes oil and biodiesel well and is less aggressive toward enzymes (Tran et al. 2013a). Researchers have also examined using ionic liquids for biodiesel production using oil from three different microalgae (*Botryococcus braunii*, *Chlorella vulgaris*, and *Chlorella pyrenoidosa*). Researchers have observed high levels of methyl ester conversion using *Penicillium expansum* lipase (PEL) and *Candida antarctica* lipase B (Novozyme 435) (Lai et al. 2012).

According to Amoah et al. (2016), researchers must also consider the phospholipids in microalgae at concentrations reaching 30 % of the total lipid content, which negatively affect enzyme catalysis because they can form water-in-oil phospholipid-based reverse micelles. The water required for enzyme activity is trapped by the micelle.

To produce biodiesel from microalgae, Huang et al. (2015) used a recombinant *Rhizomucor miehei* lipase to convert oil from *Chlorella vulgaris* with more than 90 % conversion to methyl esters and ethyl esters.

Notably, enzymatic transesterification can be advantageous for in situ biodiesel production from alternative raw materials, such as microalgae.

3.5 *In Situ Transesterification*

In situ transesterification is an efficient means to directly convert lipids from biomass to biodiesel. Because it is a simple process, various studies have used in situ transesterification with different raw materials, including oilseed plants (Hincapié et al. 2011; Martínez Arias et al. 2012) and microalgae (Hidalgo et al. 2013). This method is an alternative to the conventional transesterification process that may reduce biodiesel production costs where the lipid/oil extraction step features a high environmental or economic impact.

In this process, fatty acids are converted into alkyl esters directly inside the biomass, which eliminates the extraction step and postharvest biomass drying (Amaro et al. 2011). Microalgae lipid content can be transformed to biodiesel by enzymatic transesterification in situ, which uses methanol as raw material for the process and as solvent to assist microalgae cell wall rupture (Steriti et al. 2014), as shown in Fig. 2.

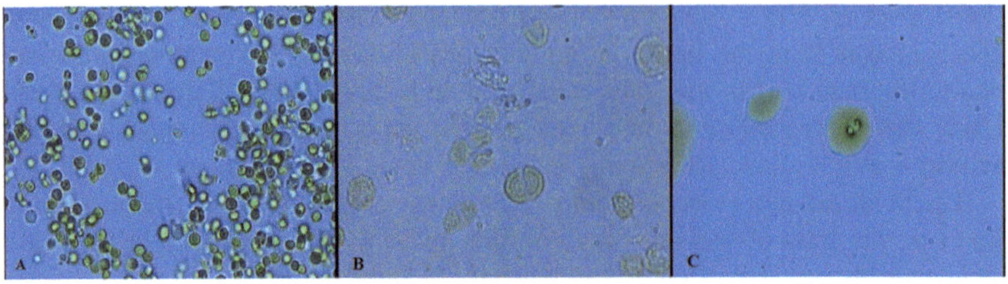

Fig. 2 Optical microscopy images of *Desmodesmus* sp. cells at different times of transesterification in situ: (**a**) intact cells at the beginning of the process (400x); (**b**) cell wall rupture after 48 h of transesterification (1000x); (**c**) the lipid content extracted from cells after transesterification in situ and evaporation of the solvent (400x)

This method may be particularly advantageous for microalgae because lipids are typically extracted using solvents, not through physical methods for oil extraction from conventional crops, as previously presented. Alcoholysis of the TG in the biomass produces better biodiesel yields compared with the conventional transesterification method with the additional advantage of reducing waste generation, power consumption, and consequently the environmental impact from these methods (Ehimen et al. 2010).

The reactions are simple and include adding alcohols, catalysts, biomass, and, occasionally, cosolvents (Rodrigues Da Silva Baumgartner et al. 2013; Atadashi et al. 2013).

For microalgae, this method has been applied using *Nannochloropsis salina*, *Chlorella* sp., *Scenedesmus* sp., *Nannochloropsis gaditana*, *Chlorella pyrenoidosa*, and *Schizochytrium limacinum*, among others (Carvalho et al. 2011; Cao et al. 2013; Jin et al. 2014; Sathish et al. 2014; Kim et al. 2015a, b; Ma et al. 2015). Tran et al. (2013b) used immobilized *Burkholderia* lipase for direct transesterification (in situ) to convert lipids into methyl esters after ultrasonically pretreating the biomass. The results showed that the ideal hexane/methanol ratio was 1.65, and different proportions of methanol relative to the lipid content yielded more than 90 % conversion to esters in the 48-h reaction period.

Biodiesel can also be produced from microalgae through in situ non-catalytic supercritical methanol transesterification, reaching 45.6 % biodiesel using the microalgae *Chlorella* sp. (native) (Jazzar et al. 2015) or *Schizochytrium limacinum* with supercritical CO_2 and adding methanol, which generates a 37.5 % yield (Bi et al. 2015). Notably, supercritical fluids are promising for in situ transesterification and should be further explored in the near future (Zeng et al. 2014).

3.6 Prospects for Industrial Production

The viability of biodiesel production using microalgae at the industrial level faces several challenges involving reduction of energy loss. The process includes a pretreatment step for free fatty acid esterification with methanol and acid catalysis followed by alkaline-catalyzed transesterification, and the cost of purifying biodiesel and glycerol through distillation was recently calculated as $0.592/L of biodiesel. This value does not consider lipid extraction from the biomass, which also impacts the production cost and may be investigated to save energy (Song et al. 2015).

In the first phase of biodiesel production from microalgae, volatile solvents are used to separate the lipids and for subsequent transesterification, which are ineffective for microalgae with high water content according to Cooney et al. (2009). Thus, the microalgae must be dried. Next, the organic solvents must be removed through reducing the pressure. The energy cost associated with the drying, distillation, and solvent recovery processes must also be considered. Modifications to the process that decrease energy consumption and optimize the use of solvents and other inputs may be the factors responsible that render biodiesel production from microalgae feasible. Peralta-Ruiz et al. (2013) performed an energy analysis for the microalgal lipid extraction pathways, and hexane was the most suitable substance for large-scale production with a 51 % maximum energy efficiency.

Lipid extraction technologies that generate a high yield are developed on a laboratory scale and are not easily converted to larger scales, such as the industrial scale. Successful industrial production is directly linked to using innovative technologies, such as in situ transesterification.

Countries such as Japan, Israel, India, and the United States have successfully ventured into this area, but the potential for using microalgae on an industrial level may be much greater when it is associated with, for example, wastewater phytoremediation. The potential for biofuel production from microalgae has driven studies and the transfer of technology to industry (http://www.algaeindustrymagazine.com).

The environmental and economic impacts of biomass production, lipid extraction, and biodiesel production have been studied in recent years (Monari et al. 2016; Delrue et al. 2012; Dassey and Theegala 2013; Collet et al. 2014; Mata et al. 2014; Sawaengsak et al. 2014; Yu et al. 2015). However, different production systems must be studied with a focus on environmental issues and the real potential of microalgae in biodiesel production to determine the actual environmental and economic gains from industrial biodiesel production using microalgae, as shown by Yu et al. (2015) in Singapore.

4 Conclusion

Biodiesel production from microalgae depends on the species, biomass production conditions, and technologies used for lipid separation and transesterification, and it is becoming feasible or is already a reality in certain countries. Several initiatives in the field have sought to use techniques that consume less energy and produce a better final biodiesel yield. Among these technologies, recent microalgal culture studies have sought greater lipid accumulation in the cells through adding different sources of carbon, metals, and vitamins and have sought to avoid using extraction solvents, such as through in situ transesterification. Technological advances from research have facilitated efficient biodiesel production from microalgae, but large-scale feasibility has not been analyzed to maximize production potential. The environmental gains from these processes should also be highlighted to render biodiesel production from microalgal biomass even more relevant in the medium and long term.

Acknowledgments The authors wish to thank FAP-UNISC, CNPq (306178/2012-5), Fapergs/Capes (DOCFIX n° 09/2012), and the Brazilian government through the Ministry of Science, Technology, and Innovation for providing necessary funds and facilities.

References

Abdelaziz AEM, Leite GB, Belhaj MA, Hallenbeck PC (2014) Screening microalgae native to Quebec for wastewater treatment and biodiesel production. Bioresour Technol 157:140–148. doi:10.1016/j.biortech.2014.01.114

Adamczak M, Bednarski W (2004) Enhanced activity of intracellular lipases from Rhizomucor miehei and Yarrowia lipolytica by immobilization on biomass support particles. Process Biochem 39:1347–1361. doi:10.1016/S0032-9592(03)00266-8

Ahmad AL, Yasin NHM, Derek CJC, Lim JK (2014) Comparison of harvesting methods for microalgae Chlorella sp. and its potential use as a biodiesel feedstock. Environ Technol 35:2244–2253. doi:10.1080/09593330.2014.900117

Amaro HM, Guedes AC, Malcata FX (2011) Advances and perspectives in using microalgae to produce biodiesel. Appl Energy 88:3402–3410. doi:10.1016/j.apenergy.2010.12.014

Amin S (2009) Review on biofuel oil and gas production processes from microalgae. Energy Covers Manage 50:1834–1840. doi:10.1016/j.enconman.2009.03.001

Amoah J, Ho SH, Hama S, Yoshida A, Nakanishi A, Hasunuma T, Ogino C, Kondo A (2016) Converting oils high in phospholipids to biodiesel using immobilized Aspergillus oryzae whole-cell biocatalysts expressing Fusarium heterosporum lipase. Biochem Eng J 105:10–15. doi:10.1016/j.bej.2015.08.007

Aravantinou AF, Theodorakopoulos MA, Manariotis ID (2013) Selection of microalgae for wastewater treatment and potential lipids production. Bioresour Technol 147:130–134. doi:10.1016/j.biortech.2013.08.024

Arias-Peñaranda MT, Cristiani-Urbina E, Montes-Horcasitas C, Esparza-Garcı AF, Torzillo G, Cañizares-Villanueva RO (2013) Scenedesmus incrassatulus CLHE-Si01: a potential source of renewable lipid for high quality biodiesel production. Bioresour Technol 140:158–164. doi:10.1016/j.biortech.2013.04.080

Ashokkumar V, Rengasamy R, Deepalakshmi S, Sivalingam A, Sivakumar P (2014) Mass cultivation of microalgae and extraction of total hydrocarbons: a kinetic and thermodynamic study. Fuel 119:308–312. doi:10.1016/j.fuel.2013.11.062

Atadashi IM, Aroua MK, Abdul Aziz AR, Sulaiman NMN (2013) The effects of catalysts in biodiesel production: a review. J Ind Eng Chem 19:14–26. doi:10.1016/j.jiec.2012.07.009

Bajaj A, Lohan P, Jha PN, Mehrotra R (2010) Biodiesel production through lipase catalyzed trans-esterification: an overview. J Mol Catal B Enzym 62:9–14. doi:10.1016/j.molcatb.2009.09.018

Bertozzini E, Galluzzi L, Penna A, Magnani M (2014) Enhancing neutral lipid content in Skeletonema marinoi through multiple phase growth in a bench photobioreactor. Algal Res 5:32–36. doi:10.1016/j.algal.2014.05.004

Bharathiraja B, Chakravarthy M, Kumar RR, Yuvaraj D, Jayamuthunagai J, Kumar RP, Palani S (2014) Biodiesel production using chemical and biological methods—a review of process, catalyst, acyl acceptor, source and process variables. Renew Sustain Energy Rev 38:368–382. doi:10.1016/j.rser.2014.05.084

Bi Z, He BB, Mcdonald AG (2015) Biodiesel production from green microalgae schizochytrium limacinum via in situ trans-esterification. Energy Fuels 29:5018–5027. doi:10.1021/acs.energyfuels.5b00559

Bligh EG, Dyer WJ (1959) A rapid method of total lipid extraction and purification. Can J Biochem Physiol 37:911–917. doi:10.1139/o59-099

Boelee NC, Temmink H, Janssen M, Buisman CJN, Wijffels RH (2014) Balancing the organic load and light supply in symbiotic microalgal-bacterial biofilm reactors treating synthetic municipal wastewater. Ecol Eng 64:213–221. doi:10.1016/j.ecoleng.2013.12.035

Brenner K, You L, Arnold FH (2008) Engineering microbial consortia: a new frontier in synthetic biology. Trends Biotechnol 26:483–489. doi:10.1016/j.tibtech.2008.05.004

Bucy HB, Baumgardner ME, Marchese AJ (2012) Chemical and physical properties of algal methyl ester biodiesel containing varying levels of methyl eicosapentaenoate and methyl docosahexaenoate. Algal Res 1:57–69. doi:10.1016/j.algal.2012.02.001

Cai T, Park SY, Li Y (2013) Nutrient recovery from wastewater streams by microalgae: status and prospects. Renew Sustain Energy Rev 19:360–369. doi:10.1016/j.rser.2012.11.030

Cao H, Zhang Z, Wu X, Miao X (2013) Direct biodiesel production from wet microalgae biomass of chlorella pyrenoidosa through in situ trans-esterification. BioMed Res Int 2013:1–6. doi:10.1155/2013/930686

Carvalho RM Jr, Vargas JVC, Ramos LP, Marino CEB, Torres JCL (2011) Microalgae biodiesel via in situ methanolysis. J Chem Technol Biotechnol 86:1418–1427. doi:10.1002/jctb.2652

Cerón Garcı AMC, Sánchez Mirón A, Fernández Sevilla JM, Molina Grima E, Garcıa Camacho F (2005) Mixotrophic growth of the microalga Phaeodactylum tricornutum: influence of different nitrogen and organic carbon sources on productivity and biomass composition. Process Biochem 40:297–305. doi:10.1016/j.procbio.2004.01.016

Cerón-García MC, Fernández-Sevilla JM, Sánchez-Mirón A, García-Camacho F, Contreras-Gómez A, Molina-Grima E (2013) Mixotrophic growth of *Phaeodactylum tricornutum* on fructose and glycerol in fed-batch and semi-continuous modes. Bioresour Technol 147:569–576. doi:10.1016/j.biortech.2013.08.092

Chandra R, Rohit MV, Swamy YV, Venkata Mohan S (2014) Regulatory function of organic carbon supplementation on biodiesel production during growth and nutrient stress phases of mixotrophic microalgae cultivation. Bioresour Technol 165:279–287. doi:10.1016/j.biortech.2014.02.102

Chao R, Mulvaney S, Huang H (1993) Effects of extraction and fractionation pressures on supercritical extraction of cholesterol from beef tallow. J Am Oil Chem Soc 70:139–143. doi:10.1007/BF02542616

Chen X, Liu T, Wang Q (2014) The growth of Scenedesmus sp. attachment on different materials surface. Microb Cell Fact 13:142. doi:10.1186/s12934-014-0142-z

Chen CL, Huang CC, Ho KC, Hsiao PX, Wu MS, Chang JS (2015) Biodiesel production from wet microalgae feedstock using sequential wet extraction/trans-esterification and direct trans-esterification processes. Bioresour Technol 194:179–186. doi:10.1016/j.biortech.2015.07.021
Cheng J, Yu T, Li T, Zhou J, Cen K (2013) Using wet microalgae for direct biodiesel production via microwave irradiation. Bioresour Technol 131:531–535. doi:10.1016/j.biortech.2013.01.045
Chinnasamy S, Bhatnagar A, Claxton R, Das KC (2010) Biomass and bioenergy production potential of microalgae consortium in open and closed bioreactors using untreated carpet industry effluent as growth medium. Bioresour Technol 101:6751–6760. doi:10.1016/j.biortech.2010.03.094
Christopher LP, Hemanathan K, Zambare VP (2014) Enzymatic biodiesel: challenges and opportunities. Appl Energy 119:497–520. doi:10.1016/j.apenergy.2014.01.017
Collet P, Lardon L, Hélias A, Bricout S, Lombaert-Valot I, Perrier B, Lépine O, Steyer J-P, Bernard O (2014) Biodiesel from microalgae—life cycle assessment and recommendations for potential improvements. Renew Energy 71:525–533. doi:10.1016/j.renene.2014.06.009
Converti A, Casazza AA, Ortiz EY, Perego P, Del Borghi M (2009) Effect of temperature and nitrogen concentration on the growth and lipid content of Nannochloropsis oculata and Chlorella vulgaris for biodiesel production. Chem Eng Proces Process Intens 48:1146–1151. doi:10.1016/j.cep.2009.03.006
Cooney M, Young G, Nagle N (2009) Extraction of bio-oils from microalgae. Sep Purif Rev 38:291–325. doi:10.1080/15422110903327919
Dassey AJ, Theegala CS (2013) Harvesting economics and strategies using centrifugation for cost effective separation of microalgae cells for biodiesel applications. Bioresour Technol 128:241–245. doi:10.1016/j.biortech.2012.10.061
Dębowski M, Zieliński M, Grala A, Dudek M (2013) Algae biomass as an alternative substrate in biogas production technologies—review. Renew Sustain Energy Rev 27:596–604. doi:10.1016/j.rser.2013.07.029
Delrue F, Setier PA, Sahut C, Cournac L, Roubaud A, Peltier G, Froment AK (2012) An economic, sustainability, and energetic model of biodiesel production from microalgae. Bioresour Technol 111:191–200. doi:10.1016/j.biortech.2012.02.020
Demirbas A (2010) Use of algae as biofuel sources. Energy Covers Manage 51:2738–2749. doi:10.1016/j.enconman.2010.06.010
Demirbas MF (2011) Biofuels from algae for sustainable development. Appl Energy 88:3473–3480. doi:10.1016/j.apenergy.2011.01.059
Dragone G, Fernandes BD, Vicente AA, Teixeira JA (2010) Third generation biofuels from microalgae. In: Méndez-Vilas A (ed) Current research, technology and education topics in applied microbiology and microbial biotechnology, pp 1355–1366. Formatex Spain. ISBN-13:978–84–614-6195-0
Efremenko EN, Nikolskaya AB, Lyagin IV, Senko OV, Makhlis TA, Stepanov NA, Maslova OV, Mamedova F, Varfolomeev SD (2012) Production of biofuels from pretreated microalgae biomass by anaerobic fermentation with immobilized Clostridium acetobutylicum cells. Bioresour Technol 114:342–348. doi:10.1016/j.biortech.2012.03.049
Ehimen EA, Sun ZF, Carrington CG (2010) Variables affecting the in situ trans-esterification of microalgae lipids. Fuel 89:677–684. doi:10.1016/j.fuel.2009.10.011
Elshahed MS (2010) Microbiological aspects of biofuel production: current status and future directions. J Adv Res 1:103–111. doi:10.1016/j.jare.2010.03.001
Fajardo AR, Cerdán LE, Medina AR, Fernández FGA, Moreno PG, Grima EM (2007) Lipid extraction from the microalga Phaeodactylum tricornutum. Eur J Lipid Sci Tech 109:120–126. doi:10.1002/ejlt.200600216
Friedrich JP, Pryde EH (1984) Supercritical CO_2 extraction of lipid-bearing materials and characterization of the products. J Am Oil Chem Soc 61:223–228. doi:10.1007/BF02678773

Gog A, Roman M, Toşa M, Paizs C, Irimie FD (2012) Biodiesel production using enzymatic trans-esterification—current state and perspectives. Renew Energy 39:10–16. doi:10.1016/j.renene.2011.08.007
Guldhe A, Singh P, Kumari S, Rawat I, Permaul K, Bux F (2016) Biodiesel synthesis from microalgae using immobilized Aspergillus niger whole cell lipase biocatalyst. Renew Energy 85:1002–1010. doi:10.1016/j.renene.2015.07.059
Gupta A, Singh D, Barrow CJ, Puri M (2013) Exploring potential use of Australian thraustochytrids for the bioconversion of glycerol to omega-3 and carotenoids production. BiochemEng J 78:11–17. doi:10.1016/j.bej.2013.04.028
Halim R, Danquah MK, Webley PA (2012) Extraction of oil from microalgae for biodiesel production: a review. Biotechnol Adv 30:709–732. doi:10.1016/j.biotechadv.2012.01.001
Hays SG, Patrick WG, Ziesack M, Oxman N, Silver PA (2015) Better together: engineering and application of microbial symbioses. Curr Opin Biotechnol 36:40–49. doi:10.1016/j.copbio.2015.08.008
Heredia-Arroyo T, Wei W, Ruan R, Hu B (2011) Mixotrophic cultivation of Chlorella vulgaris and its potential application for the oil accumulation from non-sugar materials. Biomass Bioenergy 35:2245–2253. doi:10.1016/j.biombioe.2011.02.036
Hidalgo P, Toro C, Ciudad G, Navia R (2013) Advances in direct trans-esterification of microalgal biomass for biodiesel production. Rev Environ Sci Bio 12:179–199. doi:10.1007/s11157-013-9308-0
Higgins BT, Labavitch JM, Vandergheynst JS (2015) Co-culturing Chlorella minutissima with Escherichia coli can increase neutral lipid production and improve biodiesel quality. Biotechnol Bioeng 112:1801–1809. doi:10.1002/bit.25609
Hincapié G, Mondragón F, López D (2011) Conventional and in situ trans-esterification of castor seed oil for biodiesel production. Fuel 90:1618–1623. doi:10.1016/j.fuel.2011.01.027
Hockin NL, Mock T, Mulholland F, Kopriva S, Malin G (2012) The response of diatom central carbon metabolism to nitrogen starvation is different from that of green algae and higher plants. Plant Physiol 158:299–312. doi:10.1104/pp.111.184333
Hong IK, Jeon GS, Lee SB (2014) Prediction of biodiesel fuel properties from fatty acid alkyl ester. J Ind Eng Chem 20:2348–2353. doi:10.1016/j.jiec.2013.10.011
Huang YT, Su CP (2014) High lipid content and productivity of microalgae cultivating under elevated carbon dioxide. Int J Environl Sci Tec 11:703–710. doi:10.1007/s13762-013-0251-y
Huang J, Xia J, Jiang W, Li Y, Li J (2015) Biodiesel production from microalgae oil catalyzed by a recombinant lipase. Bioresour Technol 180:47–53. doi:10.1016/j.biortech.2014.12.072
Ibáñez González MJ, Robles Medina A, Grima EM, Giménez AG, Carstens M, Cerdán LE (1998) Optimization of fatty acid extraction from *Phaeodactylum tricornutum* UTEX 640 biomass. J Am Oil Chem Soc 75:1735–1740. doi:10.1007/s11746-998-0325-z
Im H, Lee H, Park MS, Yang J-W, Lee JW (2014) Concurrent extraction and reaction for the production of biodiesel from wet microalgae. Bioresour Technol 152:534–537. doi:10.1016/j.biortech.2013.11.023
Iqbal J, Theegala C (2013) Microwave assisted lipid extraction from microalgae using biodiesel as co-solvent. Algal Res 2:34–42. doi:10.1016/j.algal.2012.10.001
Jazzar S, Quesada-Medina J, Olivares-Carrillo P, Marzouki MN, Acién-Fernández FG, Fernández-Sevilla JM, Molina-Grima E, Smaali I (2015) A whole biodiesel conversion process combining isolation, cultivation and in situ supercritical methanol trans-esterification of native microalgae. Bioresour Technol 190:281–288. doi:10.1016/j.biortech.2015.04.097
Ji M-K, Kabra AN, Salama E-S, Roh H-S, Kim JR, Lee DS, Jeon B-H (2014) Effect of mine wastewater on nutrient removal and lipid production by a green microalga Micratinium reisseri from concentrated municipal wastewater. Bioresour Technol 157:84–90. doi:10.1016/j.biortech.2014.01.087
Jiménez Callejón MJ, Robles Medina A, Macías Sánchez MD, Hita Peña E, Esteban Cerdán L, González Moreno PA, Molina Grima E (2014) Extraction of saponifiable lipids from wet

microalgal biomass for biodiesel production. Bioresour Technol 169:198–205. doi:10.1016/j.biortech.2014.06.106

Jin B, Duan P, Xu Y, Wang B, Wang F, Zhang L (2014) Lewis acid-catalyzed in situ trans-esterification/esterification of microalgae in supercritical ethanol. Bioresour Technol 162:341–349. doi:10.1016/j.biortech.2014.03.157

Jung H-C, Kwon S-J, Pan J-G (2006) Display of a thermostable lipase on the surface of a solvent-resistant bacterium, Pseudomonas putida GM730, and its applications in whole-cell biocatalysis. BMC Biotechnol 6:23–23. doi:10.1186/1472-6750-6-23

Kaiwan-Arporn P, Hai PD, Thu NT, Annachhatre AP (2012) Cultivation of cyanobacteria for extraction of lipids. Biomass Bioenergy 44:142–149. doi:10.1016/j.biombioe.2012.04.017

Kapoor M, Gupta MN (2012) Lipase promiscuity and its biochemical applications. Process Biochem 47:555–569. doi:10.1016/j.procbio.2012.01.011

Kim B, Im H, Lee JW (2015a) In situ trans-esterification of highly wet microalgae using hydrochloric acid. Bioresour Technol 185:421–425. doi:10.1016/j.biortech.2015.02.092

Kim TH, Suh WI, Yoo G, Mishra SK, Farooq W, Moon M, Shrivastav A, Park MS, Yang JW (2015b) Development of direct conversion method for microalgal biodiesel production using wet biomass of Nannochloropsis salina. Bioresour Technol 191:438–444. doi:10.1016/j.biortech.2015.03.033

Kiran B, Kumar R, Deshmukh D (2014) Perspectives of microalgal biofuels as a renewable source of energy. Energy Covers Manage 88:1228–1244. doi:10.1016/j.enconman.2014.06.022

Kirrolia A, Bishnoi NR, Singh R (2013) Microalgae as a boon for sustainable energy production and its future research & development aspects. Renew Sustain Energy Rev 20:642–656. doi:10.1016/j.rser.2012.12.003

Knothe G, Steidley KR (2011) Kinematic viscosity of fatty acid methyl esters: prediction, calculated viscosity contribution of esters with unavailable data, and carbon-oxygen equivalents. Fuel 90:3217–3224. doi:10.1016/j.fuel.2011.06.016

Krohn BJ, Mcneff CV, Yan B, Nowlan D (2011) Production of algae-based biodiesel using the continuous catalytic Mcgyan® process. Bioresour Technol 102:94–100. doi:10.1016/j.biortech.2010.05.035

Kulkarni MG, Dalai AK (2006) Waste cooking oil an economical source for biodiesel: a review. Ind Eng Chem Res 45:2901–2913. doi:10.1021/ie0510526

Lai JQ, Hu ZL, Wang PW, Yang Z (2012) Enzymatic production of microalgal biodiesel in ionic liquid [BMIm][PF6]. Fuel 95:329–333. doi:10.1016/j.fuel.2011.11.001

Lakaniemi A-M, Tuovinen OH, Puhakka JA (2013) Anaerobic conversion of microalgal biomass to sustainable energy carriers—a review. Bioresour Technol 135:222–231. doi:10.1016/j.biortech.2012.08.096

Lam MK, Lee KT (2012) Microalgae biofuels: a critical review of issues, problems and the way forward. Biotechnol Adv 30:673–690. doi:10.1016/j.biotechadv.2011.11.008

Lam MK, Lee KT (2013) Catalytic trans-esterification of high viscosity crude microalgae lipid to biodiesel: effect of co-solvent. Fuel Process Technol 110:242–248. doi:10.1016/j.fuproc.2012.12.021

Lee OK, Kim YH, Na JG, Oh YK, Lee EY (2013) Highly efficient extraction and lipase-catalyzed trans-esterification of triglycerides from Chlorella sp. KR-1 for production of biodiesel. Bioresour Technol 147:240–245. doi:10.1016/j.biortech.2013.08.037

Lee AF, Bennett JA, Manayil JC, Wilson K (2014) Heterogeneous catalysis for sustainable biodiesel production via esterification and trans-esterification. Chem Soc Rev 43:7887–7916. doi:10.1039/C4CS00189C

Leung DYC, Wu X, Leung MKH (2010) A review on biodiesel production using catalyzed trans-esterification. Appl Energy 87:1083–1095. doi:10.1016/j.apenergy.2009.10.006

Li M, Zhu Q, Hu CW, Chen L, Liu ZL, Kong ZM (2007) Cobalt and manganese stress in the microalga Pavlova viridis (Prymnesiophyceae): effects on lipid peroxidation and antioxidant enzymes. J Environ Sci 19:1330–1335. doi:10.1016/S1001-0742(07)60217-4

Liang Y (2013) Producing liquid transportation fuels from heterotrophic microalgae. Appl Energy 104:860–868. doi:10.1016/j.apenergy.2012.10.067

Likozar B, Levec J (2014) Effect of process conditions on equilibrium, reaction kinetics and mass transfer for triglyceride trans-esterification to biodiesel: experimental and modeling based on fatty acid composition. Fuel Process Technol 122:30–41. doi:10.1016/j.fuproc.2014.01.017

López Barreiro D, Prins W, Ronsse F, Brilman W (2013) Hydrothermal liquefaction (HTL) of microalgae for biofuel production: state of the art review and future prospects. Biomass Bioenergy 53:113–127. doi:10.1016/j.biombioe.2012.12.029

Ma G, Hu W, Pei H, Jiang L, Song M, Mu R (2015) In situ heterogeneous trans-esterification of microalgae using combined ultrasound and microwave irradiation. Energy Covers Manage 90:41–46. doi:10.1016/j.enconman.2014.10.061

Macario A, Giordano G (2013) Catalytic conversion of renewable sources for biodiesel production: a comparison between biocatalysts and inorganic catalysts. Catal Lett 143:159–168. doi:10.1007/s10562-012-0949-3

Maity JP, Bundschuh J, Chen CY, Bhattacharya P (2014) Microalgae for third generation biofuel production, mitigation of greenhouse gas emissions and wastewater treatment: present and future perspectives—a mini review. Energy 78:104–113. doi:10.1016/j.energy.2014.04.003

Martínez Arias EL, Fazzio Martins P, Jardini Munhoz AL, Gutierrez-Rivera L, Maciel Filho R (2012) Continuous synthesis and in situ monitoring of biodiesel production in different microfluidic devices. Ind Eng Chem Res 51:10755–10767. doi:10.1021/ie300486v

Mata TM, Martins AA, Caetano NS (2010) Microalgae for biodiesel production and other applications: a review. Renew Sustain Energy Rev 14:217–232. doi:10.1016/j.rser.2009.07.020

Mata TM, Mendes AM, Caetano NS, Martins AA (2014) Sustainability and economic evaluation of microalgae grown in brewery wastewater. Bioresour Technol 168:151–158. doi:10.1016/j.biortech.2014.04.091

Monari C, Righi S, Olsen SI (2016) Greenhouse gas emissions and energy balance of biodiesel production from microalgae cultivated in photobioreactors in Denmark: a life-cycle modeling. J Clean Prod 112:4084–4092. doi:10.1016/j.jclepro.2015.08.112

Muñoz R, Guieysse B (2006) Algal–bacterial processes for the treatment of hazardous contaminants: a review. Water Res 40:2799–2815. doi:10.1016/j.watres.2006.06.011

Mussgnug JH, Klassen V, Schlüter A, Kruse O (2010) Microalgae as substrates for fermentative biogas production in a combined biorefinery concept. J Biotechnol 150:51–56. doi:10.1016/j.jbiotec.2010.07.030

Nasir Uddin M, Daud WMAW, Abbas HF (2013) Potential hydrogen and non-condensable gases production from biomass pyrolysis: insights into the process variables. Renew Sustain Energy Rev 27:204–224. doi:10.1016/j.rser.2013.06.031

Nayak BK, Roy S, Das D (2014) Biohydrogen production from algal biomass (Anabaena sp. PCC 7120) cultivated in airlift photobioreactor. Int J Hydrogen Energy 39:7553–7560. doi:10.1016/j.ijhydene.2013.07.120

Nigam PS, Singh A (2011) Production of liquid biofuels from renewable resources. Prog Energy Combust Sci 37:52–68. doi:10.1016/j.pecs.2010.01.003

Oncel SS (2013) Microalgae for a macroenergy world. Renew Sustain Energy Rev 26:241–264. doi:10.1016/j.rser.2013.05.059

Oswald WJ (2003) My sixty years in applied algology. J Appl Phycol 15:99–106. doi:10.1023/a:1023871903434

Park J-Y, Park MS, Lee Y-C, Yang J-W (2015) Advances in direct trans-esterification of algal oils from wet biomass. Bioresour Technol 184:267–275. doi:10.1016/j.biortech.2014.10.089

Patil PD, Gude VG, Mannarswamy A, Deng S, Cooke P, Munson-Mcgee S, Rhodes I, Lammers P, Nirmalakhandan N (2011) Optimization of direct conversion of wet algae to biodiesel under supercritical methanol conditions. Bioresour Technol 102:118–122. doi:10.1016/j.biortech.2010.06.031

Peralta-Ruiz Y, González-Delgado AD, Kafarov V (2013) Evaluation of alternatives for microalgae oil extraction based on exergy analysis. Appl Energy 101:226–236. doi:10.1016/j.apenergy.2012.06.065
Persson M, Mladenoska I, Wehtje E, Adlercreutz P (2002) Preparation of lipases for use in organic solvents. Enzyme Microb Technol 31:833–841. doi:10.1016/S0141-0229(02)00184-9
Pittman JK, Dean AP, Osundeko O (2011) The potential of sustainable algal biofuel production using wastewater resources. Bioresour Technol 102:17–25. doi:10.1016/j.biortech.2010.06.035
Puna JF, Gomes JF, Correia MJN, Soares Dias AP, Bordado JC (2010) Advances on the development of novel heterogeneous catalysts for trans-esterification of triglycerides in biodiesel. Fuel 89:3602–3606. doi:10.1016/j.fuel.2010.05.035
Rawat I, Ranjith Kumar R, Mutanda T, Bux F (2011) Dual role of microalgae: phycoremediation of domestic wastewater and biomass production for sustainable biofuels production. Appl Energy 88:3411–3424. doi:10.1016/j.apenergy.2010.11.025
Rawat I, Ranjith Kumar R, Mutanda T, Bux F (2013) Biodiesel from microalgae: a critical evaluation from laboratory to large scale production. Appl Energy 103:444–467. doi:10.1016/j.apenergy.2012.10.004
Reddy HK, Muppaneni T, Patil PD, Ponnusamy S, Cooke P, Schaub T, Deng S (2014) Direct conversion of wet algae to crude biodiesel under supercritical ethanol conditions. Fuel 115:720–726. doi:10.1016/j.fuel.2013.07.090
Robles-Medina A, González-Moreno PA, Esteban-Cerdán L, Molina-Grima E (2009) Biocatalysis: towards ever greener biodiesel production. Biotechnol Adv 27:398–408. doi:10.1016/j.biotechadv.2008.10.008
Rodrigues Da Silva Baumgartner T, Mendes Burak JA, Baumgartner D, Zanin GM, Arroyo PA (2013) Biomass production and ester synthesis by in situ trans-esterification/esterification using the microalga spirulina platensis. Int J Chem Eng 2013:1–7. doi:10.1155/2013/425604
Sahu AK, Siljudalen J, Trydal T, Rusten B (2013) Utilisation of wastewater nutrients for microalgae growth for anaerobic co-digestion. J Environ Manage 122:113–120. doi:10.1016/j.jenvman.2013.02.038
Sathish A, Smith BR, Sims RC (2014) Effect of moisture on in situ trans-esterification of microalgae for biodiesel production. J Chem Technol Biotechnol 89:137–142. doi:10.1002/jctb.4125
Sawaengsak W, Silalertruksa T, Bangviwat A, Gheewala SH (2014) Life cycle cost of biodiesel production from microalgae in Thailand. Energy Sustain Dev 18:67–74. doi:10.1016/j.esd.2013.12.003
Schneider RCS, Bjerk TR, Gressler PD, Souza MP, Corbellini VA, Lobo EA (2012) Potential production of biofuel from microalgae biomass produced in wastewater. In: Fang Z (ed) Biodiesel—feedstocks, production and applications, pp 3–24. InTech. ISBN:978–953–51-0910-5 doi:10.5772/52439
Schumacher G, Blume T, Sekoulov I (2003) Bacteria reduction and nutrient removal in small wastewater treatment plants by an algal biofilm. Water Sci Technol 47:195–202
Slade R, Bauen A (2013) Micro-algae cultivation for biofuels: cost, energy balance, environmental impacts and future prospects. Biomass Bioenergy 53:29–38. doi:10.1016/j.biombioe.2012.12.019
Song C, Chen G, Ji N, Liu Q, Kansha Y, Tsutsumi A (2015) Biodiesel production process from microalgae oil by waste heat recovery and process integration. Bioresour Technol 193:192–199. doi:10.1016/j.biortech.2015.06.116
Soydemir G, Keris-Sen UD, Sen U, Gurol MD (2015) Biodiesel production potential of mixed microalgal culture grown in domestic wastewater. Bioprocess Biosyst Eng 1–7. doi:10.1007/s00449–015–1487-3
Stephens E, Ross IL, Mussgnug JH, Wagner LD, Borowitzka MA, Posten C, Kruse O, Hankamer B (2010) Future prospects of microalgal biofuel production systems. Trends Plant Sci 15:554–564. doi:10.1016/j.tplants.2010.06.003

Steriti A, Rossi R, Concas A, Cao G (2014) A novel cell disruption technique to enhance lipid extraction from microalgae. Bioresour Technol 164:70–77. doi:10.1016/j.biortech.2014.04.056

Su E, You P, Wei D (2009) In situ lipase-catalyzed reactive extraction of oilseeds with short-chained dialkyl carbonates for biodiesel production. Bioresour Technol 100:5813–5817. doi:10.1016/j.biortech.2009.06.077

Sun Y, Liu J, Xie T, Xiong X, Liu W, Liang B, Zhang Y (2014) Enhanced lipid accumulation by Chlorella vulgaris in a two-stage fed-batch culture with glycerol. Energy Fuels 28:3172–3177. doi:10.1021/ef5000326

Suwannakarn K, Lotero E, Ngaosuwan K, Goodwin JG (2009) Simultaneous free fatty acid esterification and triglyceride trans-esterification using a solid acid catalyst with in situ removal of water and unreacted methanol. Ind Eng Chem Res 48:2810–2818. doi:10.1021/ie800889w

Taher H, Al-Zuhair S, Al-Marzouqi AH, Haik Y, Farid MM (2011) A review of enzymatic trans-esterification of microalgal oil-based biodiesel using supercritical technology. Enzyme Res 2011:1–25. doi:10.4061/2011/468292

Tang X, He LY, Tao XQ, Dang Z, Guo CL, Lu GN, Yi XY (2010) Construction of an artificial microalgal-bacterial consortium that efficiently degrades crude oil. J Hazard Mater 181:1158–1162. doi:10.1016/j.jhazmat.2010.05.033

Teo CL, Jamaluddin H, Zain NM, Idris A (2014) Biodiesel production via lipase catalysed trans-esterification of microalgae lipids from Tetraselmis sp. Renew Energy 68:1–5. doi:10.1016/j.renene.2014.01.027

Toor SS, Reddy H, Deng S, Hoffmann J, Spangsmark D, Madsen LB, Holm-Nielsen JB, Rosendahl LA (2013) Hydrothermal liquefaction of *Spirulina* and *Nannochloropsis salina* under subcritical and supercritical water conditions. Bioresour Technol 131:413–419. doi:10.1016/j.biortech.2012.12.144

Tran D-T, Yeh K-L, Chen C-L, Chang J-S (2012) Enzymatic trans-esterification of microalgal oil from Chlorella vulgaris ESP-31 for biodiesel synthesis using immobilized Burkholderia lipase. Bioresour Technol 108:119–127. doi:10.1016/j.biortech.2011.12.145

Tran D-T, Chen C-L, Chang J-S (2013a) Effect of solvents and oil content on direct trans-esterification of wet oil-bearing microalgal biomass of *Chlorella vulgaris* ESP-31 for biodiesel synthesis using immobilized lipase as the biocatalyst. Bioresour Technol 135:213–221. doi:10.1016/j.biortech.2012.09.101

Tran DT, Chen CL, Chang JS (2013b) Effect of solvents and oil content on direct trans-esterification of wet oil-bearing microalgal biomass of Chlorella vulgaris ESP-31 for biodiesel synthesis using immobilized lipase as the biocatalyst. Bioresour Technol 135:213–221. doi:10.1016/j.biortech.2012.09.101

Umdu ES, Tuncer M, Seker E (2009) Trans-esterification of Nannochloropsis oculata microalga's lipid to biodiesel on Al_2O_3 supported CaO and MgO catalysts. Bioresour Technol 100:2828–2831. doi:10.1016/j.biortech.2008.12.027

Ummalyma SB, Sukumaran RK (2014) Cultivation of microalgae in dairy effluent for oil production and removal of organic pollution load. Bioresour Technol 165:295–301. doi:10.1016/j.biortech.2014.03.028

Van Den Hende S, Carré E, Cocaud E, Beelen V, Boon N, Vervaeren H (2014) Treatment of industrial wastewaters by microalgal bacterial flocs in sequencing batch reactors. Bioresour Technol 161:245–254. doi:10.1016/j.biortech.2014.03.057

Velasquez-Orta SB, Lee JGM, Harvey A (2012) Alkaline in situ trans-esterification of Chlorella vulgaris. Fuel 94:544–550. doi:10.1016/j.fuel.2011.11.045

Venkata Mohan S, Devi MP (2014) Salinity stress induced lipid synthesis to harness biodiesel during dual mode cultivation of mixotrophic microalgae. Bioresour Technol 165:288–294. doi:10.1016/j.biortech.2014.02.103

Vonortas A, Papayannakos N (2014) Comparative analysis of biodiesel versus green diesel. Wiley Interdiscip Rev Energy Environ 3:3–23. doi:10.1002/wene.78

Wahlen BD, Willis RM, Seefeldt LC (2011) Biodiesel production by simultaneous extraction and conversion of total lipids from microalgae, cyanobacteria, and wild mixed-cultures. Bioresour Technol 102:2724–2730. doi:10.1016/j.biortech.2010.11.026

Wu LF, Chen PC, Huang AP, Lee CM (2012) The feasibility of biodiesel production by microalgae using industrial wastewater. Bioresour Technol 113:14–18. doi:10.1016/j.biortech.2011.12.128

Wu YH, Yu Y, Hu HY (2013) Potential biomass yield per phosphorus and lipid accumulation property of seven microalgal species. Bioresour Technol 130:599–602. doi:10.1016/j.biortech.2012.12.116

Xu Y, Boeing WJ (2014) Modeling maximum lipid productivity of microalgae: review and next step. Renew Sustain Energy Rev 32:29–39. doi:10.1016/j.rser.2014.01.002

Xu H, Miao X, Wu Q (2006) High quality biodiesel production from a microalga Chlorella protothecoides by heterotrophic growth in fermenters. J Biotechnol 126:499–507. doi:10.1016/j.jbiotec.2006.05.002

Yang F, Long L, Sun X, Wu H, Li T, Xiang W (2014) Optimization of medium using response surface methodology for lipid production by *Scenedesmus* sp. Marine Drugs 12:1245–1257. doi:10.3390/md12031245

Yu N, Dieu LTJ, Harvey S, Lee D-Y (2015) Optimization of process configuration and strain selection for microalgae-based biodiesel production. Bioresour Technol 193:25–34. doi:10.1016/j.biortech.2015.05.101

Zeng X, Danquah MK, Chen XD, Lu Y (2011) Microalgae bioengineering: From CO_2 fixation to biofuel production. Renew Sustain Energy Rev 15:3252–3260. doi:10.1016/j.rser.2011.04.014

Zeng D, Li R, Yan T, Fang T (2014) Perspectives and advances of microalgal biodiesel production with supercritical fluid technology. RSC Adv 4:39771–39781. doi:10.1039/C4RA05766J

Zhang YM, Chen H, He CL, Wang Q (2013) Nitrogen starvation induced oxidative stress in an oil-producing green alga *Chlorella sorokiniana* C3. PLoS ONE 8:10. doi:1371/journal.pone.0069225

Zhou W, Chen P, Min M, Ma X, Wang J, Griffith R, Hussain F, Peng P, Xie Q, Li Y, Shi J, Meng J, Ruan R (2014) Environment-enhancing algal biofuel production using wastewaters. Renew Sustain Energy Rev 36:256–269. doi:10.1016/j.rser.2014.04.073

Zhu L, Hiltunen E, Shu Q, Zhou W, Li Z, Wang Z (2014) Biodiesel production from algae cultivated in winter with artificial wastewater through pH regulation by acetic acid. Appl Energy 128:103–110. doi:10.1016/j.apenergy.2014.04.039

Intensified Synthesis of Biodiesel from Sustainable Raw Materials Using Enzymatic Approach

Preeti B. Subhedar and Parag R. Gogate

Abstract The biodiesel production as a substitute fuel has been explored with great interest in recent times due to its eco-friendly advantages such as non-toxic nature and much lower emissions during usage. On the other hand, owing to an adverse economics and additional complications for design of efficient industrial scale reactor configurations, biodiesel production at large scale has not been substantially effective. The present work offers insights into the sustainable methods of production based on the waste cooking oil (WCO) and nonedible oil as the feedstock, which can lead to significant reduction in the processing costs. The focus has been kept on the enzymatic route of production, and the process intensification aspects based on the use of ultrasonic and microwave irradiations have also been discussed which is essential to overcome the inherent limitations associated with the use of enzymatic route and sustainable resources. The fundamental mechanism of intensification of biodiesel production, overview of the types of reactors and strategies for optimum design and operating factors to maximize the intensification of enzymatic biodiesel production have been presented. A detailed overview of different literature studies has also been given so as to give an idea about the range of operating parameters and the expected benefits due to the use of process intensification approaches. Overall, it has been confirmed that the application of ultrasound and microwave is able to significantly increase the biodiesel yield with significant reduction in the time as well as quantity of the starting materials besides giving eco-friendly operation.

Keywords Biodiesel • Intensification • Sustainable feedstock • Enzyme • Ultrasound • Microwave

P.B. Subhedar • P.R. Gogate (✉)
Department of Chemical Engineering, Institute of Chemical Technology, Matunga, Mumbai 400 019, India
e-mail: pr.gogate@ictmumbai.edu.in

L. Singh, V.C. Kalia (eds.), *Waste Biomass Management – A Holistic Approach*,
DOI 10.1007/978-3-319-49595-8_14

1 Introduction

Limited reserves of fossil fuels coupled with the environmental concerns such as depletion of ozone, global warming, emission of greenhouse gases, etc. have directed interest into alternative energy sources (Maceiras et al. 2011) among which biodiesel is one of the attractive options. Biodiesel has the potential to fulfil the energy requirement in the transportation as well as industrial sectors (Basha et al. 2009). Biodiesel is typically a mixture of mono alkyl esters obtained from a chemical reaction between oils including edible as well as nonedible or the animal fats and alcohols (Manh et al. 2011). Biodiesel offers superior properties as compared to diesel fuel such as non-toxic and biodegradable nature. Also, it is primarily free of sulphur components and aromatics (Lee et al. 2011). Biodiesel has substantial impact in decreasing the emissions from the engine, for example, carbon monoxide by about 44 %, partially combusted hydrocarbons by about 68 %, particulate matters by 40 % and, most importantly, sulphur oxide by almost 100 % as per the recent reports of Wu and Leung (2011) and Leduc et al. (2009). Biodiesel can be mixed with the petroleum fuel or can be used as a complete replacement successfully (Vyas et al. 2010) without any significant modifications being required in the existing setups, for example, engines in the case of transport vehicles.

The history of using biofuels dates back to the 1970s with the exploration of vegetable oils as a fuel, whereas the first production of biodiesel at laboratory scale using the chemical route dates back to 1982. A laboratory-scale plant was operated successfully in 1985 for biodiesel synthesis from rapeseed oil. Commercial production of biodiesel initially started in Europe in 1990 with over 2.7 million tonnes of biodiesel being synthesized in 2003. The current objective globally is to achieve production levels of about 20 % of entire diesel market in 2020. The strategy for production of biodiesel in the USA is to achieve a production of about 3.3 million tonnes in 2016 (Demirbas 2008). Due to this ever increasing demand for the biodiesel production, a lot of research has been directed towards developing intensified production approaches giving favourable economics, which has been a key area of concern.

The commonly applied method for the synthesis of biodiesel is based on the transesterification route using acid or base (alkaline) catalyst resulting in formation of mono alkyl esters and glycerol (Marchetti et al. 2007). The transesterification method for the production of biodiesel is in use for years; however, it has numerous difficulties such as difficult catalyst separation, huge energy consumption, unwanted side reactions and requirement of huge excess of alcohols (Fjerbaek et al. 2009). Numerous attempts have been directed in terms of developing alternative approaches for the synthesis of biodiesel with an objective of making the biodiesel production economically attractive. Recently, biodiesel synthesis using enzymatic approach involving the use of biocatalyst such as lipases has received vast consideration with moderate operating conditions and easy glycerol recovery being the key advantages (Cesarini et al. 2014). The use of enzymes gives additional advantage of conversion of free fatty acids (FFAs) into fatty acid methyl esters (biodiesel), without generation of soap, giving enhanced yield of biodiesel and reducing the costs required for the product purification. The feature of the enzymes allowing the use of materials with higher free fatty acid or water content, e.g. waste cooking oils, nonedible oils and industrial waste oil, has prompted towards an approach which can reduce the cost of production as the feedstock contributes a major fraction in the overall production cost. A major drawback of the enzymatic processes is however the requirement of large reaction times and significant cost of the enzymes, which points towards the use of suitable process intensifying approach. The present chapter provides an overview of the biodiesel production based on the enzymatic route along with the possible intensification of the process using sonochemical reactors and microwave reactors. The specific topics discussed in the present chapter comprise types of feedstocks, types of biocatalysts used for the enzymatic biodiesel production, fundamental mechanism of enhancement in the biodiesel yield due to the use of ultrasound or microwave, types of reactors and essential design as well as operational guidelines for maximizing the degree of intensification.

2 Feedstocks Used for Biodiesel Production

Production of biodiesel can be based on the use of any kind of oils or fats classified mainly into three categories: (1) edible oils such as rapeseed or soybean; (2) nonedible oils, for example, mahua, kusum, karanja or jatropha; and (3) sustainable waste materials such as animal fats, off-specification products obtained during the synthesis of vegetable oils, waste cooking oil (WCO), etc. The specific feedstock being used for the production of biodiesel generally differs with the country subject to the availability of raw materials. The European Union and Canada mainly use rapeseed oil, whereas in the USA, genetically modified soybean oil is more commonly used as the feedstock. On the other hand, some of the Asian countries such as Indonesia and Malaysia utilize palm oil as a feedstock due to abundant availability. Even though China is an agriculture-dominated country, it is still facing the problem of food supply chain. Due to the food security issues, China is mainly focussing on the use of used cooking oil as a feedstock (Tan et al. 2010).We now present a brief summary of the different feedstocks that have been explored for the production of biodiesel among the broad category as edible, nonedible and waste materials.

2.1 Edible Oils

Among the edible oils, soybean oil has been very commonly used in combination with different acyl acceptors and different solvents. Soybean oil consists of five fatty acids, viz. oleic acid (approximately 13 %), palmitic acid (approximately 13%), linolenic acid (approximately 13%), stearic acid (around 4 %) and linoleic acid (about 55 %). The USA is a major producer of soybean oil accounting for near about 40 % of the total soybean production with more than 30 million ha soybean harvest (Kinney and Clemente 2005). Xie and Wang (2014) studied the enzymatic biodiesel production using soybean oil catalysed by the immobilized form of lipase supported on the magnetic particles of Fe_3O_4/poly(styrene-methacrylic acid). It was reported that about 86 % of oil conversion is obtained in 24 h at 35 °C temperature. Xu et al. (2003) studied the enzymatic production of biodiesel using methyl acetate and soybean oil as starting materials catalysed by Novozyme 435 (immobilized *Candida antarctica* lipase). A maximum yield of 92 % was reported under optimum conditions as 30 % (w/w) enzyme loading, 1:12 molar ratio of oil to methyl acetate, 10 h of reaction time and 40 °C temperature. The production of the triacetin instead of glycerol allowed efficient recycling of the enzyme.

Sunflower oil is another commonly used raw material among the edible oils for biodiesel synthesis. Belafi-Bako et al. (2002) studied the transesterification of sunflower oil using methanol as the acyl acceptor and commercial immobilized lipase Novozyme 435 as a biocatalyst for biodiesel production. Continuous feeding of methanol into the reactor in absence of any solvent resulted in a conversion of 97 % in 16 h. Ognjanovic et al. (2009) also investigated the solvent-free biodiesel production from sunflower oil with the commercial Novozyme 435 lipase. Optimum conditions were reported to be temperature of 45 °C, enzyme loading of 3 % and molar ratio of methanol to oil of 3:1, with >99 % of oil conversion to biodiesel in 50 h of reaction time. Rapeseed oil, coconut oil and olive oil have also been utilized as a starting material in many countries. Typically, the price of vegetable oils in the edible form is greater than the price of petroleum diesel itself, and the use of vegetable oils also results into food versus fuel issue. Considering these issues, edible oils are not that commonly recommended as feedstock, and the nonedible oils are being looked as a cheaper alternative.

2.2 *Nonedible Oils*

Several oils derived from nonedible plants can also be possible raw materials for biodiesel. The significantly used nonedible crops include jatropha, karanja (Tiwari et al. 2007), tobacco (Veljković et al. 2006), mahua (Jena et al. 2010), castor oil (de Oliveira et al. 2004), cotton (Georgogianni et al. 2008), rubber (Ramadhas et al. 2005), etc. *Jatropha* is a prominent feedstock for the production of biodiesel commonly available in Southeast Asian countries, India as well as Central and South America. Currently, in India, the annual production of *Jatropha* is about 15,000 tonnes (Jain and Sharma 2010). Oil content in *Jatropha* differs depending on the species type; however, it is around 46–58 % in the kernels and 40–60 % in the seeds (Kumar and Sharma 2011). Lee et al. (2013) studied the enzymatic biodiesel production using *Jatropha* oil, and a conversion of almost 98 % was reported in 4 h of reaction time under optimized conditions of temperature of 45 °C, 250 rpm agitation speed, 10 wt% of water and 20 % of immobilized lipases on the basis of weight of oil. Karanja seeds consist of significant oil content. Karanja oil contains several toxic constituents restricting the use as cooking oil and hence offers potential for biodiesel production. In India, annual production of karanja oil is 55,000 tonnes (Jain and Sharma 2010), of which merely 6 % is presently being used for the biodiesel synthesis (Naik et al. 2008). Castor plant has comparable environmental necessities as that of *Jatropha* and is mainly cultivated as weed in many countries like India, China, Brazil and Thailand. In India approximately 0.73 Mt castor seeds are produced yearly which accounts for the 60 % of the worldwide production of castor (Kumar and Sharma 2011). Oil derived from castor is almost entirely soluble in alcohols as compared to the other vegetable oils (Meneghetti et al. 2006) offering much higher reaction rates. Transesterification of castor oil with immobilized lipase for biodiesel production was investigated by de Oliveira et al. (2004), and it was reported that conversion of about 98 % is obtained at temperature of 65 °C, 20 % of enzyme loading and 3:1 molar ratio of ethanol to oil.

Mahua is a tree which is found in most of the parts of India and Burma, and the seeds again are a substantial source of oils. The production of mahua oil in India is 180,000 t/annum (Jain and Sharma 2010). Mahua kernels comprise about 50 % of oil (No 2011). Ghadge and Raheman (2006) studied the synthesis of biodiesel from the oil of mahua (*Madhuca indica*) and reported that under optimized conditions of oil to methanol molar ratio of 1:6, 0.7 % w/v loading of KOH as catalyst and temperature of 60 °C, maximum biodiesel yield of 98 % was achieved in reaction time of 1.26 h.

2.3 Waste Oil and Animal Fats

The fundamental approach to lower the cost of production of biodiesel is to decrease the price of feedstock which can be achieved using the sustainable sources as waste oils or animal fats. Waste cooking oil (WCO) is produced after the vegetable oils are used at high temperatures during frying of food. The processing at higher temperatures leads to several hydrolysis, polymerization and oxidation reactions in oil, altering their physicochemical characteristics. The problem of no efficient technique for processing used oils from households creates significant environmental concerns due to the possible discharge into surface waters (Alcantara et al. 2000). The production of biodiesel using WCO instead of virgin refined vegetable oil is an effective approach of reducing feedstock costs, also bringing considerable eco-friendly benefits as it offers a substitute for the final discarding of the oil that creates environmental concerns (Enweremadu and Mbarawa 2009). The price of waste cooking oil is typically two- to threefold lower than the virgin edible oils, and also the usage for biodiesel decreases the price associated with the waste disposal (Phan and Phan 2008).

Andrade et al. (2013) investigated biodiesel production from WCO using enzymatic catalysis process and reported maximum biodiesel yield of 95.5 % at optimal conditions of 45 °C temperature, oil to methanol molar ratio of 1:8 and enzyme loading of 8 %. Yu et al. (2013) also studied the production of biodiesel from WCO using immobilized lipase and reported 80 % conversion under optimum conditions of 45 °C temperature and molar ratio of alcohol to oil as 5:1.

The other sustainable feedstock for the synthesis of biodiesel is animal fats which mainly comprise tallow and lard. Beef tallow is a nonedible and inexpensive animal fat which is abundantly available with limited industrial use. Therefore, due to the benefits of sustainability, the beef tallow turns out to be the second most significant raw material after soybeans for the production of biodiesel in many countries (da Cunha et al. 2009). The fatty acid composition of tallow is mostly 17–37 % palmitic acid, 26–50 % oleic acid, 6–40 % stearic acid, 1–8 % myristic acid, 0–5 % linoleic acid and other branched-chain acids (Ma et al. 1998). The efficacy of enzymatic biodiesel production from animal fat in solvent-free condition has been reported by Kumar et al. (2015) with maximum biodiesel yield being obtained at molar ratio of alcohol to oil of 4:1 at 45 °C and 16 h of reaction time. It was also reported that the enzyme activity in the presence of methanol without solvent at the optimum conditions remained comparatively constant for 10 cycles and then reduced slowly getting to zero after 50 cycles.

The composition of fatty acids in lard (Yang et al. 2003) is primarily oleic acids (41.6 %), palmitic acid (23.9 %) and some linoleic acid (13.2 %). Synthesis of biodiesel from lard based on the enzymatic approach involving the use of immobilized form of *Candida* sp. 99–125 has been investigated by Lu et al. (2007). The optimum conditions for maximum biodiesel yield of 87.4 % as established in the study were immobilized lipase quantum of 0.2 g, 20 % water by weight of fat, 8 ml n-hexane as solvent, reaction temperature of 40 °C, and addition of methanol in three stages. The immobilized lipase was observed to be stable for 180 h after repetitive use.

3 Types of Catalyst

After analysing the different feedstocks being used for the biodiesel synthesis, the various catalysts being used for the biodiesel production process have been discussed also highlighting the potential of enzymes as biocatalyst.

3.1 *Base Catalyst*

The generally used alkaline catalyst for the production of biodiesel is potassium hydroxide or sodium hydroxide (Felizardo et al. 2006) due to advantages such as higher reaction rates, ability to catalyse the process under near-ambient conditions as well as ease of availability and lower costs (Lotero et al. 2005). Also base-catalysed reaction is significantly faster as compared to acid-catalysed reactions (Kulkarni and Dalai 2006). A major limitation of using base catalyst is the possibility of soap formation for the feedstock with acid value more than 1 to 2 mg KOH/g (Felizardo et al. 2006) or when the FFA content exceeds 0.5 wt% (Wang et al. 2006). Higher water content or the higher FFA can lead to the saponification giving reduced biodiesel yield as well as significant operational problems (Liu 1994). Therefore, for the feedstock such as nonedible oils with an average FFA content of more than 6 wt. % or the waste cooking oils, alkali catalyst is certainly not appropriate for biodiesel production (Lotero et al. 2005). Alkaline alkoxides such as sodium alkoxide can also be used as an alternative as it is highly reactive and reported to give significant biodiesel yield of even more than 98 % in only 30 min (Thanh et al. 2012). Alkaline alkoxides are typically expensive than the alkaline hydroxides and hence not so commonly used especially at larger scales of operation. Also, the use of higher quantum of alkaline hydroxides instead of using more costly alkaline alkoxides can give similar biodiesel yields (Schuchardt et al. 1998).

Meng et al. (2008) investigated the biodiesel synthesis from WCO with NaOH as a catalytic agent and methanol as acyl acceptor. The various operating parameters evaluated in the study were molar ratio of methanol to oil, NaOH loading, time of reaction and the temperature. The highest biodiesel yield of 86 % was attained at the optimal reaction conditions of 6:1 methanol to oil molar ratio, NaOH loading of 0.7 %, 90 min reaction time and 50 °C temperature. Darnoko and Cheryan (2000) studied the application of two-stage approach for biodiesel production from WCO. Maximum biodiesel yield of 90 % was obtained using 1 % KOH at 60 °C temperature in 30 min of reaction time. It has been established that a two-stage, base-catalysed reaction is a cost-effective technique for the biodiesel synthesis. Tomasevic and Siler-Marinkovic (2003) investigated the biodiesel synthesis from used sunflower oil studying the effect of type of catalyst based on the use of two alkali catalysts as KOH and NaOH. The maximum yield was achieved using oil to methanol molar ratio of 1:6, 1 % KOH loading and 25 °C temperature in 30 min of reaction time. Refaat et al. (2007) studied the biodiesel synthesis from WCO with KOH and NaOH as catalysts. The optimum conditions giving maximum biodiesel yield of 98 % were oil to methanol molar ratio of 1:6, KOH loading of 1.0 % and 65 °C temperature. Also, it has been reported that use of KOH as a catalyst gave higher conversion as compared to NaOH.

3.2 Acid Catalysts

For feedstocks with relatively higher free fatty acid (FFA) content, alkaline catalysts are not appropriate as they form soap with these acids creating operational issues. For such feedstock, acid catalysts are more feasible (Ma and Hanna 1999) and offer great benefit offered is that no processing is required even for the cost-effective sustainable raw materials such as WCO and greases usually containing high FFA concentrations (more than 6 %). The overall process is economically feasible when compared to the alkali-catalysed processes using virgin oils (Zhang et al. 2003). Generally sulphuric acid (H_2SO_4), phosphoric acid (H_3PO_4), hydrochloric acid (HCl) or organic sulphonic acids are used as a catalyst (Fukuda et al. 2001). During acid-catalysed biodiesel production, in addition to transesterification reaction, esterification also takes place in which FFA reacts with methanol to form methyl ester. However, the disadvantages of the acid-catalysed reactions are the lower rate of reactions as compared to the alkali-catalysed reactions and requirement of comparatively higher temperatures to get high product yields. The use of acidic conditions can also lead to the corrosion of the reactors, and also the process requires significantly higher molar excess of alcohol, which makes subsequent downstream processing very difficult and sometimes cost intensive (Thanh et al. 2012). Wang et al. (2006) investigated the acid-catalysed transesterification of WCO based on the use of H_2SO_4 as the catalyst. It was established that higher biodiesel yield is obtained with an increase in the molar ratio of oil to methanol as well as H_2SO_4 loading. More than 90 % conversion of WCO was obtained in 10 h using oil to methanol molar ratio of 1:20 and catalyst loading of 4.0 %.

3.3 Lipases as a Catalyst

The development of green and efficient processes for biodiesel production has been mainly based on the usage of heterogeneous catalysts as either enzyme (Adamczak et al. 2009) or chemical (Bournay et al. 2005). The use of enzymatic approach offers substantial benefits over the chemical process. Lipase enzyme (triacylglycerol lipase, E.C. 3.1.1.3) is generally used for the synthesis of biodiesel having catalytic action in an aqueous solution as well as under solvent-free conditions. Lipases not only can catalyse hydrolysis reactions but also take part in several reversible reactions such as esterification or transesterification (Gupta et al. 2004). Active sites of lipase comprises three amino acid residues: (1) nucleophilic residue (cysteine, serine or aspartate), (2) acidic residue (aspartate or glutamate) and (3) histidine residue (Jaeger et al. 1994). Existence of a peptide sequence in α-helix covering the active site, called lid, is the structural feature of most forms of lipases. In the absence of a hydrophobic interface, lid restricts the access of substrate to the catalytic triad, retaining the enzyme in the closed conformation (Jaeger and Reetz 1998). Interfacial activation, i.e. the activation of lipases in the presence of hydrophobic surfaces, is a vital requirement of these enzymes that also incorporates functionality (Derewenda et al. 1992). Enzyme-catalysed biodiesel production can be carried out using both intracellular and extracellular lipases. In the case of intracellular lipases, the enzyme remains inside the cell, and in the other case, lipase is recovered from the culture broth and subsequently purified (Gog et al. 2012). But generally immobilized enzyme is used which can be easily recovered and possibly reused with some activation. Also, the immobilized form offers higher stability towards chemical and shear-induced denaturation as well as temperature shocks. Considering these aspects, immobilized enzymes have been commonly recommended as biocatalyst. In the case of lipases as catalyst, the FFAs present in the feedstock can also be transformed to alkyl esters directly allowing an easy one-step production. Thus enzymes have an interesting prospect for biodiesel production (even from sustainable resources) possibly giving reduction in the production costs based on optimized synthesis protocol. The commercially available immobilized forms of lipases such as Lipozyme or Novozyme are recommended as more appropriate catalytic agent for industrial application, as compared to the usage of free form of lipase from *Pseudomonas* or any other lipases due to the ease of recovery (Ognjanovic et al. 2009).

Many lipases from various origins have been employed for the production of biodiesel. Lipase from *Candida antarctica* has been the mostly used biocatalyst for synthesis of biodiesel. Acyl transfer reactions between oils and various acyl acceptors are catalysed by *Candida antarctica* lipase, and also it exhibits great activity in organic solvents and offers wide reactant selectivity. The lipase from *Pseudomonas* has several benefits in view of its endurance in reaction systems containing water. Kaieda et al. (2001) established that lipase enzyme obtained from *P. cepacia* represented superior resistance for inhibition by methanol than the lipases obtained from other origins such as *Candida lipolytica*, *Penicillium roqueforti*, *Rhizopus oryzae* and also *Candida antarctica*, offering it as a better choice for the synthesis of biodiesel.

4 Overview of Conventional Enzymatic Biodiesel Production Approaches

Table 1 provides a summary of the literature on enzymatic route for biodiesel production in terms of the different catalysts and feedstock used along with optimum reaction conditions and important results. A brief discussion has now been offered on the different transesterification studies involving WCO and nonedible oil with emphasis on highlighting the substrate and optimum conditions for maximum production. Hsu et al. (2001) studied the biodiesel synthesis from waste grease based on the use of two different acyl acceptors, viz. methanol and ethanol, in the presence of immobilized form of lipase obtained from *Pseudomonas cepacia*. It was reported that the use of ethanol as an acyl acceptor resulted in more conversion (84–94 %) as compared to methanol (47–89 %) in 24 h. Also, the immobilized lipase was easily recovered and could be recycled at least five times without loss in the enzyme activity. Lee et al. (2002) reported a novel approach of stepwise addition of methanol during the process of biodiesel synthesis from restaurant grease and fractionated lard. The stepwise addition was shown to give higher conversion as compared to the use of entire methanol at the start of reaction as the lipase activity is decreased by the presence of excess quantities of methanol. It was reported that using three-step reaction, 74 % conversion was achieved. Also the use of a porous material, such as silica gel, enhanced the conversion for the conventional approach of using all the amount of methanol initially. Another approach to improve the yields was based on purification of the restaurant grease

Table 1 Enzymatic production of biodiesel using various sources of lipases

Feed oil	Enzyme source	Reaction conditions					References
		Oil to acyl acceptor ratio	Enzyme conc.	Temp. (°C)	Time (h)	Yield (%)	
Cotton seed oil	*Candida antarctica*	1:6	0.017 g/g of oil	50	24	97	Royon et al. (2007)
Regenerated restaurant grease	*Candida antarctica*	1:3	10 %	30	24	96	Lee et al. (2002)
Waste cooking oil	*Candida antarctica*	1:4	3.3 g over packed bed height of 10.53 cm	40	4	80.3	Halim et al. (2009)
Waste frying oil	*Candida antarctica*	1:4	15 %	45	12	90	Azocar et al. (2010)
Lard	*Candida antarctica*, *Thermomyces lanuginosus*	1:5	0.03 % and 0.08 %	50	20	92.7	Huang et al. (2010)
Waste cooking oil	*Candida* sp. 99–125	1:3	25 %	45	(continuous operation)	91.08	Chen et al. (2009)
Jatropha oil	*Chromobacterium viscosum*	1:4	50 mg	40	8	71	Shah et al. (2004)
Jatropha oil	*Enterobacter aerogenes*	1:4	50 U of immobilized lipase/g of oil	55	48	94	Kumari et al. (2009)
Waste oil	*Penicillium expansum*	1:3	84 U/g of oil	35	24	92.8	Li et al. (2009)
Tallow and recycled restaurant grease	*Pseudomonas cepacia*	1:4	100 mg	50	8	90	Hsu et al. (2001)
Madhuca indica oil	*Pseudomonas cepacia*	1:4	50 mg	40	2.5	99	Kumari et al. (2007)
Jatropha oil	*Pseudomonas cepacia*	1:3	4 %	40	48	72.1	Aransiola (2013)
Waste frying oil	*Pseudomonas fluorescens*	1:3	5 %	45	24	63.8	Charpe and Rathod (2011)
Waste cooking oil	*Rhizopus oryzae*	1:4	30 wt% of oil	40	20	90	Chen et al. (2006)
Waste oil	*Thermomyces lanuginosus*	1:6	5 mg	40	105	92.8	Yagiz et al. (2007)

based on the removal of FFA using column chromatography and subsequent use for biodiesel production. It was reported that the conversion in 24 h increased from 25 % for the crude grease to 96 % for the case of purified grease.

Shah et al. (2004) investigated the usage of three lipases from different sources as *Candida rugosa*, *Chromobacterium viscosum* and porcine pancreas for the production of biodiesel from *Jatropha* oil under solvent-free conditions. It was reported that lipase from *C. viscosum* gave considerably higher yield as compared to other forms. The use of immobilized form of lipase from *C. viscosum* on Celite 545 enhanced the yield to 71 % in 8 h of reaction time at 40 °C. Chen et al. (2006) studied the biodiesel synthesis from WCO using immobilized lipase from *Rhizopus oryzae* and reported that highest biodiesel yield of 90 % was achieved under optimal reaction conditions of enzyme loading of 30 % (wt/wt), methanol to oil molar ratio of 4:1 and temperature of 40 °C. To counter the irreversible inactivation of the lipase due to the excess methanol, it was suggested to add methanol in steps so as to result in decreased inactivation of immobilized lipases.

The oil of *Madhuca indica* tree consists of higher free fatty acids, and therefore, it is very challenging to obtain biodiesel using a chemical approach. Kumari et al. (2007) investigated the biodiesel production from *M. indica* oil using commercial lipase obtained from *Pseudomonas cepacia*. It was reported that the use of free enzyme in the powder form resulted in 98 % conversion in 6 h requiring 50 mg of lipase under optimized conditions. Modified biocatalyst formulations, such as protein-coated microcrystals (PCMCs) and cross-linked enzyme aggregates (CLEAs), resulted in 99 % and 92 % conversion, respectively, in 2.5 h using an equal amount of enzyme. For the case of PCMCs, 99 % conversion could also be achieved using only 25 mg of enzyme in 16 h of reaction. Royon et al. (2007) investigated the enzymatic approach for the biodiesel synthesis using cottonseed oil. Immobilized form of lipase obtained from *Candida antarctica* and *t*-butanol as solvent was used in the reaction, and a biodiesel yield of 97 % was reported at 50 °C in 24 h of reaction time using 54 % oil, 13.5 % methanol, 32.5 % *t*-butanol and 0.017 g of enzyme. Yagiz et al. (2007) investigated the biodiesel production from sustainable source as waste oil using lipase immobilized on the hydrotalcite and zeolite supports. It has been reported that the maximum yield of 95 % was achieved under the optimal reaction conditions of molar ratio of oil to methanol as 1:4, 4 % catalyst loading in 5 h. Additionally, immobilized form of lipase was reported to be stable for seven cycles without any substantial loss in the enzyme activity during the reuse.

Kumari et al. (2009) investigated the synthesis of biodiesel from *Jatropha* oil in the presence of t-butanol as the solvent and immobilized lipase obtained from *Enterobacter aerogenes* as the catalyst. The use of t-butanol was reported to considerably decrease the undesirable effects incurred by the presence of methanol and glycerol giving maximum biodiesel yield as 94 % under conditions of molar ratio of oil to methanol as 1:4, enzyme loading as 50 U/g of oil, oil to t-butanol ratio of 1:0.8 v/v and 55 °C temperature in a reaction time of 48 h. A low-cost immobilized lipase from *Penicillium expansum* has also been reported to be an effective catalyst for biodiesel synthesis from WCO having high content of free fatty acids in the presence of organic solvent media (Li et al. 2009). It has been reported that water formed in the course of the esterification reaction of FFAs with alcohol restricts the biodiesel yield, and hence the use of adsorbents to remove water from the reaction can help in obtaining an enhanced biodiesel yield. Silica gel is generally reported to be the best adsorbent giving significant degree of intensification due to the efficient removal of water.

Enzymatic biodiesel production in a packed bed reactor in continuous mode from waste palm oil has also been reported which is an important investigation to examine the prospect of industrial scale production of biodiesel (Halim et al. 2009). The optimal reaction conditions were reported as packed bed height of 10.53 cm and substrate flow rate as 0.57 ml/min giving a maximum biodiesel yield of 80.3 %. The work also depicts development of models for prediction of the biodiesel yield considering the reaction control and mass transfer control. Comparison of the obtained experimental results with the models revealed that the biodiesel production process in packed bed reactor was mass transfer controlled.

Collective application of two lipases in immobilized form having reciprocal site specificity as a substitute to single lipase is a promising method that can result in higher production and reduced costs of synthesis. The combined use of two lipases, viz. Novozyme 435 which is non-specific in nature and Lipozyme TLIM which is 1,3 specific in nature, has been reported for biodiesel production from lard with optimization studies based on the response surface methodology (Huang et al. 2010). The optimum reaction conditions reported in the work were 0.04 (w/w) of lipase, 0.49 fraction of Novozyme 435 in total lipases, 0.55 (v/v) quantity of tert-butanol/oil, molar ratio of oil to methanol as 1:5 and 20 h of reaction time. Under these optimum conditions, 97.2 % of biodiesel yield was achieved, which was in well agreement to the value obtained from RSM as 97.6 %. It was also established that the mixture of lipases can be recycled 20 times with similar enzyme activity when the regeneration was performed using organic solvents.

Optimization of the production of biodiesel using a mixture of rapeseed oil and used waste frying oil in the presence of lipase as a catalyst has been studied by Azocar et al. (2010). The optimal conditions obtained by RSM analysis to give maximum biodiesel yield were molar ratio of methanol to oil as 3:1, 100 % WFO (w/v), 15 % (w/v) Novozyme 435 loading, temperature of 44.5 °C, 12 h of reaction time and agitation speed of 200 rpm. The use of WFO improved the biodiesel yield due to the higher contents of monoacylglycerols, diacylglycerols and FFA as compared to rapeseed oil. It was thus established that the substitution of rapeseed oil with WFO could reduce the biodiesel production costs due to lower price of WFO as compared to rapeseed oil.

The use of waste sunflower frying oil has also been studied for production of biodiesel based on the enzymatic transesterification approach (Charpe and Rathod 2011). Several microbial lipases were used for biodiesel production so as to establish an optimum lipase. It was reported that the *Pseudomonas fluorescens* gave maximum yield under optimum reactions conditions of 5.0 % enzyme loading, molar ratio of oil to methanol as 1:3 and 45 °C temperature. In order to avoid an inhibitory influence, three step addition of methanol was recommended giving intensification in the biodiesel yield obtained in 24 h from 56.0 % to 64.0 %. Furthermore, the effect of various solvents on the biodiesel yield was also examined, and it was reported that the non-polar solvents for example n-hexane and n-heptane gave better yield as compared to the polar solvents such as acetone. A maximum conversion of 66.12 % was obtained with n-hexane as the solvent.

Ethanolysis of *Jatropha* oil using immobilized lipase from *P. cepacia* with n-hexane as the solvent has been also reported for the biodiesel production (Aransiola 2013). Again, ethanol was added in a stepwise manner three times in order to avoid enzyme inhibition. For the case of crude *Jatropha* oil, optimized conditions of oil to ethanol ratio of 1:4 and reaction temperature of 40 °C resulted in a maximum yield as 72.1 % in 48 h of reaction time. Pretreatment of oil performed by means of acid esterification process in order to reduce the FFA content, resulted in higher biodiesel yield as 96.1 %. Reusability studies revealed that enzyme could be used for 13 cycles with only negligible loss in enzyme activity.

From the literature, it can be said that there are several points which needs to be considered so as to make the enzymatic biodiesel production economically viable, such as addition of alcohol in stepwise manner, usage of immobilized enzyme and its reusability after following proper regeneration process. Addition of alcohol in a stepwise manner facilitates minimum inhibition of the enzyme by alcohol (Vasudevan and Briggs 2008). For alcohol to be used initially, replacing methanol by ethanol as an acyl acceptor in a solvent-free condition allows the use of higher molar ratios of alcohol to oil (Antczak et al. 2009). Application of a specific solvent in the reaction system is another effective approach to overcome the problem of inhibition of lipase by methanol (Tan et al. 2010). A major hurdle for commercial use of enzymes is the cost factor. The use of immobilized form, which permits reusability, is an essential thing for large-scale biodiesel synthesis in an economical manner. Immobilization also gives improved characteristics of lipases as thermal stability, improved handling and reusability of enzyme. Reuse of the recovered immobilized enzymes significantly reduces the biodiesel production cost (Lukovic et al. 2011).

5 Intensification of Enzymatic Biodiesel Production

For the biodiesel production based on the enzymatic transesterification approach, the main drawbacks are requirement of significantly longer reaction times attributed to immiscible nature of the reactants giving higher mass transfer resistances (Ji et al. 2006) and much higher costs of the enzymes. The difference in the densities of the reactants can also result in reduced contact among the reactants as well as with the biocatalyst used in the process. Presently, production of biodiesel from lipase (enzymes) is not economically competitive as compared to the conventional fuels and hence can be only utilized for a lesser portion of prevailing demand for transport fuels (Saka and Kusdiana 2001). Lots of commercially available enzymes are too expensive considering the required scales of operation for the biodiesel production, although there is a possibility that enzymes can be recovered and reused based on the approach of immobilization.

The traditional approach of enhancing the reaction rate by increasing temperature of the reaction is not feasible in the case of enzymatic reactions due to the possibility of deactivation of the enzymes at higher operating temperatures (Marchetti and Errazu 2008). There has been significant interest in using process intensification approaches for overcoming the mass transfer limitations by way of achieving enhanced interfacial area or introducing turbulence so as to increase the intrinsic transfer coefficients (Furukawa et al. 2010). In addition, the use of process intensification technologies should also target to achieve maximum fatty acid methyl ester yields at perhaps lesser requirement of the reactants, catalyst quantum or solvents as well as lower temperatures which can yield substantial commercial benefits. The current chapter has focused on two such process intensification approaches for enzymatic reactions in terms of using ultrasound and microwave irradiations.

5.1 Sonochemical Reactors

Sonochemical reactors are based on the use of ultrasound for inducing the cavitational events in the liquid medium. For a better understanding into the expected mechanisms contributing to the increase in the yield of biodiesel, a brief discussion on the fundamentals of cavitation including the key effects driving the intensification and the reactor designs that can be utilized for the biodiesel processing has been presented.

5.1.1 Basics of Cavitation

Cavitation is defined as the formation, subsequent growth and finally intense collapse of the generated cavities induced by the pressure fluctuations due to the passage of ultrasound in the reaction media. The cavitational effects obtained in the reactor can be classified into chemical and physical effects. Generation of local hot spots with extremely high temperatures and pressures locally, which can induce the radical formation by pyrolytic dissociation of the compounds present at the site of cavity collapse are considered as the chemical effects. In addition to these chemical effects, cavitation also results in generation of strong liquid circulation currents described as acoustic streaming and local micro-scale turbulence which are more commonly described as the physical effects of cavitation. The extent of chemical and physical effects depends on the operating conditions, and it is possible to tailor a specific controlling effect depending on the requirement of the controlling mechanisms. Particularly for the biodiesel synthesis mainly restricted by the mass transfer resistances and insufficient contact of the reactants and/or catalyst, physical effects of cavitation in terms of turbulence and microstreaming are more important and hence should be maximized. The collapse of cavitating bubbles at/or near the interface of liquids results in disruption of the fluid layer forming very fine emulsions giving higher surface area for the reaction resulting in an increase in rates of reaction (Gogate et al. 2011). Intense turbulence and liquid circulation at microscale generated by cavitation also helps in intensifying the transfer processes giving enhanced yields from the process.

Sonochemical reactors are generally categorized into direct type (direct contact of transducer with the liquid) and indirect type of irradiation (transducers are typically attached to the wall or base of the vessel). The most common of the direct type reactor is the ultrasonic horn, which is submerged right into the liquid medium. In the case of ultrasonic horn, the cavitational intensity is typically restricted to an area close to the horn tip (3–5 cm distance depending on the operating power dissipation and the frequency of irradiation), and reduced activity is observed at locations away from the horn tip. Direct type of ultrasonic irradiation using probe is appropriate for laboratory-scale applications. In the indirect type of irradiation, application of ultrasound is executed in a reaction vessel submerged in ultrasonic cleaning tank-type reactors or flow cell-type configurations. In bath type of reactor configuration, ultrasonic transducers are located at the base of the reactor. Also there is no direct contact of the reaction mixture with the transducers eliminating the risk of possible metal contamination due to the possible erosion of the surface of transducers. The most active cavitational zone is found to be concentrated around the transducer and in the plane perpendicular to the transducer indicating vertical propagation of the incident energy. The different arrangements of the transducers in the case of ultrasonic bath include triangular, wedge shaped, square or hexagonal (Gogate 2008). Flow cell-type arrangement provides flexibility for both the modes of operation, i.e. batch and continuous mode. Moreover, since the transducers are situated at different sides of the flow cell, multi-frequency irradiation is also possible, which can give higher degree of cavitational effects at constant power dissipation (Kumar et al. 2007).

The two most important parameters deciding the extent of cavitational effects obtained in the case of ultrasonic reactors are the ultrasonic frequency and the power dissipation. Frequency of irradiation in particular is important in controlling the extent of physical and chemical effects being generated in the reactor. The use of higher frequencies (200−500 kHz) generates mostly chemical effects, whereas lower-frequency irradiation (20−50 kHz) gives dominant effects in terms of turbulence and/or liquid streaming beneficial for physical processing applications or reactions limited by mass transfer. For intensification of biodiesel production, as intense turbulence and microstreaming are essential, lower frequencies in the range of 20−50 kHz are recommended (Gole and Gogate 2012). Ultrasonic power influences the number of cavitational events as well as the collapse intensity (Stavarache et al. 2005) which in turn affects the magnitude of microstreaming and degree of turbulence. Typically higher ultrasonic power would be helpful in increasing the cavitational effects, but in the case of enzymatic reactions, optimum power dissipation needs to be used due to possible deactivation of the enzyme at higher power dissipation levels (Subhedar and Gogate 2013).The requirement of optimum power is also justified on the basis of decoupling effects which are dependent on the geometry of the reactor (Gole and Gogate 2012). In general, it is recommended to establish the actual value of the optimum power based on laboratory-scale investigations or literature analysis. It is also recommended to use higher areas of transducers for dissipating same power so as to obtain higher extent of cavitationally active volumes in the reactor.

5.1.2 Overview of Ultrasound-Assisted Enzymatic Biodiesel Production from Sustainable Feedstock

Several studies are available in the literature dealing with ultrasound-assisted enzymatic biodiesel production, and a brief summary has been given in Table 2. Important considerations have been also discussed now for providing a better understanding into the effects of ultrasound and its operating parameters.

Enzymatic biodiesel synthesis from sustainable source as waste oil having high content of FFA with intensification studies based on the use of ultrasound assisted approach was investigated by Wang et al. (2007). It was reported that the activity of Novozyme 435 used as the catalyst in the work was improved due to the use of ultrasound and rate of the enzyme-catalysed transesterification was also enhanced. It was also reported that low frequency and mild intensity are the crucial factors in improving enzyme activity, emulsifying oil-propanol system in a desired manner and enhancing the rate of diffusion in the reaction. Maximum conversion of 94.86 % was reported under the optimum operating parameters as 8 % lipase loading, oil to propanol molar ratio of 1:3, 28 kHz frequency, 100 W power and 40 °C temperature with a reaction time of 50 min. The maximum conversion obtained under the conventional mechanical agitation conditions was only 84.43 % confirming the degree of intensification due to ultrasound. It was also reported that several short-chain alcohols (in the range of C1–C5) resulted in good biodiesel yield under the optimum ultrasound -assisted approach. Novozyme 435 could also be reused after ten batches and did not show any structural or activity changes. The reported results confirmed that ultrasound was beneficial in enhancing the rates of reaction and at the same time did not give any adverse effect on the enzyme.

Combination of ultrasound with vibration for intensification of the biodiesel synthesis from soybean oil using Novozyme 435 as catalyst has also been investigated (Yu et al. 2010). The comparison of the biodiesel synthesis was also performed with approach of only ultrasonic irradiation and only vibration. The reported results revealed that ultrasonic irradiation improved the activity of Novozyme 435 to a higher extent as compared with the vibration. The combined operation of ultrasonic irradiation with vibration (UIV) resulted in further increase in the reaction rate giving maximum biodiesel yield of 96 % at the optimal conditions of 50 rpm as the speed of vibration, ultrasonic power of 220 W, water content of 0.5 %, 1:1 (v/v) ratio of tert-amyl alcohol/oil, molar ratio of methanol to oil of 6:1, Novozyme 435 loading of 6 % and temperature of 40 °C in 4 h of reaction time. Recycling of Novozyme 435 for five cycles was reported to be efficient without any loss in the enzyme activity which confirmed no adverse effect of UIV on the enzyme.

Table 2 Ultrasound-assisted enzymatic production of biodiesel

Feed oil	Enzyme source	Reaction conditions						References
		Oil to acyl acceptor ratio	Enzyme (wt %)	Frequency (kHz)	Power (W)	Time (min)	Yield (%)	
Waste oil	*Candida antarctica*	1:3	8	28	100	50	94.86	Wang et al. (2007)
Soybean oil	*Candida antarctica*	1:6	6	40	220	240	96	Yu et al. (2010)
Waste cooking oil	*Candida antarctica*	1:6	10	25	200	240	86.61	Gharat and Rathod (2013)
Nag Champa oil	*Candida antarctica*	1:3	3.4	20	50	450	96	Jadhav and Gogate (2014)
Soybean oil	*Candida antarctica*	1:6	10	40	132	180	81	Santin et al. (2014)
Jatropha curcas	*Chromobacterium viscosum*	1:5	5	24	100	30	84.5	Kumar et al. (2011)
Soybean oil	*Rhizomucormiehei*	1:3	5	37	100	240	90	Batistella et al. (2012)
Waste cooking oil	*Thermomyces lanuginose*	1:4	43.6	40	150	360	91	Karimi et al. (2013)
Sunflower oil	*Thermomyces lanuginosus*	1:3	7	40	132	240	96	Subhedar et al. (2015)
Waste cooking oil	*Thermomyces lanuginosus*	1:9	6	20	80	180	96	Subhedar and Gogate (2016)

Intensification of enzyme-catalysed synthesis of biodiesel from nonedible oil (*Jatropha curcas*) catalysed by immobilized form of lipase obtained from *Enterobacter aerogenes* based on the use of ultrasound has been studied by Kumar et al. (2011). The use of optimized parameters as molar ratio of methanol to oil of 4:1, lipase loading of 5.0 %, ultrasonic power density of 100 W/m^3 and duty cycle of 70 % resulted in maximum yield of biodiesel as 84.5 ± 0.5 % in 30 min. Application of ultrasound resulted in significant intensification with decrease in the time of reaction from 6 h to 30 min. Application of ultrasonic bath for intensification of enzyme-catalysed synthesis of biodiesel from soybean oil in organic solvent media has also been investigated (Batistella et al. 2012). The obtained results confirmed significant degree of intensification with maximum biodiesel yield of 90 % attained at low ultrasonic power of 100 W and 60 °C temperature in only 4 h using Lipozyme RM IM enzyme. One of the disadvantages of the process, however, was that repetitive usage of the enzyme caused a decline in the activity of enzyme and biodiesel yield in only two cycles.

Ultrasound-assisted enzyme-catalysed biodiesel synthesis from WCO using dimethyl carbonate (DMC) as acyl acceptor has also been investigated (Gharat and Rathod 2013) under solvent-free conditions using lipase obtained from *Candida antarctica* (Novozyme 435). Three different modes of operation, namely, stirring, irradiation using ultrasonic bath (UI) combined with stirring and UI without stirring, were used in the investigations. It was reported that the conventional stirring method resulted in 38.69 % conversion in 4 h, whereas only UI approach gave higher conversion as 57.68 % in 4 h. The use of the combined approach further increased the rate of reaction giving conversion of 86.6 % in the same time. The optimum conditions were reported as 25 kHz frequency, 200 W ultrasonic power, 100 rpm stirrer speed, molar ratio of DMC to oil as 6:1, Novozyme 435 loading of 10 % and 60 °C temperature. Reusability study confirmed reduction in the activity of enzyme as well as the obtained biodiesel yield.

Ultrasound-assisted biodiesel synthesis from WCO using lipase immobilized on the mesoporous silica/iron oxide nanoparticles has been reported with optimization studies (Karimi et al. 2013). The important parameters considered for initial analysis based on the response surface methodology (RSM) were methanol to WCO molar ratio, lipase loading, quantum of water and reaction time. The obtained data was subsequently used for genetic algorithm analysis, and the hybrid model established that the optimized conditions of oil to methanol molar ratio of 1:4, lipase loading of 43.6 %, water content of 10.22 % and time of 6 h would give maximum yield of 91 %. Experimental investigations confirmed this maximum expected yield and also established the reuse of immobilized enzyme for four cycles.

Jadhav and Gogate (2014) investigated the ultrasound-assisted enzymatic conversion of nonedible oils into biodiesel and established the comparison with the conventional approach. A two-stage processing of hydrolysis followed by esterification was used as a sustainable approach for processing feedstock with high free fatty acid content. A detailed study of reaction parameters revealed maximum yields at oil to water ratio of 1:1 (w/w) for Lip Z and 1:3 (w/w) for Lip 2 enzymes, 400 units loading for Lip Z and 800 mg loading for Lip 2 and 6 h of reaction time.

For esterification reaction, molar ratio of methanol to oil of 2:1, 1000 mg of enzyme loading and 20 h of reaction time were found to be the optimum conditions. Pretreated enzyme, i.e. enzyme subjected to ultrasonic irradiations, was observed to enhance the esterification yield from 75 % to 92.5 % though the time of reaction was not significantly altered. It has been also reported that application of ultrasonic irradiation in the reaction considerably enhanced the process reducing the reaction time from 20 h for the approach without ultrasound to 7.5 h for ultrasound-assisted approach. It was thus established that use of ultrasound during the reaction is more beneficial as compared to only pretreatment of enzyme using ultrasound.

Batch esterification of fatty acid charge using *Candida antarctica* B immobilized on polyurethane foam in the presence of ultrasound was investigated for biodiesel production so as to check the viability of polyurethane foam as a low-cost support for enzyme immobilization (Santin et al. 2014). For establishing the efficacy of low-cost support, batch experiments were performed under ultrasonic irradiation. Maximum biodiesel yield of 81 % was obtained at optimized conditions of 132 W output power, molar ratio of soybean fatty acid charge to ethanol as 1:6 and 10 wt% enzyme loading in 180 min reaction time. The obtained result confirmed that the use of polyurethane as a low-cost support for lipase immobilization was effective in giving high yields of biodiesel and this support can be a good alternative to the commercially available immobilized form (Novozyme 435) especially due to lower overall costs of the enzyme.

Subhedar et al. (2015) reported that application of ultrasound suppressed the need of using excess of methanol during the biodiesel production from sunflower oil in the presence of Lipozyme TL-IM enzyme. Also in comparison to the conventional approach, it has been reported that the application of ultrasound considerably increased the yield of biodiesel. Optimal reaction conditions for the conventional method were methanol to oil molar ratio of 4:1, 7 % (w/v) enzyme loading, 40 °C temperature and reaction time of 24 h, and at these optimum conditions, 93.3 % yield of biodiesel was achieved. For ultrasound-assisted method, optimal conditions were molar ratio of methanol to oil of 3:1, enzyme loading of 3 % (w/v) and reaction time of 4 h, and 96 % yield of biodiesel was achieved under these conditions. The results confirmed that application of ultrasound offered great benefits in terms of decrease in reaction time, enzyme loading and excess of reactants.

Application of ultrasonic irradiation for the interesterification of waste cooking oil with acyl acceptor as methyl acetate in the presence of immobilized lipase obtained from *Thermomyces lanuginosus* (Lipozyme TLIM) has also been recently reported by Subhedar and Gogate (2016). The work also reported experiments using the conventional approach based on stirring in order to establish the degree of intensification attained due to the use of ultrasonic irradiation. For conventional approach, maximum biodiesel yield of 90.1 % in 24 h was reported using methyl acetate to oil molar ratio of 12:1, 6 % (w/v) enzyme loading and reaction temperature of 40 °C. The optimum operating conditions for the ultrasound-assisted method were methyl acetate to oil molar ratio of 9:1, 3 % (w/v) of enzyme loading, 80 W of ultrasonic power and 3 h of reaction time, and the biodiesel yield achieved under these conditions was 96.1 %. Application of ultrasound resulted in substantial reduction in the reaction time with higher yields and lesser requirement of the enzyme, which can help in decreasing the costs of production. Enzyme reusability studies were also conducted; however, it has been reported that reuse of the Lipozyme TLIM at the optimal reaction conditions gave decreased activity of enzyme and biodiesel yield.

The discussion on the ultrasound-assisted enzymatic biodiesel production from various feedstocks has confirmed the efficacy and outstanding benefits of the ultrasonic reactors. The critical parameters that must be appropriately chosen for maximizing the degree of intensification consist of reactor configuration, ultrasonic power, frequency of irradiation and sonication time. Also, it is important to properly decide the reaction parameters of oil to alcohol molar ratio, enzyme type and loading, operating temperature, etc. The main advantages of optimized operation using ultrasound are reduction in reaction time and enzyme loading as well as reduced requirement of excess alcohol for driving the reaction in the desired direction. Mild reaction conditions for shorter reaction time also improve the quality of the produced biodiesel due to the reduction in the side reactions (Sharma et al. 2008). The overall effect would be an economically feasible operation with reduced operating costs for the efficient production of biodiesel.

5.2 *Microwave Reactors for Intensification of Biodiesel Production*

Microwaves are fundamentally nonionizing radiations with frequency in the range of 300 MHz and 300 GHz comprising of both magnetic and electric constituents which work perpendicular to one another and to the path of transmission. Microwaves lie in the middle of radio waves and infrared emissions having wavelengths between 1 mm and 1 m (Lidström et al. 2001). In microwave-assisted reactions, irradiation of microwave aids in transmitting the energy right to the starting materials at a significant rate and enhances the reaction rate, decreasing the necessary reaction time from several hours to few minutes. Microwaves can directly interact with the reactants by two mechanisms which are dipole rotation and ion migration to cause confined and quick heating of the reaction ingredients (Gude et al. 2013). Once irradiated with microwave, dipoles in the reactants arrange in the direction of application of electric field. Since the applied electric field swiftly fluctuates, the dipole attempts to quickly rearrange to the applied electric field by means of revolution. Generation of high temperature occurs through the force of friction among the arbitrarily spinning polar particles as well as the adjacent medium. In the case of ion migration, back and forth oscillation of the dissolved charged particles occurs due to the effect of microwave. As soon as the direction of electrical field alters, the bigger ions disperse their kinematic energy as temperature shock due to the fact that they get retarded and alter the directions creating frictions at molecular levels. Both these mechanisms result in high temperature and pressure gradients (Marra et al.

2010) giving enhanced rates of reaction. For biodiesel synthesis, the transesterification reaction mixture comprises oil and alcohol having polar as well as ionic constituents. Application of microwave can result in significant heating of the substrates to the necessary temperature rapidly and effectively (Barnard et al. 2007). The heating times required in the case of microwave heating are significantly lower than that required for the conventional heating where the transfer of heat to the reactants is by the conventional modes of conduction, convection and radiation from the surface. Due to this reduction in the heating time and obtained uniform heating, the required reaction times for the microwave processing are reduced (Yaakob et al. 2009).

Da Rós et al. (2012) studied the intensification of enzymatic transesterification of beef tallow using the microwave-assisted approach with an objective to decrease the reaction time. Lipase obtained from *Burkholderia cepacia* and immobilized on silica-PVA was used as a catalyst, and the effect of microwave irradiations over power dissipation range of 8–15 W on the time required to reach the desired reaction temperature has been investigated. Under optimum conditions of molar ratio of ethanol to beef tallow of 6:1 and temperature of 50 °C, nearly complete conversion of the fatty acid into biodiesel was observed in 8 h. Almost sixfold intensification in the rate of biodiesel production was obtained by using microwave irradiation as compared to conventional heating. It has been also reported that the process requires low energy and at the same time gives higher biodiesel productivity. The other reported benefits were that use of microwave irradiation accelerates the enzyme-catalysed reactions, reduces the negative effects on the enzyme such as exposure to localized higher temperatures and specificity to its substrate and allows consistent heating of the whole reaction medium.

Microwave activation of enzyme for the synthesis of biodiesel has been optimized for the case of macauba oil as starting material and Novozyme 435 and Lipozyme IM enzymes as biocatalysts based on the use of statistical design of experiments (Nogueira et al. 2010). It has been reported that the enzyme loading played a significant role, whereas the effect of temperature influenced the reaction rate only by means of interactions with other variables. The comparison of obtained results for the enzyme activity in the presence of microwave and the conventional approach confirmed beneficial results obtained due to the use of microwave.

Microwave-intensified synthesis of biodiesel from palm oil using the lipase from *Pseudomonas fluorescens* immobilized on hybrid composites of epoxy polysiloxane-polyvinyl alcohol has been reported (Da Rós et al. 2013). The effect of power dissipation of microwave over the range of 8–15 W was investigated 97.65 % biodiesel yield was reported under optimal molar ratio of ethanol to oil of 8:1 at 43 °C in 12 h of reaction time. Sixfold enhancement in the biodiesel production rate was reported as compared to the reaction without microwave irradiation, and hence microwave has been demonstrated to be a promising tool for improving the biochemical alteration of oils and fats. Overall, benefits of the microwave-assisted processing reported are (1) microwaves accelerate the enzyme-catalysed reactions; (2) there are no negative effects on the enzyme properties, for instance, substrate specificity and stability; and (3) the microwave assistance lets the complete reaction volume to be heated consistently.

The efficacy of microwave irradiation (MW) can be further enhanced by possible combination with ionic liquids (ILs) though a real analysis in terms of cost against benefits need to be performed especially at the commercial scale of operation. Combination of microwave and ILs has also been reported to intensify the enzyme-catalysed biodiesel production (Yu et al. 2011). The use of [EMIM][PF6] in combination with microwave resulted in about 1.8-folds higher activity for the lipase as compared to the use of microwave and tert-butanol. Under the optimized conditions of molar ratio of oil to methanol of 1:6, power dissipation of 480 W, [EMIM][PF6] to oil volume ratio of 2:1, Novozyme 435 loading of 6 % based on soybean oil weight, 200 rpm as stirring speed and 60 °C temperature, maximum biodiesel yield of 92 % was obtained in a reaction time of 6 h. The use of [EMIM][PF6] without MW and only MW as solvent-free system resulted in 70 % and 28 % biodiesel yield, respectively. The obtained results indicated that the integrated approach of using MW and IL resulted in significant increase in the activity of enzyme, which can be attributed to the fact that microwave acts right on the IL owing to its outstanding microwave-absorbing capability. Additionally and most importantly, the recovered IL and enzyme were reported to be reused for five consecutive reaction cycles with almost similar enzyme activity indicating good prospect for commercial processing.

6 Conclusions and Future Prospects

Lipase-catalysed production of biodiesel offers a huge prospect to produce an environment-friendly fuel as biodiesel and most importantly in an economical manner. Different lipases from various sources immobilized on different supports have revealed a range of yields with different oil substrates and acyl acceptors. The significant operating parameters associated with the biodiesel production that need to be optimized include reactant molar ratio, operating temperature and type of enzyme and its concentration. The guidelines presented in the chapter would be useful to select a feasible processing approach with optimized parameters giving maximum benefits. Using lipase as catalysts in the immobilized form offers great advantage of catalyst recovery and reuse. The application of ultrasound for enzymatic production of biodiesel gives significant process intensification benefits and allows processing from sustainable feedstock. A detailed optimization of the ultrasound parameters which is essential for maintaining the activity of enzyme can give significant benefits in terms of operation at reduced reaction temperature, lower reaction time and lower excess of alcohol to oil ratio which can give substantial decrease in the biodiesel production cost. An overview of application of ultrasonic irradiation presented for a range of different immobilized lipases with several feedstocks should give credence to the suitability of application with confirmed benefits when compared to the enzyme-catalysed biodiesel production without ultrasonic irradiation. Similar to the use of ultrasound, microwave has also been demonstrated to give process intensification benefits due to mostly the thermal effects. In general, it has been established that the application of ultrasound as well as microwave provides significant potential for the intensified biodiesel production based on the enzymatic route. Finally, as there has not been much studies based on continuous mode of operation, it is also recommended to direct future studies to incorporate the process intensification approaches, for instance, ultrasound coupled with other methods as applicable in a continuous mode of production so as to achieve economical production of biodiesel at the required scales of operation.

References

Adamczak M, Bornscheuer UT, Bednarski W (2009) The application of biotechnological methods for the synthesis of biodiesel. Eur J Lipid Sci Technol 111:800–813. doi:10.1002/ejlt.200900078

Alcantara R, Amores J, Canoira L, Fidalgo E, Franco MJ, Navarro A (2000) Catalytic production of biodiesel from soy-bean oil, used frying oil and tallow. Biomass Bioenergy 18:515–527. doi:10.1016/S0961-9534(00)00014-3

Andrade IC, Vargas FES, Fajardo CAG (2013) Biodiesel production from waste cooking oil by enzymatic catalysis process. J Chem Chem Eng 7:993–1000

Antczak MS, Kubiak A, Antczak T, Bielecki S (2009) Enzymatic biodiesel synthesis-key factors affecting efficiency of the process. Renew Energy 34:1185–1194. doi:10.1016/j.renene.2008.11.013

Aransiola EF (2013) Lipase catalysed ethanolysis of jatropha oil for biodiesel production. Energy Environ Res 3:85–92. doi:10.5539/eer.v3n1p8

Azocar L, Ciudad G, Heipieper HJ, Munoz R, Navia R (2010) Improving fatty acid methyl ester production yield in a lipase-catalyzed process using waste frying oils as feedstock. J Biosci Bioeng 109:609–614. doi:10.1016/j.jbiosc.2009.12.001

Barnard TM, Leadbeater NE, Boucher MB, Stencel LM, Wilhite BA (2007) Continuous-flow preparation of biodiesel using microwave heating. Energy Fuels 21:1777–1781. doi:10.1021/ef0606207

Basha SA, Gopal KR, Jebaraj S (2009) A review on biodiesel production, combustion, emissions and performance. Renew Sustain Energy Rev 13:1628–1634. doi:10.1016/j.rser.2008.09.031

Batistella L, Lerin LA, Brugnerotto P, Danielli AJ, Trentin CM, Popiolski A, Treichel H, Oliveira JV, De Oleveira D (2012) Ultrasound-assisted lipase-catalyzed transesterification of soybean oil in organic solvent system. Ultrason Sonochem 19:452–458. doi:10.1016/j.ultsonch.2011.11.018

Belafi-Bako K, Kovacs F, Gubicza L, Hancsok J (2002) Enzymatic biodiesel production from sunflower oil by Candida Antarctica Lipase in a solvent-free system. Biocatal Biotransformation 20:437–439. doi:10.1080/1024242021000040855

Bournay L, Casanave D, Delfort B, Hillion G, Chodorge JA (2005) New heterogeneous process for biodiesel production: a way to improve the quality and the value of the crude glycerin produced by biodiesel plants. Catal Today 106:190–192. doi:10.1016/j.cattod.2005.07.181

Cesarini S, Infanzon B, Pastor FIJ, Diaz P (2014) Fast and economic immobilization methods described for non-commercial Pseudomonas lipases. BMC Biotechnol 14:1–9. doi:10.1186/1472-6750-14-27

Charpe TW, Rathod VK (2011) Biodiesel production using waste frying oil. Waste Manag 31: 85–90. doi:10.1016/j.wasman.2010.09.003

Chen G, Ying M, Li W (2006) Enzymatic conversion of waste cooking oils into alternative fuel-biodiesel. Appl Biochem Biotechnol 129–132:911–921. doi:10.1385/ABAB:132:1:911

Chen Y, Xiao B, Chang J, Fu Y, Lv P, Wang X (2009) Synthesis of biodiesel from waste cooking oil using immobilized lipase in fixed bed reactor. Energy Convers Manage 50:668–673. doi:10.1016/j.enconman.2008.10.011

da Cunha ME, Krause LC, Moraes MSA, Faccini CS, Jacques RA, Almeida SR, Rodrigues MRA, Cramao EB (2009) Beef tallow biodiesel produced in a pilot scale. Fuel Process Technol 90: 570–575. doi:10.1016/j.fuproc.2009.01.001

Da Rós PCM, de Castro HF, Carvalho AKF, Soares CMF, de Moraes FF, Zanin GM (2012) Microwave-assisted enzymatic synthesis of beef tallow biodiesel. J Ind Microbiol Biotechnol 39:529–536. doi:10.1007/s10295-011-1059-8

Da Rós PCM, Freitas L, Perez VH, de Castro HF (2013) Enzymatic synthesis of biodiesel from palm oil assisted by microwave irradiation. Bioprocess Biosyst Eng 36:443–451. doi:10.1007/s00449-012-0801-6

Darnoko D, Cheryan M (2000) Kinetics of palm oil transesterification in a batch reactor. J Am Oil Chem Soc 77:1263–1267. doi:10.1007/s11746-000-0198-y

de Oliveira D, Di Luccio M, Faccio C, Rosa CD, Bender JP, Lipke N, Menoncin S, Amroginski C, de Oliveira JV (2004) Optimization of enzymatic production of biodiesel from castor oil in organic solvent medium. Appl Biochem Biotechnol 113–116:771–780. doi:10.1385/ABAB:115:1-3:0771

Demirbas A (2008) Comparison of transesterification methods for production of biodiesel from vegetable oils and fats. Energy Conserv Manage 49:125–130. doi:10.1016/j.enconman.2007.05.002

Derewenda U, Brzozowski AM, Lawson DM, Derewenda ZS (1992) Catalysis at the interface: the anatomy of a conformational change in a triglyceride lipase. Biochemistry 31:1532–1541. doi:10.1021/bi00120a034

Enweremadu CC, Mbarawa MM (2009) Technical aspects of production and analysis of biodiesel from used cooking oil-a review. Renew Sustain Energy Rev 13:2205–2224. doi:10.1016/j.rser.2009.06.007

Felizardo P, Neiva Correia MJ, Raposo I, Mendes JF, Berkmeier R, Bordado JM (2006) Production of biodiesel from waste frying oils. Waste Manage 26:487–494. doi:10.1016/j.wasman.2005.02.025

Fjerbaek L, Christensen KV, Norddahl B (2009) A review of the current state of biodiesel production using enzymatic transesterification. Biotechnol Bioeng 102:1298–1315. doi:10.1002/bit.22256

Fukuda H, Kondo A, Noda H (2001) Biodiesel fuel production by transesterification of oils. J Biosci Bioeng 92:405–416. doi:10.1016/S1389-1723(01)80288-7

Furukawa S, Uehara Y, Yamasaki H (2010) Variables affecting the reactivity of acid-catalyzed transesterification of vegetable oil with methanol. Bioresour Technol 101:3325–3332. doi:10.1016/j.biortech.2009.12.086

Georgogianni KG, Kontominas MG, Pomonis PJ, Avlonitis D, Gergis V (2008) Conventional and in situ transesterification of sunflower seed oil for the production of biodiesel. Fuel Process Technol 89:503–509. doi:10.1016/j.fuproc.2007.10.004

Ghadge SV, Raheman H (2006) Process optimization for biodiesel production from mahua (*Madhuca indica*) oil using response surface methodology. Bioresour Technol 97:379–384. doi:10.1016/j.biortech.2005.03.014

Gharat N, Rathod VK (2013) Ultrasound assisted enzyme catalyzed transesterification of waste cooking oil with dimethyl carbonate. Ultrason Sonochem 20:900–905. doi:10.1016/j.ultsonch.2012.10.011

Gog A, Roman M, Toşa M, Paizs C, Irimie FD (2012) Biodiesel production using enzymatic transesterification—current state and perspectives. Renew Energy 39:10–16. doi:10.1016/j.renene.2011.08.007

Gogate PR (2008) Cavitational reactors for process intensification of chemical processing applications: a critical review. Chem Eng Process Process Intensif 47:515–527. doi:10.1016/j.cep.2007.09.014

Gogate PR, Sutkar VS, Pandit AB (2011) Sonochemical reactors: important design and scale up considerations with a special emphasis on heterogeneous systems. Chem Eng J 166:1066–1082. doi:10.1016/j.cej.2010.11.069

Gole VL, Gogate PR (2012) A review on intensification of synthesis of biodiesel from sustainable feed stock using sonochemical reactors. Chem Eng Process Process Intensif 53:1–9. doi:10.1016/j.cep.2011.12.008

Gude V, Patil P, Martinez-Guerra E, Deng S, Nirmalkhandan N (2013) Microwave energy potential for biodiesel production. Sustain Chem Process 1:1–31. doi:10.1186/2043-7129-1-5

Gupta R, Gupta N, Rathi P (2004) Bacterial lipases: an overview of production, purification and biochemical properties. Appl Microbiol Biotechnol 64:763–781. doi:10.1007/s00253-004-1568-8

Halim SFA, Kamaruddin AH, Fernando WJN (2009) Continuous biosynthesis of biodiesel from waste cooking palm oil in a packed bed reactor: optimization using response surface methodology (RSM) and mass transfer studies. Bioresour Technol 100:710–716. doi:10.1016/j.biortech.2008.07.031

Hsu AF, Jones K, Marmer WN, Foglia TA (2001) Production of alkyl esters from tallow and grease using lipase immobilized in a phyllosilicate sol-gel. J Am Oil Chem Soc 78:585–588. doi:10.1007/s11746-001-0308-x

Huang Y, Zheng H, Yan Y (2010) Optimization of lipase-catalyzed transesterification of lard for biodiesel production using response surface methodology. Appl Biochem Biotechnol 160: 504–515. doi:10.1007/s12010-008-8377-y

Jadhav SH, Gogate PR (2014) Ultrasound assisted enzymatic conversion of non edible oil to methyl esters. Ultrason Sonochem 21:1374–1381. doi:10.1016/j.ultsonch.2013.12.018

Jaeger KE, Ransac S, Dijkstra BW, Colson C, van Heuvel M, Misset O (1994) Bacterial lipases. FEMS Microbiol Rev 15:29–63. doi:10.1016/0168-6445(94)90025-6

Jaeger KE, Reetz MT (1998) Microbial lipases form versatile tools for biotechnology. Trends Biotechnol 16:396–403. doi:10.1016/S0167-7799(98)01195-0

Jain S, Sharma MP (2010) Prospects of biodiesel from Jatropha in India: a review. Renew Sustain Energy Rev 14:763–771. doi:10.1016/j.rser.2009.10.005

Jena PC, Raheman H, Prasanna Kumar GV, Machavaram R (2010) Biodiesel production from mixture of mahua and simarouba oils with high free fatty acids. Biomass Bioenergy 34: 1108–1116. doi:10.1016/j.biombioe.2010.02.019

Ji J, Wang J, Li Y, Yu Y, Xu Z (2006) Preparation of biodiesel with the help of ultrasonic and hydrodynamic cavitation. Ultrasonics 44:e411–e414. doi:10.1016/j.ultras.2006.05.020

Kaieda M, Samukawa T, Kondo A, Fukuda H (2001) Effect of methanol and water contents on production of biodiesel fuel from plant oil catalyzed by various lipases in a solvent-free system. J Biosci Bioeng 91:12–15. doi:10.1016/S1389-1723(01)80103-1

Karimi M, Keyhani A, Akram A, Rahman M, Jenkins B, Stroeve P (2013) Hybrid response surface methodology-genetic algorithm optimization of ultrasound-assisted transesterification of waste oil catalysed by immobilized lipase on mesoporous silica/iron oxide magnetic core-shell nanoparticles. Environ Technol 34:2201–2211. doi:10.1080/09593330.2013.837939

Kinney AJ, Clemente TE (2005) Modifying soybean oil for enhanced performance in biodiesel blends. Fuel Process Technol 86:1137–1147. doi:10.1016/j.fuproc.2004.11.008

Kulkarni MG, Dalai AK (2006) Waste cooking oil—an economic source for biodiesel: a review. Ind Eng Chem Res 45:2901–2913. doi:10.1021/ie0510526

Kumar A, Gogate PR, Pandit AB (2007) Mapping the efficacy of new designs for large scale sonochemical reactors. Ultrason Sonochem 14:538–544. doi:10.1016/j.ultsonch.2006.11.005

Kumar A, Sharma S (2011) Potential non-edible oil resources as biodiesel feedstock: an Indian perspective. Renew Sustain Energy Rev 15:1791–1800. doi:10.1016/j.rser.2010.11.020

Kumar G, Kumar D, Poonam JR, Singh CP (2011) Enzymatic transesterification of *Jatropha curcas* oil assisted by ultrasonication. Ultrason Sonochem 18:923–927. doi:10.1016/j.ultsonch.2011.03.004

Kumar S, Ghaly AE, Brooks MS (2015) Fundamentals of renewable energy and applications production of biodesiel from animal tallow via enzymatic transesterification using the enzyme catalyst Ns88001 with methanol in a solvent-free system. Fundam Renew Energy Appl 5:1–8. doi:10.4172/20904541.1000156

Kumari A, Mahapatra P, Garlapati VK, Banerjee R (2009) Enzymatic transesterification of Jatropha oil. Biotechnol Biofuels 2:1–7. doi:10.1186/1754-6834-2-1

Kumari V, Shah S, Gupta MN (2007) Preparation of biodiesel by lipase-catalyzed transesterification of high free fatty acid containing oil from Madhuca indica. Energy Fuel 21:368–372. doi:10.1021/ef0602168

Leduc S, Natarajan K, Dotzauer E, McCallum I, Obersteiner M (2009) Optimizing biodiesel production in India. Appl Energy 86:S125–S131. doi:10.1016/j.apenergy.2009.05.024

Lee JH, Kim SB, Yoo HY, Suh YJ, Kang GB, Jang WI, Kang J, Park C, Kim SW (2013) Biodiesel production by enzymatic process using *Jatropha* oil and waste soybean oil. Biotechnol Bioprocess Eng 18:703–708. doi:10.1007/s12257-012-0805-8

Lee KT, Foglia TA, Chang KS (2002) Production of alkyl ester as biodiesel from fractionated lard and restaurant grease. J Am Oil Chem Soc 79:191–195. doi:10.1007/s11746-002-0457-y

Lee SB, Lee JD, Hong IK (2011) Ultrasonic energy effect on vegetable oil based biodiesel synthetic process. J Ind Eng Chem 17:138–143. doi:10.1016/j.jiec.2010.12.012

Li NW, Zong MH, Wu H (2009) Highly efficient transformation of waste oil to biodiesel by immobilized lipase from *Penicillium expansum*. Process Biochem 44:685–688. doi:10.1016/j.procbio.2009.02.012

Lidström P, Tierney J, Wathey B, Westman J (2001) Microwave assisted organic synthesis—a review. Tetrahedron 57:9225–9283. doi:10.1016/S0040-4020(01)00906-1

Liu KS (1994) Preparation of fatty acid methyl esters for gas-chromatographic analysis of lipids in biological materials. J Am Oil Chem Soc 71:1179–1187. doi:10.1007/BF02540534

Lotero E, Liu Y, Lopez DE, Suwannakarn K, Bruce DA, Goodwin JG (2005) Synthesis of biodiesel via acid catalysis. Ind Eng Chem Res 44:5353–5363. doi:10.1021/ie049157g

Lu J, Nie K, Xie F, Wang F, Tan T (2007) Enzymatic synthesis of fatty acid methyl esters from lard with immobilized *Candida* sp. 99-125. Process Biochem 42:1367–1370. doi:10.1016/j.procbio.2007.06.004

Lukovic N, Kneevic-Jugovic Z, Bezbradic D (2011) Biodiesel fuel production by enzymatic transesterification of oils: recent trends, challenges and future perspectives. In: Manzanera M (ed) Alternative fuel, pp 47–72. doi:10.5772/21905

Ma F, Clements LD, Hanna MA (1998) Biodiesel fuel from animal fat. Ancillary studies on transesterification of beef tallow. Ind Eng Chem Res 5885:3768–3771. doi:10.1021/ie980162s

Ma F, Hanna MA (1999) Biodiesel production: a review. Bioresour Technol 70:1–15. doi:10.1016/S0960-8524(99)00025-5

Maceiras R, Rodríguez M, Cancela A, Urrejola S, Sanchez A (2011) Macroalgae: raw material for biodiesel production. Appl Energy 88:3318–3323. doi:10.1016/j.apenergy.2010.11.027

Manh DV, Chen YH, Chang CC, Chang MC, Chang CY (2011) Biodiesel production from Tung oil and blended oil via ultrasonic transesterification process. J Taiwan Inst Chem Eng 42: 640–644. doi:10.1016/j.jtice.2010.11.010

Marchetti JM, Errazu AF (2008) Technoeconomic study of supercritical biodiesel production plant. Energy Convers Manage 49:2160–2164. doi:10.1016/j.enconman.2008.02.002

Marchetti JM, Miguel VU, Errazu AF (2007) Possible methods for biodiesel production. Renew Sustain Energy Rev 11:1300–1311. doi:10.1016/j.rser.2005.08.006

Marra F, De Bonis MV, Ruocco G (2010) Combined microwaves and convection heating: a conjugate approach. J Food Eng 97:31–39. doi:10.1016/j.jfoodeng.2009.09.012

Meneghetti SMP, Meneghetti MR, Wolf CR, Silva EC, Lima GES, Silva L, Serra TM, Cauduro F, de Oliveira LG (2006) Biodiesel from castor oil: a comparison of ethanolysis versus methanolysis. Energy Fuels 20:2262–2265. doi:10.1021/ef060118m

Meng X, Chen G, Wang Y (2008) Biodiesel production from waste cooking oil via alkali catalyst and its engine test. Fuel Process Technol 89:851–857. doi:10.1016/j.fuproc.2008.02.006

Naik M, Meher LC, Naik SN, Das LM (2008) Production of biodiesel from high free fatty acid Karanja (*Pongamia pinnata*) oil. Biomass Bioenergy 32:354–357. doi:10.1016/j.biombioe.2007.10.006

No SY (2011) Inedible vegetable oils and their derivatives for alternative diesel fuels in CI engines: a review. Renew Sustain Energy Rev 15:131–149. doi:10.1016/j.rser.2010.08.012

Nogueira BM, Carretoni C, Cruz R, Freitas S, Melo PA Jr, Costa-Felix R, Pinto JC, Nele M (2010) Microwave activation of enzymatic catalysts for biodiesel production. J Mol Catal B Enzym 67:117–121. doi:10.1016/j.molcatb.2010.07.015

Ognjanovic N, Bezbradica D, Knezevic-Jugovic Z (2009) Enzymatic conversion of sunflower oil to biodiesel in a solvent-free system: process optimization and the immobilized system stability. Bioresour Technol 100:5146–5154. doi:10.1016/j.biortech.2009.05.068

Phan AN, Phan TM (2008) Biodiesel production from waste cooking oils. Fuel 87:3490–3496. doi:10.1016/j.fuel.2008.07.008

Ramadhas AS, Jayaraj S, Muraleedharan C (2005) Biodiesel production from high FFA rubber seed oil. Fuel 84:335–340. doi:10.1016/j.fuel.2004.09.016

Refaat AA, Attia NK, Sibak HA, Sheltawy E, ElDiwani GI (2007) Production optimization and quality assessment of biodiesel from waste vegetable oil. Int J Environ Sci Technol 5:75–82. doi:10.1007/BF03325999

Royon D, Daz M, Ellenrieder G, Locatelli S (2007) Enzymatic production of biodiesel from cotton seed oil using t-butanol as a solvent. Bioresour Technol 98:648–653. doi:10.1016/j.biortech.2006.02.021

Saka S, Kusdiana D (2001) Biodiesel fuel from rapeseed oil as prepared in supercritical methanol. Fuel 80:225–231. doi:10.1016/S0016-2361(00)00083-1

Santin CMT, Scherer RP, Nyari NLD, Rosa CD, Dallago RM, de Oliveira D, Oliveira JM (2014) Batch esterification of fatty acids charges under ultrasound irradiation using *Candida antarctica* B immobilized in polyurethane foam. Biocatal Agric Biotechnol 3:90–94. doi:10.1016/j.bcab.2014.02.005

Schuchardt U, Sercheli R, Matheus R (1998) Transesterification of vegetable oils: a review. J Braz Chem Soc 9:199–210. doi:10.1590/S0103-50531998000300002

Shah S, Sharma S, Gupta MN (2004) Biodiesel preparation by lipase-catalyzed transesterification of Jatropha oil. Energy Fuel 18:154–159. doi:10.1021/ef030075z

Sharma YC, Singh B, Upadhyay SN (2008) Advancements in development and characterization of biodiesel: a review. Fuel 87:2355–2373. doi:10.1016/j.fuel.2008.01.014

Stavarache C, Vinatoru M, Nishimura R, Maeda Y (2005) Fatty acids methyl esters from vegetable oil by means of ultrasonic energy. Ultrason Sonochem 12:367–372. doi:10.1016/j.ultsonch.2004.04.001

Subhedar PB, Botelho C, Ribeiro A, Castro R, Pereira MA, Gogate PR, Cavaco-Paulo A (2015) Ultrasound intensification suppresses the need of methanol excess during the biodiesel production with Lipozyme TL-IM. Ultrason Sonochem 27:530–535. doi:10.1016/j.ultsonch.2015.04.001

Subhedar PB, Gogate PR (2013) Intensification of enzymatic hydrolysis of lignocellulose using ultrasound for efficient bioethanol production: a review. Ind Eng Chem Res 52:11816–11828. doi:10.1021/ie401286z

Subhedar PB, Gogate PR (2016) Ultrasound assisted intensification of biodiesel production using enzymatic interesterification. Ultrason Sonochem 29:67–75. doi:10.1016/j.ultsonch.2015.09.006

Tan T, Lu J, Nie K, Deng L, Wang F (2010) Biodiesel production with immobilized lipase: a review. Biotechnol Adv 28:628–634. doi:10.1016/j.biotechadv.2010.05.012

Thanh LT, Okitsu K, Van Boi L, Maeda Y (2012) Catalytic Technologies for biodiesel fuel production and utilization of glycerol: a review. Catalysts 2:191–222. doi:10.3390/catal2010191

Tiwari AK, Kumar A, Raheman H (2007) Biodiesel production from jatropha oil (Jatropha curcas) with high free fatty acids: an optimized process. Biomass Bioenergy 31:569–575. doi:10.1016/j.biombioe.2007.03.003

Tomasevic AV, Siler-Marinkovic SS (2003) Methanolysis of used frying oil. Fuel Process Technol 81:1–6. doi:10.1016/S0378-3820(02)00096-6

Vasudevan PT, Briggs M (2008) Biodiesel production—current state of the art and challenges. J Ind Microbiol Biotechnol 35:421–430. doi:10.1007/s10295-008-0312-2
Veljković VB, Lakićević SH, Stamenković OS, Todorovic ZB, Lazic ML (2006) Biodiesel production from tobacco (*Nicotiana tabacum* L.) seed oil with a high content of free fatty acids. Fuel 85:2671–2675. doi:10.1016/j.fuel.2006.04.015
Vyas AP, Verma JL, Subrahmanyam N (2010) A review on FAME production processes. Fuel 89: 1–9. doi:10.1016/j.fuel.2009.08.014
Wang JX, De Huang Q, Huang FH, Wang JW, Huang QJ (2007) Lipase-catalyzed production of biodiesel from high acid value waste oil using ultrasonic assistant. Chin J Biotechnol 23: 1121–1128. doi:10.1016/S1872-2075(07)60067-3
Wang Y, Ou S, Liu P, Xue F, Tang S (2006) Comparison of two different processes to synthesize biodiesel by waste cooking oil. J Mol Catal A Chem 252:107–112. doi:10.1016/j.molcata. 2006.02.047
Wu X, Leung DYC (2011) Optimization of biodiesel production from camelina oil using orthogonal experiment. Appl Energy 88:3615–3624. doi:10.1016/j.apenergy.2011.04.041
Xie WL, Wang JL (2014) Enzymatic production of biodiesel from soybean oil by using immobilized lipase on Fe3O4/Poly(styrene-rnethacrylic acid) magnetic microsphere as a biocatalyst. Energy Fuels 28:2624–2631. doi:10.1021/Ef500131s
Xu Y, Du W, Liu D, Zeng J (2003) A novel enzymatic route for biodiesel production from renewable oils in a solvent-free medium. Biotechnol Lett 25:1239–1241. doi:10.1023/ A:1025065209983
Yaakob Z, Ong BH, Satheesh Kumar MN, Kamarudin SK (2009) Microwave-assisted transesterification of jatropha and waste frying palm oil. Int J Sustain Energy 28:195–201. doi:10. 1080/14786450903161006
Yagiz F, Kazan D, Akin AN (2007) Biodiesel production from waste oils by using lipase immobilized on hydrotalcite and zeolites. Chem Eng J 134:262–267. doi:10.1016/j.cej.2007.03.041
Yang T, Xu X, He C, Li L (2003) Lipase-catalyzed modification of lard to produce human milk fat substitutes. Food Chem 80:473–481. doi:10.1016/S0308-8146(02)00315-1
Yu CY, Huang LY, Kuan IC, Lee SL (2013) Optimized production of biodiesel from waste cooking oil by lipase immobilized on magnetic nanoparticles. Int J Mol Sci 14:24074–24086. doi:10. 3390/ijms141224074
Yu D, Tian L, Wu H, Wang S, Wang Y, Ma D, Fang X (2010) Ultrasonic irradiation with vibration for biodiesel production from soybean oil by Novozym 435. Process Biochem 45:519–525. doi:10. 1016/j.procbio.2009.11.012
Yu D, Wang C, Yin Y, Zhang A, Gao G, Fang X (2011) A synergistic effect of microwave irradiation and ionic liquids on enzyme-catalyzed biodiesel production. Green Chem 13:1869–1875. doi:10. 1039/c1gc15114b
Zhang Y, Dubé MA, McLean DD, Kates M (2003) Biodiesel production from waste cooking oil: 1. Process design and technological assessment. Bioresour Technol 89:1–16. doi:10.1016/ S0960-8524(03)00040-3

An Integrated Approach for Efficient Energy Recovery Production from Livestock and Agro-Industrial Wastes

Sandra Silva, Ana Cristina Rodrigues, Ana Ferraz, and Joaquim Alonso

Abstract The supply and food safety needs of the growing population, particularly in (peri) urban areas, promote the intensification, concentration, and specialization of human, agricultural, livestock, and agro-industrial activities. As a result, important quantities of waste and wastewater, with significant organic and nutrient loads, are generated and need to be managed properly in order to protect the soil quality and fertility, as well as to prevent water and air pollution. Among the models, processes, treatment, and recovery technologies, anaerobic digestion can be highlighted due to the production of biogas, a renewable energy source.

The understanding of the spatiotemporal patterns of the resources and the conditions of waste and wastewater production and use provided by the analysis of these complex and adaptive socio-ecological systems, together with the knowledge of the biological process of biogas production, will support the selection and optimization of technologies, as well as the sizing and location of biogas plants for the promotion of renewable energy and local level economies. This integrated, hierarchical, and multidisciplinary approach takes a critical nature in the definition of technical and organizational solutions that contribute to sustainable development, for the effective quality of life, the local environment, and economy.

S. Silva (✉)
School of Agriculture, Polytechnic Institute of Viana do Castelo, Refóios, 4990-706 Ponte de Lima, Portugal

Institute for Systems Engineering and Computers, Rua Antero de Quental, Nº199, 3000-033 Coimbra, Portugal
e-mail: sandrasilva@esa.ipvc.pt

A.C. Rodrigues • A. Ferraz
School of Agriculture, Polytechnic Institute of Viana do Castelo, Refóios, 4990-706 Ponte de Lima, Portugal

Centre of Biological Engineering, Universidade do Minho, Campus de Gualtar, 4710-057 Braga, Portugal

J. Alonso
School of Agriculture, Polytechnic Institute of Viana do Castelo, Refóios, 4990-706 Ponte de Lima, Portugal

Research Centre in Biodiversity and Genetic Resources, Rua Padre Armando Quintas, nº 7, 4485-661 Vairão, Portugal

L. Singh, V.C. Kalia (eds.), *Waste Biomass Management – A Holistic Approach*,
DOI 10.1007/978-3-319-49595-8_15

Keywords Biogas plants • Systems analysis • Geographic information systems • Multi-criteria spatial decision • Sustainability

1 Introduction

The social needs, market dynamics, and the socioeconomic competitive advantages of local areas with adequate natural conditions for primary production and near (peri)urban areas promote intensification, concentration, and specialization of agricultural, livestock, and associated agro-industrial activities (Lefebvre et al. 2012). These processes, in the perspective of complex socio-ecological systems analysis (Kelly et al. 2015), may contribute to the increase of exogenous inputs and outputs (Lima et al. 2015) and, thus, to changes on the conditions and local natural resources (Rammel et al. 2007), as well as an increased potential conflict with the local urban elements and human activities (Steinhäußer et al. 2015). It is therefore important to evaluate the spatiotemporal patterns of biomass production and use balances, namely, waste and wastewater from agricultural, livestock, and agro-industrial origin in order to identify, select, implement, and manage treatment and recovery processes (Bardi et al. 2010; Sheets et al. 2015) related to the selection of local areas for reuse, treatment, and waste and wastewater recovery (Ma et al. 2005; Farahani et al. 2010; Höhn et al. 2014).

Anaerobic digestion is considered an important environmental technology in the bio-based economy, allowing energy production from organic waste and reduction of environmental impacts (Chen et al. 2008). The economic and environmental benefits arising from the use of this biological process to treat organic waste results from the reduction of (1) total solid concentration, by converting part of the volatile solid fraction to biogas with positive energy yield, (2) pathogens, (3) odors, and (4) greenhouse gas (GHG) emissions through fossil fuel replacement (Chynoweth et al. 2001) and biogas burning with consequent release of carbon-neutral carbon dioxide (Khalid et al. 2011). Moreover, the produced slurry (digestate) contains stabilized solids and available forms of nutrients, such as ammonia nitrogen, and can be used as soil fertilizer and/or organic amendment (Appels et al. 2011; Esposito et al. 2012; Molino et al. 2013).

At local level, the needs and challenges of management and recovery of waste and wastewaters from livestock and agro-industrial activities (Fodor and Klemeš 2012) can be transformed in critical renewable energy production for the development and qualification of local areas (Kousksou et al. 2015). The analysis of socio-ecological systems can help to understand the functioning of local areas (Mastrangelo et al. 2015). In addition, data on production systems and on the biomethanation process and technologies (Silva et al. 2014; Höhn et al. 2014) will allow to indicate, locate, size, and optimize the operation of anaerobic digestion (AD) plants for biogas production (Zubaryeva et al. 2012; Thompson et al. 2013; Perpiña et al. 2013).

Two main models can be chosen for the implementation of biogas plants in order to use livestock and agro-industrial waste as a biomass resource (Madsen et al.

2011; Wilkinson 2011): (1) centralized plants, which co-digest manures and other organic residues produced in several farms and agro-industries, and (2) on-farm plants, which co-digest manure with other farm waste and, increasingly, energy crops. There are already a significant number of examples of successful large-scale systems installed for the anaerobic processing of municipal and industrial waste in Europe, specifically in Northern Europe. Germany is the main reference in the application of on-farm anaerobic digesters, with over 4000 plants installed (Wilkinson 2011). Nevertheless, co-digestion systems have also been widely implemented in other countries. For instance, Sweden has about 10 centralized plants and more than 200 farm-scale plants, Denmark has approximately 20 centralized plants and more than 60 farm-scale plants, and Italy has around 500 AD plants (Lönnqvist et al. 2013; Nielsen and Angelidaki 2008). On-farm AD plants may include digesters with a daily biogas production ranging from 100 m^3 to 500 m^3 (Akinbami et al. 2001). On the other hand, centralized plants are usually fed with a mixture of manure and industrial organic waste (10–30 %), with organic loads ranging from 50 to 500 tons/day and a daily biogas production in the range of 1000–15,000 m^3 (Maeng et al. 1999).

The local context of agricultural system production and challenges in implementing biogas plants outlines the interest to develop a review and analysis of literature approach and present strategic and technical proposal in order to (1) determine the importance of analyzing the structure and functioning of territorial systems, as socio-ecological adaptive complex systems, with regard to the needs and possibilities implied in the management options of waste and wastewaters (Sect. 2.1); (2) establish a framework for analysis, methods, and spatially explicit multi-criteria assessment models for the location and design of biogas plants (Sect. 2.2); and (3) establish a review of the processes, techniques, and technologies for biogas production (Sect. 3). Throughout the work, it is assumed that an integrated, multidisciplinary, and hierarchical perspective from global to local scale and that the location of biogas plants and the selection of treatment processes related to an innovative/effective technological and organizational solution framework imply multidisciplinary and integrated approaches to socio-ecological complex and adaptive systems analysis.

2 Complex and Hierarchical System Analysis in Efficient Bioenergy Recovery

2.1 The Local Socio-Ecological Systems Analysis and Agro-Industrial Waste and Wastewater Treatment Challenges

The markets, associated with an increasing population, demand an intensification of the productive systems, concentration, and specialization of livestock and agro-industries in local areas with the best natural conditions for primary production

(Huang et al. 2006). This increase in the number and size of livestock and agro-industrial production plants reinforce the challenges related with land use planning and management (Schutter and Vanloqueren 2011) implicit to the high density of elements and human activities in (peri) urban and adjacent rural areas (Padgham et al. 2015).

The primary production activities depend on conditions (e.g., climate) and natural resources (e.g., soil, water, and biodiversity) highly related to the biogeochemical cycles in the context of operating environmental and territorial systems (Powley et al. 2014). Highest land suitability areas to primary production are associated to climate amenity and deep soils located at valley areas, estuaries, and coastal areas are close to (peri)urban areas along the main centralities and population concentrations that lack supply, security and food quality (Turral et al. 2011), waste and wastewater, and increasing energy production/supply needs (Fernandez-Mena et al. 2016).

The intensification of crop and livestock production activities along with the concentration of activities, plants, and human elements in (peri)urban areas contribute to the increase in exogenous inputs as well as (in)direct and cumulative impacts on the:

(1) Local natural conditions and resources (Jiao et al. 2015)
(2) Land use, land cover, and landscape structure and functioning (Winarso et al. 2015)
(3) Physical, chemical, and biological structure and functioning of the soil (Squire et al. 2015)
(4) Atmosphere and air quality (Ward et al. 2008)
(5) Hydrological and hydrogeological processes (Thackway and Specht 2015)
(6) Availability and physical-chemical quality of superficial and groundwater bodies (Davis et al. 2015)
(7) Selection and the spatial-temporal patterns and evolution of spontaneous and wild, domestic, and cultivated biodiversity (Jeliazkov et al. 2016)

These facts justify the social , political, technical, and scientific challenges as the development of ambitious and demanding political and legislative frameworks such as the Water Framework Directive (2000), the Nitrates Directive (1991), and the Habitats Directive (1992) in the European Union with implications of adaptive needs of livestock and agro-industrial units.

Simultaneously, in these (peri)urban areas, the density of elements and human activity cause:

(1) Sectorial and social conflicts (Winarso et al. 2015)
(2) The definition and implementation of environmental and legal conditions, with implications at spatial planning, environmental management, and nature conservation (Demartini et al. 2015)
(3) The implementation of collective infrastructures and individual equipments (Yangang and Jisheng 2014), including waste and wastewater treatment and recovery facilities (Appels et al. 2008; Greer 2011)

(4) Efficient and adaptive governance models related to development of spatially explicitly decision support tools (Davis et al. 2015)

In fact, these dynamics and social and productive trends in (peri)urban areas demand answers at a strategic level, including programs and policy instruments (Water Framework Directive 2000) related to spatial and environmental planning processes and tools (Alder et al. 2015), to operational activities, namely, agricultural, industrial, or commercial activities licensing, certification of livestock and agro-industrial plants (Manos et al. 2014) and associated improvement of agro-environmental management practices (Rocha et al. 2015), stakeholders' training, and collective capacity building development in a hierarchical and integrated approach from global or (inter)national, regional, and local production unit and biochemical process level.

Political and technical solutions may involve analysis of biophysical and human dimensions at local level as a socio-ecological system (Park et al. 2015) and the decision-action as an analysis process (Carlman et al. 2015). At agro-environmental systems, livestock units, and agro-industrial plants, the development and optimization of waste management processes and techniques include:

(1) Minimization of sub-product and output quantities (Gaudino et al. 2014)
(2) Increase of the product economic and commercial value
(3) Improved production techniques and technologies including the development of waste and wastewater management technologies and plans
(4) The allocation of production and external inputs quotas at the production unit level (Water Framework Directive 2000)
(5) The implementation of management systems implied to the environmental and social responsibility (Cofie and Amede 2015)

The analysis of socio-ecological systems can help to understand the composition, structure, and functioning of these complex adaptive systems. It is therefore important to evaluate the spatiotemporal production and use of biomass patterns and balances by different human activities in (peri)urban and rural areas, specifically waste and wastewater from animal husbandry and agro-industrial origin (Brito et al. 2008). The quantity and diversity of effluents and organic waste associated with all human activities present at local level, including urban and industrial effluents or solid waste management, can contribute to the definition and feasibility of waste and wastewater systems (Appels et al. 2008; Rico et al. 2015). Thus, the challenges of managing high quantities of waste and effluents from human activities can become a critical source of renewable energy production for the sustainable development and environmental qualification of local areas (Jingura and Matengaifa 2009; Neves et al. 2009; Gómez et al. 2010).

The analysis , management, and governance of territorial, environmental, organizational, and technological complex adaptive systems gain with the development and application of spatial/territorial information systems supported in web-based geographic information systems (WEBGIS). The gathering of reference and thematic spatial databases (INSPIRE Directive 2007; Alonso et al. 2013) associated with different user profile management systems can help in the collection, analysis, and simulation of data, spatial decision support systems for the development of technical and political decision-making (e.g., location problems) and operation (e.g., logistics) (Rajabifard 2007; Goodchild and Li 2012), as well as facilitates the communication and collaboration under knowledge network development (Alonso et al. 2013) as it assumes a collaborative governance tool and innovation processes (Rajabifard and Eagleson 2013). These information systems and spatial data infrastructures shall provide data on the natural environment, human dimensions, and socioeconomic dynamics (INSPIRE Directive 2007), that is, spatiotemporal patterns of waste and wastewater generation sources as other criteria and critical factors to the location and operation of treatment and recovery plants (Marchettini et al. 2007; Zubaryeva et al. 2012; Thompson et al. 2013; Höhn et al. 2014) offer collaborative instruments between the collective and individual entities involved (Zhang et al. 2012).

The implementation and operation of AD plants for energy recovery from waste and wastewater should take into consideration natural conditions, socioeconomic, and land use planning criteria, particularly the (1) socio-ecological organization (e.g., productive systems), (2) utilities and energy distribution systems, and (3) environmental policies, infrastructure, and governance tools. In order to optimize biogas plant operation, there is a clear need for adapting production systems into operating models suitably anchored to the reality of the local areas or productive units where they are implemented.

2.2 *Spatial Assessment Tools and Multi-criteria Decision Models for the Location and Sizing of Biogas Plants*

Nowadays, it is recognized that the use of livestock and agro-industrial waste as a biomass resource brings economic, environmental, and social benefits (Taleghani and Kia 2005; Ma et al. 2005). Consequently, there is a growing attention and investment on biogas plants. On the other hand, due to its undesirable nature, the population often opposes that these facilities are constructed and located near their communities, claiming health, safety, and environmental risks, economic loses (due to devaluation of property), and quality of life degradation. This social phenomenon is often dubbed as "not in my backyard" (NIMBY).

The selection of the appropriate undesirable facility location demands an evaluation process considering legal requirements (e.g., municipal regulations, governmental laws, and environmental directives) and, at the same time, economic, environmental, and social costs that are conflicting. The evaluation process must be extensive and detailed, requiring expertise in different fields such as soil science, engineering, hydro-geology, topography, land use, sociology, and economics.

Decision-making on the location of undesirable facilities is a problem that demands taking into account multiple criteria together with geographical information to achieve a satisfactory solution (Malczewski 1999, 2006; Church 2002). Multi-criteria decision aid (MCDA) supports the decision-making considering conflicting criteria and various participants, such as experts and stakeholders. MCDA allows to design, evaluate, and prioritize the decision alternative and, for such provides, a range of procedures, techniques, and algorithms to assist decision-makers in structuring decision problems.

In location problems, different multi-criteria approaches have widely been used as analytic hierarchical process (AHP), analytic network process (ANP), weighted linear combination (WLC) or simple additive method (SAM), multi-attribute utility (MAUT), Technique for Order Preference by Similarity to Ideal Solution (TOPSIS), outranking methods (ELECTRE and PROMETHEE) and fuzzy logic and multi-objective mixed-integer linear programming (MMILP), goal program, and meta-heuristic (Farahani et al. 2010; Scott et al. 2012). The literature can be found providing several studies of bioenergy location, in which MCDA methodologies were applied (see, e.g., Tuzkaya et al. 2008; Panichelli and Gnansounou 2008; Sultana and Kumar 2012; Perpiña et al. 2013; Höhn et al. 2014; Silva et al. 2014).

It is known that biomass and energy plants are geographically dependent, whereas both biomass suppliers and biogas demanders (energy plants) are dispersedly distributed and have too a strong number of spatial variables involved. To locate biogas plant, it should be taken into account the protection of the environment and the reduction of the risk of contamination by imposing to be distant from water lines and ecological and agricultural areas. Additionally, it should be considered the distance to the electricity network and roads and the slope of the terrain, to determine the economic feasibility of a candidate site. Furthermore, due to the impact that such facilities bring to the near population, it should be located as distant as possible from residential and commercial areas. To accomplish the task such as bioenergy facility, we need to prepare a number of maps (layers), each with a different theme. Geographic information system (GIS) supports the process of location modeling and analysis, since it can collect, filter, organize, and analyze the original data, and, in addition, it can present model outputs to assist decision-makers to comprehend spatial requirements and outcomes (Alçada-Almeida et al. 2009).

Geographical data and preferences of the decision-makers are integrated and transformed into a decision with the support of GIS-based MCDA. There are already case studies of the use of GIS-based MCDA to locate bioenergy facilities (e.g., Ma et al. 2005; Tavares et al. 2011; Zubaryeva et al. 2012; Ferretti and Pomarico 2012; Sánchez-Lozano et al. 2014; Silva et al. 2014). More concretely, regarding the location of biogas plants, several studies have been conducted in the last decade, aiming at the development of methodologies to optimize the location and operation of the biogas production systems. Batzias et al. (2005) developed a GIS-based biomass resource assessment application using a relational database management system to estimate biogas production from livestock manures. Zubaryeva et al. (2012) is based on AHP combined with GIS-based analysis and allowed to create a geographical information system to assist biogas planning, study feedstock from different potential sources, and present suitable biogas territorial clusters. Ma et al. (2005) developed a GIS model using AHP to assess suitability of the land for energy systems containing an anaerobic digester and to produce a suitability map which exhibits the most suitable areas to install bioenergy plants using dairy manure. Silva et al. (2014) proposed a spatial multi-criteria approach to support the location process of biogas plants using dairy manure as feedstock and using outranking ELECTRE TRI method for classified the potential locations in categories of suitability. Höhn et al. (2014) demonstrated that a GIS-based methodology, which included transport optimization, can identify the most suitable locations and calculate kernel density (KD) maps to locate high biomass concentration areas.

The complexity of the decision of finding a location for biogas facilities lies mostly in the existence of a lot of constraints (or restrictive criteria) and factors (or weighting) and conflicting objectives relating to cost, risk, equity, social rejection, and so on. Silva et al. (2014) proposed this methodology to develop the location-allocation analysis: after a preliminary analysis for selecting adequate sites to locate biogas plants by performing a land use suitability analysis, through a combination of GIS and an MCDA method (Fig. 1).

In the next phase, to answer to questions "How many biogas plants should be located?", "How large should each biogas plant be?", and "How much livestock and

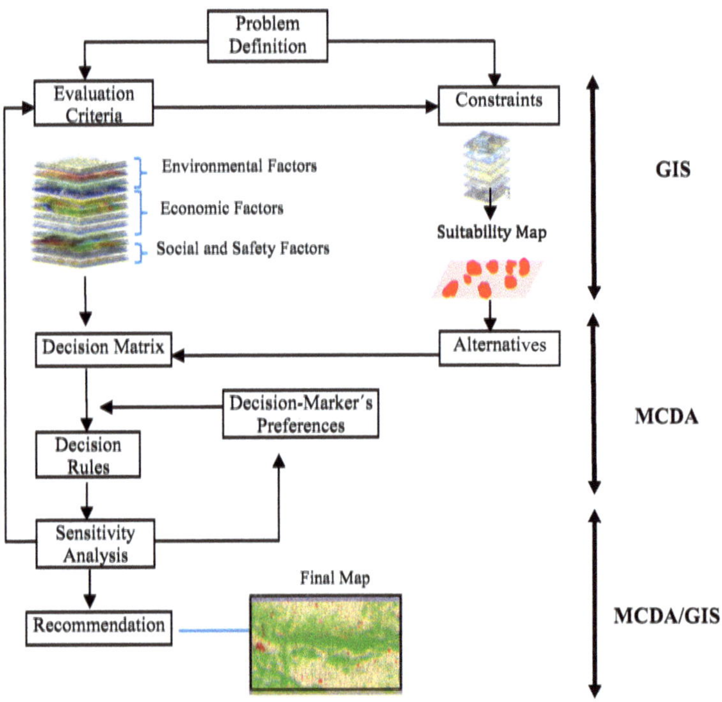

Fig. 1 Decision flowchart for spatial multi-criteria analysis (Silva et al. 2014)

agro-industrial waste should be assigned to each biogas plant?", a multi-objective mixed-integer linear programming (MMILP) formulation is developed where the objectives are to minimize initial investment and operation and maintenance costs, minimize transportation cost, and minimize social rejection. In multi-objective formulation, it is necessary to consider the characteristics related to operation, type of biomass, technical structure, dimensions, etc.

There are many types of biogas plants categorized according to the type of digested substrates, according to the technology applied, or according to their size (Holm-Nielsen et al. 2009). Al Seadi et al. (2008) classified biogas plants according to their relative size, function, and location: family-scale biogas plants (very small scale), farm-scale biogas plants (small or medium to large scale), and centralized/joint co-digestion plants (medium to large scale). For each type of biogas plant, it is necessary to define the minimum and maximum amount of manure for each type of biogas plant to ensure that it is economically and technically feasible. According to the type of biogas plant, it is necessary that the suitable site has a minimum area, which will be greater than the central. The investment and the lifetime operation and maintenance costs of facilities depend of different types (sizes) of biogas plants. It is required that the dairy farms have to be at a certain fixed maximum distance from the biogas plant assigned to it for the biomass to preserve its properties and to reduce the risks associated with transportation. The transport cost of manure from dairy farms to biogas plants should be taken into consideration in the location of these facilities.

There are various studies related to biogas plants about different aspects that are important to analyze regarding its location, such as (1) biogas potential analyses using different agro-biomasses and waste materials (Batzias et al. 2005; Murphy and Power 2009; Rosúa and Pasadas 2012); (2) the situation and potential in production and utilization of biogas (Holm-Nielsen et al. 2009; Tricase and Lombardi 2009; Salomon and Lora 2009; Gómez et al. 2010); (3) on technical analysis, economic, and environmental sustainability and the use of biogas (Maeng et al. 1999; Murphy et al. 2004; Taleghani and Kia 2005; Jingura and Matengaifa 2009; Neves et al. 2009; Rao et al. 2014); (4) performance and expansion of biogas plants (Madlener et al. 2009; Nzila et al. 2012; Bojesen et al. 2015); and (5) on the role of GIS in the mapping and evaluation of the localization of biogas plants (Dagnall et al. 2000; Voivontas et al. 2001; Batzias et al. 2005; Panichelli and Gnansounou 2008; Madlener and Schmid 2009; Sultana and Kumar 2012; Thompson et al. 2013; Höhn et al. 2014).

In this context, urges develop and implement integrated solutions that might be both technical and economically feasible with a perspective of increasing the energy recovery from agricultural livestock and agro-industrial wastes. For that, it is still necessary to consider and define the process and technologies for biogas production.

3 Anaerobic Digestion: The Process and Technologies for Biogas Production

3.1 Anaerobic Digestion Process

Anaerobic digestion may be considered a synergistic multistage biochemical process involving a bacterial consortium taking place in sequential steps, each one performed by a different group of microorganisms with specific nutritional needs, habitat requirements, growth kinetics, and ability to tolerate environment stresses (Appels et al. 2008; Khalid et al. 2011; De Vrieze et al. 2015; Li et al. 2015). The metabolic pathways involved are described as four sequential stages: hydrolysis, acidogenesis, acetogenesis, and methanogenesis (Tchobanoglous et al. 2003; Appels et al. 2008; Liu et al. 2010; Zhang et al. 2014; De Vrieze et al. 2015; Li et al. 2015).

The anaerobic digestion process is a microbial integrated system (Mao et al. 2015). The overall process can be described as the hydrolysis of high molecular weight materials and granular organic substrates (e.g., lipids, polysaccharides, proteins) into small molecular compounds and soluble organic substrates that are subsequently converted into short-chain volatile fatty acids (VFA), ammonia (NH_3), hydrogen (H_2), carbon dioxide (CO_2), and other by-products by fermentative bacteria.

The VFA are further degraded into acetate, H_2, and CO_2 by acetogenic bacteria, which are used to produce methane (CH_4) by methanogenic bacteria, converting acetate into CH_4 and CO_2 (acetoclastic methanogenesis) or using H_2 and CO_2, respectively, as electron donor and acceptor, to produce CH_4 (hydrogenotrophic methanogenesis) contributing with 60–75 % of total methane production (Fig. 2). Formate, carbinol, and carbon monoxide could also be converted to CH_4. The hydrolysis process for particulate substrate solubilization, mediated by extracellular hydrolytic enzymes, is assumed to be the rate-limiting step governing the overall AD process and conversion efficiency of the biomass feedstock (Kleerebezem and Loosdrecht 2007; Appels et al. 2008; Liu et al. 2010; Li et al. 2011; Zhang et al. 2014).

Microbial consortium involved in biomethanation includes hydrolytic, fermentative, acetogenic, and methanogenic microorganisms. Hydrolytic bacteria such as *Acetivibrio, Bacteroides, Bacillus, Cellulomonas, Clostridium, Erwinia, Microbispora, Ruminococcus, Streptomyces,* and *Thermomonospora* are able to hydrolyze cellulolytic biomass. The major genera of fermentative bacteria are *Acetobacterium*, *Clostridium*, *Sporomusa*, *Saccharomyces*, *Butyribacterium*, *Lactobacillus*, and *Streptococcus* (Li et al. 2011). The acetogen group includes the genera *Syntrophomonas* and *Syntrophobacter* bacteria that convert the acid-phase products into acetates and hydrogen (Shah et al. 2014). Acetogens are not able to subsist in high partial hydrogen pressure. Therefore, a symbiotic relationship is established with methanogens, hydrogen consumers (Li et al. 2011). The most commonly found methanogens genera are *Methanobacterium*,

Fig. 2 Anaerobic digestion process: organic matter conversion into biogas (Chen et al. 2014; Zhang et al. 2014)

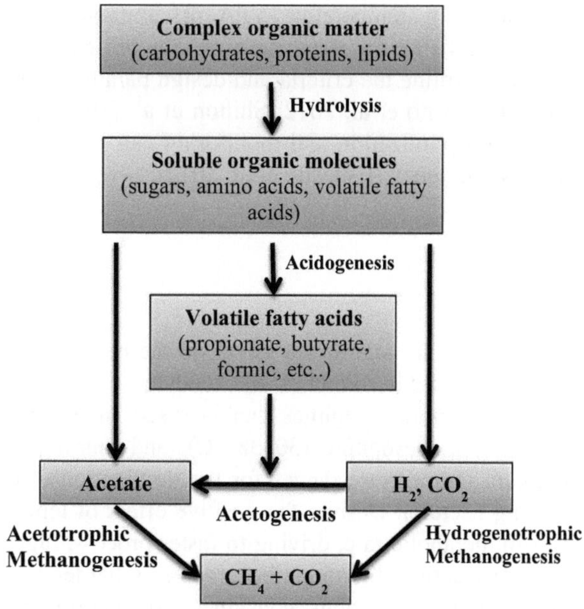

Methanothermobacter (formerly *Methanobacterium*), *Methanobrevibacter*, *Methanosarcina*, and *Methanosaeta* (formerly *Methanothrix*) (Kondusamy and Kalamdhad 2014).

Biogas consists of methane (50–75 %) and carbon dioxide (25–45 %), with trace gases such as water vapor (1–2 %), nitrogen (1–5 %), hydrogen (0–3 %), hydrogen sulfide (0.1–0.5 %), and carbon monoxide (0–0.03 %) (AEBIOM 2009). To use the produced biogas as fuel, e.g., through natural gas distribution network, it is necessary to achieve a final quality comparable to the natural gas, with CH_4 concentration greater than 95 %. Therefore, it is necessary to remove contaminants such as H_2O, sulfur compounds, halogenated organic molecules, CO_2, O_2, and metals (Molino et al. 2013). Carbon dioxide removal is accomplished by passing the gas into lime water (Kondusamy and Kalamdhad 2014).

3.2 Anaerobic Digestion Technologies

3.2.1 Design Criteria and Operational Parameters

In addition to understand the complex interactions of microorganisms involved in the anaerobic digestion process for better process control (Maspolim et al. 2015), several guideline operational parameters for digester design can be established, such as volumetric solid load, solid retention time, volatile solid reduction, and biogas production. It is important to ensure the control of key environmental conditions, e.g., temperature, VFA, pH, ammonia, nutrients, carbon/nitrogen/phosphorus (C/N/P) ratio, trace elements, the presence of inhibitors, and particle size, and to determine the criteria and design parameters which lead to process optimization (Esposito et al. 2012; Shilton et al. 2013; Zhang et al. 2014). Operational condition optimization and monitoring can prevent system failures and improve process efficiency (Esposito et al. 2012).

Temperature

Temperature has significant effect on the bacterial community, as well as on the anaerobic digestion kinetics and methane production yield (Khalid et al. 2011). Three temperature regimes can be used in anaerobic digesters: psychrophilic (20–25 °C), mesophilic (30–38 °C), and thermophilic (50–57 °C) with varied optimum temperature ranges for the domination of different strains of methane-forming bacteria. Despite the positive effect of temperature increase on microorganism metabolic rate, driving to faster process, higher solid reduction, and pathogen destruction, thermophilic process is harder to control, is more sensible to environmental conditions, consumes more energy, produces low-quality supernatant, has more odor risk, and demands great care to ensure process stability (Appels et al. 2008; Khalid et al. 2011; Kwietniewska and Tys 2014; Mao et al. 2015). Chae et al. (2008) studying swine manure digestion within the range from 25 to 35 °C reported a nonlinear increment on CH_4 yields with temperature. As methane production was similar at 30–35 °C, and 13–17 % higher than at 25 °C, authors concluded that higher temperature may not be the optimal condition due to the large energy requirements being necessary to balance the increase in methane yield against the energy consumption to heat the reactor. Therefore, optimal digestion temperature relies on the local climate, feedstock used and its volatile solid content, and the economic efficiency of heating (Chae et al. 2008; Deng et al. 2014). Most of the high-rate digesters are operated in the mesophilic range (Mao et al. 2015).

Most common methods used for sludge heating include heat exchangers, where the recirculating sludge is mixed with raw sludge prior to heating, also providing its inoculation with anaerobic microorganisms and steam injection. Usually, the digester biogas is used to produce the hot water needed (Appels et al. 2008).

The amount of heat required for sludge to reach the digester temperature, Q_1 (J/d), and to compensate heat losses, Q_2 (J/s), can be calculated from Eqs. (1 and 2):

$$Q_1 = W_f C_P (T_2 - T_1) \tag{1}$$

$$Q_2 = UA(T_2 - T_a) \tag{2}$$

where W_f represents the feeding sludge rate (kg/d), C_p its specific heat (4200 J/kg°C), T_2 the digester operating temperature (°C), T_1 the feed sludge temperature (°C), U the heat transfer coefficient (W/m^2°C), A the reactor surface area through which heat loss occur (m^2), and T_a the environmental temperature (°C) (Appels et al. 2008).

pH

The operational pH affects the digestive progress and efficiency. The ideal pH range for anaerobic digestion has been reported to be 6.8–7.4 (Mao et al. 2015), 6.8–7.2 (Ward et al. 2008; Kwietniewska and Tys 2014), and 6.8–7.5 (Khalid et al. 2011). Within this wide pH range, the optimal value varies with substrate and digestion technique (Kondusamy and Kalamdhad 2014). Methanogenesis is most efficient at pH 6.5–8.2, with optimal pH 7.0. The growth rate of methanogenic bacteria decreases significantly at pH below 6.6. Acid-forming bacteria are less sensitive and tolerate pH in the range of 4.0–8.5, but the optimal pH for hydrolysis and acidogenesis is between 5.5 and 6.5 (Kwietniewska and Tys 2014; Mao et al. 2015). The pH varies because of parameters such as VFA, bicarbonate, and alkalinity concentration and also by the fraction of CO_2 produced (Kondusamy and Kalamdhad 2014). At the beginning of the fermentation, acidogens and acetogens produce acids and CO_2 resulting in pH decrease. With acid consumption by methanogens the pH increases and then stabilizes. The pH also affects the proportion of methanogenesis inhibitors, such as fatty acids and hydrogen sulfide at pH below 7 and ammonium-nitrogen at pH values higher than 7 (Kwietniewska and Tys 2014).

C/N Ratio and Biogas Production

The concentrations of carbon and nitrogen indicate AD process performance. Carbon acts as the energy source, and nitrogen is the major nutrient for microbial growth, contributing to enhance microbial population. The optimal C/N ratio for organic waste AD is 20–35 (Kondusamy and Kalamdhad 2014; Kwietniewska and Tys 2014), with a ratio of 25 being the most commonly used (Mao et al. 2015). Nitrogenous compounds, usually proteins, are converted to ammonium that is assimilated by microbial biomass for new cell production (Khalid et al. 2011). When C/N ratio is low (nitrogen rich substrates), anaerobic digestion results in increased content of free ammonia released, and its toxic effect also causes a large amount of VFA accumulation and pH fluctuations (Kwietniewska and Tys 2014). In contrast, high C/N ratio provides insufficient nitrogen to maintain cell biomass and leads to fast nitrogen degradation by microbes, resulting in lower biogas production. Thus, ammonia inhibition may be avoided by optimizing the C/N ratio (Mao et al. 2015).

Co-digestion of organic mixtures is used to balance the C/N ratio, allowing to buffer the organic loading rate and ammonia production (Khalid et al. 2011). Other advantages include toxic compound dilution, providing pH and moisture content adjustment, increasing available biodegradable material, and widening the bacterial community involved in the process (Khalid et al. 2011; Esposito et al. 2012). In this context, co-digestion synergistic effects may favor an increased production of methane exceeding that obtained on single substrates (Smith and Almquist 2013) and improve process stability and performance (Esposito et al. 2012).

Waste from agriculture activities and agro-industries accounts with large amounts of potential feedstock for AD (Kothari et al. 2014). The composition and characteristics of different feedstock significantly affects process performance and operational parameters including optimal retention time, biogas production yield, and solid conversion ratio (Li et al. 2011), and single feedstocks rarely have the ideal composition for optimal AD.

In fact, sewage sludge are poor in organic matter and may have inhibitory concentrations of heavy metals; animal manures have low C/N ratios; agro-industrial wastes have seasonality constraints and have high C/N ratio; and, on the contrary, slaughterhouse wastes have low C/N ratio and/or long-chain fatty acids (Mata-Alvarez et al. 2014). Using co-substrates with low lipid and N content increases biogas production and reduces operational problems arising from intermediate volatile compound accumulation and excessive NH_3 concentrations (Khalid et al. 2011). Manures are often associated to poor methane yields (Mata-Alvarez et al. 2014). Cattle manure, with high contents of nonbiodegradable substances and low C/N ratio, results in low methane yield in anaerobic mono-digestion (Agyeman and Tao 2014). Co-digestion of agricultural waste with manure waste provides positive synergistic effects and can potentially dilute toxic compounds (Mao et al. 2015) and make plants economically feasible (Mata-Alvarez et al. 2014).

Gas yield and composition vary with the operating conditions of the digester and the type of feedstock (Surendra et al. 2014), in particular regarding its carbohydrate, protein, and lipid contents. Despite providing higher biogas production yields, lipids need higher retention times because of their slow biodegradability (Esposito et al. 2012). A wide variety of livestock and agro-industrial wastes may be used as feedstock, for instance, pig manure, cow manure, chicken manure, and fruit waste. For mono-digestion and co-digestion processes using manure or fruit waste as feedstock, the volatile solids (VS) usually correspond to 70–85 % of the total solids (TS), with a C/N ratio from 3 to 35, leading to a biogas yield of 0.2–0.6 (m^3 kg^{-1} VS) and a CH_4 content of 55–80 %, in digestors operating with HRT of 8–40 d (Kothari et al. 2014). According to Kondusamy and Kalamdhad (2014), when considering waste from animal origin, several factors can affect biogas quality and quantity, namely, the animal type, age, feeding and living conditions, and waste storage conditions.

Methane is the only component of biogas that contributes to the heating value (Surendra et al. 2014). Raw biogas, with 60 % of CH_4, has a net calorific value of 21.48 MJ/m^3 (Braun 2007).

Equations (3) and (4) may be used to estimate methane production rate (Appels et al. 2008):

$$V_{CH4} = 0.35(ES_0 - 1.42\Delta X) \tag{3}$$

$$\Delta X = \frac{YES_0}{1 + k_d\theta_c} \tag{4}$$

where V_{CH4} represents methane production volume (m^3/d), 0.35 the theoretical conversion factor for CH_4 production from 1 kg BOD, E the substrate utilization efficiency, S_0 the influent flow ultimate BOD (kg/d), 1.42 the conversion factor for cell biomass into BOD, ΔX the cell production net mass (kg/d), Y the yield coefficient (g/g), k_d the endogenous coefficient (d^{-1}), and θ_c the mean cell residence time (d), also reported as solids retention time (SRT).

Organic Loading Rate (OLR)

The OLR represents the amount of volatile solids fed into a digester per day under continuous feeding. With increasing OLR, the biogas yield increases to an extent, but the equilibrium and productivity of the digestion process can be disturbed, and above the optimal OLR, the volatile solid degradation and biogas yield decreases due to overloading. High OLR results in hydrolysis/acidogenesis bacterial activity higher than methanogenesis bacterial activity, leading to VFA accumulation and consequently an irreversible acidification that inhibits methanogenic bacteria. Thermophilic AD and effluent recirculation may contribute to control overloading inhibition (Kwietniewska and Tys 2014; Mao et al. 2015). Optimal OLR depends on feedstock used for co-digestion, reactor temperature, and mixing conditions. According to Mao et al. (2015) thermophilic system and effluent recirculation have great potential to relieve the overloading inhibition. Li et al. (2015) studied the effect of OLR in mesophilic co-digestion of rice straw and cow manure, reporting a stable and efficient co-digestion with OLR of 6 kg VS $m^{-3}d^{-1}$, with a specific biogas production of 383.5 L/kg VS and a volumetric rate of 2.3 m^3 $m^{-3}d^{-1}$, foam formation for OLR $\geq$ 8 kg VS $m^{-3}d^{-1}$, and inhibition caused by VFA accumulation for OLR $\geq$ 12 kg VS $m^{-3}d^{-1}$.

Hydraulic Retention Time and Solids Retention Time

The time required to complete the degradation of organic matter is associated with the microbial growth rate and depends on the process temperature, OLR, and substrate composition. In biological processes, two retention times can be defined, the average biomass retention time (SRT) and the liquid fraction retention time (HRT). Under mesophilic conditions, the average HRT is 15–30 d, being reduced to nearly 2 weeks in thermophilic conditions (Kwietniewska and Tys 2014; Mao et al. 2015). If the HRT is short and bacterial loss exceeds the growth rate of bacteria, digester washout occurs and the process can fail (Kwietniewska and Tys 2014). To prevent this phenomenon, for digestion temperature of 35 °C, the minimum SRT is 10 d. Low OLR and a long HRT provide the best strategy for achieving constant and maximal methane yields (Appels et al. 2008). Pig and cattle manure, differing in their composition, require different HRT. Cattle manure, with higher cellulose and hemicellulose contents and less lipids, needs higher HRT. Average mesophilic digestion HRT for cattle manure, cattle manure with straw bedding, and pig manure were found to be, respectively, 12–18 d, 18–36 d, and 10–15 d (Kothari et al. 2014).

Inhibitors

The low methane production rate is often the main limitation for a large-scale application of anaerobic digestion. One of the major challenges of the anaerobic digestion process may be microbial population vulnerability to operational condition fluctuations and inhibition caused by inhibitory compounds currently present in wastes and wastewaters at relatively high concentrations. The methanogens are very sensible toward adverse environmental conditions, so it is very important to maintain optimal conditions for these microbes (Kondusamy and Kalamdhad 2014).

Ammonium nitrogen, sulfide, and metals are usually reported as inhibitors in anaerobic digestion processes (Chen et al. 2008). Organic overloading, sudden pH variations, and high salt and ammonium concentrations may cause process inhibition due to volatile fatty acid (VFA) accumulation (De Vrieze et al. 2012). Ammonium ion (NH_4^+) and free ammonia (NH_3) usually occur in livestock manure and other agro-industrial wastes and wastewaters (Abbasi et al. 2012; Rico et al. 2015). Both forms of inorganic ammonia nitrogen can directly and/or indirectly lead to process inhibition in an anaerobic digester (Yenigün and Demirel 2013). Table 1 summarizes the results of the effects of different ammonia concentrations on the anaerobic digestion process.

The differences in tolerance to ammonia may be related to the composition of the substrate, operational conditions (e.g., pH and temperature), inoculum biomass, and acclimation periods (De Vrieze et al. 2012; Rajagopal et al. 2013).

Another important issue is the potential negative effect of emerging organic pollutants, such as antibiotics, on methanogenic activity. Indeed, antibiotics are emerging as a new group of pollutants in wastewater treatment plants (WWTPs) as a result of their intensive use for human, veterinary, and agricultural purposes. They are generally referred as inhibitory compounds to microorganisms in activated sludge in batch assays (Shilton et al. 2013), and, since biological systems rely on the composition and activity of their microbial communities, the increasing occurrence of antibiotics in waste and wastewater might hinder the proper functioning of WWTPs.

The overall system ability to tolerate adverse conditions depends, in part, on waste composition, feeding rate, inoculum type, and digester operational conditions (Dearman et al. 2006; De Vrieze et al. 2013). Therefore, mixing different organic substrates (co-digestion), microbial acclimation to inhibitors and the adoption of

Table 1 Effects of different ammonia concentrations on the anaerobic digestion process and methane production

Ammonium nitrogen concentration (mg/L)	Result	Reference
1.7–14 g/L	50 % inhibition of methane production	Chen et al. (2008)
2.1, 2.6 and 3.1 g/L	Optimal concentrations for methanogens, in dependence on inoculum origin	Prochazka et al. (2012)
> 2.5 g/L	May lead to 100 % inhibition	Yenigün and Demirel (2013)

methods for toxic compounds removal or attenuation prior to DA process can improve its efficiency (Chen et al. 2008).

3.2.2 Anaerobic Digesters

Agro-industrial waste and wastewater handling and disposal represent a substantial part of farm operation cost and also represent environmental, economic, and social challenges to livestock operations. Therefore, developing robust and cost-effective technologies for waste-to-energy conversion can solve some of its environmental problems and improve the social acceptability of livestock operations.

Different types of digesters have been developed for the treatment of waste and wastewaters (Kleerebezem and van Loosdrecht 2007). Bioreactor design strongly influences the performance of the anaerobic digestion process (Khalid et al. 2011). Therefore, it should accomplish optimal operational and management criteria for each particular application depending on the feedstock characteristics. According to Ward et al. (2008), design requirements should consider the maximization of methane production, with a high and sustainable organic loading rate and a short hydraulic retention time.

Concerning the process configuration, anaerobic digesters can be classified as slurry-phase or solid-phase digesters depending on feedstock solids concentration. Slurry-phase bioreactors contain 10–25 % dry matter, and solid-phase systems are characterized by 30–40 % dry matter content (Khalid et al. 2011). According to Kothari et al. (2014), food waste, animal manure, and sewage sludge are usually treated in slurry-phase digesters, whereas the organic fraction of municipal solid waste, crop residues, and energy crops (lignocellulosic biomass) are treated by solid-phase systems.

Digesters can operate in batch or continuous mode. Batch reactors, operated with simple and inexpensive equipment, can be used to perform anaerobic digestion processes with high organic matter removal efficiency and relatively low hydraulic retention time, despite having limitations such as biogas volume and quality fluctuations (Khalid et al. 2011). Common applications are the anaerobic sequencing batch reactors (ASBR), operating in repetitive cycles: (a) fill, (b) reaction, (c) settling, and (d) drawing (Fig. 3). Despite its simplicity, ASBR provides efficient quality control of the effluent, low input process and mechanical requirements, cost-effectiveness, high biogas yields, and flexibility in operation (Mao et al. 2015) even under less favorable conditions, such as operating at low temperature or with low organic matter concentration (Rodrigues et al. 2009), also contributing to overcome problems related with overdesign capacity. In fact, they have been successfully applied in the treatment of the liquid fraction of dairy cattle manure by anaerobic digestion (Rodrigues et al. 2010). Furthermore, SBR operational flexibility can enhance nutrient removal from alternating between anaerobic and oxic/anoxic conditions for each cycle (Zhu et al. 2006). According to Tauseef et al. (2013), ASBR digesters can be used to treat manure with a high organic matter removal efficiency, particularly if the solid concentration is lower than 1 %, being an ideal treatment process for wastewater resulting from cleaning operations in dairies.

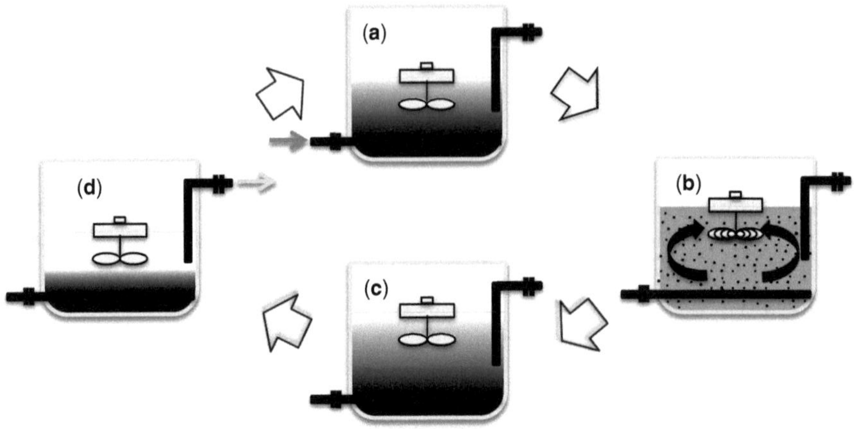

Fig. 3 ASBR operating cycles: (**a**) fill, (**b**) reaction, (**c**) settling, and (**d**) drawing

Continuous stirred tank reactors (CSTR) are the first generation of high-rate anaerobic reactors, widely used to treat wastewater containing high levels of suspended solids, e.g., high-strength liquid animal manure and organic industrial wastes. Microorganisms are suspended in the digester through intermittent or continuous mixing. Complete mixing offers good substrate-sludge contact with slight mass transfer resistance but consumes considerable energy and is labor-intensive (Mao et al. 2015). In these reactors, solids retention time (SRT) is identical to hydraulic retention time (HRT) and biomass washout occurs, but due to the intense mixing and temperature control, the anaerobic digestion occurs with a high rate (Tauseef et al. 2013).

The most common continuous fed systems operate in one-stage reactor, with all the metabolic stages occurring in a single reactor or in a two-stage reactor where the hydrolysis/acidification and acetogenesis/methanogenesis processes are separated, in order to provide optimum environmental conditions for each process and, therefore, to increase overall process performance (Ward et al. 2008; Schievano et al. 2014).

Acidogenic and methanogenic bacteria differ in nutritional requirements, growth kinetics, and sensitivity to environmental conditions. Therefore, the balance between them can take a long time in a single reactor. In two-stage reactors, it is possible to establish the optimal environmental conditions for each microorganism group, optimizing acid formation in the first phase and methane production in the second and enhancing overall process stability and control (Liu et al. 2010).

A controlled acidogenic fermentation that allows efficient hydrogen production has been considered the best pathway to pretreat raw biomass and enhance methanogenic process (Schievano et al. 2014). The control of the acidification phase through the HRT optimization is particularly important for the sensitive methanogenic bacteria, since it may prevent overloading and the production of toxic compounds (Khalid et al. 2011). For instance, in single-stage reactors, the methanogenic bacteria can be washed out at short SRT operation due to their slow growing rates, causing the accumulation of VFA produced during acidogenesis, resulting in system pH decrease and cause process failure. Nevertheless, reactor operation at extended SRT and neutral pH may suppress the acidogenic bacteria's capacity to produce VFA (Maspolim et al. 2015). The two-stage reactors can be able to maintain pH at 6 and 7, for optimal operation of both hydrolytic acidogenic and methanogenic bacteria, respectively (Kondusamy and Kalamdhad 2014). The first digester is usually heated in order to achieve the optimal temperatures for the hydrolytic and acidogenic bacteria performance. On the contrary, the second digester is not normally heated because of the exothermic nature of the methane formation reaction. The higher capital costs and the operational and maintenance requirements can be overcome by higher process performance, with increase degradation yield and biogas production (Ward et al. 2008; Khalid et al. 2011).

Anaerobic filters are attached growth reactors operating under up-flow or down-flow condition, developed to provide a support medium for the intimate contact between the influent and the bacterial mass through biofilm formation, allowing for a biomass retention time longer than the hydraulic retention time (HRT) (Mao et al. 2015). Biofilm formation improves the degradation rate by increasing cell density and decreasing the hydraulic retention time. These systems have excellent adaptability to carbon source and organic load fluctuations. One of the major operational problems of such reactors may be filter media clogging. Therefore, anaerobic filters could be more suitable to treat wastewaters with low suspended solid concentration and a particle size less than 0.75 mm. On the other hand, substrates with a high solid content should be submitted to a pretreatment in order to induce solubilization of particulate organic matter before they can feed these high rate systems (Mao et al. 2015; Aslanzadeh et al. 2014). For dairy wastewater treatment, a removal of 1–5 % of the total solids is usually needed before feeding the digester (Tauseef et al. 2013).

3.2.3 Pretreatments

Several physical, thermochemical, biological, or combined pretreatments are described to enhance AD performance by promoting particulate substrate hydrolysis and solubilization, the limiting steps for complex substrate biodegradation (Kleerebezem and van Loosdrecht 2007; Esposito et al. 2012; Zhang et al. 2014). Particle size reduction and solubilization of organic matter will improve biodegradability and also microbial access to the substrate through release, exposure, or transformation of organic material. Consequently, such pretreatments will improve the rate and extent of substrate degradation, both of which are directly correlated to the methane yield (Carlsson et al. 2012) and biogas plant performance (Harris and McCabe 2015). However, pretreatments may favor the formation of refractory compounds, counteracting the positive effects on biodegradability, and the increase of solubilized COD may have inhibitory effects by overloading the AD system (Carlsson et al. 2012).

Chemical treatments include acid (e.g., H_2SO_4 and HCl) and alkali (e.g., NaOH) hydrolysis and oxidative pretreatments (e.g., ozonation and peroxidation). Thermal treatments involve substrate exposure to high temperatures (150–220 °C) at 600–2500 kPa to promote chemical reactions and solubilize larger biomolecules. Mechanical pretreatments, used to enhance digestion of cellular wastes (e.g., sludge) and cellulosic wastes, require mechanisms of solids physical disintegration such as high-pressure homogenizers, microwaves, and ultrasonicators. Biological treatments use enzymes and biosurfactants (Harris and McCabe 2015).

Livestock manure and agro-industrial wastes are rich in cellulose, hemicellulose, and lignin, components not readily available for immediate fermentation (Kothari et al. 2014). Thermal and chemical pretreatments have been successfully applied to energy crops and/or harvesting residues and agro-industrial waste (Carlsson et al. 2012).

Despite the overall improvement in waste stabilization and methane production, the viability of pretreatments application should consider the additional processing costs (Esposito et al. 2012). The assessment of the potential improvement in methane production should consider mass and energy balances, as well as the effect of the pretreatment on different substrates and on the overall system performance, addressing technical, energy, and economic aspects (Carlsson et al. 2012).

For biogas plant optimization and process sustainability, a number of factors need to be considered when selecting AD technologies, pretreatments, and their respective operational conditions. Dealing with different feedstock origins, with a variety of physicochemical and biodegradability characteristics, and considering local environmental particularities (e.g., climate) are additional challenges to project design that should not follow standard parameterization. In fact, the final option should result from individual technical and economical balance based on process performance, advantages, and disadvantages of each technology and the operational and maintenance costs.

4 Conclusions

The development of livestock and husbandry production activities, the agro-industrial processing, and production inputs in growth contexts of (peri)urban areas result in losses and challenges of environmental and quality of life. Therefore, there is a need to develop and implement technological and organizational solutions, particularly the development of infrastructures by private and public entities, including waste and wastewater treatment plants, as well as develop adequate governance models and tools.

In the location and sizing of biogas plant decisions, it is essential to develop an integrated approach, based on the territorial reality, as well as on the characteristics and estimated quantities of waste and wastewater generated at regional, local, and production unit level. A multi-criteria and multi-objective analysis, including spatially explicit environment (GIS), should be carried out to define different scenarios for the location and dimension of waste and wastewater treatment plants for biogas production.

Apart from the methane production improvement, the enhancement of the agriculture-based AD plants profitability include (1) gains from the digestate sale as fertilizer; (2) location of the power plants near to the consumers; (3) integration of the power plants in the farms, reducing operational costs; or (4) integration of the biomass transport in the business of, either biomass producers or power plant operators.

Environmental and social problems related to excess of biomass production may become an opportunity, when there is a knowledge of the local socio-ecological systems, the productive systems as well as the biological, physical, and chemical processes for biogas production. Such knowledge will contribute to the selection of technologies, location, and design of waste and wastewater treatment and recovery systems, integrated in territorial dynamics and energy production innovation, as a critical resource for local sustainable and smart development.

References

Abbasi T, Tauseef SM, Abbasi SS (2012) Anaerobic digestion for global warming control and energy generation—an overview. Renew Sustain Energy Rev 16:3228–3242. doi:10.1016/j.rser.2012.02.046

AEBIOM–European Biomass Association (2009) A biogas roadmap for Europe. www.aebiom.org

Agyeman FO, Tao W (2014) Anaerobic co-digestion of food waste and dairy manure: effects of food waste particle size and organic loading rate. J Environ Manage 133:268–274. doi:10.1016/j.jenvman.2013.12.016

Akinbami JFK, Ilori MO, Oyebisi TO, Akinwumi IO, Adeoti O (2001) Biogas energy use in Nigeria: current status, future prospects and policy implications. Renew Sust Energy Rev 5:97–112. doi:10.1016/S1364-0321(00)00005-8

Al Seadi T, Rutz D, Prassl H, Köttner M, Finsterwalder T (2008) Biogas handbook. University of Southern Denmark, Esbjerg, 126 p. ISBN:978-87-992962-0-0

Alçada-Almeida L, Coutinho-Rodrigues J, Current J (2009) A multiobjective modeling approach to locating incinerators. Socioecon Plann Sci 43:111–120. doi:10.1016/j.seps.2008.02.008

Alder A, Prasuhn V, Liniger H, Herweg K, Hurni H, Candinas A, Ulrich H, Gujer (2015) A high-resolution map of direct and indirect connectivity of erosion risk areas to surface waters in Switzerland—a risk assessment tool for planning and policy-making. Land Use Policy, 48:236–249. doi:10.1016/j.landusepol.2015.06.001

Alonso J, Castro P, Guerra C, Goncalves J, Pôças I, Marcos B, Honrado J (2013) Novel tools to improve the management of spatial data quality in the context of ecosystem and biodiversity monitoring, in: GI_Forum 2013—Creating the GISociety, Salzbug, pp 7. doi:10.1553/giscience2013

Appels L, Baeyens J, Degrève J, Dewil R (2008) Principles and potential of the anaerobic digestion of waste-activated sludge. Prog Energy Combust 34:755–781. doi:10.1016/j.pecs.2008.06.002

Appels L, Lauwers J, Degrève J, Helsen L, Lievens B, Willems K, Impe JV, Dewil R (2011) Anaerobic digestion in global bio-energy production: potential and research challenges. Renew Sust Energy Rev 15:4295–4301. doi:10.1016/j.rser.2011.07.121

Aslanzadeh S, Rajendran K, Taherzadeh MJ (2014) A comparative study between single- and two-stage anaerobic digestion processes: effects of organic loading rate and hydraulic retention time. Int Biodeter Biodegr 95:181–188. doi:10.1016/j.ibiod.2014.06.008

Bardi L, Malusà E, Zoppellari F, Bosco F, Bergesio B, Bertin B (2010) Production of renewable energies and biomolecules from livestock and agro-industrial waste: the BIOMOLENER Project. J Biotechnol 150:172. doi:10.1016/j.jbiotec.2010.08.449

Batzias FA, Sidiras DK, Spyrou EK (2005) Evaluating livestock manures for biogas production: a GIS based method. Renew Energy 30:1161–1176. doi:10.1016/j.renene.2004.10.001

Bojesen M, Boerboom L, Skov-Petersen H (2015) Towards a sustainable capacity expansion of the Danish biogas sector. Land Use Policy 42:264–277. doi:10.1016/j.landusepol.2014.07.022

Braun R (2007) Anaerobic digestion: a multi-faceted process for energy, environmental management and rural development. In: Ranalli P (ed) Improvement of crop plants for industrial end uses. Springer, pp 335–416. ISBN:13 978–1–4020-5486-0

Brito LM, Coutinho J, Smith SR (2008) Methods to improve the composting process of the solid fraction of dairy cattle slurry. Bioresour Technol 99:8955–8960. doi:10.1016/j.biortech.2008.05.005

Carlman I, Grönlund E, Longueville A (2015) Models and methods as support for sustainable decision-making with focus on legal operationalization. Ecol Model 306:95–100. doi:10.1016/j.ecolmodel.2014.10.010

Carlsson M, Lagerkvist A, Morgan-Sagastume F (2012) The effects of substrate pre-treatment on anaerobic digestion systems: a review. Waste Manage 32:1634–1650. doi:10.1016/j.wasman.2012.04.016

Chae KJ, Jang A, Yim SK, Kim IS (2008) The effects of digestion temperature and temperature shock on the biogas yields from the mesophilic anaerobic digestion of swine manure. Bioresour Technol 99:1–6. doi:10.1016/j.biortech.2006.11.063

Chen Y, Cheng JJ, Creamer KS (2008) Inhibition of anaerobic digestion process: a review. Bioresour Technol 99:4044–4064. doi:10.1016/j.biortech.2007.01.057

Chen JL, Ortiz R, Steele TWJ, Stuckey DC (2014) Toxicants inhibiting anaerobic digestion: a review. Biotechnol Adv 32:1523–1534. doi:10.1016/j.biortech.2007.01.057

Church RL (2002) Geographical information systems and location science. Comput Oper Res 29:541–562. doi:10.1016/S0305-0548(99)00104-5

Chynoweth DP, Owens JM, Legrand R (2001) Renewable methane from anaerobic digestion of biomass. Renew Energy 22:1–8. doi:10.1016/S0960-1481(00)00019-7

Cofie O, Amede T (2015) Water management for sustainable agricultural intensification and smallholder resilience in sub-Saharan Africa. Water Resour Rural Dev 6:3–11. doi:10.1016/j.wrr.2015.10.001

Dagnall S, Hill J, Pegg D (2000) Resource mapping and analysis of farm livestock manures—assessing the opportunities for biomass-to-energy schemes. Bioresour Technol 71:225–234. doi:10.1016/S0960-8524(99)00076-0

Davis J, O'Grady A, Dale A, Arthington A, Gell P, Driver P, Bond N, Casanova M, Finlayson M, Watts R, Capon S, Nagelkerken I, Tingley R, Fry B, Page T, Specht A (2015) When trends intersect: the challenge of protecting freshwater ecosystems under multiple land use and hydrological intensification scenarios. Sci Total Environ 534:65–78. doi:10.1016/j.scitotenv.2015.03.127

De Vrieze J, Hennebel T, Boon N, Verstraete W (2012) Methanosarcina: the rediscovered methanogen for heavy duty biomethanation. Bioresour Technol 112:1–9. doi:10.1016/j.biortech.2012.02.079

De Vrieze J, De Lathouwer L, Verstraete W, Boon N (2013) High-rate iron-rich activated sludge as stabilizing agent for the anaerobic digestion of kitchen waste. Water Res 47:3732–3741. doi:10.1016/j.watres.2013.04.020

De Vrieze J, Gildemyn S, Vilchez-Vargas R, Jáuregui R, Pieper DH, Verstraete W, Boon N (2015) Inoculum selection is crucial to ensure operational stability in anaerobic digestion. Appl Microbiol Biot 99:189–199. doi:10.1007/s00253-014-6046-3

Dearman B, Marschner P, Bentham RH (2006) Methane production and microbial community structure in single-stage batch and sequential batch systems anaerobically co-digesting food waste and biosolids. Appl Microbiol Biotechnol 69:589–596. doi:10.1007/s00253-005-0076-9

Demartini E, Gaviglio A, Bertoni D (2015) Integrating agricultural sustainability into policy planning: a geo-referenced framework based on Rough Set Theory. Environ Sci Policy 54:226–239. doi:10.1016/j.envsci.2015.07.006

Deng L, Yang H, Liu G, Zheng D, Chen Z, Liu Y, Pu X, Song L, Wang Z, Lei Y (2014) Kinetics of temperature effects and its significance to the heating strategy for anaerobic digestion of swine wastewater. Appl Energy 134:349–355. doi:10.1016/j.apenergy.2014.08.027

Esposito G, Frunzo L, Giordano A, Liotta F, Panico A, Pirozzi F (2012) Anaerobic co-digestion of organic wastes. Rev Environ Sci Biotechnol 11:325–341. doi:10.1007/s11157-012-9277-8

Farahani RZ, SteadieSeifi M, Asgari N (2010) Multiple criteria facility location problems: a survey. Appl Math Model 34:1689–1709. doi:10.1016/j.apm.2009.10.005

Fernandez-Mena H, Nesme T, Pellerin S (2016) Towards an Agro-Industrial Ecology: a review of nutrient flow modelling and assessment tools in agro-food systems at the local scale. Sci Total Environ 543:467–479. doi:10.1016/j.scitotenv.2015.11.032

Ferretti V, Pomarico S (2012) Integrated sustainability assessments: a spatial multi-criteria evaluation for siting a waste incinerator plant in the Province of Torino (Italy). Environ Dev Sustain 14:843–867. doi:10.1007/s10668-012-9354-8

Fodor Z, Klemeš JJ (2012) Waste as alternative fuel—minimising emissions and effluents by advanced design. Process Saf Environ Protect 90(3):263–284. doi:10.1016/j.psep.2011.09.004

Gaudino S, Goia I, Grignani C, Monaco S, Sacco D (2014) Assessing agro-environmental performance of dairy farms in northwest Italy based on aggregated results from indicators. J Environ Manage 140:120–134. doi:10.1016/j.jenvman.2014.03.010

Gómez A, Zubizarreta J, Rodrigues M, Dopazo C, Fueyo N (2010) Potential and cost of electricity generation from human and animal waste in Spain. Renew Energy 35:498–505. doi:10.1016/j.renene.2009.07.027

Goodchild M, Li L (2012) Assuring the quality of volunteered geographic information. Spatial Stat 1:110–120. doi:10.1016/j.spasta.2012.03.002

Greer D (2011) Funding anaerobic digestion facilities. BioCycle 52:70–73

Habitats Directive (1992) Council Directive Habitats (92/43/EEC) on the conservation of natural habitats and of wild fauna and flora

Harris PW, McCabe BK (2015) Review of pre-treatments used in anaerobic digestion and their potential application in high-fat cattle slaughterhouse wastewater. Appl Energy 155:560–575. doi:10.1016/j.apenergy.2015.06.026

Höhn J, Lehtonen E, Rasi S, Rintala J (2014) A Geographical Information System (GIS) based methodology for determination of potential biomasses and locals for biogas plants in southern Finland. Appl Energy 113:1–10. doi:10.1016/j.apenergy.2013.07.005

Holm-Nielsen JB, Al Seadi T, Oleskowicz-Popiel P (2009) The future of anaerobic digestion and biogas utilization. Bioresour Technol 100:5478–5484. doi:10.1016/j.biortech.2008.12.046

Huang B, Shi X, Yu D, Öborn I, Blombäck K, Pagella T, Wang H, Sun W, Sinclair F (2006) Environmental assessment of small-scale vegetable farming systems in peri-urban areas of the Yangtze River Delta Region, China. Agric Ecosyst Environ 112:391–402. doi:10.1016/j.agee.2005.08.037

INSPIRE Directive (2007) 2007/2/EC of the European Parliament and of the Council of 14 March 2007 establishing an Infrastructure for Spatial Information in the European Community

Jeliazkov A, Mimet A, Chargé R, Jiguet F, Devictor V, Chiron F (2016) Impacts of agricultural intensification on bird communities: new insights from a multi-level and multi-facet approach of biodiversity. Agric Ecosyst Environ 216:9–22. doi:10.1016/j.agee.2015.09.017

Jiao W, Ouyang W, Hao F, Lin C (2015) Anthropogenic impact on diffuse trace metal accumulation in river sediments from agricultural reclamation areas with geochemical and isotopic approaches. Sci Total Environ 536:609–615. doi:10.1016/j.scitotenv.2015.07.118

Jingura RM, Matengaifa R (2009) Optimization of biogas production by anaerobic digestion for sustainable energy development in Zimbabwe. Renew Sust Energy Rev 13:1116–1120. doi:10.1016/j.rser.2007.06.015

Kelly C, Ferrara A, Wilson G, Ripullone F, Nolè A, Harmer N, Salvati L (2015) Community resilience and land degradation in forest and shrubland socio-ecological systems: evidence from Gorgoglione, Basilicata, Italy. Land Use Policy 46:11–20. doi:10.1016/j.landusepol.2015.01.026

Khalid A, Arshad M, Anjum M, Mahmood T, Dawson L (2011) The anaerobic digestion of solid organic waste. Waste Manage 31:1737–1744. doi:10.1016/j.wasman.2011.03.021

Kleerebezem R, van Loosdrecht MCM (2007) Mixed culture biotechnology for bioenergy production. Curr Opin Biotechnol 18:207–212. doi:10.1016/j.copbio.2007.05.001

Kondusamy D, Kalamdhad AS (2014) Pre-treatment and anaerobic digestion of food waste for high rate methane production—a review. J Environ Chem Eng 2:1821–1830. doi:10.1016/j.jece.2014.07.024

Kothari R, Pandey AK, Kumar S, Tyagi VV, SK T (2014) Different aspects of dry anaerobic digestion for bio-energy: an overview. Renew Sustain Energy Rev 39:174–195. doi:10.1016/j.wasman.2011.03.021

Kousksou T, Allouhi A, Belattar M, Jamil A, El Rhafiki T, Arid A, Zeraouli Y (2015) Renewable energy potential and national policy directions for sustainable development in Morocco. Renew Sust Energy Rev 47:46–57. doi:10.1016/j.rser.2015.02.056

Kwietniewska E, Tys J (2014) Process characteristics, inhibition factors and methane yields of anaerobic digestion process, with particular focus on microalgal biomass fermentation. Renew Sust Energy Rev 34:491–500. doi:10.1016/j.rser.2014.03.041

Lefebvre M, Espinosa, Gomez M, Paloma S (2012) The influence of the Common Agricultural Policy on agricultural landscape, JRC Scientifics and Policy Reports. European Commission. Report EUR 25459 EN. 75 pp and Annexes

Li Y, Park SY, Zhu J (2011) Solid-state anaerobic digestion for methane production from organic waste. Renew Sust Energy Rev 15:821–826. doi:10.1016/j.rser.2010.07.042

Li L, He Q, Ma Y, Wang X, Peng X (2015) Dynamics of microbial community in a mesophilic anaerobic digester treating food waste: relationship between community structure and process stability. Bioresour Technol 189:113–120. doi:10.1016/j.biortech.2015.04.015

Lima M, Romanelli A, Massone H (2015) Assessing groundwater pollution hazard changes under different socio-economic and environmental scenarios in an agricultural watershed. Sci Total Environ 530–531:333–346. doi:10.1016/j.scitotenv.2015.05.026

Liu XY, Ding HB, Wang JY (2010) Food waste to bioenergy. In: Khanal SK, Surampalli RY, Zhang TC, Lamsal BP, Tyagi RD, Kao CM (ed) Bioenergy and biofuel from biowastes and biomass. American Society of Civil Engineers, pp 43–70. ISBN:978–0–7844-1089-9. doi:10.1061/9780784410899.ch03

Lönnqvist T, Silveira S, Sanches-Pereira A (2013) Swedish resource potential from residues and energy crops to enhance biogas generation. Renew Sust Energy Rev 21:298–314. doi:10.1016/j.rser.2012.12.024

Ma J, Scott NR, DeGloria S, Lembo AJL (2005) Siting analysis of farm-based centralized anaerobic digester systems for distributed generation using GIS. Biomass Bioenergy 28:591–600. doi:10.1016/j.biombioe.2004.12.003

Madlener R, Schmid C (2009) Spatial diffusion of biogas technology in Switzerland: a GIS-based multi-agent simulation approach. Int J Environ Pollut 39:28–43. doi:10.1504/IJEP.2009.027141

Madlener R, Hengeller Antunes C, Dias LC (2009) Assessing the performance of biogas plants with multi-criteria and data envelopment analysis. Eur J Oper Res 197:1084–1094. doi:10.1016/j.ejor.2007.12.051

Madsen M, Holm-Nielsen JB, Esbensen KH (2011) Monitoring of anaerobic digestion processes: a review perspective. Renew Sust Energy Rev 15:3141–3155. doi:10.1016/j.rser.2011.04.026

Maeng H, Lund H, Hvelplund F (1999) Biogas plants in Denmark: technological and economic developments. Appl Energy 64:195–206. doi:10.1016/S0306-2619(99)00067-7

Malczewski J (1999) GIS and multi-criteria decision analysis. Wiley, New York, NY, 408 p. ISBN:978–0–471-32944-2

Malczewski J (2006) GIS-based multi-criteria decision analysis: a survey of the literature. Int J Geogr Inf Sci 20:703–726. doi:10.1080/13658810600661508

Manos B, Bartocci P, Partalidou M, Fantozzi F, Arampatzis S (2014) Review of public–private partnerships in agro-energy districts in Southern Europe: the cases of Greece and Italy. Renew Sust Energy Rev 39:667–678. doi:10.1016/j.rser.2014.07.031

Mao C, Feng Y, Wang X, Ren G (2015) Review on research achievements of biogas from anaerobic digestion. Renew Sust Energy Rev 45:540–555. doi:10.1016/j.rser.2015.02.032

Marchettini N, Ridolfi R, Rustici M (2007) An environmental analysis for comparing waste management options and strategies. Waste Manage 27:562–571. doi:10.1016/j.wasman.2006.04.007

Maspolim Y, Zhou Y, Guo C, Xiao K, Ng WJ (2015) Comparison of single-stage and two-phase anaerobic sludge digestion systems—performance and microbial community dynamics. Chemosphere 140:54–62. doi:10.1016/j.rser.2015.02.032

Mastrangelo M, Weyland F, Herrera L, Villarino S, Barral M, Auer A (2015) Ecosystem services research in contrasting socio-ecological contexts of Argentina: critical assessment and future directions. Ecosyst Serv 16:63–73. doi:10.1016/j.ecoser.2015.10.001

Mata-Alvarez J, Dosta J, Romero-Güiza MS, Fonoll S, Peces M, Astals S (2014) A critical review on anaerobic co-digestion achievements between 2010 and 2013. Renew Sust Energy Rev 36:412–427. doi:10.1016/j.rser.2014.04.039

Molino A, Nanna F, Ding Y, Bikson B, Braccio G (2013) Biomethane production by anaerobic digestion of organic waste. Fuel 103:1003–1009. doi:10.1016/j.fuel.2012.07.070

Murphy JD, Power N (2009) Technical and economic analysis of biogas production in Ireland utilizing three different crop rotations. Appl Energy 10:3–15. doi:10.1016/j.apenergy.2008.03.015

Murphy JD, McKeogh E, Kiely G (2004) Technical/economic/environmental analysis of biogas utilization. Appl Energy 77:407–427. doi:10.1016/j.apenergy.2003.07.005

Neves LCM, Converti A, Vessoni TC (2009) Biogas production: new trends for alternative energy sources in rural and urban zones. Chem Eng Technol 32:1147–1153. doi:10.1002/ceat.200900051

Nielsen HB, Angelidaki I (2008) Strategies for optimizing recovery of the biogas process following ammonia inhibition. Bioresour Technol 99:7995–8001. doi:10.1016/j.biortech.2008.03.049

Nitrates Directive (1991) Council Directive (91/676/EEC) Council Directive of 12 December 1991 concerning the protection of waters against pollution by nitrates from agricultural sources

Nzila C, Dewulf J, Spanjers H, Tuigong D, Kiriamiti H, van Langenhove H (2012) Multi criteria sustainability assessment of biogas production in Kenya. Appl Energy 93:496–506. doi:10.1016/j.apenergy.2011.12.020

Padgham J, Jabbour J, Dietrich K (2015) Managing change and building resilience: a multi-stressor analysis of urban and peri-urban agriculture in Africa and Asia. Urban Climate 12:183–204. doi:10.1016/j.uclim.2015.04.003

Panichelli L, Gnansounou E (2008) GIS-based approach for defining bioenergy facilities location: ya case study in Northern Spain based on marginal delivery costs and resources competition between facilities. Biomass Bioenergy 32:289–300. doi:10.1016/j.biombioe.2007.10.008

Park YS, Baehr C, Larocque G, Sánchez-Pérez J, Sauvage S (2015) Special issue: ecological modelling for ecosystem sustainability. 19th ISEM Conference, 28–31 October 2013, Toulouse, France Ecological Modelling, p 306. doi:10.1016/j.ecolmodel.2015.04.008

Perpiña C, Martínez-Llario JC, Pérez-Navarro Á (2013) Multi-criteria assessment in GIS environments for siting biomass plants. Land Use Policy 31:326–335. doi:10.1016/j.landusepol.2012.07.014

Powley H, Krom M, Emeis KC, Van Cappellen P (2014) A biogeochemical model for phosphorus and nitrogen cycling in the Eastern Mediterranean Sea: part 2. Response of nutrient cycles and primary production to anthropogenic forcing: 1950–2000. J Marine Syst 139:420–432. doi:10.1016/j.jmarsys.2014.08.017

Prochazka J, Dolejs P, Máca J, Dohanyos M (2012) Stability and inhibition of anaerobic processes caused by insufficiency or excess of ammonia nitrogen. Appl Microbiol Biotechnol 93:439–447. doi:10.1007/s00253-011-3625-4

Rajabifard A (2007) Towards a spatially enabled society. The University of Melbourne, 400 p. ISBN:978–0–7325-1620–8

Rajabifard A, Eagleson S (2013) Spatial enablement from an international context—a vision for the North and West Melbourne Corridor. In: Rajabifard A, Eagleson S (eds) Spatial data access and integration to support liveability: a case study in North and West Melbourne. University of Melbourne, Victoria, Australia, pp 1–10. ISBN:978–0–9922918-0-8

Rajagopal R, Massé DI, Singh G (2013) A critical review on inhibition of anaerobic digestion process by excess ammonia. Bioresour Technol 143:632–641. doi:10.1016/j.biortech.2013.06.030

Rammel C, Stagl S, Wilfing H (2007) Managing complex adaptive systems—a coevolutionary perspective on natural resource management. Ecol Econ. doi:10.1016/j.ecolecon.2006.12.014

Rao B, Mane A, Rao AB, Sardeshpande V (2014) Multi-criteria analysis of alternative biogas technologies. Energy Procedia 54:292–301. doi:10.1016/j.egypro.2014.07.272

Rico C, Muñoz N, Fernández J, Rico JL (2015) High-load anaerobic co-digestion of cheese whey and liquid fraction of dairy manure in a one-stage UASB process: limits in co-substrates ratio and organic loading rate. Chem Eng J 262:794–802. doi:10.1016/j.cej.2014.10.050

Rocha J, Roebeling P, Rial-Rivas M (2015) Assessing the impacts of sustainable agricultural practices for water quality improvements in the Vouga catchment (Portugal) using the SWAT model. Sci Total Environ 536:48–58. doi:10.1016/j.scitotenv.2015.07.038

Rodrigues A, Ferraz A, Mamede J, Alonso J (2009) Tratamento de efluentes agro-pecuários por digestão anaeróbia num Reactor Descontínuo Sequencial à temperatura ambiente. In: Proceedings of 3rd International congress of energy, environmental engineering and management, pp 83. ISBN:13 978–84–92669-15-8

Rodrigues A, Ferraz A, Alonso J (2010) Evaluation of the potential of anaerobic digestion in a sequencing batch reactor for the treatment of dairy wastewaters. In: Cordovil C, Ferreira L (eds) Proceedings of the 14th Ramiran International conference on treatment and use of organic residues in agriculture: challenges and opportunities towards sustainable management. ISBN:978–972–8669-47-8

Rosúa JM, Pasadas M (2012) Biomass potential in Andalusia, from grapevines, olives, fruit trees and poplar, for providing heating in homes. Renew Sust Energy Rev 16:4190–4195. doi:10.1016/j.rser.2012.02.035

Salomon KR, Lora EES (2009) Estimate of the electric energy generating potential for different sources of biogas in Brazil. Biomass Bioenergy 33:1101–1107. doi:10.1016/j.biombioe.2009.03.001

Sánchez-Lozano JM, Hengeller Antunes C, García-Cascales MS, Dias LC (2014) GIS-based photovoltaic solar farms site selection using ELECTRE-TRI: evaluating the case for Torre Pacheco, Murcia, Southeast of Spain. Renew Energy 66:478–494. doi:10.1016/j.renene.2013.12.038

Schievano A, Tenca A, Lonati S, Manzini E, Adani F (2014) Can two-stage instead of one-stage anaerobic digestion really increase energy recovery from biomass? Appl Energy 124:335–342. doi:10.1016/j.apenergy.2014.03.024

Schutter O, Vanloqueren G (2011) The New Green Revolution: how twenty-first-century science can feed the world. Solutions 2(4):33–44

Scott JA, Ho W, Dey PK (2012) A review of multi-criteria decision-making methods for bioenergy systems. Energy 42:146–156. doi:10.1016/j.energy.2012.03.074

Shah FA, Mahmood Q, Shah MM, Pervez A, Asad SA (2014) Microbial ecology of anaerobic digesters: the key players of anaerobiosis. Sci World J. doi:10.1155/2014/183752
Sheets JP, Yang L, Ge X, Wang Z, Li Y (2015) Beyond land application: emerging technologies for the treatment and reuse of anaerobically digested agricultural and food waste. Waste Manage 44:94–115. doi:10.1016/j.wasman.2015.07.037
Shilton A, Powell N, Broughton A, Pratt C, Pratt S, Pepper C (2013) Enhance biogas production using cow manure to stabilize co-digestion of whey and primary sludges. Environ Technol 34 (17):2491–2496. doi:10.1080/09593330.2013.774032
Silva S, Almeida-Alçada L, Dias L (2014) Biogas plants site selection integrating multi-criteria decision aid methods and GIS techniques: a case study in a Portuguese region. Biomass Bioenergy 71:58–68. doi:10.1016/j.biombioe.2014.10.025
Smith DB, Almquist CB (2013) The anaerobic co-digestion of fruit and vegetable waste and horse manure mixtures in a bench-scale, two-phase anaerobic digestion system. Environ Technol 35:859–867. doi:10.1080/09593330.2013.854398
Squire G, Hawes C, Valentine T, Young M (2015) Degradation rate of soil function varies with trajectory of agricultural intensification. Agric Ecosyst Environ 202:60–167. doi:10.1016/j.agee.2014.12.004
Steinhäußer R, Siebert R, Steinführer A, Hellmich M (2015) National and regional land-use conflicts in Germany from the perspective of stakeholders. Land Use Policy 49:183–194. doi:10.1016/j.landusepol.2015.08.009
Sultana A, Kumar A (2012) Optimal siting and size of bioenergy facilities using geographic information system. Appl Energy 94:192–201. doi:10.1016/j.apenergy.2012.01.052
Surendra KC, Takara D, Hashimoto AG, Khanal SK (2014) Biogas as a sustainable energy source for developing countries: opportunities and challenges. Renew Sust Energy Rev 31:846–859. doi:10.1016/j.rser.2013.12.015
Taleghani G, Kia AS (2005) Technical-economical analysis of the Saveh biogas power plant. Renew Energy 30:441–446. doi:10.1016/j.renene.2004.06.004
Tauseef SM, Premalatha M, Abbasi T, Abbasi SA (2013) Methane capture from livestock manure. J Environ Manage 117:187–207. doi:10.1016/j.jenvman.2012.12.022
Tavares G, Zsigraiová Z, Semiao V (2011) Multi-criteria GIS-based siting of an incineration plant for municipal solid waste. Waste Manage 31:1960–1972. doi:10.1016/j.wasman.2011.04.013
Tchobanoglous G, Burton FL, Stensel HD (2003) Wastewater engineering: treatment and reuse, 4th edn. Metcalf & Eddy. ISBN:0–07–112250–8
Thackway R, Specht A (2015) Edited by Specht A, Gordon I, Groves R, Lambers H, Phinn S. Reprint of: Synthesising the effects of land use on natural and managed landscapes. In: Catalysing transdisciplinary synthesis in ecosystem science and management. Sci Total Environ 534:14–30. doi:10.1016/j.scitotenv.2015.06.093
Thompson E, Wang Q, Li M (2013) Anaerobic digester systems (ADS) for multiple dairy farms: a GIS analysis for optimal site selection. Energy Policy 61:114–124. doi:10.1016/j.enpol.2013.06.035
Tricase C, Lombardi M (2009) State of the art and prospects of Italian biogas production from animal sewage: technical-economic considerations. Renew Energy 34:477–485. doi:10.1016/j.renene.2008.06.013
Turral H, Burke J, Faurès JM (2011) Climate Change, water and food security. FAO Water Report 36. ISBN:978–92–5-106795-6
Tuzkaya G, Önüt S, Tuzkaya UR, Gülsün B (2008) An analytic network process approach for locating undesirable facilities: an example from Istanbul, Turkey. J Environ Manage 88:970–983. doi:10.1016/j.jenvman.2007.05
Voivontas D, Assimacopoulos D, Koukios EG (2001) Assessment of biomass potential for power production: a GIS based method. Biomass Bioenergy 20:101–112. doi:10.1016/S0961-9534(00)00070-2
Ward AJ, Hobbs PJ, Holliman PJ, Jones DL (2008) Optimisation of the anaerobic digestion of agricultural resources. Bioresour Technol 99:7928–7940. doi:10.1016/j.biortech.2008.02.044

Water Framework Directive (2000) Council Directive Directive 2000/60/EC of the European Parliament and of the Council establishing a framework for the Community action in the field of water policy

Wilkinson KG (2011) A comparison of the drivers influencing adoption of on-farm anaerobic digestion in Germany and Australia. Biomass Bioenergy 35:1613–1622. doi:10.1016/j.biombioe.2011.01.013

Winarso H, Hudalah D, Firman T (2015) Peri-urban transformation in the Jakarta metropolitan area. Habitat Int 49:221–229. doi:10.1016/j.habitatint.2015.05.024

Yangang F, Jisheng L (2014) The modification of North China quadrangles in response to rural social and economic changes in agricultural villages: 1970–2010s. Land Use Policy 39:266–280. doi:10.1016/j.landusepol.2014.02.009

Yenigün O, Demirel B (2013) Ammonia inhibition in anaerobic digestion: a review. Process Biochem 48:901–911. doi:10.1016/j.procbio.2013.04.012

Zhang H, Yi S, Wu Y (2012) Decision support system and monitoring of eco-agriculture based on WebGIS in Shule Basin. Energy Procedia 14:382–386. doi:10.1016/j.egypro.2011.12.946

Zhang C, Su H, Baeyens J, Tan T (2014) Reviewing the anaerobic digestion of food waste for biogas production. Renew Sust Energy Rev 38:383–392. doi:10.1016/j.rser.2014.05.038

Zhu J, Zhang Z, Miller C (2006) A laboratory scale sequencing batch reactor with the addition of acetate to remove nutrient and organic matter in pig slurry. Biosyst Eng 93(4):437–446. doi:10.1016/j.biosystemseng.2006.01.010

Zubaryeva A, Zaccarelli N, Del Giudice C, Zurlini G (2012) Spatially explicit assessment of local biomass availability for distributed biogas production via anaerobic co-digestion –Mediterranean case study. Renew Energy 39:261–270. doi:10.1016/j.renene.2011.08.021

Microbial Fuel Cells: Types and Applications

Ravinder Kumar, Lakhveer Singh, and A.W. Zularisam

Abstract Microbial fuel cells (MFCs) are bioelectrochemical devices that convert the chemical energy present in organic or inorganic compounds into electric current by using microorganisms as the catalysts. MFCs are of different types; however, the basic designs used in the laboratories for its applications include double-chamber MFC, single-chamber MFC, upflow MFC and stacked MFC. Moreover, some other designs have also been used for the studies. The type of electrode materials and proton exchange membrane (PEM) used in MFCs has most significant role for its outcomes for different applications such as bioelectricity generation, wastewater treatment, bioremediation of toxic compounds, biohydrogen production and biosensors. Furthermore, MFCs are operated at the optimized parameters such as thermophilic temperatures, neutral pH, etc. to obtain more significant results for respective application. This chapter explores the various types of MFCs, the operational parameters to improve its performance and the most studied applications of the MFCs.

Keywords Microbial fuel cells • Catalysts • Wastewater treatment • Bioelectricity generation • Biosensors • Proton exchange membrane

1 Introduction

The microbial fuel cell (MFC) technology is one of the most attractive technologies at present for renewable energy production and simultaneous wastewater treatment. MFCs are the bioelectrochemical devices that utilize microorganisms as the biocatalysts to convert the chemical energy present in organic or inorganic compounds into electric current (Aelterman et al. 2006; Bermek et al. 2014; Kumar et al. 2016). A typical double-chamber MFC is made up of two chambers, i.e. the anode and the cathode. Usually a proton exchange membrane (PEM) is placed between these two chambers that allows the protons produced at the anode to pass

R. Kumar • L. Singh (✉) • A.W. Zularisam
Faculty of Engineering Technology, Universiti Malaysia Pahang (UMP), Lebuhraya Tun Razak, Gambang, 26300 Kuantan, Pahang, Malaysia
e-mail: lucki.chem09@gmail.com; lakhveer@ump.edu.my

L. Singh, V.C. Kalia (eds.), *Waste Biomass Management – A Holistic Approach*,
DOI 10.1007/978-3-319-49595-8_16

through itself to the cathode. The cathode and the anode are connected by an electrical circuit (e.g. with titanium wires or copper wires) to make it a complete system. The organic substrates are oxidized by the microorganisms at the anode chamber and produce electrons, protons and carbon dioxide. The electrons generated from the microbial metabolic activity are firstly transferred to the anode surface by redox-active proteins or cytochromes and then passed to the cathode through electrical circuit (Borole et al. 2011; Kumar et al. 2015). At the cathode chamber, the reduction of electrons takes place. Generally an electron acceptor is provided at the cathode, e.g. oxygen or ferricyanide. Subsequently, electrons combine with protons and oxygen at the cathode and form water. This reaction can be further facilitated by a catalyst such as platinum. The general working process and the components of an MFC are showed in Fig. 1.

$$\text{Anode reaction}: CH_3COO^- + 2H_2O \quad 2CO_2 + 7H^+ + 8e^-$$
$$\text{Cathode reaction}: O_2 + 4e^- + 4H^+ \quad 2H_2O$$

MFC technology is an encouraging technology for electric current generation from diverse materials, such as natural organic matter and complex organic waste, and can be beneficially united with applications in wastewater treatment (Chaudhuri and Lovely 2003; Inoue et al. 2010; Jiang et al. 2014). There are some unique characteristics of MFCs that make this technology more advantageous than other technologies. (1) The MFCs provide comparatively higher conversion proficiency for chemical energy to electric current. (2) The MFCs can produce fruitful results at varying temperature conditions (from 20 to 40 °C) that makes MFC technology unique to other present bioenergy practices. (3) During operation of MFCs, they do not require an external electric for aeration to provide oxygen (as electron acceptor) as the cathode can be passively aerated. MFCs are constructed in different designs using diverse materials. These systems are usually operated at optimized conditions to extract more energy from the system, but they can be also operated at varying conditions such as at low or high temperatures; acidic or basic pH, with different electron acceptors; etc. (Amend and Shock 2001; Logan 2004; Oh et al. 2010; Patil et al. 2011; Tang et al. 2015).

2 Materials for Construction of MFC

Anode. The materials used to make it as anode need to be a conductive material. Also the material should be environment-friendly and chemically inert to the electrolyte (anolyte) in the anode chamber. The electrodes made up of carbon materials are widely used in the MFCs, for examples, graphite plates, graphite rods, graphite felt electrodes, graphite granules, carbon cloth, carbon brush and stainless steel, etc. (Logan and Regan 2006; Liu et al. 2014). Generally, the electrode materials that are cheaper and exhibit higher surface area are of great interest in MFCs (Bergel et al. 2005). The bare electrodes that have low surface area

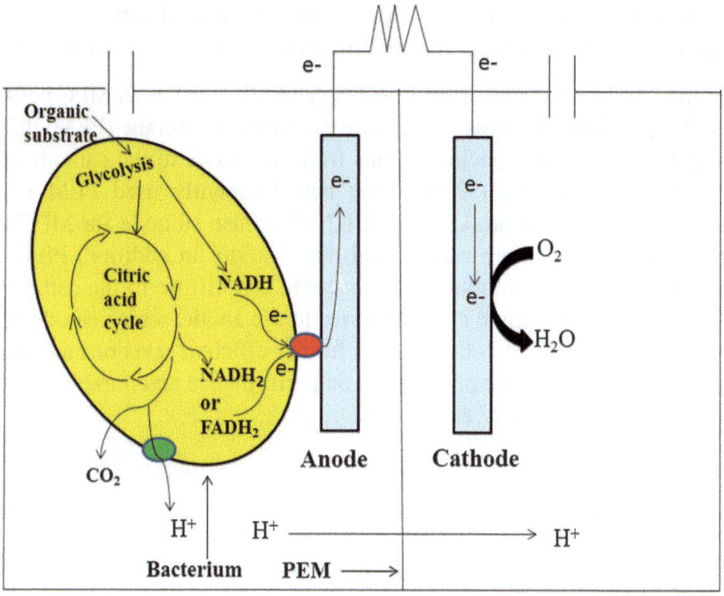

Fig. 1 General principle of a microbial fuel cell (Kumar et al. 2016)

can be easily modified with conductive nanomaterials of higher surface area (e.g. graphene) for improved MFC performance (Tang et al. 2015). There are many techniques to coat the nanomaterial on the bare electrodes. The anodic modifications with the nanoparticles have shown fruitful results so far in the MFCs. Such nanomodifications usually favour the biofilm formation on the anode, decrease the MFC start-up time, facilitate the electron transfer mechanism, also decrease the internal resistance of the system and consequently increase the overall performance of the MFC (Bergel et al. 2005; Jiang et al. 2014; Shen et al. 2014).

Cathode The cathode compartment contains the cathode material, a catalyst to increase the reduction of electrons and an electron acceptor (Wang et al. 2014; Zhang et al. 2012). The electrode materials used as the anode as mentioned above are/can be used as the cathode. Moreover, a catalyst (e.g. platinum) is employed to the cathode electrode to increase the rate of oxygen reduction when oxygen is used as the electron acceptor. Oxygen is one of the best alternatives for the electron acceptor as it is cheapest, abundantly available and has high redox potential (Bond and Lovley 2003; Rhoads et al. 2005; Huang et al. 2012). In contrast, ferricyanide (K_3 [Fe $(CN)_6$]) is a precisely common electron acceptor and has been widely used in MFCs due to its decent enactment. A catalyst is not required at the cathode when ferricyanide is used as the electron acceptor in MFCs. This is particularly because it exhibits a small overpotential with a plain carbon electrode (Huang et al. 2012). But the use of ferricyanide also has some limitations in MFCs, e.g. the inadequate reoxidation by oxygen, which needs the catholyte to be frequently changed.

Moreover, ferricyanide can also be diffused into the anode chamber through the ion exchange membrane; therefore, its use can decrease the performance of MFCs.

Membrane An ion exchange membrane is generally used in an MFC between the anode and the cathode chamber, e.g. proton exchange membrane (PEM) that allows the passage of protons or specific cations from the anode to the cathode compartment (Heilmann and Logan 2006). The best frequently used PEM is Nafion. However, in place of Nafion, Ultrex CMI-7000 is also suitable for MFC applications and is ominously more economical than Nafion. In addition, PEM may be leaky to oxygen, and the anolyte or the bacteria can diffuse to the cathode, while catholyte such as ferricyanide can also move to the anode, which can decrease the performance of an MFC effectively. But, further efficient revisions are indispensable to appraise the influence of the PEM on performance and durable permanence (Min and Logan 2004; Jafary et al. 2013).

3 Types of MFC

3.1 *Double-Chamber MFC*

Double-chamber MFC (see Fig. 2a) is the simplest design among all MFCs (Niessen et al. 2004; Phung et al. 2004; Kumar et al. 2016). In a typical design, one bottle (can be of different designs) is used as anode while the other one as cathode, separated by PEM. Usually in two-chamber MFC, defined medium (or substrate) in the anode and defined catholyte solution are used to generate energy. In other words, the double-chamber MFC is often operated in batch mode. The double-chamber MFC may be in the shape of bottles or cube. The choice of catholyte in the MFC can define the nomenclature of the design. For example, if the air is used in the cathode to provide the electron acceptor, i.e. oxygen, then the MFC can be called as two-chamber air-cathode MFC (Ringeisen et al. 2006; Shantaram et al. 2005). Such MFCs may prove valuable to generate electricity in remote sensing regions.

3.2 *Single-Chamber MFC (SCMFC)*

This type of MFC is made up of one chamber only that contains both the anode and the cathode (see Fig. 2b), which was introduced by Doo Hyun Park and J. Gregory Zeikus (2003). The anode is either positioned far or near to the cathode separated by PEM. It has been stated that by decreasing interelectrode spacing, internal ohmic resistance can also be diminished. This can be achieved by evading the use of catholyte as a result of joining two chambers and thus raises the power density. Such MFC is simple and economical and also produces much power in rival to double-chamber MFC (Chaudhuri and Lovley 2003; Ringeisen et al. 2006).

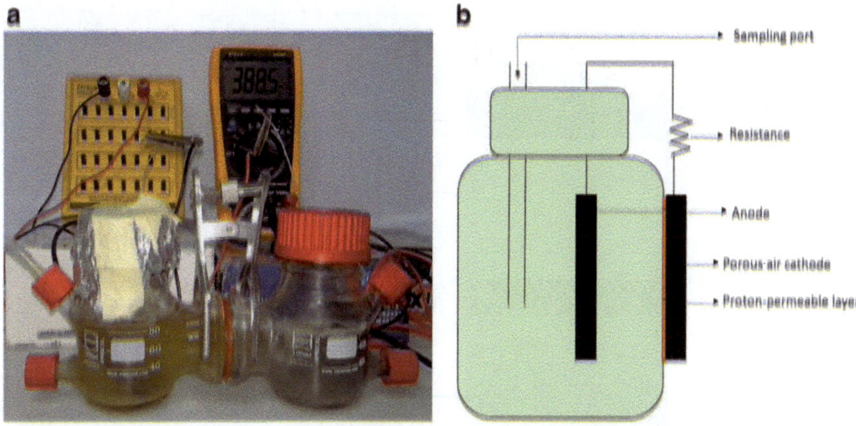

Fig. 2 (**a**) Simple design of double-chamber microbial fuel cell. (**b**) Schematic design of single-chamber microbial fuel cell. (**c**) Schematic design of upflow microbial fuel cell. (**d**) Design of stacked microbial fuel cell

However, in SCMFC the major problems such as microbial adulteration and reverse passage of oxygen from cathode to anode occur normally. SCMFCs propose simpler and economic designs. Such MFCs generally have simply an anodic chamber with no requisite of air in a cathodic chamber (Rabaey et al. 2004; Rabaey et al. 2005).

3.3 Upflow MFC

The new design came into existence with increase in the interest in MFC research. The upflow MFC is cylinder-shaped MFC (He et al. 2006). The MFC is made up of the cathode chamber at the top and the anode at the bottom. Both the chambers are apportioned by glass wool and glass bead layers. The substrate is provided from the bottom of the anode that moves upward to the cathode and leaves at the top (see Fig. 2c). A gradient is formed between the electrodes which also help in the favourable action of the fuel cell (Cheng et al. 2006a). In his design, there are no distinct anolyte and catholyte. Moreover, it does not have any physical parting. Therefore, proton transmission-related difficulties are very less (Zhou et al. 2013; Venkata Mohan et al. 2014).

Upflow mode MFCs are fascinating for wastewater treatment since they can be easily scaled up as compared to other designs of MFCs. However, the major drawback of the fuel cell is the energy costs to pump the substrate that are highly greater than their power outcome (Zhou et al. 2013). So we can conclude that the prime purpose of upflow MFC is wastewater treatment instead of power generation (Brutinel and Gralnick 2012). These kinds of MFCs are commonly employed in basic research, and the studies propose that the power densities are low due to high internal resistance, electrode-based losses and complex design.

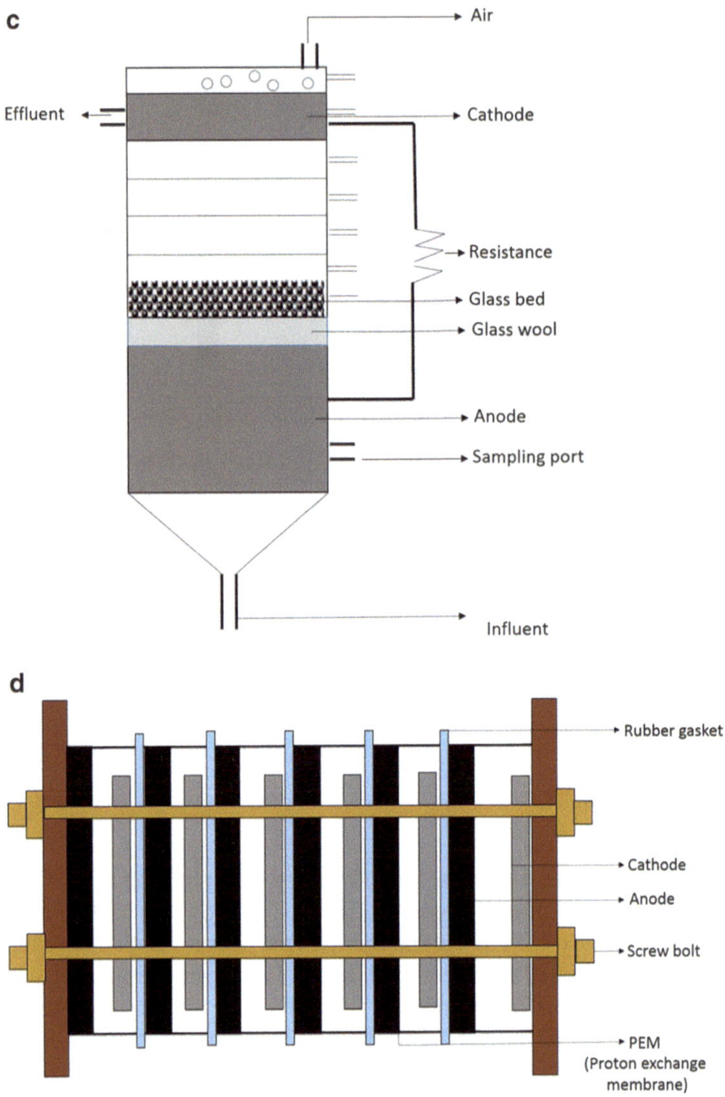

Fig. 2 (continued)

3.4 *Stacked MFC*

A stacked MFC is generally a combination of some MFCs that are either coupled in series or in parallel to enhance the power output (Logan and Regan 2006; Aelterman et al. 2006; Sun et al. 2012). The output of MFC is increased by connecting some MFCs by multiplying individual power output or current output. Generally a single unit of MFC (with oxygen as an electron acceptor) can generate a maximum open-circuit voltage (OCV) of 0.8 V (Logan et al. 2005). Therefore, a number of MFC unit cells can be stacked in a series or parallel connection, and their individual power output can be multiplied to get the resultant power output. However, some other operational factors also play an important role in current generation in a stacked MFC, which can either decrease or increase the overall performance of the MFC (Zhou et al. 2013; Venkata Mohan et al. 2014). Moreover, after connecting the individual MFC in a stacked MFC, the final voltage may not be exactly the total of the individual cell voltages because there will be a loss of voltage when every cell is connected either in series or parallel. The parallel-connected stack MFCs have proven to produce more current as in rival when the MFCs are stacked in a series connection. Therefore, we can say that in parallel-connected stack MFCs, higher bioelectrochemical reaction rate is achieved than in series-connected stack MFCs. Moreover, a parallel connection is favoured to maximize chemical oxygen demand (COD) removal for enhanced wastewater treatment efficiency, if the MFC units are not autonomously functioned. In a successful study, six MFCs were connected in series or in parallel with copper wires. The MFC was fed with acetate as substrate in the anode and ferricyanide as the catholyte, and graphite rods were used as the anode and the cathode. The results of the study demonstrated that the stacked MFC in a series connection produced a volumetric power density of 59 W/m^3 and when connected in parallel generated 51 W/m^3 of power density (Aelterman et al. 2006).

The coulombic efficiency (CE) in MFCs can give the estimation of the electrons transferred (from the total electrons generated theoretically from substrate oxidation) from the anode that help to generate the current (Kumar et al. 2015). In stacked MFCs, the different rates of CE can be achieved while linking the cells in different connections. The former study exhibited that the stacked MFC achieved higher rate of CE when operated in parallel, i.e. 78 %, than in series that produced the CE of only 12 %. The major obstacle in stack MFC to achieve higher voltage outputs is the voltage reversal. The voltage reversal may be due to the depletion of the substrate in the cell, pointing to the diminished ability of the bacteria to produce higher voltage (Logan and Regan 2006; Aelterman et al. 2006).

3.5 *Other Designs*

The basic designs of MFCs are overwhelmingly used for the studies, and less intention is paid to develop new designs to overcome the drawbacks of the existing designs. In order to diminish one obstacle, a flat-plate MFC (FPMFC) in 2004 was designed by Min and Logan to reduce the ohmic resistance that is caused due to more interelectrode spacing (Min and Logan 2004). This kind of design is generally used in chemical fuel cells, also generating more power than the former designs. In FPMFC, the anode and the cathode were made up of flat plates (each plate with a projected surface area of 225 cm^2). A Nafion membrane was placed between the two plates. This reactor generated the power density of 56 mW/m^2 with domestic wastewater as a substrate, and 58 % COD was achieved in the study (Min and Logan 2004). The reactor was also used to produce power with other substrates like acetate, glucose, starch, etc. but produced less power output than the other designs, i.e. cube reactor. It may be due to the too closely joined electrodes, and oxygen may pass through the membrane to the bacteria in the anode chamber, therefore affecting the growth of the microbial community (Phung et al. 2004; Patil et al. 2011).

4 Factors Affecting the Performance of MFCs

4.1 Electrode Material

In the MFCs , the anode and the cathode are made up of the electrode material that should be conductive in nature, non-corrosive, non-fouling to the bacteria (in case of anode) and cost-effective. The electrode material with high surface area also increases the performance of the MFC (Zhang et al. 2012; Alatraktchi et al. 2014). The development of the electrodes for the MFC has reached to the higher level. Moreover, many electrodes with modification of nanoparticles have produced electricity manifold than the plain electrodes. In the anode, the electrode modification with the nanomaterial or catalyst that can support the biofilm formation and increase the electron transfer rates is highly beneficial for increased power output. For example, magnetite nanoparticles increased the current production in the MFC using *G. sulfurreducens* as an inoculum (Alatraktchi et al. 2014). The study found that the nanoparticles increased the electrical conductivity of biofilm-electrode interface, thus boosted the electron transfer mechanism (Kumar et al. 2016). In the cathode, Pt or Pt-coated cathodes produced higher electric current as compared to plain cathodes containing no catalyst. Pt has been regarded the best catalyst for oxygen (when used as electron acceptor) reduction in MFCs (Bond and Lovely 2003), but its higher cost is also one of the obstacles for the technology to get launched at large-scale applications.

4.2 Proton Exchange System

The proton exchange system of an MFC consists of proton exchange membrane (PEM) between the anode chamber and the cathode chamber. The membrane contains the pores with charged side walls that help the movement of protons from anode to cathode. Hence this membrane has a pivotal function in the MFCs that can affect the power output of the MFCs (Gil et al. 2003; Cheng et al. 2006b). During the operation of the MFC, there might be a diffusion of the anolyte through the membrane to the catholyte. This can cause fouling of the membrane, subsequently can block the passage of protons to the cathode and consequently will decrease the power output of the MFC. Alternatively, the passage of catholyte to the anode chamber will affect the performance of the bacteria. For example, if the electron acceptor is oxygen, its diffusion to the anode can affect the biofilm growth of the anaerobic bacteria, consequently affecting the whole performance of the MFC (Reguera et al. 2005; Shen et al. 2014). Moreover, the membrane increases the internal resistance of the MFC, and the diffusion of ions or the electrolytes through it causes concentration polarization loss of the system, which has a direct effect on the current generation of the fuel cell. Nafion membrane is the most commonly used PEM in the MFC technology. This is because the membrane is selectively permeable to the protons (Gil et al. 2003). Though, some other cations also transported through Nafion during the MFC operation. But, the use of membrane is highly beneficial to maintain the balance of the charge between the anolyte and the catholyte (Raghavulu et al. 2013). Moreover, it is also found that the ratio of membrane surface area to volume of MFC is vital for electrical output of the system. The results of a study suggested that MFCs with PEM of large surface area generate less internal resistance and hence the higher power output. Some studies have also shown that the MFCs without a PEM produce more power output up to a specific period than membrane-less MFC (Gil et al. 2003). The higher cost of the PEM is also one of its disadvantages. Therefore, we can imagine membrane-less MFCs in the future at large-scale applications.

4.3 Operational Condition

This section includes only the basic parameters for the MFC operation, such as pH, temperature, organic loading, feed rate and shear stress. It is found that performances of MFCs at the small scale such as in the laboratory are still very low than the ideal performances. Therefore, it is highly beneficial to optimize the operating conditions to improve the overall performance of the MFCs.

4.3.1 Effect of pH

In the anodic chamber of the MFCs, the microorganisms oxidize the substrates and produce protons that move to the cathode through the PEM where they combine with electrons and an electron acceptor (if oxygen) to form water. After a long period of the MFC operation, the concentration of protons increases in the anolyte due to the slow or restricted flow of the protons via the PEM. This accumulation of the protons makes the anode chamber more acidic and much unfavourable for the bacterial growth though bacteria are more active at nearly neutral pH (Venkata Mohan et al. 2014). On the second hand, the cathode chamber becomes more alkaline due to the deficiency of replacement of protons from the oxidation reaction and constant diminution of protons by the reduction reaction (He et al. 2006). This pH concentration gradient across the two chambers leads to an electrochemical/thermodynamic drawback on MFC performance.

The high pH in the cathode chamber can substantially reduce the current generation. The potential of the oxygen reduction increases with a decrease in pH value (according to the Nernst equation). Therefore, the low operational pH is advantageous for the oxygen reduction and subsequently to achieve higher electric current from MFCs (Niessen et al. 2004; Phung et al. 2004; Kumar et al. 2016). Usually, bacteria need a neutral pH for their optimum growth, and bacteria counter to the variations in internal and external pH by regulating their activities. The bacterial growth requirements can change the pH in the anode chamber which can further cause some variations in primary physiological parameters, such as membrane potential, concentration of ions, biofilm formation and proton-motive force (Kumar et al. 2016).

The pH maintenance in the anodic compartment is one of the significant factors, which affects microbial metabolic activity and subsequently disturbs the electron and proton generation mechanism (Kumar et al. 2015). The lower pH in the anolyte decreases the bacterial activity and therefore affects the biofilm formation and current output of the MFC. A study revealed that MFCs worked at low anodic pH exhibited higher proton transfer rates; therefore, higher amount of protons accumulated at the cathode side that further diminished the current density (Bermek et al. 2014). However, an alternative study validated that the pH ranging from 6 to 9 is appropriate for microbial growth and achieving comparatively higher power outputs.

It is more convenient to sustain two different pH conditions in a dual-chamber MFC to optimize the anodic and cathodic reactions as compared to a single-chamber MFC (Huang et al. 2012). On the other hand, it is more difficult to maintain the pH in air-cathode MFC, because of the presence of one electrolyte only that is present in the anode chamber, referred to as anolyte. The electrolyte present in the cathode is referred to as catholyte (in two-chamber MFC). The air-cathode MFCs (the cathode electrode is exposed to the air) exhibit simpler configuration and have proven to generate more power outputs than the conventional MFCs (Bermek et al. 2014). A study demonstrated that the air-cathode MFC can be operated with an anolyte of pH range between 8 and 10. The results exhibited that the anodic bacterial activity was optimum at a neutral pH, while the oxygen reduction reaction was amended at a higher pH.

4.3.2 Effect of Temperature

In MFCs, the temperature is one of the most important factors that affect the kinetics of the whole system (Amend and Shock 2001). Any large deviation in the temperature during the operation of MFC may affect its performance to a great extent. It mainly affects the microbial metabolism, mass transfer and thermodynamics (electrode potentials and Gibbs free energy). The temperature has been included a vital parameter for the MFC's performance toward electricity generation and wastewater treatment efficiency (Amend and Shock 2001; Logan 2004; Oh et al. 2010; Patil et al. 2011; Tang et al. 2015). Usually, the MFCs are operated at the room temperature nearly in the range between 25 and 30 °C. However, MFCs have shown good results (with respect to both, i.e. power density and COD removal) at higher temperatures as well (Jafary et al. 2013). The increase in power density may be due to the augmentation of the bacterial metabolism and membrane permeability. Moreover, the high temperature can increase the conductivity of the anolyte as well as the catholyte thus decreasing the ohmic resistance of the MFC and hence rising the power density (Phung et al. 2004). It has been observed that the slight increase in the temperature does not the membrane permeability. The studies have shown that the effect of temperature on bacterial activity follows an exponential trend. Therefore, the enhanced power output due to increase in temperature can be directly attributed to the increased bacterial activity. The bacterial activity can be determined in terms of biofilm growth at the anode which has an impact on the current production (Wang et al. 2014; Yang et al. 2015). The studies have shown that the temperature plays a vital role in the start-up of the MFC and hence in the initial biofilm formation. It has been observed that the higher temperatures decrease the start-up time of the MFC operation and lead to the stable biofilm formation (Phung et al. 2004).

A study found that the temperature ranges between 30 and 45 °C are more beneficial for the operation of MFCs to obtain higher power outputs because the bacterial biofilms showed maximum catalytic activity between the mentioned temperature ranges (Yong et al. 2014). However, some particular bacterial species can work effectively in a particular temperature range; it is useful to adjust the temperature range during the MFC operation to achieve the maximum outputs from the system (Kumar et al. 2015). Moreover, the variation in the temperature can result in the different microbial communities in the anodic chamber of the MFC (Phung et al. 2004). After the establishment of a stable biofilm, the bacteria can adjust their metabolism according to the small variations in the temperature.

4.3.3 Feed Rate and Shear Stress

The MFCs can be operated in two modes, the first is batch mode and the second is continuous mode. In the batch mode, the substrate is provided in the initiation of the cycle; on the other hand, in the continuous mode, the substrate is provided after short intervals in the cycle. The operation of MFCs in the continuous mode exhibits hydrodynamic problems that further affect the whole performance of the system. Therefore, the flow rate and the subsequent hydraulic retention time (HRT) and the shear stress are important parameters that should be optimized for MFC operation to obtain the maximum output from the fuel cell (Zhou et al. 2013; Venkata Mohan et al. 2014). It has been found that flow rate affects the performance of MFCs for both power density and COD removal. The studies suggest that higher flow rates decrease the power output as well as COD removal efficiency and coulombic efficiency (Logan et al. 2005; Sun et al. 2012). In practice, the higher flow rates decrease the HRT. It means the bacteria get less time to oxidize the substrate, therefore waning the COD removal efficiency of the MFC. Moreover, another vital parameter in MFCs is the hydrodynamic strength. It affects the bacterial adhesion and biofilm formation on the anode. The studies have demonstrated that low shear rates lead the formation of thicker biofilms (Logan et al. 2005). The formation of denser biofilms may be attributed to stable bacterial attachment on the electrode (anode). The operation of MFC at different shear rates can also result in different bacterial communities at the anode. It has been seen that higher shear rates in the MFCs decline the microbial diversity, resulting mostly in the homogeneous biofilm formation (Ringeisen et al. 2006; Raghavulu et al. 2013).

5 Applications

In laboratory MFCs have been already experimented for many applications such as electricity generation, wastewater treatment, biosensing and hydrogen production. Each application is discussed in detail further in the chapter.

5.1 *Electricity Generation*

It is quite evident that most of the studies of MFCs are performed for the electricity generation, and it is the prime application of the technology (Orellana et al. 2013). Some examples of MFC performance for electricity generation are given in Table 1. In the anode chamber of the MFC, the microorganisms oxidize the substrate into protons and electrons that are passed through PEM and electrical connection, respectively, to the cathode (Aelterman et al. 2006; Bermek et al. 2014; Kumar et al. 2015). The two chambers of the MFC can be electrically connected to a multimeter and with an external resistor box, to measure the voltage, and

Table 1 Performance of MFCs for bioelectricity generation

Type of MFC	Substrate	Power density	References
Single-chamber MFC	Glucose	68 mW/m^2	Logan (2004)
	Acetic acid	835 mW/m^2	Oh et al. (2010)
	Ethanol	820 mW/m^2	Logan and Regan (2006)
	Domestic wastewater	114 mW/m^2	Jiang et al. (2014)
Double-chamber MFC	Glucose	855 mW/m^2	Chaudhuri and Lovely (2003)
	Acetate	1926 mW/m^2	Inoue et al. (2010)
	Acetate	1.9 mW/m^2	Tang et al. (2015)
	Acetate	1200 mW/m^2	Patil et al. (2011)
	Cellulose	188 mW/m^2	Liu et al. (2014)
	Wastewater	2485 mW/m^2	Amend and Shock (2001)

subsequently the power can be calculated using Ohm's law. The substrates that can be completely oxidized into electrons are of great importance in MFCs to achieve higher coulombic efficiency and subsequently the power output of the MFCs. A study has shown that *Geobacter sulfurreducens* can reduce the acetate completely into electrons and protons (Reguera et al. 2005). The electrical output of the MFC depends on many factors mainly including the design of MFC, electrode materials, inoculum (pure culture or mixed culture), proton exchange membrane and the operational conditions (Sun et al. 2012). Many approaches are already employed to increase the electrical output in the MFCs. The amendments in MFCs are basically focused on new MFC designs to reduce the internal resistance of the system, cost-effective electrode materials with high surface area, cheaper cation exchange membranes, modifications of the electrode material with nanomaterials (e.g. gold nanoparticles, nickel nanoparticles) and other physical (e.g. heat treatment of stainless steel electrode) or chemical (nitrogen-doped electrodes) treatment methods (Zhou et al. 2013; Venkata Mohan et al. 2014; Kumar et al. 2016).

5.2 *Wastewater Treatment*

The MFCs have shown the potential to treat different industrial, urban or domestic wastewaters (Rhoads et al. 2005; Oh et al. 2010; Zhou et al. 2013). Some examples of MFC performance for wastewater treatment are given in Table 2. Though, the highly toxic wastewaters cannot be completely treated in MFCs, however MFCs are able to reduce the COD of wastewaters much enough to meet discharge regulations before it is released into the environment. The MFCs have proved up to 98 % COD removal from the wastewater (Oh et al. 2010). Alternatively, the wastewaters rich in organic materials (carbohydrates, proteins, lipids, minerals, fatty acids, etc.)

Table 2 Performance of MFCs for wastewater treatment/bioremediation

Type of MFC	Wastewater/heavy metals	% COD removal	References
Single-chamber MFC	Olive mill wastewaters	65	He et al. (2006)
	Biodiesel wastes	90	Brutinel and Gralnick (2012)
	Brewery wastewater	98	
	Azo dye Congo red	98	Rhoads et al. (2005)
	Cadmium	90	Orellana et al. (2013)
	Chromium (VI)	99	Yong et al. (2014)
Double-chamber MFC	Domestic wastewater	88	Cheng et al. (2006a)
	Chemical wastewater	63	Zhou et al. 2013
	Real urban wastewater	70	Logan et al. (2005)
	Food waste leachate	85	Liu et al. (2004)
	Cyanide	88	Min and Logan (2004)

provide the substrate for microbial metabolism to produce electrons and protons. Moreover, wastewaters are also the source of inoculum. The basic wastewater treatment assays (COD, BOD, total solids, nitrogen removal) can be employed to measure the treatment efficiency of the MFCs before and after the MFC operation (Zhou et al. 2013). The COD removal in MFCs can be further improved by operating the MFCs at optimized conditions such as mesophilic temperatures which have shown to increase the COD removal. Moreover, the MFC operation in fed-batch mode is advantageous to obtain high COD removal rate. Usually, the MFC studies operated for wastewater treatment are coupled with power generation; however the coulombic efficiency obtained in such cases is quite low varying from 10 % to 30 % only (Liu et al. 2004).

5.3 *Biosensor*

The application of MFC technology besides electricity generation and wastewater treatment is its use as a biosensor for pollutant detection in water (Shantaram et al. 2005; Zhou et al. 2013). The linear relationship between the coulombic yield of MFC and wastewater strength appoints MFC as a BOD sensor. MFC-based biosensor has advantages over conventional biosensors. Such biosensors are comparatively cheaper because they don't need transducer which is generally used in conventional biosensors. Moreover, they can be operated for very long period such as 5 years without any maintenance. Therefore, MFC-based biosensors have more stability and reliability. Several studies have shown that on the basis of linear correlation, wide BOD ranges (low/high) can be measured in the MFC-based biosensors.

5.4 *Biohydrogen*

The typical double-chamber MFC can be amended to microbial electrolysis cell (MEC) for hydrogen production (Rhoads et al. 2005; Zhou et al. 2013; Kumar et al. 2015). The basic principle of an MEC remains quietly similar; instead electric current is provided at the cathodic chamber. An MEC is also made up of two chambers, i.e. the anode and the cathode. Like MFC, an ion exchange membrane separates both chambers of MEC (Zhou et al. 2013). In the anode chamber, the exoelectrogens metabolize the substrate and produce electrons and protons. The protons are moved to the cathode similarly in MFCs. However, the reaction between protons and electrons to produce hydrogen at the cathode is thermodynamically not possible. To accomplish this reaction, electric current is provided at the cathode. Usually, >0.3 V is enough to fulfil the electrical requirement. Such low voltages can be easily obtained in the MFCs. Therefore, the MFCs employed to generate electricity can be coupled with MEC to fulfil the electrical requirement.

The hydrogen produced from the MEC can be easily stored and subsequently can be used to produce electricity.

6 Future Directions

The MFC technology is still not commercialized despite of 10 years of intensive research on the MFC studies. There is yet to solve many problems in the technology to launch the MFCs in the real-world applications. The main drawback of the MFCs is the insufficient power output. The other limitations are related to the high cost of the electrode materials, membranes and the cathode catalyst. The power output in MFCs can be improved in the future by providing electrode materials of high surface area, while the absence of PEM in futuristic MFC (at large scale) can make the MFC more economical. The MFCs used for wastewater treatment still need effective amendment to completely purify the water. A less work has been done for biosensor application. Because the MFCs are biofilm-based biosensors, the response time is longer in MFC-based biosensor.

Acknowledgments The authors are thankful to the Universiti Malaysia Pahang Research Scheme (Grant No. RDU-140379) for financial support.

References

Aelterman P, Rabaey K, Pham TH, Boon N, Verstraete W (2006) Continuous electricity generation at high voltages and currents using stacked microbial fuel cells. Environ Sci Technol 40:3388–3394. doi:10.1021/es0525511

Alatraktchi FA, Zhang Y, Angelidaki I (2014) Nanomodification of the electrodes in microbial fuel cell: Impact of nanoparticle density on electricity production and microbial community. Appl Energy 116:216–222. doi:10.1016/j.apenergy.2013.11.058

Amend JP, Shock EL (2001) Energetics of overall metabolic reactions of thermophilic and hyperthermophilic Archaea and Bacteria. FEMS Microbiol Rev 25:175–243. doi:10.1111/j.1574-6976.2001.tb00576.x

Bergel A, Feron D, Mollica A (2005) Catalysis of oxygen reduction in PEM fuel cell by seawater biofilm. Electrochem Commun 7:900–904. doi:10.1016/j.elecom.2005.06.006

Bermek H, Catal T, Akan SS, Ulutas MS, Kumru M, Ozguven M, Liu H, Ozcelik H, Akarsubasi AT (2014) Olive mill wastewater treatment in single-chamber air-cathode microbial fuel cells. World J Microbiol Biotechnol 30:1177–1185. doi:10.1007/s11274-013-1541-8

Bond DR, Lovley DR (2003) Electricity production by *Geobacter sulfurreducens* attached to electrodes. Appl Environ Microbiol 69:1548–1555. doi:10.1128/AEM.69.3.1548-1555.2003

Borole AP, Reguera G, Ringeisen B, Wang ZW, Feng Y, Kim BH (2011) Electroactive biofilms: current status and future research needs. Energy Environ Sci 4:4813–4834. doi:10.1039/C1EE02511B

Brutinel E, Gralnick J (2012) Shuttling happens: soluble flavin mediators of extracellular electron transfer in *Shewanella*. Appl Microbiol Biotechnol 93:41–48. doi:10.1007/s00253-011-3653-0

Chaudhuri SK, Lovley DR (2003) Electricity generation by direct oxidation of glucose in mediator less microbial fuel cells. Nat Biotechnol 21:1229–1232. doi:10.1038/nbt867

Cheng S, Liu H, Logan BE (2006a) Increased performance of single chamber microbial fuel cells using an improved cathode structure. Electrochem Commun 8:489–494. doi:10.1016/j.elecom.2006.01.010

Cheng S, Liu H, Logan BE (2006b) Increased power generation in a continuous flow MFC with advective flow through the porous anode and reduced electrode spacing. Environ Sci Technol 40:2426–2432. doi:10.1021/es051652w

Gil GC, Chang IS, Kim BH, Kim M, Jang JK, Park HS, Kim HJ (2003) Operational parameters affecting the performance of a mediator-less microbial fuel cell. Biosen Bioelect 18:327–334. doi:10.1016/S0956-5663(02)00110-0

He Z, Wagner N, Minteer SD, Angenent LT (2006) The upflow microbial fuel cell with an interior cathode: assessment of the internal resistance by impedance spectroscopy. Environ Sci Technol 40:5212–5217. doi:10.1021/es060394f

Heilmann J, Logan BE (2006) Production of electricity from proteins using a single chamber microbial fuel cell. Water Environ Res 78:531–537. doi:10.2175/106143005X73046

Huang L, Chai X, Quan X, Logan BE, Chen G (2012) Reductive dechlorination and mineralization of pentachlorophenol in biocathode microbial fuel cells. Bioresour Technol 111:167–174. doi:10.1016/j.biortech.2012.01.171

Inoue K, Leang C, Franks AE, Woodard TL, Nevin KP, Lovley DR (2010) Specific localization of the c-type cytochrome OmcZ at the anode surface in current-producing biofilms of *Geobacter sulfurreducens*. Environ Microbiol Rep 3:211–217. doi:10.1111/j.1758-2229.2010.00210.x

Jafary T, Ghoreyshi AA, Najafpour G, Fatemi S, Rahimnejad M (2013) Investigation on performance of microbial fuel cells based on carbon sources and kinetic models. Int J Energy Res 37:1539–1549. doi:10.1002/er.2994

Jiang X, Hu J, Lieber AM, Jackan CS, Biffinger JC, Fitzgerald LA, Ringeisen BR, Lieber CM (2014) Nanoparticle facilitated extracellular electron transfer in microbial fuel cells. Nano Lett 14:6737–6742. doi:10.1021/nl503668q

Kumar R, Singh L, Wahid ZA (2016) Exoelectrogens: recent advances in molecular drivers involved in extracellular electron transfer and strategies used to improve it for microbial fuel cell applications. Renew Sustain Energy Rev 56:1322–1336. doi:10.1016/j.rser.2015.12.029

Kumar R, Singh L, Wahid ZA, Fadhil M (2015) Exoelectrogens in microbial fuel cells towards bioelectricity generation: a review. Int Journal Energy Res 39:1048–1067. doi:10.1002/er.3305

Liu H, Ramnarayanan R, Logan BE (2004) Production of electricity during wastewater treatment using a single chamber microbial fuel cell. Environ Sci Technol 38:2281–2285. doi:10.1021/es034923g

Liu X, Huang Y, Sun X, Sheng G, Zhao F, Wang S, Yu H (2014) Conductive carbon nanotube hydrogel as a bioanode for enhanced microbial electrocatalysis. Appl Mater Interf 6:8158–8164. doi:10.1021/am500624k

Logan BE (2004) Extracting hydrogen and electricity from renewable resources. Environ Sci Technol 38:160–167. doi:10.1021/es040468s

Logan BE, Murano C, Scott K, Gray ND, Head IM (2005) Electricity generation from cysteine in a microbial fuel cell. Water Res 39:942–952. doi:10.1016/j.watres.2004.11.019

Logan BE, Regan JM (2006) Microbial fuel cells—challenges and applications. Environ Sci Technol 40:5172–5180. doi:10.1021/es0627592

Min B, Logan BE (2004) Continuous electricity generation from domestic wastewater and organic substrates in a flat plate microbial fuel cell. Environ Sci Technol 38:5809–5814. doi:10.1021/es0491026

Niessen J, SchroÈder U, Scholz F (2004) Exploiting complex carbohydrates for microbial electricity generation-a bacterial fuel cell operating on starch. Electrochem Commun 6:955–958. doi:10.1016/j.elecom.2004.07.010

Oh S, Kim J, Premier G, Lee T, Changwon K, Sloan W (2010) Sustainable wastewater treatment: how might microbial fuel cells contribute. Biotechnol Adv 28:871–881. doi:10.1016/j.biotechadv.2010.07.008

Orellana R, Leavitt JJ, Comolli LR, Csencsits R, Janot N, Flanagan KA, Gray AS, Leang C, Izallalen M, Mester T (2013) U (VI) reduction by a diversity of outer surface c-type cytochromes of *Geobacter sulfurreducens*. Appl Environ Microbiol 79:6369–6374. doi:10.1128/AEM.02551-13

Park DH, Zeikus JG (2003) Improved fuel cell and electrode designs for producing electricity from microbial degradation. Biotechnol Bioeng 81:348–355. doi:10.1002/bit.10501

Patil SA, Harnisch F, Koch C, Hubschmann T, Fetzer I, Carmona-Martinez AA, Muller S, Schroder U (2011) Electroactive mixed culture derived biofilms in microbial bioelectrochemical systems: the role of pH on biofilm formation, performance and composition. Bioresour Technol 102:9683–9690. doi:10.1016/j.biortech.2011.07.087

Phung NT, Lee J, Kang KH, Chang IS, Gadd GM, Kim BH (2004) Analysis of microbial diversity in oligotrophic microbial fuel cells using 16S rDNA sequences. FEMS Microbiol Lett 233:77–82. doi:10.1016/j.femsle.2004.01.041

Rabaey K, Boon N, Hofte M, Verstraete W (2005) Microbial phenazine production enhances electron transfer in biofuel cells. Environ Sci Technol 39:3401–3408. doi:10.1021/es048563o

Rabaey K, Boon N, Siciliano SD, Verhaege M, Verstraete W (2004) Biofuel cells select for microbial consortia that self-mediate electron transfer. Appl Environ Microbiol 70:5373–5382. doi:10.1128/AEM.70.9.5373-5382.2004

Raghavulu SV, Annie Modestra J, Amulya K, Reddy CN, Venkata Mohan S (2013) Relative effect of bioaugmentation with electrochemically active and nonactive bacteria on bioelectrogenesis in microbial fuel cell. Bioresour Technol 146:696–703. doi:10.1016/j.biortech.2013.07.097

Reguera G, McCarthy KD, Mehta T, Nicoll JS, Tuominen MT, Lovley DR (2005) Extracellular electron transfer via microbial nanowires. Nature 435:1098–1101. doi:10.1038/nature03661

Rhoads A, Beyenal H, Lewandowski Z (2005) Microbial fuel cell using anaerobic respiration as an anodic reaction and biomineralized manganese as a cathodic reactant. Environ Sci Technol 39:4666–4671. doi:10.1021/es048386r

Ringeisen BR, Henderson E, Wu PK, Pietron J, Ray R, Little B, Biffinger JC, Jones-Meehan JM (2006) High power density from a miniature microbial fuel cell using *Shewanella oneidensis* DSP10. Environ Sci Technol 40:2629–2634. doi:10.1021/es052254w

Shantaram A, Beyenal H, Raajan R, Veluchamy A, Lewandowski Z (2005) Wireless sensors powered by microbial fuel cells. Environ Sci Technol 39:5037–5042. doi:10.1021/es0480668

Shen HB, Yong XY, Chen YL, Liao ZH, Si RW, Zhou J, Zheng T (2014) Enhanced bioelectricity generation by improving pyocyanin production and membrane permeability through sophorolipid addition in *Pseudomonas aeruginosa*-inoculated microbial fuel cells. Bioresour Technol 167:490–494. doi:10.1016/j.biortech.2014.05.093

Sun M, Reible DD, Lowry GV, Gregory KB (2012) Effect of applied voltage, initial concentration, and natural organic matter on sequential reduction/oxidation of nitrobenzene by graphite electrodes. Environ Sci Technol 46:6174–6181. doi:10.1021/es300048y

Tang J, Chen S, Yuan Y, Cai X, Zhou S (2015) In situ formation of graphene layers on graphite surfaces for efficient anodes of microbial fuel cells. Biosens Bioelectron 71:387–395. doi:10.1016/j.bios.2015.04.074

Venkata Mohan S, Velvizhi G, Modestra JA, Srikanth S (2014) Microbial fuel cell: critical factors regulating bio-catalyzed electrochemical process and recent advancements. Renew Sustain Energy Rev 40:779–797. doi:10.1016/j.rser.2014.07.109

Wang Y, Wang A, Zhou A, Liu W, Huang L, Xu M, Tao H (2014) Electrode as sole electrons donor for enhancing decolorization of azo dye by an isolated WYZ-2. Bioresour Technol 152:530–533. doi:10.1016/j.biortech.2013.11.001

Yang Y, Ding Y, Hu Y, Cao B, Rice SA, Kjelleberg S, Song H (2015) Enhancing bidirectional electron transfer of *Shewanella oneidensis* by a synthetic flavin pathway. ACS Synth Biol 7:815–823. doi:10.1021/sb500331x

Yong XY, Feng J, Chen YL, Shi DY, Xu XY, Zhou J, Wang SY, Xu S, Yong YC, Sun YM, Shi CL, OuYang PK, Zheng T (2014) Enhancement of bioelectricity generation by cofactor manipulation in microbial fuel cell. Biosens Bioelect 56:19–25. doi:10.1016/j.bios.2013.12.058

Zhang Y, Sun J, Hu Y, Li S, Xu Q (2012) Bio-cathode materials evaluation in microbial fuel cells: a comparison of graphite felt, carbon paper and stainless steel mesh materials. Int J Hydrogen Energy 37:16935–16942. doi:10.1016/j.ijhydene.2012.08.064

Zhou M, Wang H, Hassett JD, Gu T (2013) Recent advances in microbial fuel cells (MFCs) and microbial electrolysis cells (MECs) for wastewater treatment, bioenergy and bioproducts. J ChemTechnol Biotechnol 88:508–518. doi:10.1002/jctb.4004

Index

L. Singh, V.C. Kalia (eds.), *Waste Biomass Management – A Holistic Approach*,
DOI 10.1007/978-3-319-49595-8

MIX
Papier aus verantwortungsvollen Quellen
Paper from responsible sources
FSC® C105338

If you have any concerns about our products,
you can contact us on
ProductSafety@springernature.com

In case Publisher is established outside the EU,
the EU authorized representative is:
Springer Nature Customer Service Center GmbH
Europaplatz 3, 69115 Heidelberg, Germany

Printed by Libri Plureos GmbH
in Hamburg, Germany